# Functional Glasses and Glass-Ceramics

# Functional Glasses and Glass-Ceramics

## Processing, Properties, and Applications

*Basudeb Karmakar*

ELSEVIER

Butterworth-Heinemann
An imprint of Elsevier

Butterworth-Heinemann is an imprint of Elsevier
The Boulevard, Langford Lane, Kidlington, Oxford OX5 1GB, United Kingdom
50 Hampshire Street, 5th Floor, Cambridge, MA 02139, United States



**Library of Congress Cataloging-in-Publication Data**
A catalog record for this book is available from the Library of Congress

**British Library Cataloguing-in-Publication Data**
A catalogue record for this book is available from the British Library

ISBN: 978-0-12-805056-9

For information on all Butterworth-Heinemann publications
visit our website at https://www.elsevier.com/books-and-journals



*Publisher:* Matthew Deans
*Acquisition Editor:* Christina Gifford
*Editorial Project Manager:* Ana Claudia A. Garcia
*Production Project Manager:* Anusha Sambamoorthy
*Cover Designer:* Alan Studholme

Typeset by SPi Global, India

# Dedication

Dedicated to
My beloved parents,
*Late Kunjalal Karmakar (Father)*
*and*
*Late Tarubala Karmakar (Mother)*
My wife,
*Gouri Karmakar*
and
My son,
*Dr. Supratim Karmakar, MBBS*

# Contents

# Author's biography

**Basudeb Karmakar** is now the senior principal scientist and head of Glass Science & Technology Section, CSIR-Central Glass & Ceramic Research Institute, Kolkata, India. He is also the professor of the Academy of Scientific and Innovative Research (AcSIR) and Fellow of the Indian Institute of Ceramics (FIICeram). He received his postgraduate degree in Inorganic Chemistry from the Calcutta University, Kolkata, and earned his PhD in Materials Science from the Jadavpur University, Kolkata. He has over 35 years of research and teaching experience. His major research areas are on the preparation, characterization, and property evaluation of different types of glasses (viz., silica, silicate, borate, borosilicate, phosphate, fluorophosphate, antimony oxide, and bismuth oxide), glass-ceramics, ceramic oxide powders, gels, etc. synthesized employing either by the melt-quenching or sol-gel techniques. His notable R&D contributions are in the area of metal-glass nanocomposites, semiconductor-glass nanocomposites, rare-earth doped glass and glass-ceramic nanocomposites, photoluminescent glasses, radiophotoluminescent glass, lead-free environmentally friendly glass powders and pastes for plasma TV, thermally cyclable glass sealant for solid oxide fuel cell, high density radiation-shielding window glass, Nd-doped phosphate laser glass, ultralow expansion transparent glass-ceramics, machinable glass-ceramics, chalcogenide glasses, high purity silica glass, etc. He guided nearly 30 graduate and postgraduate students to accomplish their project or thesis work on glass, glass nanocomposites, and glass-ceramic nanocomposites. Under his supervision, five students received the PhD degree, and currently, three more students are working for their PhD dissertation on glass, glass nanocomposites, or glass-ceramic nanocomposites. He is the recipient of 12 different prestigious awards. He is a DAAD (Germany) fellow and visited Germany, Russia, and Czech Republic in connection with research. He is the peer reviewer of about 35 international journals and life member of 7 different professional institutions. He has published more than 125 research papers in International peer-reviewed journals (Citation, 1495; *h*-index, 21), 95 contributions in international and national conferences/symposia and invited presentations, 10 book chapters, and 11 national/international patents. He is the editor-in-chief of the book on "*Glass Nanocomposites: Synthesis, Properties, and Applications*," B. Karmakar, K. Rademann, and A.L. Stepanov (Eds.), Elsevier, Amsterdam, 2016. He was involved and completed more than 25 R&D projects on glass and glass-ceramics. His current research interests include the areas such as efficiency enhancement of silicon solar cell using plasmonic metal-glass hybrid nanocomposites, semiconductor-glass hybrid nanocomposites, different functional glasses and glass-ceramic nanocomposites, chalcogenide glasses, and thermally cyclable glass sealant for solid oxide fuel cell.

# Foreword

Prof. Dr. Basudeb Karmakar is the head of the Glass Science & Technology Section of the Glass Division at the Central Glass and Ceramic Research Institute (CGCRI) in Kolkata, India. He has long been engaged in intensive research and development of various types of oxide and nonoxide glasses and glass-ceramics, rare-earth and dye-doped luminescent glasses and glass-ceramics, nanometal-glass hybrid nanocomposites, and lead-free and low-softening point glasses, particularly for applications in solar cell technology, light emitting diodes, and photonics. The results of Prof. Karmakar's research activities have been published in 125 articles in indexed periodicals, starting in 1985.

I am pleased and honored to be invited to write the foreword for the book "*Functional Glasses and Glass-Ceramics: Processing, Properties, and Applications*," by Prof. Karmakar. Given the rapid growth of the field of glassy materials, a wide range of synthesis techniques, properties, and applications for glassy materials are already available, but progress in this field continues to be made at a remarkable rate. This book will be a useful mine of state-of-the-art information for those engaged in research and education in this domain.

The book covers different types of synthesis, properties, and applications for a wide spectrum of glasses and glass-ceramics, including oxide, nonoxide, and metallic glasses. It describes various synthesis techniques, such as melt-quenching, sol-gel, and laser irradiation, and discusses physical characteristics such as thermal, optical, nonlinear, optoelectronic, spectroscopic, microscopic, electric, and magnetic properties.

The monograph begins by introducing the basic concepts of functional glasses and glass-ceramics, as well as structure-property correlations of these materials. It then goes on to discuss the functionalization of glasses by transition metals and rare-earth metal doping, functional glass-ceramics, functionalization of glasses by laser irradiation, microfabrication, functionalization by incorporation of semiconductors, glasses and glass-ceramics for biomedical applications, glass and glass-ceramics for solid oxide fuel cell (SOFC) sealants, functional chalcogenide and halide glasses and glass-ceramics, and functional bulk metallic glasses.

There is an undeniable and ongoing demand for the development of novel functional glasses and glass-ceramics with improved properties for future technological utilization. This demand has stimulated a rapid growth in these materials in recent years, particularly in view of the availability of new strategies for synthesis, tools for characterization, and the search for new applications. By including most of these developments, the author has sought to create this book as a knowledge base for functional glasses and glass-ceramics. Given the rapidly increasing diversity of this field, further progress is likely to depend to a great extent on the availability of overviews that present the wide spectrum of the growing number of new results critically. In fact, the crucial tasks of these overviews are threefold: to summarize existing knowledge, to focus critically on new insights gained through recently published papers, and to create the basis upon which the next generation of researchers can safely stand when they seek to look beyond the existing limits. Hence, this book will most likely represent a valuable source of information not only for scientists working on glassy materials but also for many interdisciplinary scientists involved in chemistry, physics, ceramics, and other engineering disciplines. I look forward to receiving a copy of this fascinating book signed by its illustrious author Prof. Karmakar.

**Prof. Dr. Edgar D. Zanotto**
Center for Research, Technology and Education in Vitreous Materials (CeRTEV)
Department of Materials Engineering (DEMa),
Federal University of São Carlos (UFSCar), São Carlos, SP, Brazil
www.certev.ufscar.br

# Preface

The "*functional glasses and glass-ceramics*" are defined as those glasses and glass-ceramics that perform some specific functions that are used to accomplish some specific application need(s) or intelligent action(s). There are a wide range of reports in the published domain of the literature on glasses and glass-ceramics, which are used to accomplish various functions. But there is no suitable book reviewing all these developments and recent literature. This book intends to fill this gap. This is as a consequence of the fact: a rapid development in a particular field requires a book, which covers the forefront research in a widespread range including material synthesis, property evaluation, and applications.

This book contains 12 chapters that are divided into three parts. Part I is the introduction to functional glasses and glass-ceramics. It describes the basic concept and structure-property correlation of functional glasses and glass-ceramics (Chapter 1). Part II covers oxide glass-based different functional glasses and glass-ceramics. They are functionalized oxide glasses by 3d transition metal ion doping (Chapter 2), functionalized oxide glasses by 4f rare-earth metal ion doping (Chapter 3), functional glass-ceramics (Chapter 4), functionalization of glasses by laser irradiation microfabrication (Chapter 5), functionalization of glasses by incorporation of semiconductor (Chapter 6), glasses and glass-ceramics for biomedical applications (Chapter 7), glass and glass-ceramics for solid oxide fuel cell (SOFC) sealants (Chapter 8), and functional glasses and glass-ceramics from solid waste materials (Chapter 9). Part III discusses functional glasses and glass-ceramics of nonoxide systems. They are functional chalcogenide glasses and glass-ceramics (Chapter 10), functional halide glasses and glass-ceramics (Chapter 11), and functional bulk metallic glasses (Chapter 12). The book covers all the types of glasses starting from oxide and nonoxide to bulk metallic glasses (BMGs), diverse synthesis techniques (e.g., melt-quenching, sol-gel, and laser irradiation), various properties (e.g., physical, thermal, optical, nonlinear, photonic, optoelectronic, spectroscopic, microscopic, electrical, and magnetic), and numerous applications.

This is a timely book. The chapters have been written by the author based on his knowledge and work experience accumulated through research and teaching in the area of various glasses and glass-ceramics. We have tried to create this book as a treasure of up-to-date understanding and information on various types of functional glasses and glass-ceramics.

In spite of our best effort to make this book perfect, there may be some mistakes. If you find anything that needs to be corrected, please feel free to contact us.

This book is envisioned for the postgraduate students, researchers, educators, scientists, professionals, practitioners in industry, and R&D managers and planners not only of the glass and materials science and engineering but also for many interdisciplinary areas like chemistry, physics, ceramics, metallurgy, electronics, optoelectronics, photonics, and electrical engineering.

We believe that this book will be very helpful to the readers to broaden their knowledge and insight into the different subjects of this interdisciplinary area. We also hope that this book will certainly enhance the rapid progress of science and technology of functional glasses and glass-ceramics.

We wish you a very pleasant journey on your way to discovering the fascinating world of functional glasses and glass-ceramics.

**Basudeb Karmakar**
CSIR-Central Glass and Ceramic Research Institute, Kolkata, India
January 2017

# Acknowledgments

This book emerged as a result of experience accumulated during many years of research and teaching in the area of glass science and technology.

I would like to thank the authors whose research results and ideas are presented here. I am indebted to them. I would also like to thank the individuals, trustees, and publishers who have permitted the reproduction of their photographs and figures.

I am very grateful to Prof. Dr. E.D. Zanotto for his contribution of the foreword of this book.

I thank the referees for their critical reading, suggestions, and constructive criticism.

I also wish to express my great appreciation to my publisher, Elsevier Inc., for their trust and encouragement. Finally, I would like to thank many unnamed editorial and production members of Elsevier Inc. for their expert work.

I would like to thank Dr. K. Muraleedharan, director, CSIR-CGCRI, for his kind encouragement and support; Mr. Kamal Dasgupta, then acting director, CSIR-CGCRI, for his permission to write this book and encouragement; and Dr. Ranjan Sen, head, Glass Division of the institute, for encouragement and support. I wish to thank my scientist colleagues Dr. K. Annapurna, Dr. Ashis Kumar Mandal, Dr. Atiar Rahaman Molla, Dr. Anal Tarafder, Dr. Kaushik Biswas, and Mr. S. Balaji for their many fruitful scientific discussion and encouragement. I also thank my PhD students, Tirtha, Shiv, Atiar, Anal, Mrinmoy, Nibedita, Chirantan, and Nilanjana for working on many advanced topics on functional glasses and glass-ceramics, which promoted me to think of such a book on "*functional glasses and glass-ceramics*."

Last but certainly not the least, I am deeply indebted to my family. Without their sacrifice, support, and inspiration, this book is impossible.

# Part One

# Introduction

# Introduction to functional glasses and glass-ceramics

**1**

## Chapter Outline



## 1.1 Introduction

Glass is an important ancient material. It is being used by mankind from the early days of human civilization to accomplish various functions. Nowadays, glass and its associated products are a part of our everyday life starting from early morning to late at night whether we are at home, at work, or almost everywhere. This is due to the fact that it functions silently in making our life easier. Glass-ceramics are polycrystalline materials derived from some selected glasses by controlled partial crystallization. They are used in some particular areas to get some specific functions done. There are many books that described both science and technology of these materials. A long, chronologically well-ordered list of these books is provided in the references [1–52]. All these books have several strong superior properties and excellent coverage of topics; on the other hand, they have limitations as well. The main limitations of these books are that majority of these books are old, and current topics and outlooks are obviously not available therein. It has been observed that there is a rapidly growing domain of published literature, on glass and its related products, particularly for the last two decades, which reported much new information on various aspects such as composition, processing, functional properties, and functional applications. There is no book reviewing all these up-to-date progress and literature particularly with respect to the functionality of glasses and glass-ceramics. The present book is intended

Functional Glasses and Glass-Ceramics. http://dx.doi.org/10.1016/B978-0-12-805056-9.00001-5

to fill this gap. To accomplish this mission, it is planned to describe the basic concept and classification of functional glasses and glass-ceramics in this chapter first.

## 1.2 Basic concept of functional glasses and glass-ceramics

Glass products are numerous. A nonexhaustive list of these glass products and their purposes is function-wise provided in Table 1.1. It is seen that glass products are used in many fields, for example, optics, electrical, electronics, semiconductor industry, medicine, biotechnology, aviation, and astronomy, to accomplish some functions. According to this logic, all glasses are functional. Hence, the question of "*What is actually meant by functional glasses and glass-ceramics*?" remains unsettled. Let us discuss different aspects of this question.

**Table 1.1 Function-wise nonexhaustive list of glass products and their purposes**

| Serial no. | Function | Glass product and/or purpose |
|---|---|---|
| 1. | Optical | • Optical glass lens, prism, and mirror for microscope, telescope, spectrophotometer, camera, etc.<br>• Colored signal glasses for traffic control<br>• Colored glass filters<br>• Optical fibers for communication (e.g., telephones) |
| 2. | Photonic | • Laser glass for generation of laser beams of different wavelengths<br>• Optical fiber amplifiers<br>• Gradient index glass for reprography<br>• Nonlinear optical and magneto-optical glasses |
| 3. | Ophthalmic | • Spectacle lens for vision correction<br>• Photochromic glass, sun goggle, welder's glass, etc. for eye protection |
| 4. | Lighting | • Tubing for neon signs, fluorescent lamps, mercury and sodium vapor lamps<br>• Envelope for incandescent and projection lamps<br>• Ultraviolet (UV) transmitting glasses for photographic purpose<br>• Phototropic glass for illumination control<br>• Photosensitive glass for reproduction of picture in glass |
| 5. | Structural | • Flat glasses for housing and buildings (e.g., windows, facades, conservatory, insulation, and reinforcement structures)<br>• Glass mirrors for looking glass and decoration<br>• Laminated safety glasses<br>• Toughened or tempered glasses for wind-shields in automobiles, aircraft, etc.<br>• Double-walled window glass for insulation against heat and sound, and controlled illumination |

**Table 1.1 Continued**

| | | |
|---|---|---|
| 6. | Thermal | • Heat-absorbing and fire-proof glasses for windows<br>• Glasses for thermometers and thermoflasks<br>• Windows for high temperature furnaces<br>• Foamed glasses for thermal insulation<br>• Thermal shock resistance opal and borosilicate glasses for microwave oven<br>• Thermal shock resistance glass-ceramics for cook top panel |
| 7. | Chemical | • Container glasses for packaging, for example, bottle for drink, jar for food, flacon for cosmetics and pharmaceuticals, tank, and ampoule<br>• Table wares, for example, drinking glasses, plate, cups, and bowls<br>• Laboratory glasswares for research, testing, and diagnosis<br>• Glass-lined metallic containers for chemical processing<br>• Tubing and piping for chemical transportation and heat exchangers<br>• Cylinders for fractionating columns<br>• Gage glass in boilers and other vessels to observe liquid level<br>• Glass electrodes for pH meters, ion sensitive meters, etc.<br>• Sintered glass for liquid filtration<br>• Glass wool for air and liquid filtration<br>• Glasses as fertilizers |
| 8. | Electrical | • Insulator and lightening arrester in power lines<br>• X-ray and cathode-ray generating tubes<br>• Insulator for radio antennae (low dielectric loss glass) and in high frequency circuits |
| 9. | Electronic | • Envelopes and other components for electronic valves<br>• Sealing glasses for electronic components (e.g., transistor, diodes, and relays)<br>• Electrically conducting glass in microcircuits<br>• Glass plates for printed circuit boards<br>• Aircraft nose cones |
| 10. | Mechanical | • Torsion fibers, jewel bearings, and optically flat surface plates for high precision instruments<br>• Molds for fabrication of rubber and plastics<br>• Diffusion vacuum pumps<br>• Rollers and thread guides for textile industry<br>• E-glass fiber yarns and textiles<br>• Glass lubricant for extrusion of metals |

The meaning of the word "*function*" in English, as a verb, is to "*work or operate in a proper or particular way*" and that of the word "*functional*," as an adjective, is "*of or having a special activity, purpose, or task.*" The synonyms of "*functional*" are useful, serviceable, purposeful, efficient, practical, well designed, etc. Consequently, the "*functional glasses and glass-ceramics*" are defined as those glasses and glass-ceramics that perform some specific functions, which are used to accomplish some specific application need(s) or intelligent action(s). With respect to properties, they could be defined as "*functional glasses and glass-ceramics*" that possess at least one or more inherent or inborn particular properties and functions independently based on this specific property. With respect to preparation, they could be defined as "*functional glasses and glass-ceramics*" that are prepared using a "*target-motivated*" approach and possess certain designed properties to achieve the purpose(s).

## 1.3 Structure-property correlation of glasses and glass-ceramics

From the above definitions, it is revealed that the origin of functionalities of the glasses and glass-ceramics is in their properties. On the other hand, the properties are controlled by the structures of the glasses and glass-ceramics. Therefore, there is a strong correlation between the structure and property. Since this is not the right place to discuss the structure-property correlation of the glasses and glass-ceramics elaborately, readers are referred to consult the references [5,7–9,12,14,16,18,25,27,28,35–37,43–47,52] to understand this fact of the glasses and glass-ceramics in greater detail. A very short outline is given as follows. Gaskell [53] has described the glass properties relating to their sensitivity (or insensitivity) to the structural characteristics on various length scales such as local structure (0.0–0.5 nm), medium-range structure (0.6–1.5 nm), and long-range structure (>1.5 nm). This is schematically shown in Fig. 1.1. This representation gives us a clear idea about the dependence of different properties of glasses on the length scales of the structures. However, the glass properties are fundamentally composition-dependent and are controlled largely by the first neighbor shell of the network formers or less often by the local structure around the network modifiers. Different types of structure-property correlation of the glasses and glass-ceramics would be described case-wise in the subsequent chapters.

## 1.4 Publication trend and growth on glasses and glass-ceramics

An attempt was made using the Web of Science (Version 5.17) to provide a global publication trend and growth quantitatively in the domain of glasses and glass-ceramics. Here, data were extracted by searching using the topic keyword "*glass*" followed by refining the document type as "*article*" published in the journals and meetings. The topic keyword "*glass*" was found either in the title of the article or in its abstract or both. It is observed that a total of 123,477 articles were published during period of 1950–2014 (65 years) and 112,633 articles during period of 1990–2014 (25 years). As a result, it is seen that majority (91.2%) of the articles were published during the last 25 years (1990–2014), and only a minor amount (8.8%) was

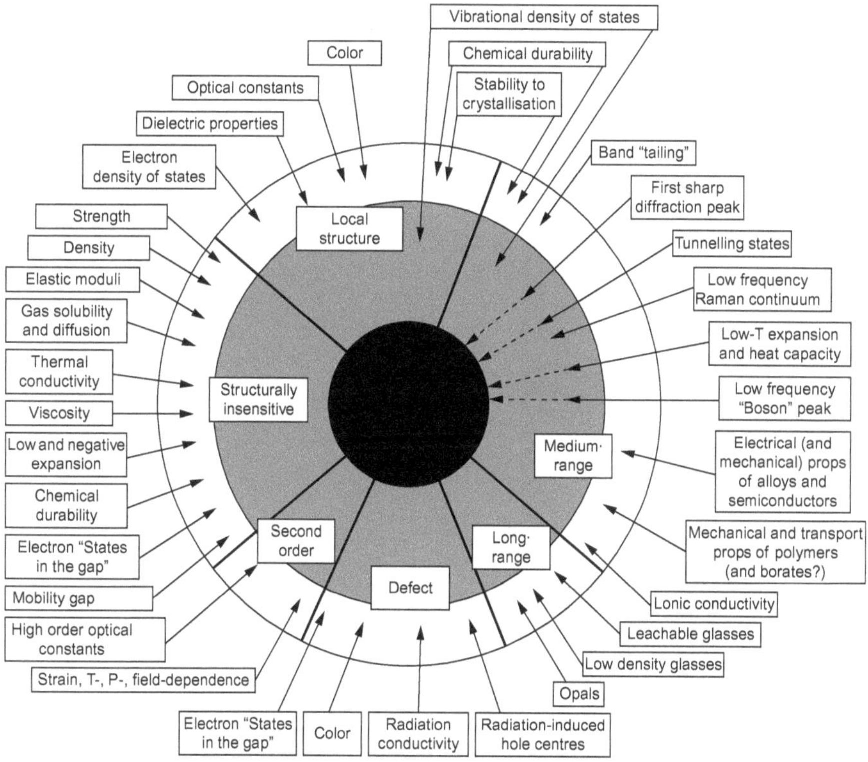


**Fig. 1.1** Schematic representations of glass properties relating to their sensitivity (or insensitivity) to the structural characteristics on various length scales: local (0.0–0.5 nm), medium (0.6–1.5 nm), and long (>1.5 nm) ranges. *Arrows* pointing into the outer ring represent "exoteric" or "materials" properties in contrast to the second ring *(lightly mottled)* that are more "esoteric" properties. The central ring represents the possible Unified Theory of Properties in Amorphous (UTOPIA) *(solids)*.
Reproduced with permission from P.H. Gaskell, Structure and properties of glasses—how far do we need to go?, J. Non-Cryst. Solids 222 (1997) 1–12, Copyright © Elsevier.

published during the period of 1950–89 (previous 40 years). For this reason, for discussion on the global publication trend and growth on glasses and glass-ceramics in the rest parts of the book, we shall consider only the last 25 years (1990–2014). Accordingly, the number of paper published per year on glasses and glass-ceramics and functional glasses and glass-ceramics during the last 25 years (1990–2014) are shown in Fig. 1.2A and B, respectively. In the latter case, the keyword "functional" was found either in the title of the article or in its abstract or both. It is observed that in both cases, the global general trends are increasing with time. Plausible reasons are increase in (i) number of researchers, (ii) funding for research, and (iii) facility for research in the area of glass and glass-ceramics.

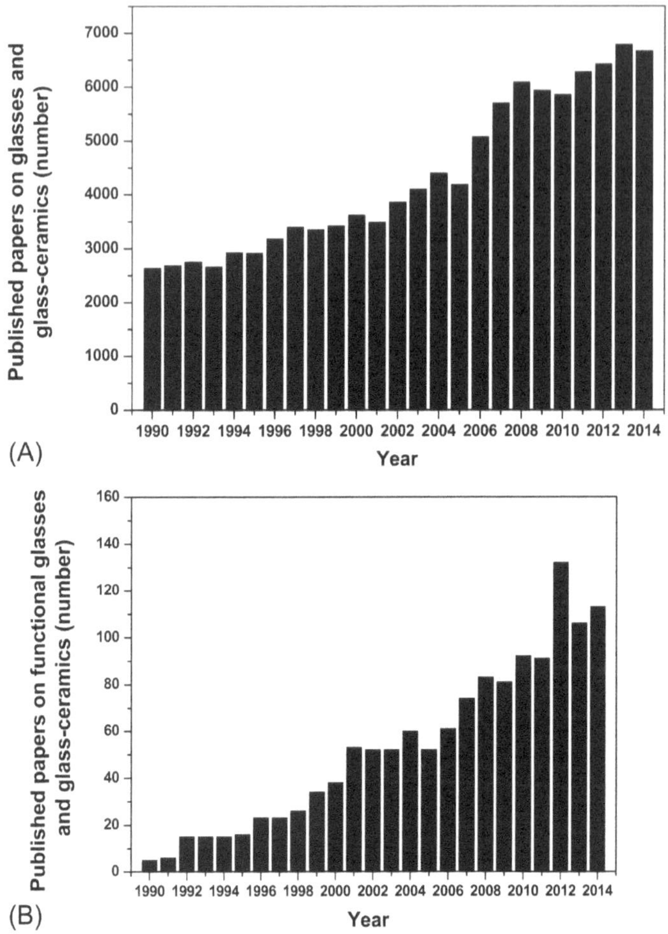


**Fig. 1.2** (A) Number of paper published per year on glasses and glass-ceramics during the last 25 years (1990–2014). Data were extracted using Web of Science (Version 5.17) by searching using the topic keyword "glass" followed by refining the document type as "article" published in journals and meetings. (B) Number of paper published per year on functional glasses and glass-ceramics during the last 25 years (1990–2014). Data were extracted from the data as shown in (A) by further refining using the keyword "functional."

## 1.5 Classification of functional glasses and glass-ceramics

The definition of "*functional glasses and glass-ceramics*," as described in Section 1.2, reflects the ability of a glass or glass-ceramics to perform a certain predestined "*task*." In this general definition, a wide range of glasses and glass-ceramics can be included together with an ample range of properties and applications. Therefore, it would be recognizable to identify the class of functional glasses and glass-ceramics with properties and applications. However, all the properties are directly dependent on the chemical composition, singularities of the crystallographic structure, and manufacturing process. On the other hand, for use of a glass or glass-ceramics in a particular application, it is essential to understand the relationships between its composition, structure, processing, and properties. Taking into consideration of all these facts, functional glasses and glass-ceramics are broadly classified into four types based on (i) composition, (ii) processing, (iii) property, and (iv) application. These are described in the subsequent sections.

### 1.5.1 Composition-based functional glasses and glass-ceramics

Functional glasses and glass-ceramics based on composition are of two types: (i) oxide and (ii) nonoxide. This is illustrated in Fig. 1.3. The oxide-based glasses and glass-ceramics may be of silicate, borate, phosphate, and germinate type in the context of glass forming oxide components such as silica ($SiO_2$), boric oxide ($B_2O_3$), phosphorus oxide ($P_2O_5$), and germanium oxide ($GeO_2$), respectively. These materials may also be created from the combination of these oxides along with other suitable oxides. The nonoxide glasses and glass-ceramics may be of chalcogenide, halide, and metallic type depending on their constituent components. For example, chalcogenide glasses are prepared from chalcogen elements (e.g., sulfur (S), selenium (Se), and tellurium (Te)) and other elements such as arsenic (As), germanium (Ge), gallium (Ga), and antimony (Sb). $As_2S_3$, $As_2Se_3$, $Ge_{22}As_{20}Se_{58}$, and $Ge_{20}Sb_{15}Se_{65}$ are some of the examples of the chalcogenide glasses. Halide glasses are made of inorganic halide (e.g., fluoride ($F^-$), chloride ($Cl^-$), bromide ($Br^-$), and iodide ($I^-$)) salts. Heavy metal halides (HMH) are the major constituents of this family of glasses. The glass systems $ZrF_4$–$BaF_2$–$LaF_3$–$AlF_3$ (ZBLA), $ZrF_4$–$BaF_2$–$LaF_3$–$AlF_3$–NaF (ZBLAN), $ZrF_4$–$BaF_2$–$YF_3$–$AlF_3$ (ZBYA), etc. are the examples of the halide family glasses. Metallic glasses are glasses of alloys of pure metals or the combinations of metals and metalloids. These glasses may be binary (e.g., $Ni_{60}Nb_{40}$, $Cu_{50}Zr_{50}$, $Pd_{80}Si_{20}$, $Au_{73}Ge_{27}$, and $Co_{75}P_{25}$), ternary (e.g., $Pd_{77}Au_5Si_{18}$, $Ni_{80}Si_8B_{12}$, $Fe_{81}Cr_2B_{17}$, and $Ti_{72}Fe_{12}Si_6$), quaternary (e.g., $Fe_{40}Ni_{40}P_{14}B_6$, $Ni_{68}Cr_{10}Si_{10}B_{12}$, and $Fe_{75}P_{15}C_6Al_4$), etc. systems. It is worth mentioning that here

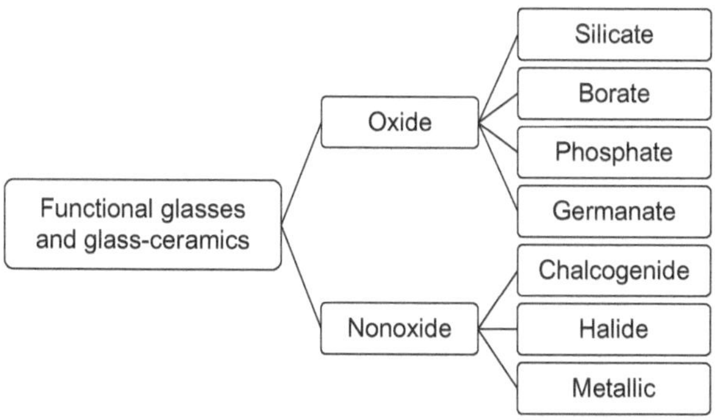


**Fig. 1.3** Composition-based classification of functional glasses and glass-ceramics.

the term "*composition*" indicates not only the components but also their quantity in the glasses and glass-ceramics.

Fig. 1.4A shows the number of paper published per year on glasses and glass-ceramics based on composition during the last 25 years (1990–2014). Data were extracted from the data as shown in Fig. 1.2A by further refining using the keyword

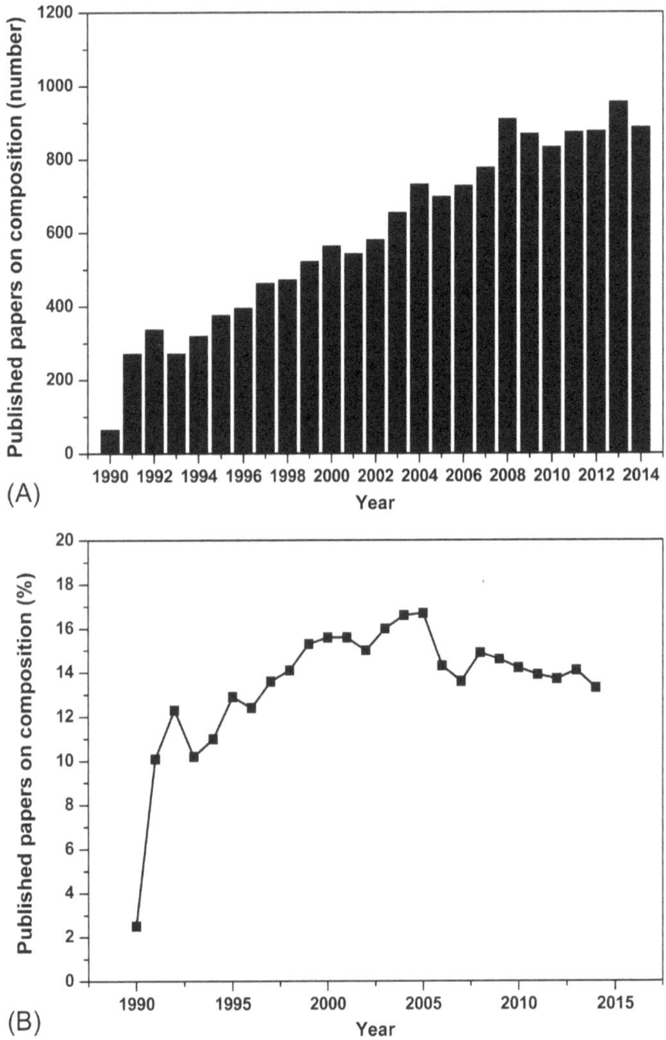


**Fig. 1.4** (A) Number of paper published per year on glasses and glass-ceramics based on composition during the last 25 years (1990–2014). Data were extracted from the data as shown in Fig. 1.2A by further refining using the keyword "composition." (B) Year-wise plot of percentage of papers published on composition of glasses and glass-ceramics calculated from the data as shown in Fig. 1.2A.

"*composition*." It is seen that there is an increasing trend of paper published per year on composition-based glasses and glass-ceramics with time, that is, the area is growing. Fig. 1.4B shows the year-wise plot of percentage of papers published on composition of glasses and glass-ceramics calculated from the data as shown in Fig. 1.2A. From this plot, it is seen that there are three types of trends of paper publication on composition of glasses and glass-ceramics. The steep rise in percentage of paper publication is in the period 1990–92, a gradual increasing trend is in the period 1993–2005, and a nearly horizontal with a little bit of decreasing trend is in the period 2006–14. It is, therefore, disclosed that at present, the researchers are giving little less attention on composition oriented investigation on glasses and glass-ceramics.

### 1.5.2 Processing-based functional glasses and glass-ceramics

Functional glasses and glass-ceramics can be prepared by various techniques. Some of these processing methods are shown in Fig. 1.5. Melt quenching, sol-gel, chemical vapor deposition, physical vapor deposition, sputtering, plasma discharge, etc. are some of these methods. Readers are referred to consult the references [6,11,37,43,45,52] for details of these methods.

From the foregoing discussion, it is clear that the functional glasses and glass-ceramics can be generated by creation of functional properties in the glasses and glass-ceramics by adopting suitable technique(s) and protocol(s). This is shown in Fig. 1.6. The creation of functional properties in glasses and glass-ceramics to yield functional glasses and glass-ceramics may be done by various approaches. Some of these approaches are illustrated in Fig. 1.7. They are doping with $3d^{1-9}$ transition and $4f^{1-13}$ rare-earth metal ions, crystallization by thermal or laser treatment, incorporation of metal and semiconductor nanoparticles, adjustment of chemical composition, applying coatings of thin films, shaping (e.g., lens and prism), and sizing.

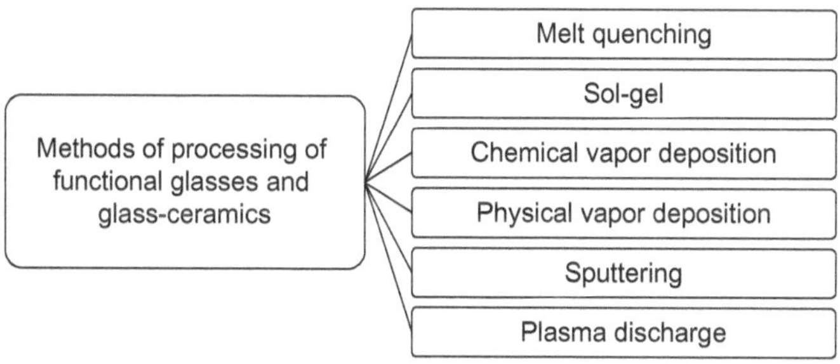


**Fig. 1.5** Different methods of processing of functional glasses and glass-ceramics.

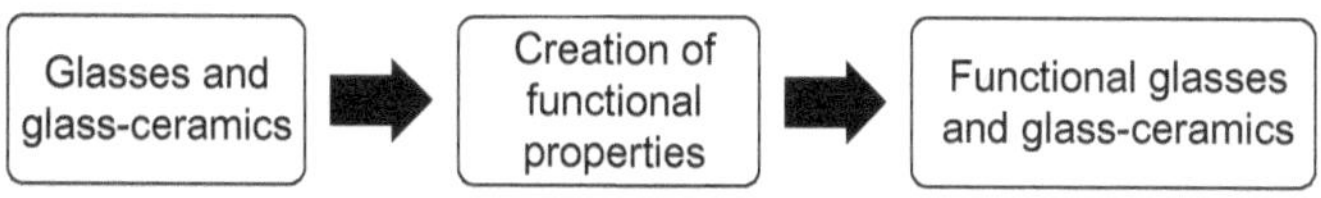


**Fig. 1.6** General approach of creation of functional glasses and glass-ceramics.

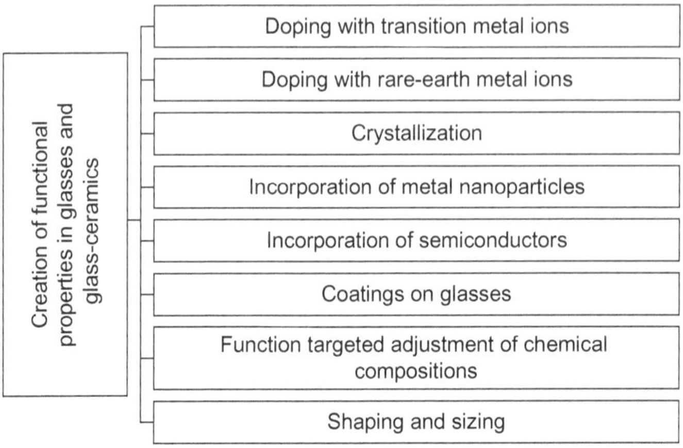


**Fig. 1.7** Some approaches for creation of functional properties in glasses and glass-ceramics to yield functional glasses and glass-ceramics.

Fig. 1.8A shows the number of paper published per year on glasses and glass-ceramics based on processing during the last 25 years (1990–2014). Data were extracted from the data as shown in Fig. 1.2A by further refining using the keyword "*processing*." It is seen that there is an increasing trend of paper published per year on processing-based glasses and glass-ceramics with time, that is, the field is growing. Fig. 1.8B shows the year-wise plot of percentage of papers published on processing of glasses and glass-ceramics calculated from the data as shown in Fig. 1.2A. From this plot, it is seen that there are three types of trends of paper publication on processing of glasses and glass-ceramics. There is a steep rise in the percentage of paper publication during the period 1990–91, a gradual increasing trend during the period 1993–99, and a nearly horizontal with a little bit of decreasing trend during the period 2000–14. It is, therefore, disclosed that at present the researchers are giving little less attention on processing oriented investigation on glasses and glass-ceramics.

### 1.5.3 Property-based functional glasses and glass-ceramics

It is already mentioned that the properties are at the central point for exhibiting different functionalities of the glasses and glass-ceramics. These properties are in the categories of optical, photonic, thermal, electrical, electronic, microstructure, mechanical, chemical, etc. A correlation between the properties and functions of glasses and glass-ceramics is summarized in Table 1.2. The readers may find it useful to review the references [8,9,12,27,35,37,42,43,45–47].

Fig. 1.9A shows the number of paper published per year on glasses and glass-ceramics based on property during the last 25 years (1990–2014). Data were extracted from the data as shown in Fig. 1.2A by further refining using the keyword "*property*." It is seen that there is an increasing trend of paper published per year on property-based glasses and glass-ceramics with time, that is, the field of research is growing. Fig. 1.9B shows the year-wise plot of percentage of papers published on

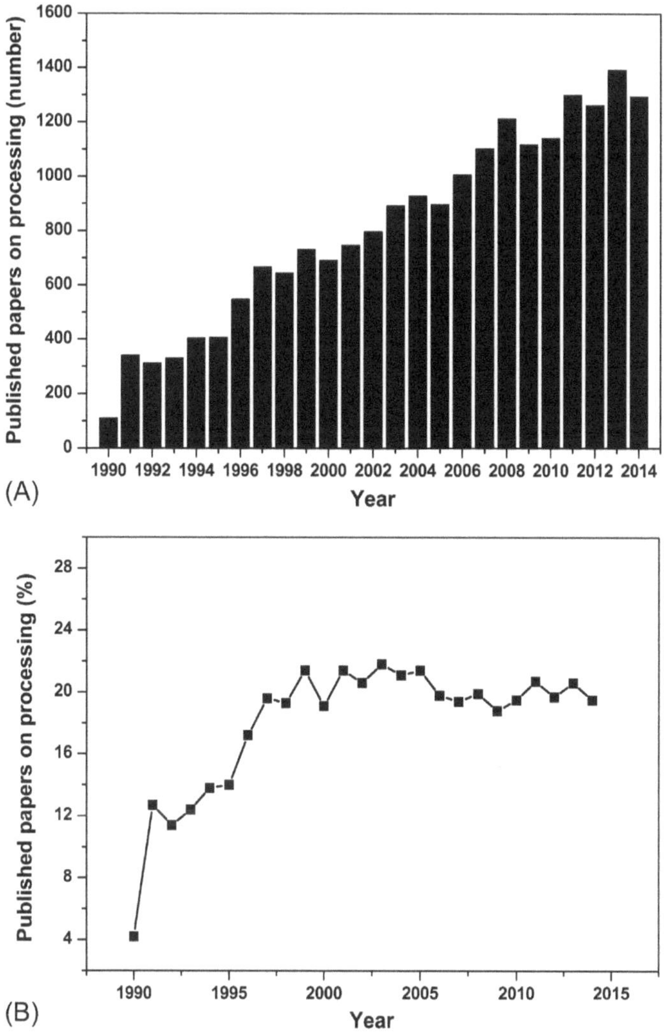


**Fig. 1.8** (A) Number of paper published per year on glasses and glass-ceramics based on processing during the last 25 years (1990–2014). Data were extracted from the data as shown in Fig. 1.2A by further refining using the keyword "processing." (B) Year-wise plot of percentage of papers published on processing of glasses and glass-ceramics calculated from the data as shown in Fig. 1.2A.

property of glasses and glass-ceramics calculated from the data as shown in Fig. 1.2A. From this plot, it is seen that there are two distinct types of trends of paper publication on property of glasses and glass-ceramics. The initial steep rise in percentage of paper publication lies during the period 1990–92, and thereafter, the trend is gradual increasing during the period 1993–2014. It is, therefore, disclosed that at present, the researchers are giving greater attention on property oriented investigation on glasses and glass-ceramics.

**Table 1.2 Correlation between property and function of glasses and glass-ceramics**

| Property category | Property | Function |
|---|---|---|
| Physical | Density | A fundamental property influences directly the refractive index and other properties |
| Optical | Refractive index | Control light path |
| | Dispersion | Control chromatic aberration, improve image quality |
| | Transmission and absorption | Control transparency window of undoped glasses and selective absorption of doped (3d and 4f metal ions, nanometal, semiconductor, etc.) glasses from ultraviolet (UV) to far infrared (FIR) regions of the electromagnetic spectrum |
| Photonic | Nonlinear optical (NLO) property | Optical switching and modulation |
| | Photoluminescence and emission | Absorption, luminescence, and lasing action |
| | Magneto-optical property | Control of optical effect under magnetic field |
| Thermal | Coefficient of thermal expansion | Control of dimensional changes with temperature |
| | Thermal conductivity | Transport of heat energy through the material |
| Electrical | Electrical conductivity | Resistance to mobility of ions or free electrons |
| | Dielectric property: permittivity (dielectric constant) and dielectric loss | Extent of polarizability of ions and loss of electrical energy under electric field |
| Electronic | Electronic conduction | Conduction of electrical charge by electrons ($e^-$) or holes ($h^+$) |
| Microstructure | Morphology (size and shape) and orientation of grain | Control properties of the glasses and glass-ceramics |
| Mechanical | Elastic property: Young's and bulk moduli | Control breakage of the glass |
| | Hardness: Knoop and Vickers | Control strength of the glass |
| | Fracture toughness | Resistance under load |
| Chemical | Durability | Inertness to environment or chemicals |

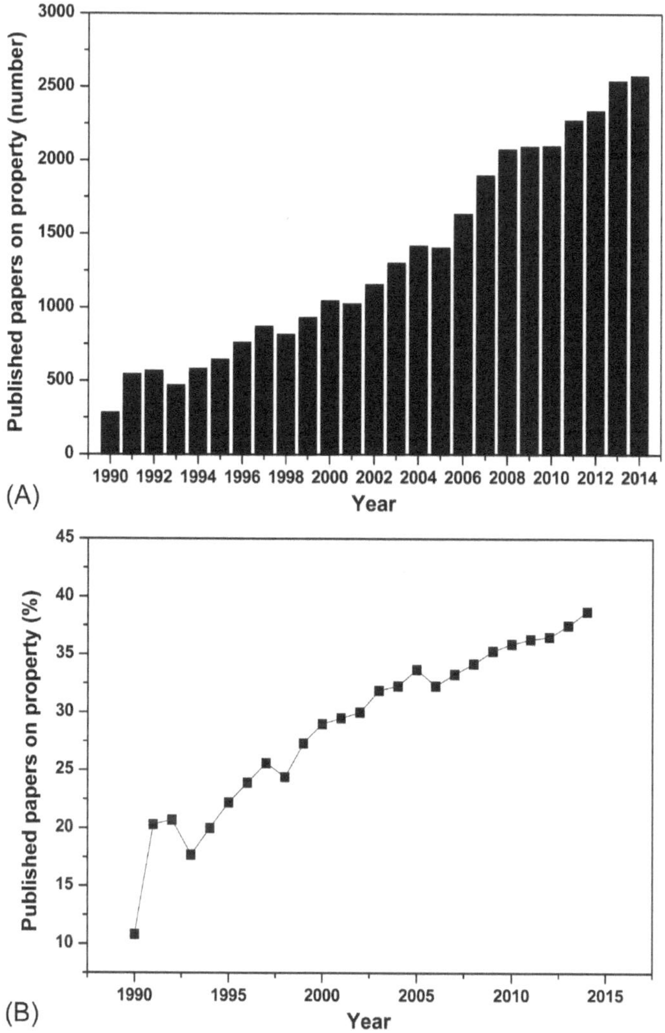


**Fig. 1.9** (A) Number of paper published per year on glasses and glass-ceramics based on property during the last 25 years (1990–2014). Data were extracted from the data as shown in Fig. 1.2A by further refining using the keyword "property." (B) Year-wise plot of percentage of papers published on property of glasses and glass-ceramics calculated from the data as shown in Fig. 1.2A.

## *1.5.4 Application-based functional glasses and glass-ceramics*

The ultimate destinations of the products of functional glasses and glass-ceramics are their use in various technological applications by the way of fabrication of different devices. To provide a clear idea about this fact, some application area-wise lists of glass products and their purposes are reviewed in Table 1.3. These application areas are energy, display, health care, nuclear, defense, space, etc.

**Table 1.3 Application area-wise lists of glass products and their purposes**

| Serial no. | Area of application | Glass product and/or purpose |
|---|---|---|
| 1. | Energy | • Insulating sealing glasses for solid oxide fuel cell (SOFC)<br>• Low-iron cover glasses for photovoltaic solar cells<br>• Solar power plants with glass made parabolic trough technology<br>• Glass fibers for wind energy<br>• Glass-based solid electrolytes for lithium batteries<br>• Glass microspheres for hydrogen storage that is used as fuel in SOFC |
| 2. | Display | • Glass substrates for display devices, for example, liquid crystal display (LCD), thin film transistors (TFTs), and mobile phones<br>• Sealing glasses for plasma display panel (PDP)<br>• Thin film (e.g., heat reflecting, electrochromic, and transparent conducting glasses) coated display glass substrates |
| 3. | Health care | • Bioactive glass and glass-ceramics for restoration of bone and teeth<br>• Microarray glass substrates for genetic research<br>• Glass image guides for pathological microscopes<br>• Flexible optical fiber light guides for medical examination by endoscopy (e.g., gastroscope and colonoscope)<br>• Rod-shaped glass lens system in telescope of laparoscopic surgical instruments<br>• Contact lens and spectacle lens for vision correction<br>• Glasses for medical technology, biotechnology, and life science engineering<br>• Infrared (IR) transmitting glasses for medical diagnosis and treatment<br>• Ultraviolet (UV) transmitting glasses for germicidal lamps<br>• Silver containing glasses for germicidal food containers<br>• Glass made packaging for pharmaceuticals<br>• Glass for solar-driven water purification |
| 4. | Nuclear | • Glass for fixation of radioactive wastes<br>• X-ray and $\gamma$-ray shields<br>• Dosimeter for high-energy X- and $\gamma$-rays and neutron<br>• Study of nuclear reactions and detection of nuclear tracks |
| 5. | Defense | • Glasses for modern navigation and surveillance equipment<br>• Infrared (IR) transmitting glasses for imaging infrared seekers for missiles, thermal imagers, and night vision cameras<br>• Glass and glass-ceramics for armor<br>• Hermetic glass packaging for protecting sensitive electronic components |

**Table 1.3 Continued**

| | | |
|---|---|---|
| 6. | Space | • Bulletproof safety glasses<br>• Rocket and aircraft nose cones<br>• Windows for spacecrafts<br>• Glass made satellite navigation systems<br>• Ultralow expansion glass-ceramics for high precision telescope mirror blanks<br>• Machinable glass-ceramics for high temperature ultrahigh vacuum sealing |

Fig. 1.10A shows the number of paper published per year on glasses and glass-ceramics based on application during the last 25 years (1990–2014). Data were extracted from the data as shown in Fig. 1.2A by further refining using the keyword "application." It is seen that there is an exponential increasing trend of paper published per year on application-based glasses and glass-ceramics with time, that is, the field of research is growing very fast. Fig. 1.10B shows the year-wise plot of percentage of papers published on application of glasses and glass-ceramics calculated from the data as shown in Fig. 1.2A. From this plot, it is seen that there is a continuous increasing trend of paper publication on application of glasses and glass-ceramics for the last 25 years (1990–2014). It is, therefore, disclosed that the researchers are always giving greater attention on application oriented investigation on glasses and glass-ceramics.

Majority of the aforementioned functional glasses and glass-ceramics including both oxide and nonoxide systems are described in the subsequent chapters of this book.

## 1.6 Conclusions and future outlook

In the beginning of this chapter, a complete chronologically well-ordered and state-of-the-art list of books published on glass and glass-ceramics and function-wise non-exhaustive list of glass products and their purposes are provided. This is followed by the description of the topics such as basic concept and definition of functional glasses and glass-ceramics, structure-property correlation, publication trend, and growth on glasses and glass-ceramics as well as their functional counterparts. Classification of functional glasses and glass-ceramics has been made based on composition, processing, property, and application.

Current research trend on glasses and glass-ceramics has also been disclosed by analyzing the research article publication trends. It is observed from the plot of percentage of published articles (on composition, processing, property, and application) versus the year that currently the researchers are giving less attention on composition and processing oriented investigation and greater attention on property and application oriented investigation on glasses and glass-ceramics. This observation clearly reflects the present day need on glass and glass-ceramics research, that is,

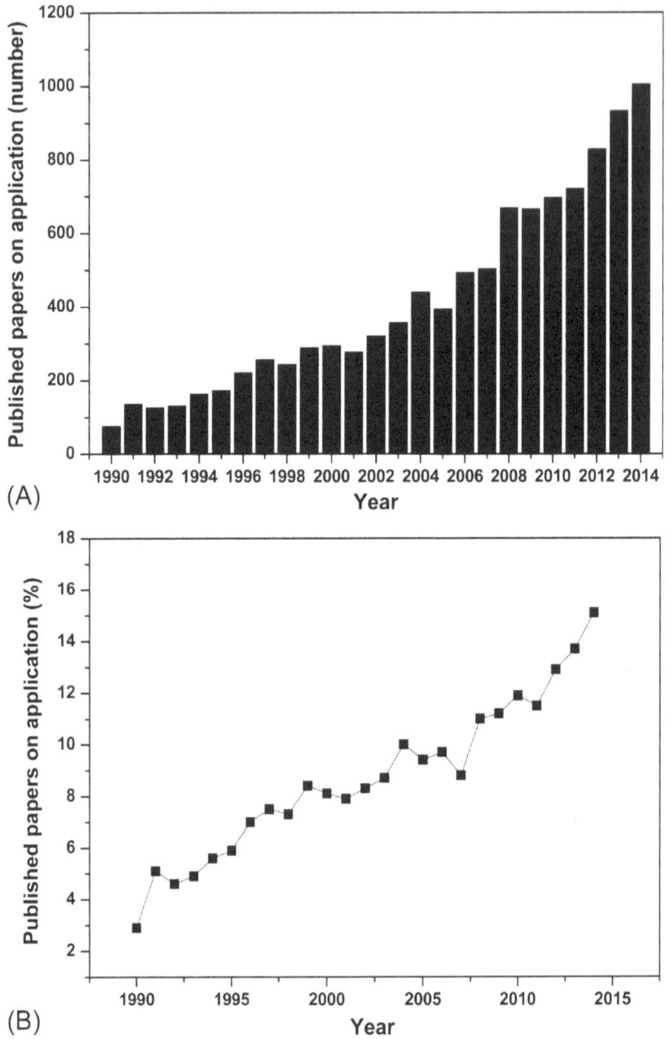


**Fig. 1.10** (A) Number of paper published per year on glasses and glass-ceramics based on application during the last 25 years (1990–2014). Data were extracted from the data as shown in Fig. 1.2A by further refining using the keyword "application." (B) Year-wise plot of percentage of papers published on application of glasses and glass-ceramics calculated from the data as shown in Fig. 1.2A.

the improvement of properties and finding out newer applications of glasses and glass-ceramics. It is also realized that the researchers should give attention not only to property and application oriented investigations but also to improved device making investigations, because of the ultimate destination or use of glasses and glass-ceramics is in proper functioning for a "*targeted action*" in a device.

## Exercises

**1.1** What are the definitions of functional glasses and glass-ceramics?
**1.2** What are the effects of composition and structure on the properties and functions of silicate, borate, germanate, and phosphate glasses? Explain with suitable examples. (Hint: consult references [27,36,37,43].)
**1.3** What are the effects of composition and structure on the properties and functions of lithium aluminosilicate and magnesium boroaluminosilicate glass-ceramics? Explain with suitable examples. (Hint: consult references [12,39,47].)
**1.4** Name different kinds of nonoxide glass systems with their most important and functional characteristics.
**1.5** Name different functions and the corresponding glass and glass-ceramics products.
**1.6** Explain the correlation between different properties and functions of glasses and glass-ceramics.
**1.7** Name several application area-wise different glass products and their purposes.
**1.8** Mention with proper justification the classification of functional glasses and glass-ceramics.
**1.9** What are the current research trends on glasses and glass-ceramics? Analyze the trends.
**1.10** In your opinion, what would be the future research motive on glasses and glass-ceramics?
**1.11** Explain your answer with proper justification.

## References

[1] G.W. Morey, The Properties of Glass, Reinhold Publishing Corp., New York, NY, 1938.
[2] W.A. Weyl, Coloured Glasses, Society of Glass Technology, Sheffield, 1951.
[3] S.R. Scholes, Modern Glass Practice, Industrial Publications, Chicago, IL, 1952.
[4] G.O. Jones, Glass, Methuen & Co., London, 1956.
[5] J.D. Mackenzie (Ed.), In: Modern Aspects of the Vitreous State, Vols. I, II, and III, Butterworth & Co. (Publishers) Ltd., London, 1960.
[6] F.V. Tooley, Handbook of Glass Manufacture, Vols. I and II, Ogden Publishing Co., New York, NY, 1960.
[7] K. Pátek, Glass Lasers, Butterworth & Co (Publishers) Ltd., London, 1970.
[8] R.H. Doremus, Glass Science, John Wiley & Sons, New York, NY, 1973.
[9] J. Wong, C.A. Angell, Glass Structure by Spectroscopy, Marcel Dekker, Inc., New York, NY, 1976.
[10] C.R. Bamford, Colour Generation and Control in Glass, Elsevier, Amsterdam, 1977.
[11] C.L. Babcock, Silicate Glass Technology Methods, John Wiley & Sons, New York, NY, 1977.
[12] P.W. McMillan, Glass-Ceramics, second ed., Academic Press, London, 1979.
[13] H. Rawson, Properties and Applications of Glass, Elsevier, Amsterdam, 1980.
[14] A. Paul, Chemistry of Glasses, Chapman and Hall, London, 1982.
[15] V. Fanderlik, Optical Properties of Glass, Elsevier, Amsterdam, 1983.
[16] M.B. Volf, Chemical Approach to Glass, Elsevier, Amsterdam, 1984.
[17] H.K. Pulker, Coatings on Glass, Elsevier, Amsterdam, 1984.
[18] D.R. Uhlmann, N.J. Kreidl (Eds.), Glass: Science and Technology, Vol. 2: Processing I, Academic Press, New York, NY, 1984.

[19] G.W. McLelland, E.B. Shand, Glass Engineering Handbook, McGraw-Hill, New York, NY, 1984.
[20] M. Tomozawa, R.H. Doremus (Eds.), Glass, Academic Press, New York, NY, 1985.
[21] T.S. Izumitani, Optical Glass, American Institute of Physics, New York, NY, 1986.
[22] Z. Strnad, Glass-Ceramic Materials, second ed., Elsevier Science Publishers, Amsterdam, 1986.
[23] A.E. Comyns (Ed.), Fluoride Glasses, John Wiley & Sons, New York, NY, 1989.
[24] M.H. Lewis, Glasses and Glass-Ceramics, Chapman and Hall, London, 1989.
[25] S.R. Elliott, Physics of Amorphous Materials, Longman, London, 1990.
[26] P.W. France, M.G. Drexhage, J.M. Parker, M.W. Moore, S.F. Carter, J.V. Wright, Fluoride Glass Optical Fibres, Blackie, London, 1990.
[27] H. Scholze, Glass: Nature, Structure, and Properties, Springer-Verlag, New York, NY, 1991.
[28] J. Zarzycki, Glasses and the Vitreous State, Cambridge University Press, Cambridge, 1991.
[29] I.D. Aggarwal, G. Lu (Eds.), Fluoride Glass Fiber Optics, Academic Press, Boston, MA, 1991.
[30] N.F. Borrelli, D.W. Hall, D.R. Uhlmann, N.J. Kreidl, Optical Properties of Glass, The American Ceramic Society, Westerville, OH, 1991.
[31] K.M. Nair (Ed.), Glasses for Electronic Applications, The American Ceramic Society, Westerville, OH, 1991.
[32] I. Fanderlik, Silica Glass and Its Application, Elsevier Science Publishers, Amsterdam, 1991.
[33] H. Rawson, Glasses and Their Applications, Institute of Materials Publications, London, 1991.
[34] M. Cable, J.M. Parker, High Performance Glasses, Blackie, Glasgow, 1991.
[35] G. Fuxi, Optical and Spectroscopic Properties of Glass, Springer-Verlag, Berlin, 1992.
[36] W. Vogel, Glass Chemistry, second ed., Springer-Verlag, Berlin, 1994.
[37] A.K. Varshneya, Fundamentals of Inorganic Glasses, Academic Press, Boston, MA, 1994.
[38] J.E. Shelby (Ed.), Rare Elements in Glasses, Trans Tech Publications, Zurich, 1994.
[39] H. Bach (Ed.), Low Thermal Expansion Glass Ceramics, Springer-Verlag, Berlin, 1995.
[40] H. Bach, N. Neuroth (Eds.), The Properties of Optical Glass, Springer, Berlin, 1995.
[41] H.G. Pfaender, Schott Guide to Glass, Chapman & Hall, London, 1996.
[42] V.F. Kokorina, Glasses for Infrared Optics, CRC Press, New York, NY, 1996.
[43] J.E. Shelby, Introduction to Glass Science and Technology, The Royal Society of Chemistry, Cambridge, 1997.
[44] H. Bach, D. Krause (Eds.), Analysis of the Composition and Structure of Glass and Glass Ceramics, Springer, Berlin, 1999.
[45] M. Yamane, Y. Asahara, Glasses for Photonics, Cambridge University Press, Cambridge, 2000.
[46] R.A.H. El-Mallawany, Tellurite Glasses Handbook: Physical Properties and Data, CRC Press, Boca Raton, FL, 2002.
[47] W. Höland, G. Beall, Glass-Ceramic Technology, The American Ceramic Society, Westerville, OH, 2002.
[48] R. Fairman, B. Ushkov (Eds.), Semiconducting Chalcogenide Glass I: Glass Formation, Structure, and Simulated Transformations in Chalcogenide Glasses, Academic Press, New York, NY, 2004.
[49] R. Fairman, B. Ushkov (Eds.), Semiconducting Chalcogenide Glass II: Properties of Chalcogenide Glasses, Academic Press, New York, NY, 2005.

[50] R. Fairman, B. Ushkov (Eds.), Semiconducting Chalcogenide Glass III: Applications of Chalcogenide Glasses, Academic Press, New York, NY, 2005.
[51] V. Marghussian, Nano-Glass Ceramics: Processing, Properties and Applications, Elsevier, Amsterdam, 2015.
[52] B. Karmakar, K. Rademann, A.L. Stepanov (Eds.), Glass Nanocomposites: Synthesis, Properties and Applications, Elsevier, Amsterdam, 2016.
[53] P.H. Gaskell, Structure and properties of glasses—how far do we need to go? J. Non-Cryst. Solids 222 (1997) 1–12.

# Part Two

# Functional glasses and glass-ceramics of oxide systems

# Functionalization of glasses by 3d transition metal ion doping

2

## Chapter Outline



## 2.1 Introduction

Oxide glasses exhibit a wide range of transparency in the wavelength range of 0.2–4.5 μm of the electromagnetic spectrum. This transparency is largely governed by two factors, namely, chemical composition and impurity. The chemical composition mainly controls the ultraviolet (UV) and infrared (IR) absorption edges in the wavelength ranges of 0.2–0.4 and 3.0–4.5 μm, respectively. In the absence of any impurities in the glasses, there is no absorption band in the range 0.3–3.0 μm. The common impurities in glass, which exhibit absorption bands in between this range, are hydroxyl group (—OH group), 3d transition metal ions, 4f-rare-earth metal ions, etc. Therefore, these impurities must be absent in the glasses when we are dealing with the colorless clear transparent glasses. Nevertheless, these colorless clear transparent glasses could be functionalized for certain specific purposes by doping (i.e., adding in small quantity) with 3d transition metal (scandium to copper) ions in the form of oxide and exploiting their specific or combination of absorption bands in the wavelength range of 0.3–3.0 μm. Atomic number, symbol, name, valences, and outer shell electronic configuration of the 3d transition metal elements are depicted in Fig. 2.1. In the visible

**Functional Glasses and Glass-Ceramics. http://dx.doi.org/10.1016/B978-0-12-805056-9.00002-7**

**3d transition metal elements**

| | | | | | | | | |
|---|---|---|---|---|---|---|---|---|
| 21 | 22 | 23 | 24 | 25 | 26 | 27 | 28 | 29 |
| Sc | Ti | V | Cr | Mn | Fe | Co | Ni | Cu |
| **Scandium** | **Titanium** | **Vanadium** | **Chromium** | **Manganese** | **Iron** | **Cobalt** | **Nickel** | **Copper** |
| +3 | +2, +3 +4 | +2 to +5 | +2, +3, +4, +6 | +2 to +7 | +2, +3 | +2, +3 | +2, +3 | +1, +2 |
| $3d^14s^2$ | $3d^24s^2$ | $3d^34s^2$ | $3d^54s^1$ | $3d^54s^2$ | $3d^64s^2$ | $3d^74s^2$ | $3d^84s^2$ | $3d^{10}4s^1$ |

26 ← Atomic number
Fe ← Symbol
**Iron** ← Name
+2, +3 ← Valences
$3d^64s^2$ ← Outer electronic configuration

**Fig. 2.1** Atomic number, symbol, name, valences, and outer shell electronic configuration of the 3d-transition-metal elements.

wavelengths (380–780 nm), they exhibit various human eye detectable colors (e.g., red, green, and blue) in the glasses. Some of the uses of the 3d transition metal ion functionalized glasses are: colored glasses, signal glasses, filter glasses, laser glasses, etc.

However, the functions (position of the absorption bands and their extent of absorption) of the 3d transition metal ions in the glasses depend on several factors such as

**(i)** chemical composition and component of the glass,
**(ii)** concentration and combination of 3d transition metal oxides in the glass,
**(iii)** glass melting atmosphere (e.g., oxidizing, reducing, and inert),
**(iv)** glass melting temperature and duration,
**(v)** sources of raw materials,
**(vi)** volume of glass melted (prepared) per melt, and
**(vii)** glass melting container (e.g., refractory or noble metal crucible and tank furnace).

Therefore, for successful functionalization of glasses by 3d transition metal ion doping, it is necessary to control and consider all of the above factors. As a consequence of this fact, there are a huge number of publications in this domain.

An attempt was made using the Web of Science (version 5.17) as described in Chapter 1 to provide a global publication trend and quantitative growth in the area of 3d transition metal ion doped glasses and glass-ceramics. Here, year-wise data were extracted for the last 25 years (1990–2014) by searching, using the topic keyword "*glass*" followed by refining the document type as "*article*" published in the journals and meetings. It was further refined using the term "*transition metal*." This plot is shown in Fig. 2.2A. It gives us a clear publication trend and quantitative growth of published papers on transition metal containing glasses and glass-ceramics during the last 25 years (1990–2014). To find out the total number of published papers on individual transition metal ion containing glasses and glass-ceramics during the last 25 years (1990–2014), the total publication of 112,633 articles was further refined by term of "*individual 3d transition metal*" (e.g., iron, copper, and titanium). This plot is illustrated in Fig. 2.2B. It gives us a clear quantitative idea of paper published on individual transition metal ion containing glasses and glass-ceramics as well as a

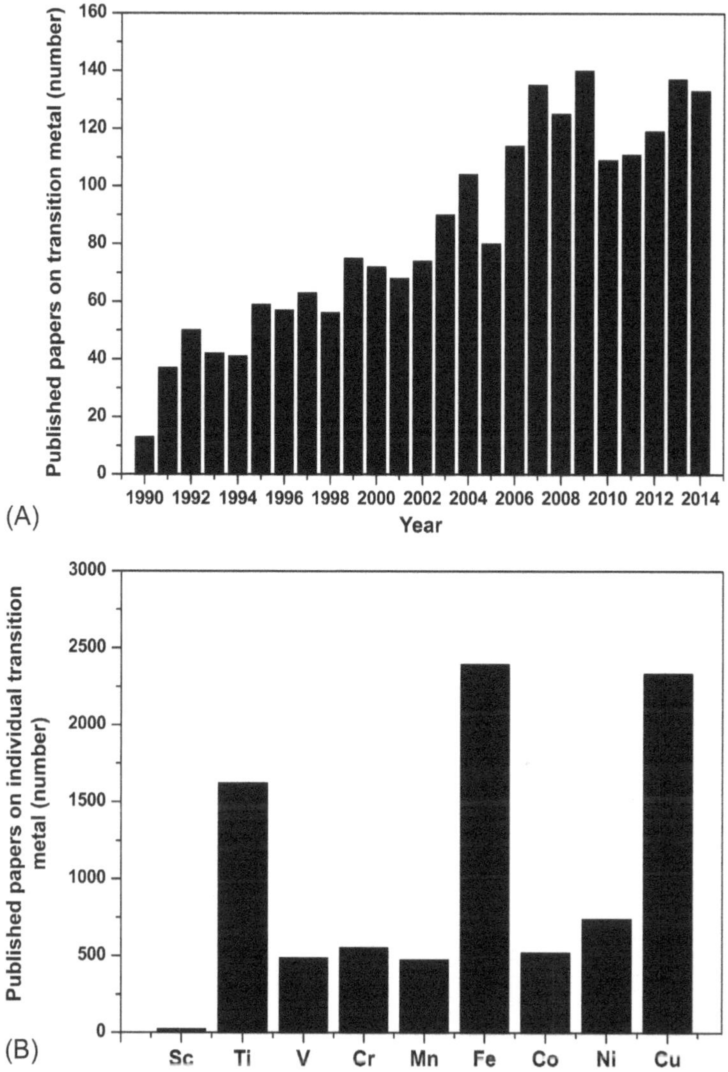


**Fig. 2.2** (A) Year-wise published papers on transition metal containing glasses and glass-ceramics during the last 25 years (1990–2014). (B) Total number of published papers on individual transition metal ion during the last 25 years (1990–2014).

comparison among them. The decreasing order of publication trend of the individual transition metal ion doped glasses and glass-ceramics is as follows:

$$Fe > Cu > Ti \gg Ni > Cr > Co > V > Mn \ggg Sc$$

It is seen that during the last 25 years (1990–2014) among the 112,633 published articles, the highest number (2394) of papers was published using iron (Fe), the second highest (2335) is with copper (Cu), the third position (1623) is with titanium (Ti), and the lowest position (22 only) is with scandium (Sc). This trend of publications also gives us the idea about the importance of doping the glasses with these ions.

It is necessary to know the underlying theories to understand the activities of these ions in glasses particularly with respect to their functionality. For this reason, the necessary theories are briefly described in the next section.

## 2.2 Theoretical considerations

There are four different theories to explain different aspects of complex ion or compound formation of 3d transition metal ions with oxygen or other ions or molecules. These theories are

**(i)** valence bond theory (VBT),
**(ii)** molecular orbital theory (MOT),
**(iii)** crystal field theory (CFT), and
**(iv)** ligand field theory (LFT).

### 2.2.1 *Valence bond theory*

The valence bond theory considers only the bonding between two atoms at a time using the electrons of the valence orbitals through hybridization (e.g., $sp^3$, $sp^3d$, and $sp^3d^2$) of atomic orbitals (e.g., s, p, d, and f ). Hybridization of the transition metal ion with oxygen ions in the complex ions can well be explained by the valence bond theory. Some examples are presented below.

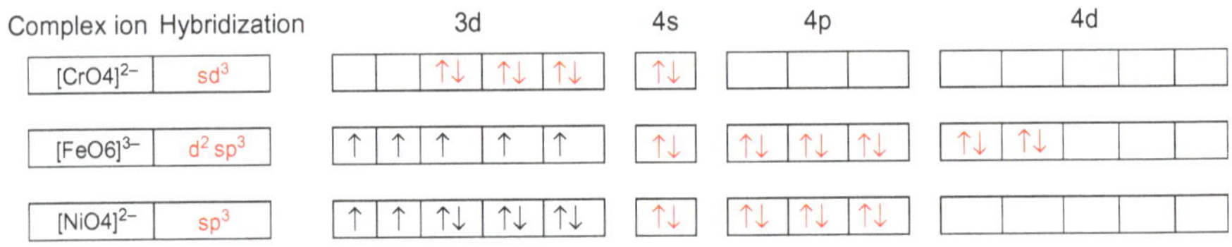


According to this hybridization, it is seen that the $[CrO_4]^{2-}$ ion is an inner-orbital complex, whereas $[FeO_6]^{3-}$ and $[NiO_4]^{2-}$ ions are outer-orbital complexes. Some general characteristics of 3d transition metal ions in glasses [1,2] are listed in Table 2.1. It depicts the electronic configuration, coordination number, complex ions, geometry, orbital configuration, and hybridization of complex ions along with the glass color and absorption region. The colors are generated due to the incompletely filled d-orbitals ($d^{1-9}$) of 3d transition metal ions and formation of tetrahedral, octahedral, square planar, etc. complex ions by them with oxygen ions. The VBT can explain the geometry of the complex ions, but it cannot explain the electronic transitions for the color generation, that is, the spectroscopic properties and magnetic properties. For this reason, it is superseded by MOT, CFT, and LFT. The MOT can explain it very well. It is a very useful theory but more sophisticated and to some extent difficult to understand and use.

Table 2.1 **General characteristics of 3d transition metal ions in glass [1,2]**

| Electronic configuration | Metal ion | Coordination number | Geometry[a] | Orbital configuration | Complex ion | Hybridization | Glass color | Absorption region |
|---|---|---|---|---|---|---|---|---|
| $3d^0$ | $Ti^{4+}$ | | | | | | Colorless | |
| | $V^{5+}$ | | | | | | Yellow to colorless | |
| | $Cr^{6+}$ | 4 | $T_d$ | | $[CrO_4]^{2-}$ | $sd^3$ | Light yellow | Intense UV absorption |
| $3d^1$ | $Ti^{3+}$ | $6^b$ | $O_h$ | $t_{2g}^1$ | $[TiO_6]^{3-}$ | $d^2sp^3$ | Violet-purple | Intense UV absorption |
| | $V^{4+}$ | | | | | | Blue | |
| | $Mn^{6+}$ | | | | | | Colorless | |
| $3d^2$ | $V^{3+}$ | 6 | $O_h$ | $t_{2g}^2$ | $[VO_6]^{3-}$ | $d^2sp^3$ | Greenish | UV absorption |
| $3d^3$ | $Cr^{3+}$ | 6 | $O_h$ | $t_{2g}^3$ | $[CrO_6]^{3-}$ | $d^2sp^3$ | Green | |
| $3d^4$ | $Cr^{2+}$ | | | | | | Light blue | |
| | $Mn^{3+}$ | $6^b$ | $O_h$ | $t_{2g}^3e_g^1$ | $[MnO_6]^{3-}$ | $d^2sp^3$ | Violet | UV absorption |
| $3d^5$ | $Mn^{2+}$ | 4 | $T_d$ | $t_2^3e^2$ | $[MnO_4]^{2-}$ | $sp^3$ | Yellowish | |
| | $Fe^{3+}$ | 6 | $O_h$ | $t_{2g}^3e_g^2$ | $[FeO_6]^{3-}$ | $d^2sp^3$ | Yellow | UV absorption |
| $3d^6$ | $Fe^{2+}$ | 6 | $O_h$ | $t_{2g}^4e_g^2$ | $[FeO_6]^{4-}$ | $d^2sp^3$ | Blue-green | IR absorption |
| | $Co^{3+}$ | | | $t_{2g}^4e_g^2$ | $[CoO_6]^{3-}$ | $d^2sp^3$ | Light yellow | |
| $3d^7$ | $Co^{2+}$ | 4 | $T_d$ | $t_2^3e^4$ | $[CoO_4]^{2-}$ | $sp^3$ | Deep blue | IR absorption |
| $3d^8$ | $Ni^{2+}$ | 4 | $T_d$ | $t_2^4e^4$ | $[NiO_4]^{2-}$ | $sp^3$ | Yellow-brown | IR absorption |
| $3d^9$ | $Cu^{2+}$ | $6^b$ | $O_h$ | $t_{2g}^6e_g^3$ | $[CuO_6]^{2-}$ | $d^2sp^3$ | Light blue | IR absorption |
| $3d^{10}$ | $Cu^+$ | | | | | | Colorless | |

[a] $O_h$, octahedral; $T_d$, tetrahedral.
[b] Tetragonal distortion: Jahn-Teller effect.

### 2.2.2 Ligand field theory

In contrast, the crystal field theory (CFT) is simpler than the MOT. The crystal field theory is based on completely ionic character of bonding, that is, interaction between metal (e.g., 3d transition metal) ion and ligand (e.g., oxygen) ion is electrostatic. The ligands are treated as point charges. Here, covalent character of bonding is completely neglected, whereas the ligand field theory (LFT) includes both ionic and covalent aspects of bonding. Hence, the LFT accepted all the assumptions of CFT including covalent characters of the ligands. These are as follows. It is worth mentioning that the LFT is more popular compared with other theories. For this reason, it is described here:

- **(i)** In an isolated 3d transition metal atom or ion, the d-orbitals are all degenerate, that is, they have identical orbital energy. This is shown in Figs. 2.3 [3] and 2.4.
- **(ii)** If a spherically symmetrical field of negative charges is placed around the metal ion, these orbitals remain degenerate, but all of them are raised in energy as a result of the repulsion between the negative charges on the ligands and in the d-orbitals. This hypothetical spherical field is known as "*barycenter*" (see Fig. 2.4).
- **(iii)** When a ligand is added, the spherical symmetry of the d-orbitals is broken.
- **(iv)** The appearance of their energy levels depends on the structure of the complex ion (e.g., octahedral and tetrahedral). This is shown in Fig. 2.4.
- **(v)** The separation of energy levels depends on the nature of the ligands. Some ligands produce larger separation than others.

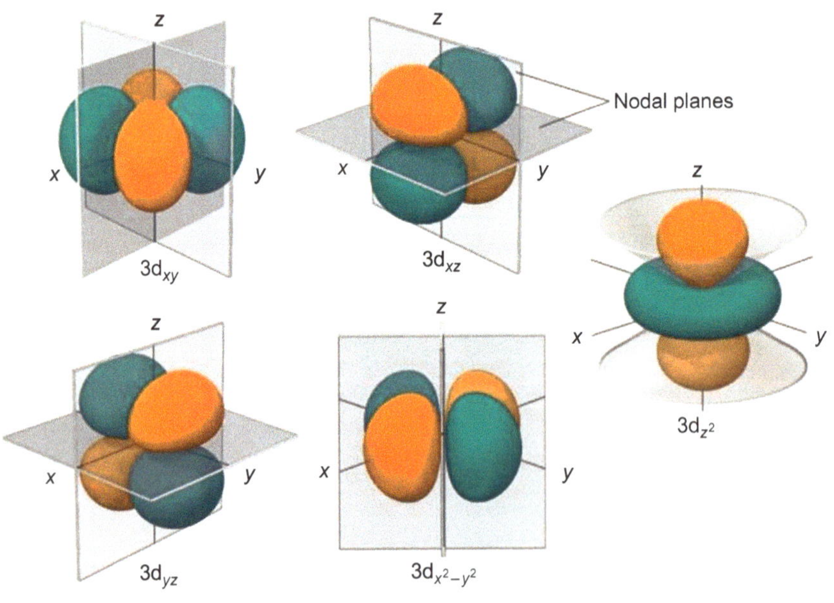


**Fig. 2.3** Five equivalent d-orbitals without any external field.
Courtesy: B. Averill, P. Eldredge, General Chemistry: Principles, Patterns, and Applications, Flat World Knowledge, Washington, DC, 2011.

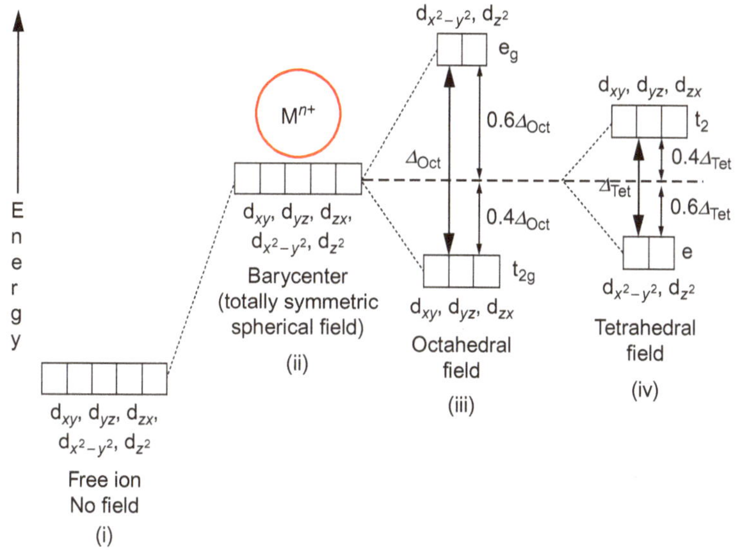


**Fig. 2.4** Energy level diagrams showing the position of all the five d-orbitals (i) without any external field and under (ii) totally symmetrical spherical field and their splitting under the (iii) octahedral and (iv) tetrahedral fields. It is seen that as a result of higher interaction the octahedral splitting energy ($\Delta_{Oct}$) is more than that of tetrahedral ($\Delta_{Tet}$).

In the octahedral ($O_h$) point group, the $d_{x^2-y^2}$ and $d_{z^2}$ orbitals belong to the $e_g$ irreducible representation, and $d_{xy}$, $d_{yz}$, and $d_{xz}$ belong to the $t_{2g}$ representation. In octahedral field, the $d_{x^2}-d_{y^2}$ and $d_{z^2}$ orbitals ($e_g$) point directly toward ligands, which are placed on the centers of the faces of a cube, whereas the $d_{xy}$, $d_{yz}$ and $d_{zx}$ orbitals ($t_{2g}$) do not. These are shown in Figs. 2.5 and 2.6A and also see Fig. 2.3. This enhances more repulsion between the electrons of the $d_{x^2}-d_{y^2}$ and $d_{z^2}$ orbitals with ligands and consequently raises their energy level relative to the other three, $t_{2g}$ (see Fig. 2.4). The magnitude to which $e_g$ and $t_{2g}$ sets of orbitals are split is denoted by $\Delta_{Oct}$. The $t_{2g}$ set must be stabilized by the energy as much as the $e_g$ set is destabilized to conserve the barycenter energy in octahedral field.

In the tetrahedral ($T_d$) point group, the $d_{x^2-y^2}$ and $d_{z^2}$ orbitals belong to the e representation, and $d_{xy}$, $d_{yz}$, and $d_{xz}$ belong to the $t_2$ representation. In tetrahedral complex (Fig. 2.6B), the ligands are on the alternate corners of two opposite faces of a cube. This is shown in Fig. 2.7. Under this situation, it is the $t_2$ orbitals that are destabilized as they point toward the incoming point charges of the ligands and the e orbitals are stabilized so that the barycenter is preserved (Fig. 2.4). The magnitude to which $t_2$ and e sets of orbitals are split is denoted by $\Delta_{Tet}$. In comparison with the octahedral ligand field, the energy of the $d_{xy}$, $d_{yz}$, and $d_{xz}$ orbitals ($t_2$) increases, while the energy of the $d_{x^2-y^2}$ and $d_{z^2}$ orbitals (e) decreases in a tetrahedral ligand field. Since there are smaller numbers of ligands and they do not line up as well with the transition metal d-orbitals (Fig. 2.7), the magnitude of the tetrahedral splitting ($\Delta_{Tet}$) is roughly 4/9 of the octahedral splitting ($\Delta_{Oct}$).

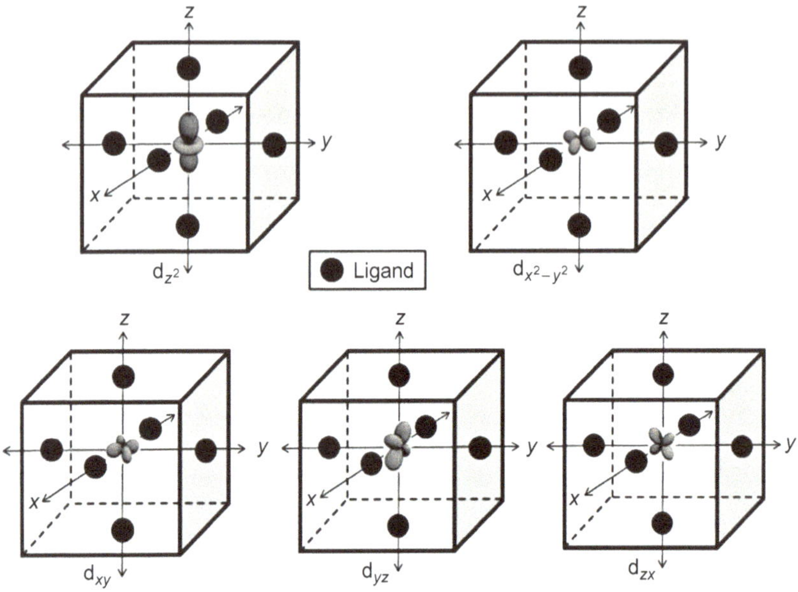


**Fig. 2.5** Interaction of all the five d-orbitals with ligands in an octahedral field. In a cube, the ligands of an octahedral complex are on the center of the cube faces.

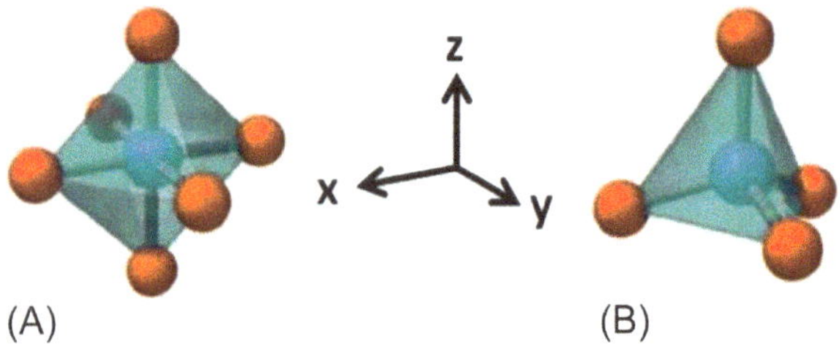


**Fig. 2.6** Geometric arrangements and metal-ligand (blue and red balls represent for 3d transition metal ion and oxygen ligand, respectively) interactions under (A) octahedral and (B) tetrahedral fields. It is seen that the interaction of d-orbitals of the metal ion with the oxygen ligands is more in the situation of octahedral field and less in the case of tetrahedral field.

The readers are referred to Bates [4], Bamford [5], and Paul [6] for detailed descriptions of these theories.

#### *2.2.2.1 Tetragonal distortion of octahedral complex (The Jahn-Teller effect)*

The Jahn-Teller (J-T) effect is very common in 3d transition metal ion doped oxide glasses. Therefore, it is necessary to describe it here. In some octahedral complexes, whenever the $e_g$ or $t_{2g}$ set of orbitals are unequally filled with electrons (e.g., one or

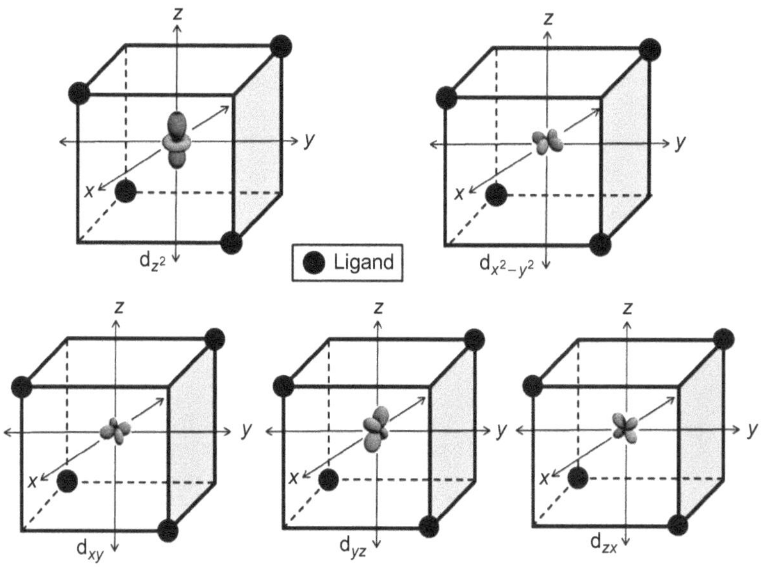


**Fig. 2.7** Interaction of all the five d-orbitals with ligands in a tetrahedral field. In a cube, the ligands of a tetrahedral complex are on the alternate corners of two opposite faces of the cube.

three electrons in the $d_{x^2} - d_{y^2}$ and $d_{z^2}$ orbitals of the $e_g$ set), some distortion occurs. This is known as "*Jahn-Teller effect or distortion.*" The J-T theorem states that "any nonlinear complex in a degenerate electronic state will be unstable and will undergo some sort of distortion to lower its symmetry and remove the degeneracy." The complexes of $[MnO_6]^{3-}$ and $[CuO_6]^{2-}$ ions have the orbital configurations $t_{2g}^3e_g^1$ and $t_{2g}^6e_g^3$, respectively, (Table 2.1) and exhibit the "*Jahn-Teller effect or distortion.*" In octahedral field, the $d_{x^2} - d_{y^2}$ and $d_{z^2}$ orbitals of the $e_g$ set are generally degenerate (of same energy, Figs. 2.4 and 2.8). If they are asymmetrically filled such as

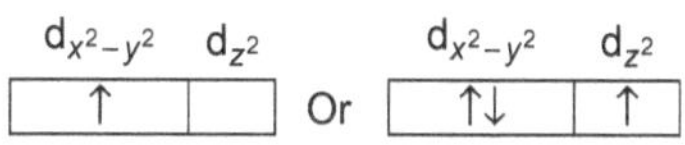


then their degeneracy is lost, that is, the two orbitals are no longer of identical in energy. Under this situation, the ligand approaching along the *z*-axis encounter greater repulsion than the other four ligands of the *x*-*y* plane (see Figs. 2.5 and 2.8). This repulsion and distortion result in elongation of the octahedron along the *z*-axis and compression in the *xy*-plane. This is shown in Fig. 2.8 (right side) and known as "*tetragonal distortion.*" Compressive J-T effects occur when the degeneracy is destroyed by the stabilization (dropping in energy) of the d-orbitals without a *z*-component, while the orbitals with a *z*-component are destabilized (greater in energy). This is due to the higher energy of the *z*-component d-orbitals have better overlap with the ligand orbitals. This is also shown in Fig. 2.8 (left side).

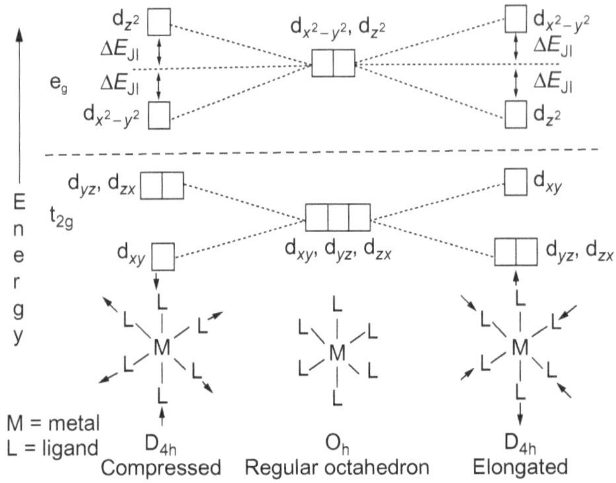


**Fig. 2.8** Energy diagram for elongation (right) and compression (left) Jahn-Teller effects. Influences on the structures of the complexes are also illustrated at the bottom.

## 2.3 Synthesis, batch calculation, and characterization of glass

### 2.3.1 Synthesis

Glasses are prepared from the mixture of raw materials by heating them to the molten liquid state at high temperature, in the range of 1200–1600°C depending on the composition and components of glasses, followed by annealing near the glass transition temperature to release the residual thermal stresses. In research laboratory practice, high pure chemicals are used as raw materials, whereas less pure minerals are used in industry for commercial production of glass objects. These raw materials eliminate the volatile materials such as adsorbed moisture ($H_2O$), water ($H_2O$) of crystallization or hydration, and carbon dioxide ($CO_2$) from carbonates on heating at high temperature and convert into resultant oxides. Each glass has a predetermined fixed oxide composition. It has also a fixed batch (a mixture of calculated raw materials of the glass) composition consisting of fixed raw materials. Therefore, it is necessary to calculate quantitatively the extent of loss from these raw materials, due to volatilization at high temperature, to keep the final product composition unchanged.

### 2.3.2 Batch calculation

In practice, the batch calculation is the chemical calculation. For successful batch calculation, it is necessary to know the following computations:

**(i)** Conversion of glass composition from weight (wt.%) to mole (mol.%) and vice versa
**(ii)** Calculation of conversion factors from raw material to oxide and vice versa

**(iii)** Calculation of batch composition (in weight) for a fixed glass composition using specified raw materials
**(iv)** Calculation of glass composition (in wt.% or mol.%) from batch composition

These are described as follows with a single example of the most simple soda lime silicate glass. The single example of simple soda lime silicate glass was considered to figure out the differences in the above cases.

(i) ***Conversion of glass composition from weight (wt.%) to mole (mol.%) and vice versa***

*Example 1.* Convert the glass composition of $15Na_2O$–$10CaO$–$75SiO_2$ (wt.%) into mole percent (mol.%).

| **Oxide component** | **Composition (wt.%)** | **Molecular weight (MW)** | **Mole fraction** | **Composition (mol.%)** |
|---|---|---|---|---|
| $Na_2O$ | 15 | 61.98 | 0.2420 | 14.50 |
| $CaO$ | 10 | 56.08 | 0.1783 | 10.69 |
| $SiO_2$ | 75 | 60.09 | 1.2481 | 74.81 |
| | | *Total* | *1.6684* | |

The following formulas are used in the above calculations:

$$\text{Mole fraction} = \frac{\text{Composition in wt.\%}}{\text{Molecular weight}}$$

$$\text{Composition in mol.\%} = \frac{\text{Individual mole fraction} \times 100}{\text{Summation of mole fractions}}$$

*Example 2.* Convert the glass composition of $15Na_2O$–$10CaO$–$75SiO_2$ (mol.%) into weight percent (wt.%).

| **Component** | **Composition (mol.%)** | **Molecular weight (MW)** | **Composition in weight** | **Composition (wt.%)** |
|---|---|---|---|---|
| $Na_2O$ | 15 | 61.98 | 929.70 | 15.50 |
| $CaO$ | 10 | 56.08 | 560.80 | 9.35 |
| $SiO_2$ | 75 | 60.09 | 4506.75 | 75.15 |
| | | *Total* | *5997.25* | |

The following formulations are used in the above calculations:

$$\text{Composition in weight} = \text{Composition in mol.\%} \times \text{Molecular weight}$$

$$\text{Composition in wt.\%} = \frac{\text{Individual composition in weight} \times 100}{\text{Summation of composition in weights}}$$

(ii) ***Calculation of conversion factors for from raw material to oxide and vice versa***

*Example 3.* Calculate the conversion factors for from soda ash ($Na_2CO_3$) to sodium oxide ($Na_2O$) and vice versa.

The stoichiometric decomposition equation is as follows:

$$\underset{105.98\,g}{Na_2CO_3} = \underset{61.98\,g}{Na_2O} + CO_2\uparrow$$

Conversion factor for soda ash ($Na_2CO_3$) to sodium oxide ($Na_2O$)

$$= \frac{\text{Wt. of } Na_2O \text{ in stoichiometric decomposition equation}}{\text{Wt. of } Na_2CO_3 \text{ in stoichiometric decomposition equation}} = 0.585$$

Conversion factor for sodium oxide ($Na_2O$) to soda ash ($Na_2CO_3$)

$$= \frac{\text{Wt. of } Na_2CO_3 \text{ in stoichiometric decomposition equation}}{\text{Wt. of } Na_2O \text{ in stoichiometric decomposition equation}} = 1.710$$

Similarly, the conversion factors for other raw materials could be calculated. Different raw materials and their conversion factors are listed in Table 2.2. In actual fact, the oxide materials of 100% purity have the conversion factors of unity. Here, the raw materials have been broadly classified into the groups of glass former, glass modifier, intermediate, refining agent, coloring agent, and opacifying agent. Now, we shall exemplify how to use these conversion factors for calculation of batch and glass composition.

**(iii)** ***Calculation of batch composition (in weight) for a fixed glass composition using specified raw materials***

*Example 4*. Calculate the batch composition for 1000 g of a glass of composition $15Na_2O$–$10CaO$–$75SiO_2$ (wt.%) using soda ash, limestone, and sand as the raw materials.

| Glass oxide component | Glass composition (wt.%) | Raw material | Conversion factor | Wt. of raw material for 100 g glass (g) | Batch composition for 1000 g glass (g) |
|---|---|---|---|---|---|
| $Na_2O$ | 15 | Soda ash | 1.710 | $1.710 \times 15 = 25.65$ | 256.5 |
| $CaO$ | 10 | Limestone | 1.786 | $1.786 \times 10 = 17.86$ | 178.6 |
| $SiO_2$ | 75 | Sand | 1.000 | $1.000 \times 75 = 75.00$ | 750.0 |

The following formula is used for the above calculation:

$$\text{Wt. of raw material for 100g glass (g)} = (\text{Conversion factor}) \times (\text{Weight of individual oxide in 100g glass})$$

Therefore, the batch composition for 1000 g of glass is 256.5 g soda ash, 178.6 g limestone, and 750.0 g sand.

**(iv)** ***Calculation of glass composition (in wt.% or mol.%) from batch composition***

*Example 5*. Calculate the glass composition in wt.% and mol.% from the batch composition of 256.5 g soda ash, 178.6 g limestone, and 750.0 g sand.

| Raw material | Batch composition (g) | Glass oxide component | Conversion factor | (Conversion factor) × (individual batch weight) | Wt. of oxide derived from batch (g) | Glass composition (wt.%) |
|---|---|---|---|---|---|---|
| Soda ash | 256.5 | $Na_2O$ | 0.585 | $0.585 \times 256.5$ | 150.05 | 15.00 |
| Limestone | 178.6 | $CaO$ | 0.560 | $0.560 \times 178.6$ | 100.02 | 10.00 |
| Sand | 750.0 | $SiO_2$ | 1.000 | $1.000 \times 750.0$ | 750.00 | 74.99 |
| | | | | *Total* | *1000.07* | |

Table 2.2 **Raw materials used in glass making and their respective conversion factors required for batch and glass composition calculations [7,8]**

| Raw material | Chemical formula | Supplied oxide | Oxide content (%) | Conversion factor | |
|---|---|---|---|---|---|
| | | | | Raw material to oxide | Oxide to raw material |
| ***Glass former*** | | | | | |
| Sand, quartz | $SiO_2$ | $SiO_2$ | 100 | 1.000 | 1.000 |
| Boric acid | $H_3BO_3$ | $B_2O_3$ | 56.3 | 0.563 | 1.776 |
| Anhydrous borax | $Na_2B_4O_7$ | $B_2O_3$ | 69.2 | 0.692 | 1.445 |
| | | $Na_2O$ | 30.8 | 0.308 | 3.247 |
| Borax (hydrated) | $Na_2B_4O_7{\cdot}10H_2O$ | $B_2O_3$ | 36.5 | 0.365 | 2.740 |
| | | $Na_2O$ | 16.3 | 0.163 | 6.135 |
| Phosphorous oxide | $P_2O_5$ | $P_2O_5$ | 100 | 1.000 | 1.000 |
| Sodium phosphate | $NaPO_3$ | $P_2O_5$ | 69.6 | 0.696 | 1.437 |
| | | $Na_2O$ | 30.4 | 0.304 | 3.290 |
| Calcium phosphate | $Ca_3(PO_4)_2$ | $P_2O_5$ | 45.8 | 0.458 | 2.183 |
| | | CaO | 54.2 | 0.542 | 1.845 |
| Germanium oxide | $GeO_2$ | $GeO_2$ | 100 | 1.000 | 1.000 |
| ***Glass modifier*** | | | | | |
| Lithium carbonate | $Li_2CO_3$ | $Li_2O$ | 40.4 | 0.404 | 2.475 |
| Soda ash | $Na_2CO_3$ | $Na_2O$ | 58.5 | 0.585 | 1.710 |
| Sodium nitrate (Chile saltpeter) | $NaNO_3$ | $Na_2O$ | 36.5 | 0.365 | 2.740 |
| Salt cake | $Na_2SO_4$ | $Na_2O$ | 43.7 | 0.437 | 2.288 |
| Potash | $K_2CO_3$ | $K_2O$ | 68.2 | 0.682 | 1.466 |
| Potassium nitrate (saltpeter) | $KNO_3$ | $Na_2O$ | 46.6 | 0.466 | 2.146 |

*Continued*

**Table 2.2 Continued**

| Raw material | Chemical formula | Supplied oxide | Oxide content (%) | Conversion factor | |
|---|---|---|---|---|---|
| | | | | Raw material to oxide | Oxide to raw material |
| Magnesium carbonate | $MgCO_3$ | MgO | 47.8 | 0.478 | 2.092 |
| Lime (burnt) | CaO | CaO | 100 | 1.000 | 1.000 |
| Lime (hydrated) | $CaO.H_2O$ | CaO | 75.7 | 0.757 | 1.321 |
| Limestone | $CaCO_3$ | CaO | 56.0 | 0.560 | 1.786 |
| Strontium carbonate | $SrCO_3$ | SrO | 70.2 | 0.702 | 1.425 |
| Barium carbonate | $BaCO_3$ | BaO | 77.7 | 0.777 | 1.287 |
| Dolomite (Burnt) | $CaO{\cdot}MgO$ | CaO | 58.2 | 0.582 | 1.718 |
| | | MgO | 41.8 | 0.418 | 2.392 |
| Dolomite | $CaO{\cdot}MgO{\cdot}2CO_2$ | CaO | 30.4 | 0.304 | 3.289 |
| | | MgO | 21.9 | 0.219 | 4.566 |
| Zinc oxide | ZnO | ZnO | 100 | 1.000 | 1.000 |
| Zinc carbonate | $ZnCO_3$ | ZnO | 64.9 | 0.649 | 1.541 |
| Red lead | $Pb_3O_4$ | PbO | 97.7 | 0.977 | 1.024 |
| Litharge | PbO | PbO | 100 | 1.000 | 1.000 |
| ***Intermediate*** | | | | | |
| Alumina (calcined) | $Al_2O_3$ | $Al_2O_3$ | 100 | 1.000 | 1.000 |
| Alumina (hydrated) | $Al_2O_3.3H_2O$ | $Al_2O_3$ | 65.4 | 0.654 | 1.529 |
| Feldspar (typical) | $K_2O{\cdot}Al_2O_3{\cdot}6SiO_2$ | $K_2O$ | 16.9 | 0.169 | 5.917 |
| | | $Al_2O_3$ | 18.3 | 0.183 | 5.464 |
| | | $SiO_2$ | 64.8 | 0.648 | 1.543 |
| Titanium oxide | $TiO_2$ | $TiO_2$ | 100 | 1.000 | 1.000 |
| Zirconium oxide | $ZrO_2$ | $ZrO_2$ | 100 | 1.000 | 1.000 |

| | | | | | |
|---|---|---|---|---|---|
| ***Refining agent*** | | | | | |
| Arsenious oxide (white arsenic) | $As_2O_3$ | $As_2O_5$ | 116.2 | 1.162 | 0.861 |
| Antimony oxide | $Sb_2O_3$ | $Sb_2O_3$ | 100 | 1.000 | 1.000 |
| Ammonium nitrate | $NH_4NO_3$ | Decomposes | – | – | – |
| ***Coloring agent*** | | | | | |
| Vanadium oxide | $V_2O_5$ | $V_2O_5$ | 100 | 1.000 | 1.000 |
| Chromium trioxide | $Cr_2O_3$ | $Cr_2O_3$ | 100 | 1.000 | 1.000 |
| Potassium chromate | $KCrO_4$ | $Cr_2O_3$ | 39.1 | 0.391 | 2.558 |
| | | $(CrO_3)$ | 51.5 | 0.515 | 1.942 |
| | | $K_2O$ | 48.5 | 0.485 | 2.062 |
| Potassium dichromate | $K_2Cr_2O_7$ | $Cr_2O_3$ | 51.7 | 0.517 | 1.934 |
| | | $(CrO_3)$ | 68.0 | 0.680 | 1.471 |
| | | $K_2O$ | 32.0 | 0.320 | 3.125 |
| Manganese dioxide (pyrolusite) | $MnO_2$ | $MnO_2$ | 85.0 (Maximum content) | 0.850 | |
| Iron oxide | $Fe_2O_3$ | $Fe_2O_3$ | 100 | 1.000 | 1.000 |
| Cobalt oxide | $Co_3O_4$ | CoO | 93.4 | 0.934 | 1.071 |
| Nickel oxide (black) | $Ni_2O_3$ | NiO | 90.3 | 0.903 | 1.107 |
| Copper oxide | CuO | CuO | 100 | 1.000 | 1.000 |
| ***Opacifying agent*** | | | | | |
| Calcium fluoride | $CaF_2$ | CaO | 71.8 | 0.718 | 1.393 |
| | | $F_2$ | 48.7 | Huge loss by volatilization | Huge loss by volatilization |
| Cryolite | $Na_3AlF_6$ | $Na_2O$ | 44.3 | 0.443 | 2.257 |
| | | $Al_2O_3$ | 24.3 | 0.243 | 4.115 |
| | | $F_2$ | 54.3 | Huge loss by volatilization | Huge loss by volatilization |

Therefore, the glass of composition is $15.00Na_2O$–$10.00CaO$–$74.99SiO_2$ (wt.%). A negligible difference in the content of $SiO_2$ is originated due to usage of the digit rounding off rule. The composition in mol.% could also be calculated following the same procedure demonstrated earlier in Example 1.

### 2.3.3 *Characterization*

Glasses are characterized by evaluation of various properties. Density is found out with the help of the Archimedes principle by using water or nonaqueous solvent (e.g., toluene and xylene) as immersion liquid. Glassy nature or amorphous character is established by recording X-ray diffraction (XRD) pattern. Thermal properties such as coefficient of thermal expansion (CTE), glass transition or transformation temperature ($T_g$), glass deformation temperature ($T_d$), and glass crystallization temperature ($T_c$) are measured by the dilatometer, differential scanning calorimeter (DSC), or differential thermal analyzer (DTA). Optical properties are measured by different instruments such as refractive index by refractometer, absorption or transmission spectrum by ultraviolet (UV)-visible (Vis)-near infrared (NIR) spectrophotometer, and photoluminescence spectrum by fluorescence spectrometer. Glass structural properties are assessed by recording the bond vibration energy by means of the Fourier transform infrared (FTIR) or Raman spectrometers.

## 2.4 Processing and functional properties

### 2.4.1 *Absorption and photoluminescence spectra of individual 3d transition metal ions in different oxide glasses*

Different oxide glasses, on doping with 3d transition metal ions, exhibit many useful functionality mainly based on their optical properties. This is as a consequence of their (titanium to copper ions) incompletely filled d-orbitals ($d^{1-9}$). For this reason, it is intended to describe their individual absorption and photoluminescence spectra in different oxide glasses in this section. It is observed that their absorption and photoluminescence spectra depend on many factors. Some of these factors are

**(i)** glass component and composition,
**(ii)** glass melting atmosphere (e.g., oxidizing, reducing, or inert),
**(iii)** glass melting temperature and duration,
**(iv)** concentration of 3d transition metal ions,
**(v)** ionic or covalent character of glass,
**(vi)** spectra recording temperature, and
**(vii)** excitation wavelength for recording photoluminescence spectra

Consequently, it is aimed to enlighten some of these factors as well. Outer shell electronic configuration, valence state, glass composition and color, absorption band position, and their assignment of different 3d transition metal ions in silicate glasses are summarized in Table 2.3 based on the investigation reported in references [9–12].

**Table 2.3** **Electronic configuration, valence state, glass composition and color, and absorption band position and their assignment of 3d transition metal ions in silicate glasses**

| Electronic configuration | Metal ion | Glass | | Band position | | Band assignment | Reference |
|---|---|---|---|---|---|---|---|
| | | Composition[a] | Color | (nm) | ($cm^{-1}$) | | |
| $3d^1$ | $Ti^{3+}$ | A | Violet | 640 | 15,625 | $^2T_{2g} \rightarrow {}^2E_g$ | [9] |
| | | | | 570 | 17,544 | $^2T_{2g} \rightarrow {}^2E_g$ | |
| $3d^2$ | $V^{3+}$ | B | Brown | 635 | 15,700 | $^3T_{1g}$ (F) $\rightarrow {}^3T_{2g}$ (F) | [10] |
| | | | | 460 | 21,700 | $^3T_{1g}$ (F) $\rightarrow {}^3T_{1g}$ (P) | |
| $3d^3$ | $Cr^{3+}$ | B | Green | 650 | 15,400 | $^4A_{2g}$ (F) $\rightarrow {}^4T_{2g}$ (F) | [10] |
| | | | | 435 | 23,000 | $^4A_{2g}$ (F) $\rightarrow {}^4T_{1g}$ (F) | |
| $3d^4$ | $Cr^{2+}$ | C | Pale blue | 625 | 1600 | $^5E_g$ (D) $\rightarrow {}^5T_{2g}$ (D) | [10] |
| | | | | 490 | 20,400 | $^5E_g$ (D) $\rightarrow {}^5T_{2g}$ (D) | |
| | $Cr^{2+}$ | B | Pale blue | 630 | 15,900 | $^5E_g$ (D) $\rightarrow {}^5T_{2g}$ (D) | [10] |
| | | | | 480 | 20,800 | $^5E_g$ (D) $\rightarrow {}^5T_{2g}$ (D) | |
| | $Mn^{3+}$ | B | Pale red | 730 | 13,700 | $^5E_g$ (D) $\rightarrow {}^5T_{2g}$ (D) | [10] |
| | | | | 467 | 21,400 | $^5E_g$ (D) $\rightarrow {}^5T_{2g}$ (D) | |
| $3d^5$ | $Mn^{2+}$ | C | Red-brown | 417 | 24,000 | $^6A_{1g}$ (S) $\rightarrow {}^4E_g$ (G)<br>$^6A_{1g}$ (S) $\rightarrow {}^4A_{1g}$ (G) | [10] |
| | $Mn^{2+}$ | B | Yellow | 415 | 24,100 | $^6A_{1g}$ (S) $\rightarrow {}^4E_g$ (G)<br>$^6A_{1g}$ (S) $\rightarrow {}^4A_{1g}$ (G) | [10] |
| | $Fe^{3+}$ | D | | 470 | 21,260 | | [11] |
| | | | | 442 | 22,600 | | |
| | | | | 414 | 24,130 | | |
| | | | | 378 | 26,450 | | |
| | | | | 223 | 44,850 | | |

*Continued*

Table 2.3 **Continued**

| Electronic configuration | Metal ion | Glass |  | Band position |  | Band assignment | Reference |
|---|---|---|---|---|---|---|---|
|  |  | Composition[a] | Color | (nm) | ($cm^{-1}$) |  |  |
|  | $Fe^{3+}$ (octahedral) | E |  | 893 | 11,200 | $^6A_{1g} \rightarrow {}^4T_{1g}$ | [12] |
|  |  |  |  | 641 | 15,600 | $^6A_{1g} \rightarrow {}^4T_{2g}$ |  |
|  |  |  |  | 376 | 26,600 | $^6A_{1g} \rightarrow {}^4E_g$ (D) |  |
|  | $Fe^{3+}$ (tetrahedral) | E |  | 595 | 16,800 | $^6A_1 \rightarrow {}^4T_1$ | [12] |
|  |  |  |  | 488 | 20,500 | $^6A_1 \rightarrow {}^4T_2$ |  |
|  |  |  |  | 435 | 23,000 | $^6A_1 \rightarrow ({}^4E_1, {}^6A_1)$ |  |
|  |  |  |  | 413 | 24,200 | $^6A_1 \rightarrow {}^4T_2$ |  |
|  |  |  |  | 376 | 26,600 | $^6A_1 \rightarrow {}^4E$ (D) |  |
| $3d^6$ | $Fe^{2+}$ | D |  | 3030 | 3300 |  | [11] |
|  |  |  |  | 2040 | 4900 |  |  |
|  |  |  |  | 1075 | 9300 |  |  |
|  |  |  |  | 420 | 23,820 |  |  |
|  |  |  |  | 190 | 52,600 |  |  |
|  | $Fe^{2+}$ | C | Green-brown | 1900 | 5260 | $^5T_{2g}$ (D) $\rightarrow {}^5E_g$ (D) | [10] |
|  |  |  |  | 1100 | 9090 | $^5T_{2g}$ (D) $\rightarrow {}^5E_g$ (D) |  |
|  | $Fe^{2+}$ | B | Blue-green | 1850 | 5400 | $^5T_{2g}$ (D) $\rightarrow {}^5E_g$ (D) | [10] |
|  |  |  |  | 1080 | 9260 | $^5T_{2g}$ (D) $\rightarrow {}^5E_g$ (D) |  |
| $3d^7$ | $Co^{2+}$ | C | Dark blue | 1680 | 5920 | $^4A_2$ (F) $\rightarrow {}^4T_1$ (F) | [10] |
|  |  |  |  | 1520 | (Barycenter) | $^4A_2$ (F) $\rightarrow {}^4T_1$ (F) |  |
|  |  |  |  | 1340 | 6580 | $^4A_2$ (F) $\rightarrow {}^4T_1$ (F) |  |
|  |  |  |  |  | (Barycenter) |  |  |
|  |  |  |  |  | 7460 |  |  |
|  |  |  |  |  | (Barycenter) |  |  |
|  |  |  |  | 664 | 15,060 | $^4A_2$ (F) $\rightarrow {}^2T_1$ (G) | [10] |
|  |  |  |  |  |  | $^4A_2$ (F) $\rightarrow {}^2E$ (G) |  |

| | | | | | | | |
|---|---|---|---|---|---|---|---|
| | | | | 600 | 16,670 | $^4A_2$ (F) → $^4T_1$ (P) | [10] |
| | | | | 495 | 20,200 | $^4A_2$ (F) → $^2T_2$ (G) | |
| | $Co^{2+}$ | B | Dark blue | 1700 | 5880 | $^4A_2$ (F) → $^4T_1$ (F) | [10] |
| | | | | 1470 | (Barycenter) | $^4A_2$ (F) → $^4T_1$ (F) | |
| | | | | 1200 | 6800 | $^4A_2$ (F) → $^4T_1$ (F) | |
| | | | | | (Barycenter) | | |
| | | | | | 8300 | | |
| | | | | | (Barycenter) | | |
| | | | | 645 | 15,500 | $^4A_2$ (F) → $^2T_1$ (G) | [10] |
| | | | | | | $^4A_2$ (F) → $^2E$ (G) | |
| | | | | 591 | 16,920 | $^4A_2$ (F) → $^4T_1$ (P) | [10] |
| | | | | 518 | 19,300 | $^4A_2$ (F) → $^2T_2$ (G) | |
| $3d^8$ | $Ni^{2+}$ | C | Brown | 1730 | 5780 | $^3A_{2g}$ (F) → $^3T_{2g}$ (F) | [10] |
| | | | | 1090 | 9170 | $^3A_{2g}$ (F) → $^3T_{1g}$ (F) | |
| | | | | 640 | 15,620 | $^3A_{2g}$ (F) → $^1E_g$ (D) | |
| | | | | 535 | 18,690 | $^3A_{2g}$ (F) → $^3T_{1g}$ (P) | |
| | | | | 460 | 21,740 | $^3A_{2g}$ (F) → $^3T_{1g}$ (P) | |
| | $Ni^{2+}$ | B | Brown | 1900 | 5260 | $^3A_{2g}$ (F) → $^3T_{2g}$ (F) | [10] |
| | | | | 980 | 10,200 | $^3A_{2g}$ (F) → $^3T_{1g}$ (F) | |
| | | | | 630 | 15,870 | $^3A_{2g}$ (F) → $^1E_g$ (D) | |
| | | | | 540 | 18,520 | $^3A_{2g}$ (F) → $^3T_{1g}$ (P) | |
| | | | | 446 | 22,420 | $^3A_{2g}$ (F) → $^3T_{1g}$ (P) | |
| $3d^9$ | $Cu^{2+}$ | C | Green | 790 | 12,660 | $^2E_g$ (D) → $^2T_{2g}$ (D) | [10] |
| | $Cu^{2+}$ | B | Blue-green | 780 | 12,820 | $^2E_g$ (D) → $^2T_{2g}$ (D) | [10] |

[a]Respective 3d transition metal oxide was added with the nominal glass composition: A = 25 $K_2O$–75$SiO_2$ (mol.%) + 10 (wt.%) $TiO_2$; B (albite-diopside glass) = 4.50$Na_2O$–7.90MgO–9.85$Al_2O_3$–63.36$SiO_2$–13.46CaO (wt.%); C (albite glass) = 9.13$Na_2O$–19.51$Al_2O_3$–69.86$SiO_2$–0.03CaO (wt.%); D = 13$Na_2O$–6MgO–71$SiO_2$–10CaO (mol.%) + 1 wt.% $Fe_2O_3$; E = 14.5$Na_2O$–11.5CaO–73.7$SiO_2$–(wt.%) + 0.3 wt.% $Fe_2O_3$.

### 2.4.1.1 Titanium

Titanium (III), having $3d^1$ electronic configuration, exhibits violet to violet-purple coloration in the glasses as a result of d-d transitions. Absorption spectra of $Ti^{3+}$ ions in silicate ($1.0K_2O$–$3.0SiO_2$, mol) glasses containing different amounts (2.5–20.0 g) of $TiO_2$ per 100 g glass are shown in Fig. 2.9. It forms the octahedral complex of $[TiO_6]^{3-}$ in oxide glass with $d^2sp^3$ hybridization (Table 2.1).The band at 570 nm is assigned to the $^2T_{2g} \rightarrow {}^2E_g$ transition. The band at 640 nm is attributed to the transition between the sublevels of $T_{2g}$ ground state and the $E_g$ excited state [9].These are also enlisted in Table 2.3. Bausa et al. [13] have reported the absorption and fluorescence

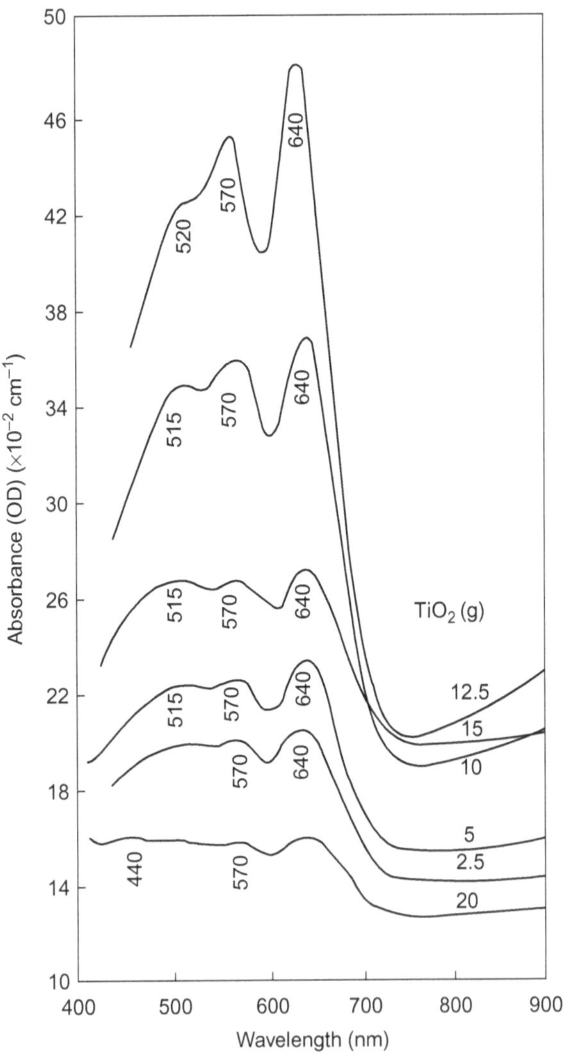


**Fig. 2.9** Absorption spectra of $Ti^{3+}$ ions in silicate glasses containing different amounts (2.5–20.0 g) of $TiO_2$ per 100 g glass.
Reproduced with permission from N. Abd El-Shafi, M.M. Morsi, Optical absorption and infrared studies of some silicate glasses containing titanium, J. Mater. Sci. 33 (1997) 5185–5189, Copyright © Springer.

spectra (excitation at $\lambda_{exc} = 632.8$ nm) of $Ti^{3+}$ ions in sodium aluminophosphate glasses at liquid nitrogen temperature (LNT). This is show in Fig. 2.10A. They observed two absorption bands peaking at 560 and 697 nm. These two peaks have been interpreted as a dynamic J-T splitting in the $E_g$ excited state resulting in tetragonal distortion and accounts for the spin-orbit interaction. The broad and structure less photoluminescence band centered at 860 nm due to the transition between the lowest J-T component of the $E_g$ excited state and the split components of the $T_{2g}$ ground state manifold (see Section 2.2, J-T effect). Very recently, Wen and Tanner [14] have reported the charge transfer emission spectra of $TiO_2$-doped sodium borosilicate glasses measured at room temperature. This is depicted in Fig. 2.10B. The emission spectra of the $TiO_2$-doped glasses are similar under excitation 317 or 309 nm, but their intensity is much higher than for undoped glasses. The emission peak is broadened, and its position is shifted to lower wavelength (from 525 nm down to 504 nm) compared with the undoped glass. This band is assigned to $Ti^{4+}$–$O^{2-}$ charge transfer emission.

### 2.4.1.2 Vanadium

Vanadium (III) is a $3d^2$ system. The absorption spectrum of $V^{3+}$ ions in silicate glass is depicted in Fig. 2.11A. The $V^{3+}$ ion forms the greenish colored octahedral complex of $[VO_6]^{3-}$ in oxide glass with $d^2sp^3$ hybridization (Table 2.1). It exhibited two absorption bands at 460 and 635 nm. The strong band at 460 nm is assigned to the ${}^3T_{1g}$ (F) → ${}^3T_{1g}$ (P) transition, and the less intense shoulder at 635 nm is attributed to

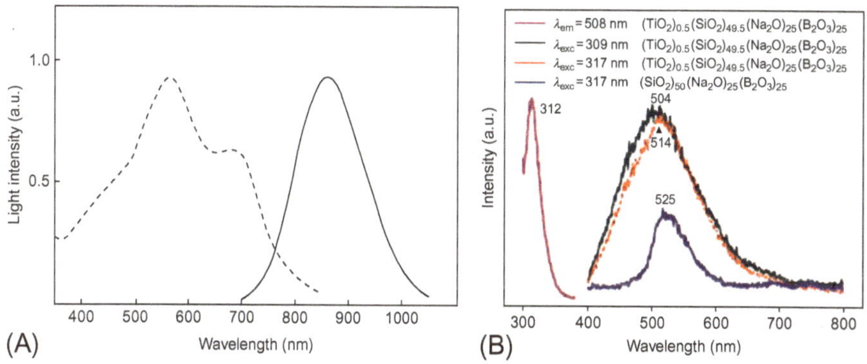


**Fig. 2.10** (A) Absorption (left) and fluorescence (right) spectra of $Ti^{3+}$ ions in aluminophosphate glasses at liquid nitrogen temperature (LNT). Excitation at $\lambda_{exc} = 632.8$ nm. (B) Excitation (left) and emission (right) spectra of $TiO_2$-doped and undoped borosilicate glasses.

(A) Reproduced with permission from L.E. Bausa, F. Jaque, J. Garcia Sole, A. Duran, Photoluminescence of $Ti^{3+}$ in $P_2O_5$–$Na_2O$–$Al_2O_3$ glass, J. Mater. Sci. 23 (1988) 1921–1922, Copyright © Springer. (B) Reproduced with permission from H. Wen, P.A. Tanner, Optical properties of 3d transition metal ion-doped sodium borosilicate glass, J. Alloys Compd. 625 (2015) 328–335, Copyright © Elsevier.

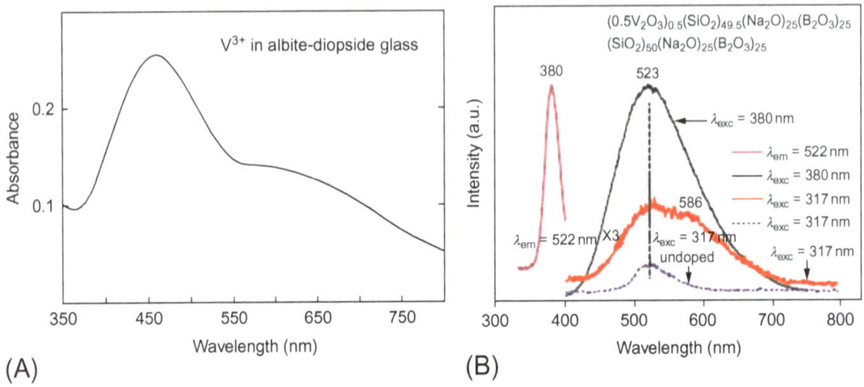


**Fig. 2.11** (A) Absorption spectrum of $V^{3+}$ ions in silicate glass. (B) Excitation (left) and emission (right) spectra of $V_2O_3$-doped and undoped borosilicate glasses.
(A) Courtesy: H. Keppler, Crystal field spectra and geochemistry of transition metal ions in silicate melts and glasses, Am. Miner. 77 (1992) 62–75. (B) Reproduced with permission from H. Wen, P.A. Tanner, Optical properties of 3d transition metal ion-doped sodium borosilicate glass, J. Alloys Compd. 625 (2015) 328–335, Copyright © Elsevier.

the $^3T_{1g}$ (F) → $^3T_{2g}$ (F) transition [10]. Very recently, Wen and Tanner [14] have reported the room temperature excitation and emission spectra of $V_2O_3$-doped sodium borosilicate glasses. This is illustrated in Fig. 2.11B. No emission band is obtained under the excitation around the wavelength of 612 nm. Therefore, it is difficult to conclude about the presence of $V^{3+}$ ions in the sodium borosilicate glass. Nevertheless, two emission bands are obtained at 525 and 586 nm under excitation at 317 nm. These two emission bands are assigned to $VO_4$ tetrahedral emission transitions of $^3T_2$, $^3T_1 \rightarrow {}^1A_1$.

### 2.4.1.3 Chromium

Trivalent chromium ($Cr^{3+}$) is a $3d^3$ system. It produces green glass containing the octahedral $[CrO_6]^{3-}$ complex ion with $d^2sp^3$ hybridization (Tables 2.1 and 2.3). Obviously, the $Cr^{2+}$ is of the $3d^4$ system, and it produces pale or light blue glasses. Fig. 2.12A illustrates the absorption spectra of $Cr^{3+}$ and $Cr^{2+}$ ions in different silicate glasses [10].The $Cr^{3+}$ ion displays two absorption bands peaking at 435 and 650 nm that are assigned to the $^4A_{2g}$ (F) → $^4T_{1g}$ (F) and $^4A_{2g}$ (F) → $^4T_{2g}$ (F) transitions, respectively. Similarly, the $Cr^{2+}$ ion demonstrates two absorption bands depending on the glass composition in the ranges of 480–490 and 625–630 nm that are assigned to the $^5E_g$ (D) → $^5T_{2g}$ (D) transition. Its spectra can satisfactorily be explained by the distorted octahedral sites. The appearance of shoulder around 480–490 nm in addition to the main band around 625–630 nm is caused by the splitting of the $^5E_g$ level due to the above said distortion (see Section 2.2, J-T effect). Choi et al. [15] have demonstrated the absorption spectra of $Cr^{3+}$, $Cr^{4+}$, and $Cr^{6+}$ ions in soda-lime-silicate and calcium-aluminate glasses. This is shown in Fig. 2.12B. The absorption band around

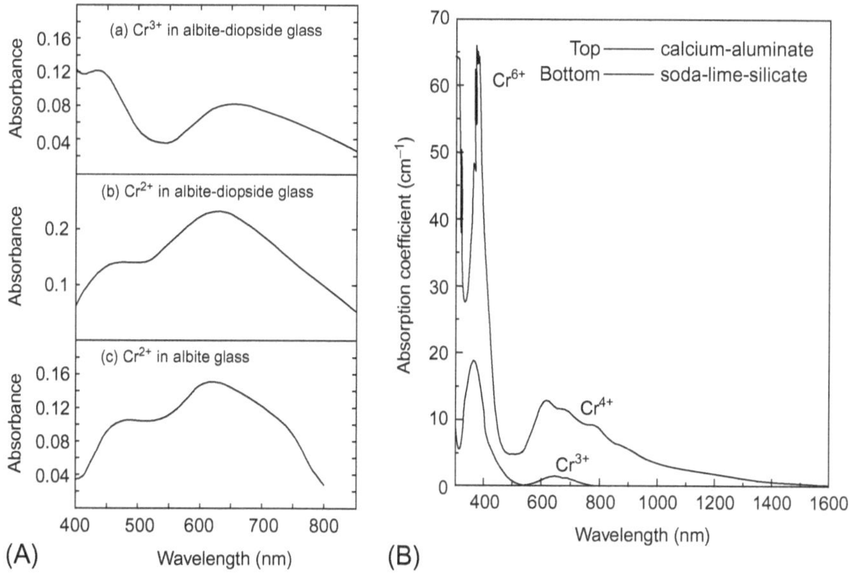


**Fig. 2.12** (A) Absorption spectra of $Cr^{3+}$ and $Cr^{2+}$ ions in different silicate glasses. (B) Absorption spectra of $Cr^{3+}$, $Cr^{4+}$, and $Cr^{6+}$ ions in silicate and aluminate glasses. (A) Courtesy: H. Keppler, Crystal field spectra and geochemistry of transition metal ions in silicate melts and glasses, Am. Miner. 77 (1992) 62–75. (B) Reproduced with permission from Y.G. Choi, K.H. Kim, Y.S. Han, J. Heo, Oxidation state and local coordination of chromium dopant in soda-lime-silicate and calcium-aluminate glasses, Chem. Phys. Lett. 329 (2000) 370–376, Copyright © Elsevier.

370 nm in both glasses is assigned to a charge transfer associated with hexavalent chromium ($Cr^{6+}$) ion. Besides, there are characteristic absorption bands corresponding to $Cr^{3+}$ and $Cr^{4+}$ in soda-lime-silicate and calcium-aluminate glasses, respectively. These spectra disclose that $Cr^{6+}$ exists in both glass samples, and the extinction coefficient of $Cr^{4+}$ species is much higher than that of $Cr^{3+}$ ions. The room temperature emission spectra, under excitation at 800 nm, of chromium ions in (a) silicate and (b) aluminate glasses are shown in Fig. 2.13A. The $Cr^{6+}$ ion being the $3d^0$ system emits no fluorescence due to d-d transition. The emission of $Cr^{3+}$ ions is likely to be reabsorbed by the absorption of the $Cr^{4+}$ ions. For this reason, it is very difficult to analyze the absorption and emission spectra due to the intra-3d-configurational transitions of chromium.

The above investigation was carried out in the year 2000; later in 2002, Koepke et al. [16] reported the low- (a, c, and e) and room temperature (b, d, and f) emission spectra of chromium ions in different glasses such as $ZrO_2$–$Al_2O_3$–$SiO_2$ (ZAS), $Li_2B_4O_7$ (LBO), and sol-gel silica ($SiO_2$) glasses under excitation at 640, 606, and 457 nm, respectively. These are illustrated in Fig. 2.13B. The room temperature emission spectra (a and b) of the ZAS glass exhibit a broad band assigned to the transition of ${}^4T_{2g} \rightarrow {}^4A_{2g}$ in the low crystal field. The spectra (c and d) of the LBO glass display

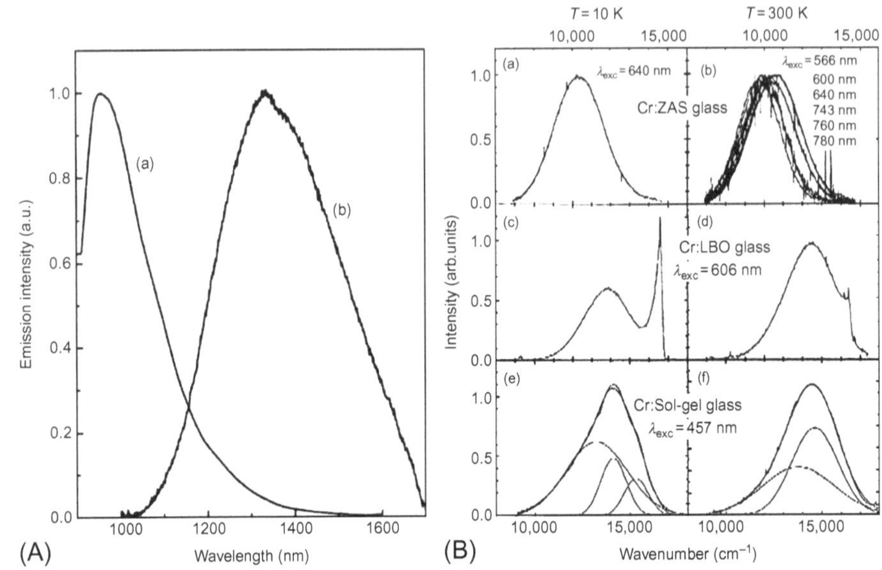


**Fig. 2.13** (A) Room temperature emission spectra of chromium ions in (a) silicate and (b) aluminate glasses. Excitation at $\lambda_{exc} = 800$ nm. (B) Low (a, c, and e) and room temperature (b, d, and f) emission spectra of chromium ions in different glasses. The bottom wavenumber scale is only for Cr: sol-gel glasses.
(A) Reproduced with permission from Y.G. Choi, K.H. Kim, Y.S. Han, J. Heo, Oxidation state and local coordination of chromium dopant in soda-lime-silicate and calcium-aluminate glasses, Chem. Phys. Lett. 329 (2000) 370–376, Copyright © Elsevier. (B) Reproduced with permission from C. Koepke, K. Wisniewski, M. Grinberg, Excited state spectroscopy of chromium ions in various valence states in glasses, J. Alloys Compd. 341 (2002) 19–27, Copyright © Elsevier.

the emission bands corresponding to the $^4T_{2g} \rightarrow {}^4A_{2g}$ and $^2E_g \rightarrow {}^4A_{2g}$ transitions at different temperatures. On the other hand, the sol-gel silica glass under excitation at 457 nm exhibits the emission band peaking around 13,260 $cm^{-1}$ at low temperature (10 K) and 13,690 $cm^{-1}$ at room temperature (300 K). It is believed that this emission band is owing to the $Cr^{6+}$ ions, that is, from the $Cr^{5+}O^{-}$ center of the $3d^12p^5$, ($t_1^5$, 2e) configuration to the ground state of the $Cr^{6+}O^{2-}$ center of the $3d^02p$, ($t_1^6$) configuration.

### *2.4.1.4 Manganese*

Bivalent ($Mn^{2+}$) and trivalent ($Mn^{3+}$) manganese are $3d^5$ and $3d^4$ systems, respectively. The $Mn^{2+}$ ion produces yellow to red-brown glasses forming tetrahedral $[MnO_4]^{2-}$ complex with $sp^3$ hybridization, whereas the $Mn^{3+}$ ion yields violet to pale red glasses creating octahedral $[MnO_6]^{3-}$ complex with $d^2sp^3$ hybridization (see Tables 2.1 and 2.3). The absorption spectrum of $Mn^{2+}$ ion in albite silicate glass

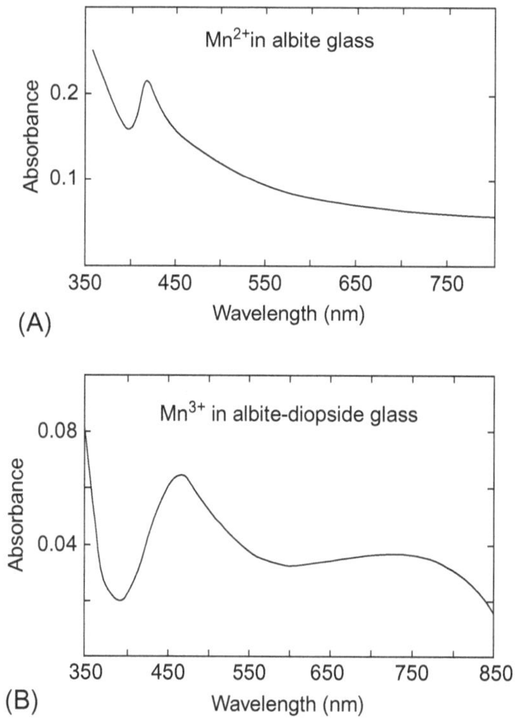


**Fig. 2.14** Absorption spectra of (A) $Mn^{2+}$ and (B) $Mn^{3+}$ ions in different silicate glasses. Courtesy: H. Keppler, Crystal field spectra and geochemistry of transition metal ions in silicate melts and glasses, Am. Miner. 77 (1992) 62–75.

is depicted in Fig. 2.14A. It displays a weak band peaking at 415 nm. The band is assigned to two energetically indistinguishable transitions of $^6A_{1g}$ (S)$\rightarrow$$^4E_g$ (G) and $^6A_{1g}$ (S)$\rightarrow$$^4A_{1g}$ (G) [10]. The absorption spectrum of $Mn^{3+}$ ion in albite-diopside silicate glass is illustrated in Fig. 2.14B. It exhibits two bands at 467 and 730 (broad) nm. They result in from the splitting of $^5E_g$ (D)$\rightarrow$$^5T_{2g}$ (D) transition because of distortion of octahedral coordination (see Section 2.2, J-T effect). This distortion causes splitting of the $^5E_g$ (D) level, which explains the appearance of the shoulder at 730 nm along with the main band at 467 nm.

Wen and Tanner [14] have reported the excitation and emission spectra of $Mn^{2+}$-doped and undoped borosilicate glasses, which are shown in Fig. 2.15A. The excitation spectrum (shown at the left side) exhibits two bands centered at 356 and 420 nm. The emission spectra of the $Mn^{2+}$-doped glasses, under excitation at 356 and 420 nm, display a broad band peaking around 580 nm. It is assigned to the $^4T_{1g}$ ($^4$G)$\rightarrow$$^6A_{1g}$ ($^6$S) transition of $Mn^{2+}$ in octahedral environment. It is also observed that the tetrahedrally coordinated $Mn^{2+}$ ion with a weak ligand field produces a green emission, whereas the $Mn^{2+}$ ion in octahedrally coordinated geometry with a strong ligand field displays an orange to red emission. Machado et al. [17] have reported the room temperature emission spectra, under excitation at $\lambda_{exc}=410$ nm, with increasing concentration (samples S-3 to S-6 contain 2.5–20 mol.% $Mn_2O_3$, prepared under reducing atmosphere of decomposed $NH_4H_2PO_4$) of $Mn^{2+}$ ion in barium phosphate

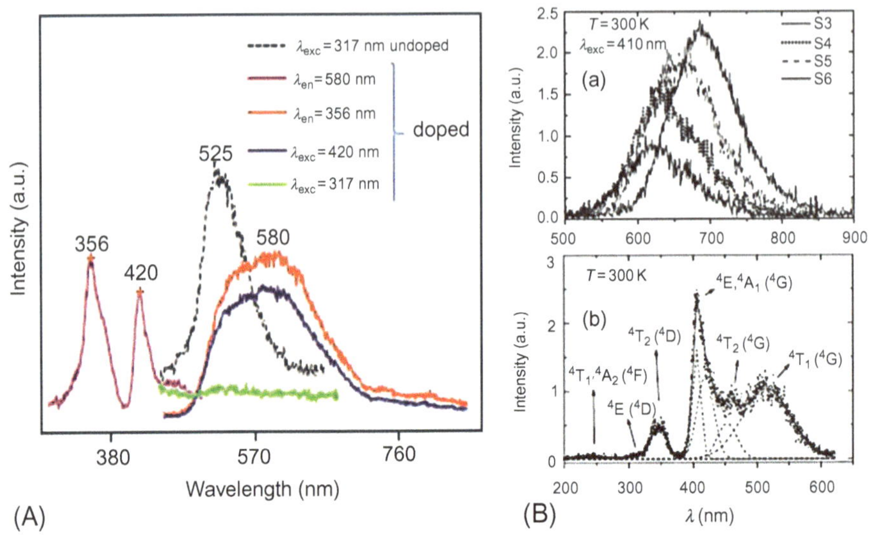


**Fig. 2.15** (A) Excitation (left) and emission (right) spectra of $Mn^{2+}$-doped and undoped borosilicate glasses. (B) (a) Emission spectra with increasing concentration of $Mn^{2+}$ ion in phosphate glasses (excitation at $\lambda_{exc} = 410$ nm) and (b) excitation spectrum obtained by monitoring at maximum of emission band.

(A) Reproduced with permission from H. Wen, P.A. Tanner, Optical properties of 3d transition metal ion-doped sodium borosilicate glass, J. Alloys Compd. 625 (2015) 328–335, Copyright © Elsevier. (B) Reproduced with permission from I.E.C. Machado, L. Prado, L. Gomes, J.M. Prison, J.R. Martinelli, Optical properties of manganese in barium phosphate glasses, J. Non-Cryst. Solids, 348 (2004) 113–117, Copyright © Elsevier.

glasses. This is shown in Fig. 2.15B (a). Only one emission band is detected at around 617 nm with a red shift (up to 690 nm) associated with increase of manganese concentration. This was explained by the magnetic interactions between the pairs of $Mn^{2+}$ ions. The excitation spectrum obtained by monitoring at maximum of the emission band is shown in Fig. 2.15B (b). The identified excitation bands around 515, 405, and 346 nm are assigned to the $^6A_1$ (S) → $^4T_1$ (G), $^6A_1$ (S) → $^4A_1$ (G), $^4E$ (G) and $^6A_1$ (S) → $^4T_2$ (D) transitions, respectively.

### *2.4.1.5 Iron*

Bivalent ($Fe^{2+}$) and trivalent ($Fe^{3+}$) iron are $3d^6$ and $3d^5$ systems, respectively. The $Fe^{2+}$ ion produces blue-green to green-brown glasses forming octahedral $[FeO_6]^{4-}$ complex with $d^2sp^3$ hybridization whereas the $Fe^{3+}$ ion yields yellow glasses creating octahedral $[FeO_6]^{3-}$ complex with $d^2sp^3$ hybridization (see Tables 2.1 and 2.3). The absorption spectra of ferric ($Fe^{3+}$) and ferrous ($Fe^{2+}$) ions in silicate ($Na_2O$-CaO-MgO-$SiO_2$) glasses are illustrated in Fig. 2.16A [11]. The absorption bands of tetrahedral ($T_d$)

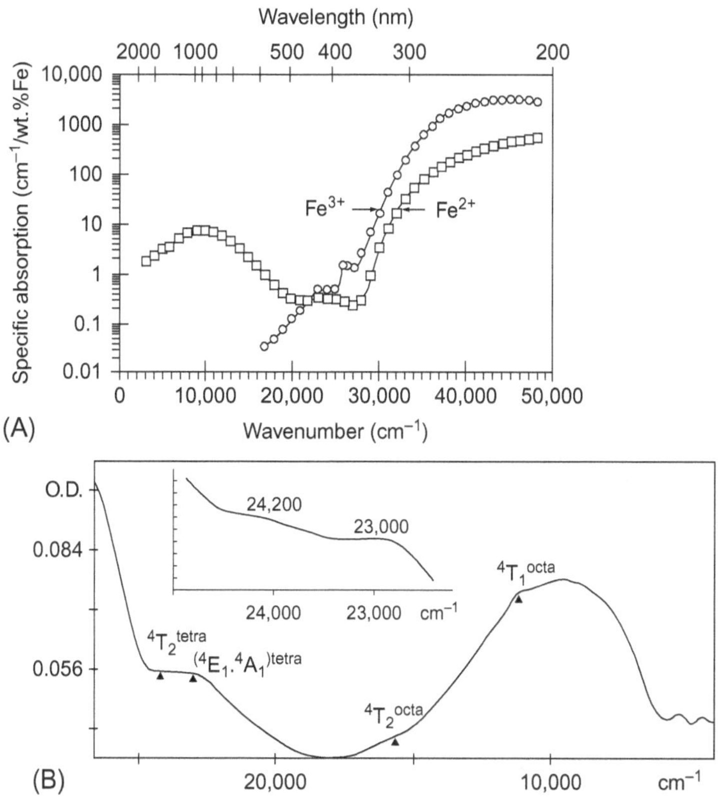


**Fig. 2.16** (A) Absorption spectra of ferric ($Fe^{3+}$) and ferrous ($Fe^{2+}$) ions in silicate glasses. (B) Absorption bands of tetrahedral ($T_d$) and octahedral ($O_h$) ferric ($Fe^{3+}$) ions in silicate glass. (A) Reproduced with permission from L.B. Glebov, E.N. Boulos, Absorption of iron and water in the $Na_2O$-CaO-MgO-$SiO_2$ glasses. II. Selection of intrinsic, ferric, and ferrous spectra in the visible and UV regions, J. Non-Cryst. Solids 242 (1998) 49–62, Copyright © Elsevier. (B) Reproduced with permission from B. Hannoyer, M. Lenglet, J. Dürr, R. Cortes, Spectroscopic evidence of octahedral iron (III) in soda-lime-silicate glasses, J. Non-Cryst. Solids 151 (1992) 209–316, Copyright © Elsevier.

and octahedral ($O_h$) ferric ($Fe^{3+}$) ions in soda-lime-silicate glass are demonstrated in Fig. 2.16B [12]. The octahedral $Fe^{3+}$ ion exhibits absorption bands at 376, 641, and 893 nm that are assigned to the $^6A_{1g} \rightarrow {}^4E_g$ (D), $^6A_{1g} \rightarrow {}^4T_{2g}$ and $^6A_{1g} \rightarrow {}^4T_{1g}$ transitions, respectively. In contrast, the tetrahedral $Fe^{3+}$ ion displays absorption bands at 376, 413, 435, 488, and 595 nm that are assigned to the $^6A_1 \rightarrow {}^4E$ (D), $^6A_1 \rightarrow {}^4T_2$, $^6A_1 \rightarrow ({}^4E_1, {}^6A_1)$, $^6A_1 \rightarrow {}^4T_2$, and $^6A_1 \rightarrow {}^4T_1$ transitions, respectively. The absorption spectrum of ferrous ($Fe^{2+}$) ions in albite-diopside silicate glass is shown in Fig. 2.17A [10]. The spectrum consists of a strong band around 1100 nm with a weak shoulder around 1900 nm. The former band around 1100 nm is assigned to the $^5T_{2g}$ (D) $\rightarrow {}^5E_g$ (D) transition of octahedral oxygen complex of $Fe^{2+}$ ion. The origin of the

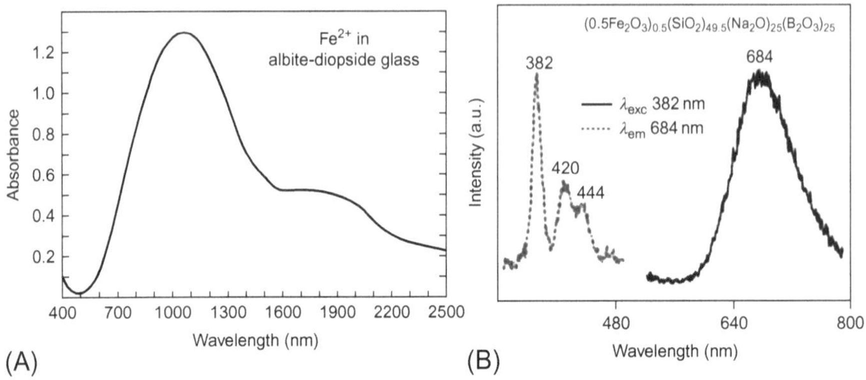


**Fig. 2.17** (A) Absorption spectrum of ferrous ($Fe^{2+}$) ions in a different silicate glass. (B) Excitation (left) and emission (right) spectra of $Fe^{3+}$ ions in borosilicate glasses. (A) Courtesy: H. Keppler, Crystal field spectra and geochemistry of transition metal ions in silicate melts and glasses, Am. Miner. 77 (1992) 62–75. (B) Reproduced with permission from H. Wen, P.A. Tanner, Optical properties of 3d transition metal ion-doped sodium borosilicate glass, J. Alloys Compd. 625 (2015) 328–335, Copyright © Elsevier.

band around 1900 nm is a component of the $^5T_{2g}$ (D) → $^5E_g$ (D) transition of octahedral oxygen complex of $Fe^{2+}$ ion that has been splitted into two components by a distortion of octahedral geometry.

The excitation and emission spectra of $Fe^{3+}$ ion in sodium borosilicate glasses are presented in Fig. 2.17B [14]. The right hand side emission spectrum exhibit a broad band around 684 nm under excitation of 420 or 382 nm. The left hand side excitation spectrum peaking at 444, 420, and 382 nm recognizes the emission of the $Fe^{3+}$ ion.

### 2.4.1.6 Cobalt

Bivalent ($Co^{2+}$) and trivalent ($Co^{3+}$) cobalt are $3d^7$ and $3d^6$ systems, respectively. The $Co^{2+}$ ion produces deep blue glasses forming tetrahedral $[CoO_4]^{2-}$ complex with $sp^3$ hybridization, whereas the $Co^{3+}$ ion yields light yellow glasses creating octahedral $[CoO_6]^{3-}$ complex with $d^2sp^3$ hybridization (see Tables 2.1 and 2.3). The absorption spectrum of $Co^{2+}$ ions in albite-diopside silicate glass is illustrated in Fig. 2.18A [10]. The absorption bands peaking around 518, 591, and 645 nm is assigned to the $^4A_2$ (F) → $^2T_2$ (G), $^4A_2$ (F) → $^4T_1$ (P), and $^4A_2$ (F) → $^2T_1$ (G) transitions, respectively. The latest band around 645 nm is assigned to the $^4A_2$ (F) → $^2E$ (G) transition as well. Three "*barycenter*" bands peaking around 1200, 1470, and 1700 nm have been assigned to the single transition of $^4A_2$ (F) → $^4T_1$ (F). It is also seen that all the above band positions alter significantly with change in glass composition (see Table 2.3). The Tanabe-Sugano is very useful for interpreting the spectra of 3d transition metal ions in a host including both weak and strong ligand fields. Consequently, the Tanabe-Sugano diagram for tetrahedral $Co^{2+}$ ion in albite-diopside silicate glass with predicted band positions for $B$ (Racah parameter) = 787 $cm^{-1}$ and $\Delta$ (ligand field splitting) = 4040 $cm^{-1}$ is shown in Fig. 2.18B [10]. The Racah interelectron repulsion parameter ($B$) is very useful for clarification of spectra and gives the measure of

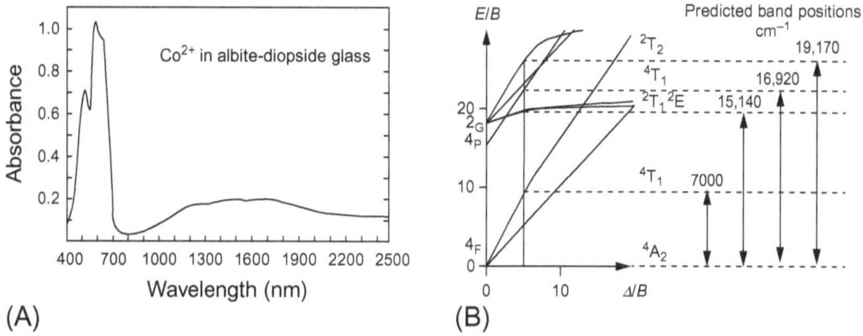


**Fig. 2.18** (A) Absorption spectrum of $Co^{2+}$ ions in silicate glass. (B) Tanabe-Sugano diagram for tetrahedral $Co^{2+}$ ion with predicted band positions for $B=787\ cm^{-1}$ and $\Delta=4040\ cm^{-1}$. Courtesy: H. Keppler, Crystal field spectra and geochemistry of transition metal ions in silicate melts and glasses, Am. Miner. 77 (1992) 62–75.

covalency arising from the delocalization of d-electrons from the transition metal onto the ligand. If $B$ value is smaller than that of the free metal ion, then the d-electrons are greatly delocalized onto the ligand with greater covalency.

Figs. 2.19(A) and (B) represent the room temperature excitation and emission spectra of $Co^{2+}$ in the $45B_2O_3$–5ZnO–49.5PbO (mol%) lead borate glass, respectively [18]. The excitation spectrum was recorded with the emission at 625 nm. This emission transition of $Co^{2+}$ ions in the visible region at 625 nm is assigned to the $^2E(^2G)\rightarrow{}^4A_2(^4F)$ transition of $Co^{2+}$ ions in the tetrahedral co-ordination.

### 2.4.1.7 Nickel

Bivalent nickel ($Ni^{2+}$) is a $3d^8$ system. The $Ni^{2+}$ ion produces yellow-brown to brown glasses forming tetrahedral $[NiO_4]^{2-}$ complex with $sp^3$ hybridization (see Tables 2.1).

The absorption spectrum of $Ni^{2+}$ ion in albite-diopside silicate glass is illustrated in Fig. 2.20A [10]. The absorption bands peaking around 446, 540, 630, 980, and 1900 nm are assigned to the $^3A_{2g}$ (F) → $^3T_{1g}$ (P), $^3A_{2g}$ (F) → $^3T_{1g}$ (P), $^3A_{2g}$ (F) → $^1E_g$ (D), $^3A_{2g}$ (F) → $^3T_{1g}$ (F) and $^3A_{2g}$ (F) → $^3T_{1g}$ (F) transitions in octahedral field, respectively. It is also seen that all the above band positions alter significantly with change in glass composition (see Table 2.3).

The room temperature photoluminescence spectra under excitation at 800 nm of $Ni^{2+}$ ions in tellurite glasses with composition of $20ZnF_2$–$30As_2O_3$–$(50-x)TeO_2$–$x$NiO ($x=0.2$–2.0 mol.%) are shown in Fig. 2.20B [19]. The spectrum of each of the glasses exhibits a broad emission band in the region 1100–1500 nm. The band is assigned to the $^3T_{2g}$ ($^3$F) → $^3A_{2g}$ ($^3$F) transition of octahedral $Ni^{2+}$ ion in tellurite glasses [19]. With the increasing content of $Ni^{2+}$ ions in the glass matrix, the full width at half maximum (FWHM) of the band is found to be increased with a shift of the peak position toward to some extent higher wavelength (from 1330 to 1354 nm). The gradual increase in optical absorption band intensity in these glasses indicates that the $Ni^{2+}$ ions occupy tetrahedral coordination in lower concentration but at higher concentration they are mostly in octahedral coordination. Generally, octahedral $Ni^{2+}$ ion

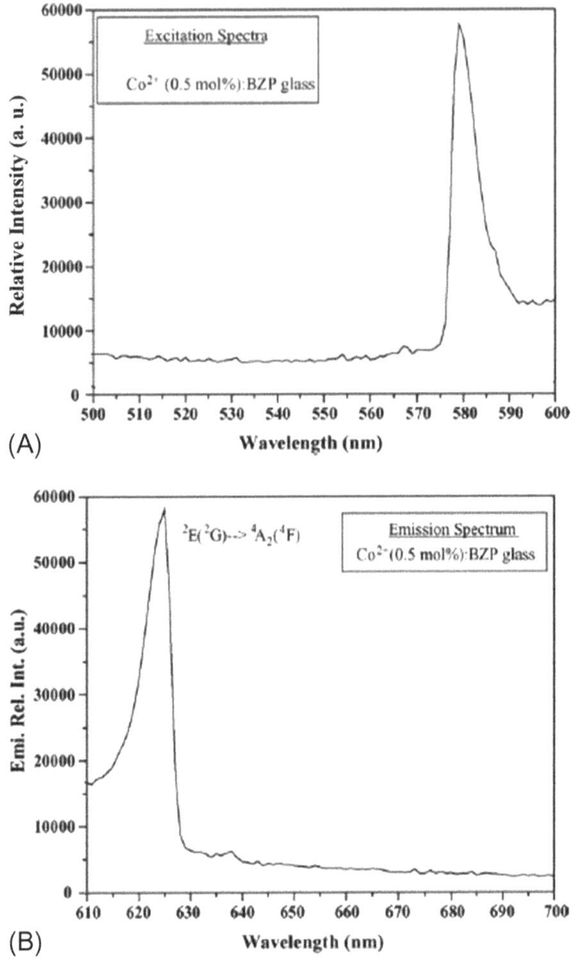


**Fig. 2.19** (A) Excitation and (B) emission spectra $Co^{2+}$ in borate glass.
Reproduced with permission from G. Lakshminarayana, S. Buddhudu, Spectral analysis of $Mn^{2+}$, $Co^{2+}$ and $Ni^{2+}$: $B_2O_3$–ZnO–PbO glasses, Spectrochimica Acta Part A: Molecular and Biomolecular Spectroscopy on ScienceDirect" Spectrochimica Acta A, 63 (2006) 295–304, Copyright © Elsevier.

exhibits luminescence emission, and so far, no emission caused by tetrahedral $Ni^{2+}$ ion has been reported either in glass or crystal hosts. For these reasons, the emission band is credited to the $^3T_{2g}$ ($^3F$) → $^3A_{2g}$ ($^3F$) transition of $Ni^{2+}$ ions in octahedral field. The width and the shape of this transition specify that there is a comparatively increased in Stokes shift between the emission and absorption band.

### 2.4.1.8 Copper

The univalent ($Cu^+$) and bivalent ($Cu^{2+}$) copper are the member of the $3d^{10}$ and $3d^9$ system, respectively. The $Cu^+$ ion produces colorless glasses, whereas the $Cu^{2+}$ ion yields light blue to green glasses creating octahedral $[CuO_6]^{2-}$ complex with $d^2sp^3$

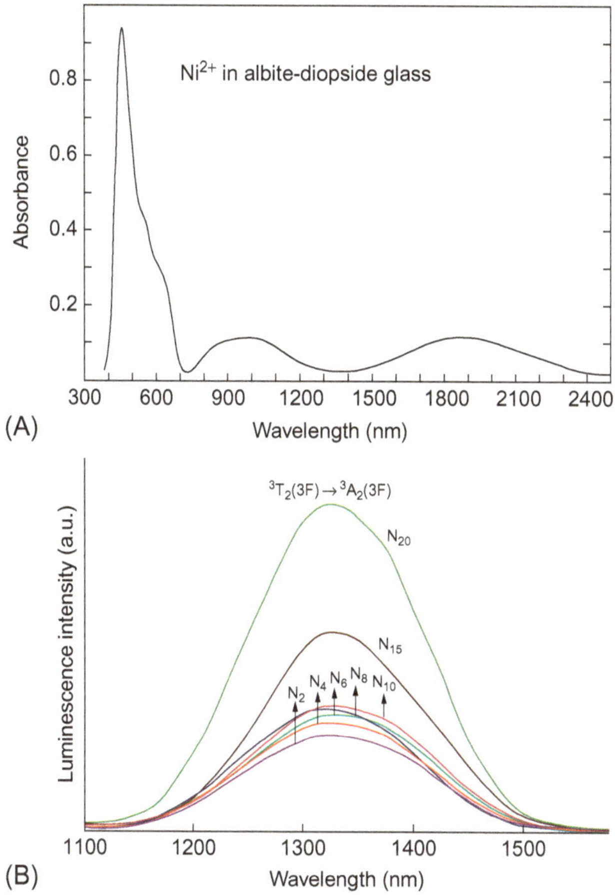


**Fig. 2.20** (A) Absorption spectrum of $Ni^{2+}$ ions in silicate glass. (B) Photoluminescence spectra of $Ni^{2+}$ ions in tellurite glasses with increasing concentration (0.2–2.0 mol.%) of NiO. Excitation at $\lambda_{exc} = 800$ nm.
(A) Courtesy: H. Keppler, Crystal field spectra and geochemistry of transition metal ions in silicate melts and glasses, Am. Miner. 77 (1992) 62–75. (B) Reproduced with permission from Y. Gandhi, N. Krishna Mohan, N. Veeraiah, Role of nickel ion coordination on spectroscopic and dielectric properties of $ZnF_2$–$As_2O_3$–$TeO_2$:NiO glass system, J. Non-Cryst. Solids 357 (2011) 1193–1202, Copyright © Elsevier.

hybridization (see Tables 2.1 and 2.3). The absorption spectrum of $Cu^{2+}$ ions in albite silicate glass is illustrated in Fig. 2.21A [10]. The asymmetrical absorption band peaking around 790 nm is assigned to the $^2E_g$ (D) → $^2T_{2g}$ (D) transition of $Cu^{2+}$ ion in octahedral coordination. The asymmetry of the band is caused by J-T effect or distortion (see Section 2.2). In consequence of this distortion, the spectrum exhibits a shoulder around 1100 nm as well. It is noteworthy that the peak position may change with alteration glass composition (see Table 2.3).

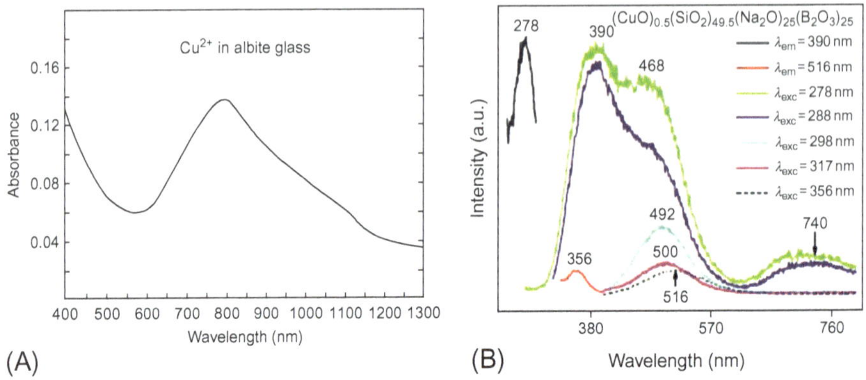


**Fig. 2.21** (A) Absorption spectrum of $Cu^{2+}$ ions in silicate glass. (B) Excitation (left) and emission (right) spectra of $Cu^{2+}$ ions in borosilicate glasses.
(A) Courtesy: H. Keppler, Crystal field spectra and geochemistry of transition metal ions in silicate melts and glasses, Am. Miner. 77 (1992) 62-75. (B) Reproduced with permission from H. Wen, P.A. Tanner, Optical properties of 3d transition metal ion-doped sodium borosilicate glass, J. Alloys Compd. 625 (2015) 328–335, Copyright © Elsevier.

The excitation (left) and emission (right) spectra of $Cu^{2+}$ ions in sodium borosilicate glasses are shown in Fig. 2.21A [14]. Two excitation bands are observed peaking at 278 and 356 nm. The emission bands in the ranges of 445–492 and 535–555 nm are caused by the $Cu^{2+}$ and $Cu^{+}$ ions, respectively. The emission bands peaking at 390, 468, and 740 nm, appeared under excitation at 278 nm, are caused by the $Cu^{2+}$ ion. The excitation (left) and emission (right) spectra, under excitation at different wavelengths, of $Cu^{+}$ ions in ion-exchanged borosilicate (BK7) glasses are shown in Fig. 2.22A [20]. Here, the excitation spectrum is splitted into three peaks at 282, 287, and 298 nm. The

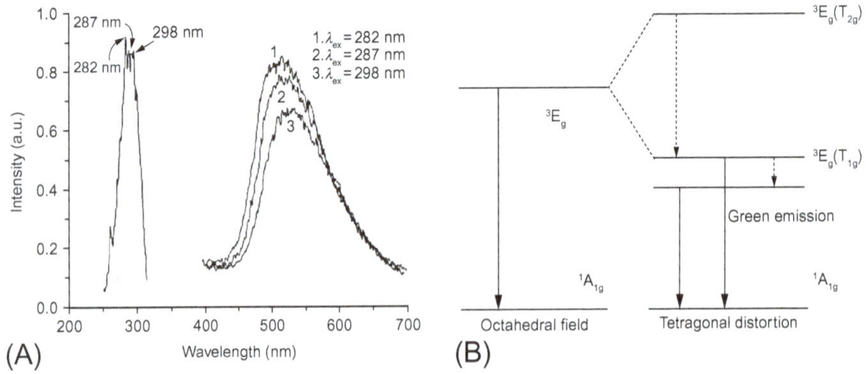


**Fig. 2.22** (A) Excitation (left) and emission (right) spectra of $Cu^{+}$ ions in borosilicate glasses. (B) Energy-level diagram for a tetragonal distortion of $Cu^{+}$ ion under the octahedral field. Reproduced with permission from Y. Ti, F. Qiu, Y. Cao, L. Jia, W. Qin, J. Zheng, G. Farrell, Photoluminescence of copper ion exchange BK7 glass planar waveguides, J. Mater. Sci. 43 (2008) 7073–7078, Copyright © Springer.

obtained emission spectra, under excitation with all these three excitation wavelengths, show that the $Cu^+$ ions exhibit a strong green emission band around 520 nm. It is seen that intensity and peak position of the emission band have been considerably changed depending on the excitation wavelengths. This was created by the variation of excited-state electronic level positions of the $Cu^+$ ion and strongly affected the transitions from the excited levels to the ground state, which resulted in emission band shift and intensity variation. In fact, the univalent copper ($Cu^+$) having completely filled d-orbital ($3d^{10}$), transitions for excitation and emission are unexpected. However, here, the excitation band is originated from the transitions of the ground state to states originating from the $3d^9 4s^1$ configuration. Even though this transition is forbidden, according to the Laporte selection rules, it may really take place in the glass for the reason that the low symmetry of the local electric field around the $Cu^+$ ions in the matrix may partly mix the orbits and permits some transitions. The shift of the emission bands may also cause by variations in the symmetry of the sites. It was found that the $Cu^+$ ions uniformly distributed in a near-pure silica glass structure occupying tetragonally distorted octahedral sites. When the ion-exchange time is increased, the distribution of impurities around the $Cu^+$ can change greatly, and it can bring on a strong electrostatic infinitesimal disturbance, which can influence the d-orbitals of $Cu^+$. In fact, the distortion may significantly lower the triplet emitter level by the splitting of the $^3E_g$ ($T_{1g}$) in the two spin-orbit components, which shows the level scheme for the $Cu^+$ ion. This is shown in Fig. 2.22B by the energy-level diagram for a tetragonal distortion of $Cu^+$ ion under the octahedral field (see also Section 2.2, J-T effect) [20].

## 2.5 Functional applications

### 2.5.1 *Colored glasses*

It is clear from the description of previous sections that different base glasses (e.g., silicate, borate, borosilicate, and phosphate) produce simple colored glasses upon introduction of various 3d transition metal (e.g., Ti, V, Cr, Mn, Fe, Co, Ni, and Cu) ions. Various technical aspects about manufacturing of different colored glasses are described by Weyl [21]. Coloring of glasses by 3d transition metal ions is essential for use in architectural purpose (as flat building glass), container glass, ophthalmic glass (prescription eye glasses), decorative glass wares, etc. In colored flat building glasses, the color renders rejection of entry of harmful ultraviolet radiation (UVR), intensity solar radiation, and solar heat [22]. In colored container glasses, it protects the contents (e.g., food materials and medicines) from the harmful effects of UV light and high intensity of visible light as well. Similarly, in eye care glasses, the color protects the eye from the harmful effects of UV light (creates cortical cataract) and high intensity of visible radiation [22]. Photo damages of skin, eye, etc. and their protection by glasses have recently been reviewed by Almutawa et al. [22]. Some of the hues of colored glasses are shown in Fig. 2.23.

There are various kinds of colored glasses, but the most important types are those manufactured by doping with 3d transition metal ions are blue, green, green-gray, and bronze glasses [5]. The general characteristic of all these glasses is that they all have

**Fig. 2.23** Some of the hues of colored glasses.

high solar radiation absorption. They transmit nearly 50% of the incident solar radiation at a thickness of 5–6 mm for use as building glass. For this reason, it is necessary to optimize the 3d transition metal ions concentration based on glass components and composition. Either cobalt oxide ($Co_3O_4$) or copper oxide (CuO) is used as colorant for production of soda lime silicate glass-based blue glasses, whereas a combination of chromium oxide ($Cr_2O_3$) and iron oxide ($Fe_2O_3$) is used for green glasses. Spectral profiles of the (A) blue and (B) green glasses are shown in Fig. 2.24 [5]. Green-gray glasses are produced using combination of nickel oxide (NiO), cobalt oxide ($Co_3O_4$), and iron oxide ($Fe_2O_3$), while bronze glasses are produced using selenium (Se), cobalt oxide ($Co_3O_4$), and iron oxide ($Fe_2O_3$) in soda lime silicate glasses. Spectral profiles of the (A) green-gray and (B) bronze glasses are illustrated in Fig. 2.25 [5].

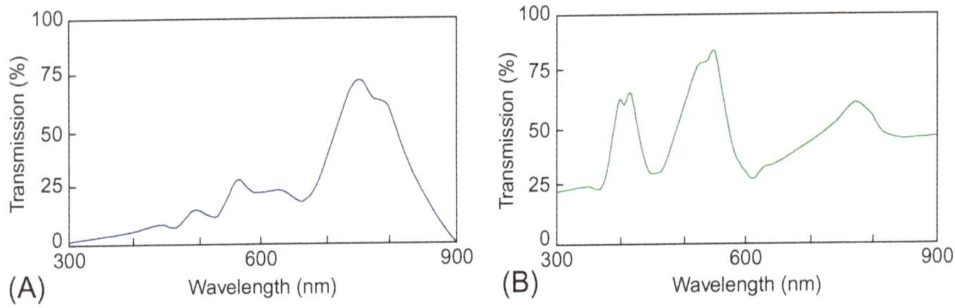


**Fig. 2.24** Spectral profiles of (A) blue and (B) green glasses.

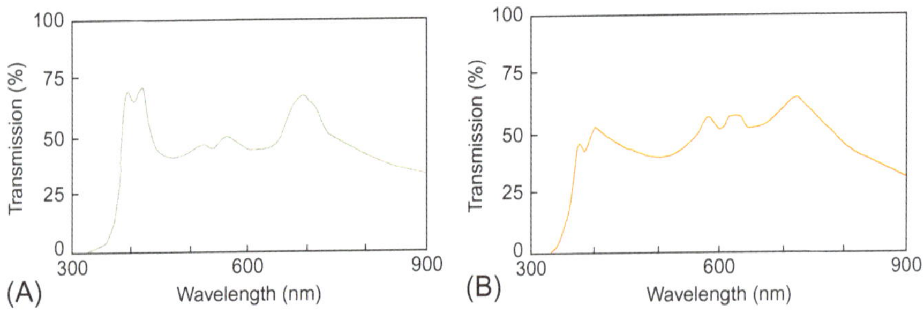


**Fig. 2.25** Spectral profiles of (A) green-gray and (B) bronze glasses.

## 2.5.2 Colored filter glasses

Colored filter glasses are those glasses that selectively transmit (pass) light of a particular range of wavelengths while blocking the other wavelengths of the electromagnetic spectrum. There are many types of colored filter glasses (e.g., absorptive, interference, or dichroic), but only the absorptive type of filter glasses are prepared by doping different base glasses (e.g., silicate, phosphate, borate, and borosilicate) with 3d transition metal ions using their selective absorption properties of the optical radiation as described in earlier sections. They exhibit colors if their absorption or filter action appears in the visible spectral range. There are various manufacturers of filter glasses such as Schott (Germany), Corning Glass Works (the United States), and Hoya Glass Works (Japan), and they produce a large number of glasses of different types. The manufacturing technologies are proprietary of each of these companies and undisclosed. A very little information is available in open literatures. Since there are a limited number of techniques of manufacturing these glasses, most of them differ from one another only due to variation in concentration of 3d transition metal ions. Therefore, it is possible to obtain same spectral transmission by using different thicknesses of a different manufacturer [23].

Colored filter glasses are used in many optical instruments such as camera, telescope, night vision devices, thermal imaging devices, medical devices, UV-Vis-NIR spectrophotometers, and fluorescence microscopes. [24]. In astronomy, filter glasses are used to limit light passed to the spectral band of interest. In fluorescence spectrometer, the absorbed light coming from the light source is separated from the emission light of the sample under investigation with the help of band-pass glass filters. Raman spectrometer uses narrow band-pass filters to separate strong incident laser light from the scattered light of the sample due to Raman effect. Digital surveillance camera requires IR cut filters to block IR light and only to pass visible light. Medical and biotechnology devices require UV band-pass and band-edge filters to increase the signal-to-noise ratio in measurements. Hence, it seen that there are different types of filter glasses. The most important types are band-pass, short-pass, and neutral-density filter glasses.

### 2.5.2.1 Bandpass filter glass

Band-pass filter glasses are categorized by their transmission characteristics. They have very high transmission (passband) within a certain wavelength range while blocking in both longer and shorter wavelength regions. The passband may be wider or narrower. The cutoff between minimum and maximum transmission may be gradual or sharp. Examples of this type of filter glasses are (A) blue-green band-pass filter glass (BG38, Schott) and (B) green band-pass filter glass (VG9, Schott) [24]. Their spectral profiles with their respective photograph (see inset) are shown in Fig. 2.26. They are used in electronic cameras such as charge-coupled device (CCD) or complementary metal-oxide semiconductor (CMOS) for conversion of light into electrons. Besides these there are several other functional uses as well.

Rayan et al. [25] have reported the optical spectroscopic analysis of band-pass filters of cupric oxide-doped barium phosphate glasses in the $CuO–BaO–ZnO–Na_2O–P_2O_5$ system. This is shown in Fig. 2.27A. It exhibits the effects of copper

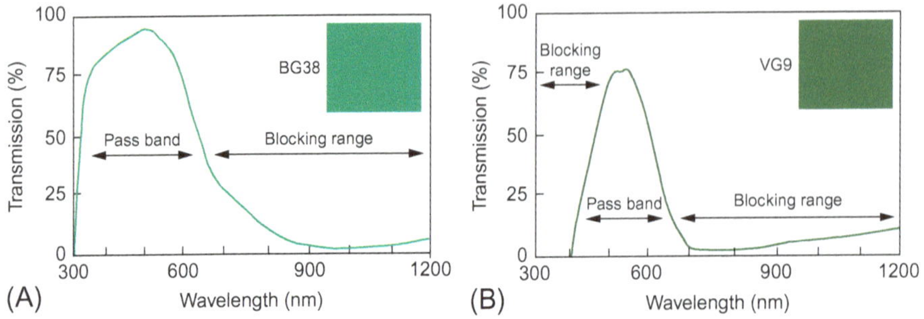


**Fig. 2.26** Spectral profiles of (A) blue-green band-pass filter glass (BG38, Schott) and (B) green band-pass filter glass (VG9, Schott).

oxide (CuO) content ($x=2$–12 mol.%) on the cutting bands in the UV and IR regions. The absorption bands around 600 and 1200 nm are assigned to the $^2B_{2g} \rightarrow {}^2B_{1g}$, $^2A_{1g} \rightarrow {}^2B_{1g}$ and $^2E_g \rightarrow {}^2B_{1g}$ transitions of cupric ($Cu^{2+}$) ions in phosphate glasses [25]. It is also seen that with increase in CuO content both the bands move and the intensity of the pass band decreases up to a certain limit. Very recently, Elhaes et al. [26] have reported the synthesis and characterization of $Cu_2O$-doped phosphate glass band-pass filters in the $P_2O_5$–ZnO–$Na_2O$–CaO–CuO ($x=2$, 4, 6, 8, and 10 mol.%) system. This is shown in Fig. 2.27B. It is seen that similarly as in the earlier case, here, all the samples exhibits high absorption both in the UV region up to 320 nm and IR region starting from 690 to 1100 nm. It is also seen that the band-pass transmission

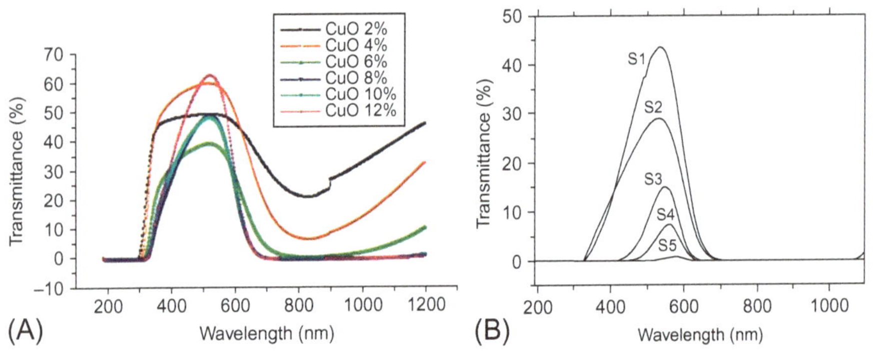


**Fig. 2.27** Alteration of spectra of band-pass filter glass with copper oxide content in (A) $x$CuO–$(20-x)$BaO–30ZnO–10$Na_2O$–40$P_2O_5$ ($x=2$–12 mol.%). (B) 42$P_2O_5$–39ZnO–$(18-x)Na_2O$–1CaO–$x$CuO ($x=2$, 4, 6, 8, and 10 mol.%) glass systems.
(A) Reproduced with permission from D.A. Rayan, Y.H. Elbashar, M.M. Rashad, A. El-Korashy, Optical spectroscopic analysis of cupric oxide doped barium phosphate glass for band pass absorption filter, J. Non-Cryst. Solids 382 (2013) 52–56, Copyright © Elsevier.
(B) Reproduced with permission from H. Elhaes, M. Attallah, Y. Elbashar, M. El-Okr, M. Ibrahim, Application of $Cu_2O$-doped phosphate glasses for bandpass filter, Physica B 449 (2014) 251–254, Copyright © Elsevier.

maxima shifts toward longer wavelengths and its intensity decreases with increase in the content of $Cu_2O$ ($x=2$, 4, 6, 8, and 10 mol.%).

### 2.5.2.2 *Short-pass filter glass*

A short-pass filter is colored glass filter that has high transmission (passes) in the shorter wavelengths (usually in the UV and Vis regions) and blocking in the longer wavelengths (NIR region). The slope of the short-pass filter is less steep at the transition between passband and blocking range. Since the short-pass filter blocks the heat producing IR light, it is regarded as "*heat protection or cold light filter.*" It is also regarded as "*eye-safe laser protective glass filter*" during practice with neodymium doped solid state lasers of 1064 nm wavelength (e.g., Nd-YAG laser). It is used as filters of operation theater (OT) room lamps. Its colorlessness in the visible wavelength range does not impart any color falsification in the operation room. In fluorescence microscopy, the short-pass filters are frequently used as excitation filters.

Example of this type of filter glasses is the KG series of Schott. The spectral profile of KG3 with its photograph (see inset) is shown in Fig. 2.28A [24]. The glass has a slight greenish-blue hue. A suitable glass system is $MgO$–$ZnO$–$B_2O_3$–$Al_2O_3$–$P_2O_5$. The filter glass is prepared using this glass system after hugely doping with ferric oxide ($Fe_2O_3$) and strong reducing agent. Concentration of ferric oxide and reducing agent is adjusted in such a way to obtain a very high concentration of ferrous ($Fe^{2+}$) ions, if possible only $Fe^{2+}$ ion, in the glass. The $Fe^{2+}$ ($d^6$ system) ion has strong absorption in the NIR region (1000–1100 nm) due to d-d transition with high visible transmission (see Fig. 2.17A and Table 2.3).

### 2.5.2.3 *Neutral-density filter glass (NG5)*

Neutral-density filter glass is a gray glass with uniform attenuation and almost constant transmission in the visible range (from 400 to 800 nm), therefore, is only slightly wavelength-dependent. Example of this type of filter glasses is the NG series of Schott. The spectral profile of NG5 with its photograph (see inset) is shown in Fig. 2.28B [24]. They are used to reduce the intensity of light by reflecting or

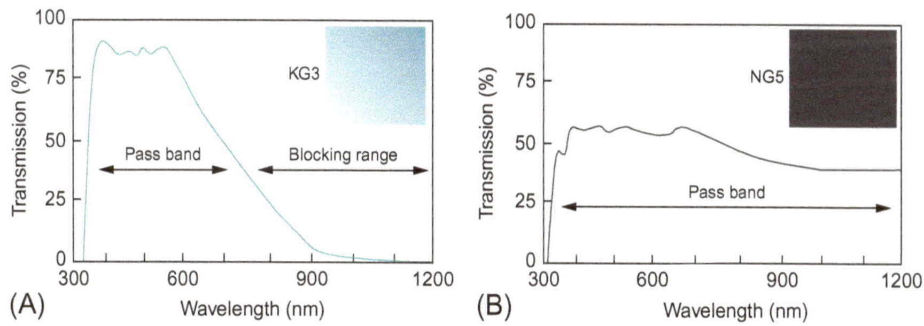


**Fig. 2.28** Spectral profiles of (A) short-pass filter glass (KG3, Schott) and (B) neutral-density filter glass (NG5, Schott).

absorbing a portion of it. A neutral-density filter glass has a predetermined optical density. The amount of desired filtering can be controlled by the use of different thickness of the filter glass. They are used in optics and photography when the user requires a defined attenuation of the intensity of radiation sources over the large visible to NIR spectral range. They are useful for taking photographs in bright light to avoid blurry and making photographic exposures longer. It is also recommended by the ophthalmologist in prescription spectacles when the brightness of light over a certain limit is harmful to the eye of the patient. This type of filter glasses is manufactured using a suitable base glass doping with a judicial combination of nickel oxide (very large amount), chromium oxide (medium amount), and iron oxide (small amount).

## 2.6 Summary and future outlook

Functionalization of different oxide glasses by 3d transition metal ion doping for various applications is described here. General methods of synthesis, batch calculation, and characterization of glasses are briefly described mainly for the beginners. It is observed that all the 3d transition metal ions exhibit their respective color in the glasses depending on glass composition and components. At the start, the underlying theories explicitly the valence bond (concisely) and ligand field (elaborately) are described for understanding the origin of coloration in these glasses on doping with 3d transition metal ions. The absorption and photoluminescence of individual 3d transition metal ion in different oxide glasses are described with proper spectra and interpretation in detailed. These glasses have various functional applications based on their colors that are caused by the absorption bands. The most important uses are as colored glasses and colored filter glasses. These two categories of glasses are described in details particularly the spectra and varieties of colorants taking commercial glass as an example.

Functionalization (e.g., absorption and photoluminescence) of glasses fundamentally depends on various manufacturing parameters such as glass component and composition, glass melting atmosphere (e.g., oxidizing, reducing, or inert), and concentration of 3d transition metal ions. It is also seen that their exhibited properties are influenced by the characterization conditions, for example, spectra recording temperature and excitation wavelength for recording of photoluminescence spectra. The process technologies of commercial colored glasses and colored filter glasses are all proprietary of the manufacturing companies. A very little is known. For example, very recently Rayan et al. [25] and Elhaes et al. [26] have reported some studies about band-pass filter glasses. Therefore, it is realized that more explorations are necessary with respect to each of the aspects indicated above and many more.

## Exercises

**2.1** What are the optical effects of different 3d-transition-metal ions in glasses? Explain the reasons.

**2.2** Describe different factors those affect the optical properties of 3d-transition-metal ions in glasses.

**2.3** Describe bonding and hybridization of $[CrO_4]^{2-}$, $[FeO_6]^{3-}$, and $[NiO_4]^{2-}$ complex ions in oxide glasses.

**2.4** Write the coordination number and hybridization of the complex ions formed by different 3d-transition-metal ion on doping in oxide glasses.

**2.5** Describe in details the ligand field theory with energy diagram for the octahedral and tetrahedral geometries of the complex ions.

**2.6** Describe in details with examples and energy diagram the different types of J-T effects in oxide glasses.

**2.7** Describe with example the absorption and photoluminescence spectra with proper interpretation of each of the 3d-transition-metal ions separately in oxide glasses.

**2.8** Although the $Cu^{+}$ ion is a $3d^{10}$ system, it exhibits absorption and photoluminescence spectrum in certain oxide glasses. Why? Explain reason in details.

**2.9** What are colored glasses? Describe with examples.

**2.10** What are colored filter glasses? Describe with examples.

# References

[1] Yu.A. Guloyan, Complete analysis of ionic staining of glasses by transition-metal compound, Glas. Ceram. 64 (2007) 153–158.

[2] A.K. Varshneya, Fundamentals of Inorganic Glasses, Academic Press, Boston, MA, 1994.

[3] B. Averill, P. Eldredge, General Chemistry: Principles, Patterns, and Applications, Flat World Knowledge, Washington, DC, 2011.

[4] T. Bates, Ligand field theory and absorption spectra of transition metal ions in glasses, in: J.D. Mackenzie (Ed.), Modern Aspects of the Vitreous State, vol. 2, Butterworths, London, 1962, , pp. 195–254.

[5] C.R. Bamford, Control of Colours and Generation in Glass, Elsevier-North Holland Publications, Amsterdam, 1977.

[6] A. Paul, Chemistry of Glasses, second ed., Chapman and Hall, London, 1990, pp. 276–348.

[7] D. Rider, Glass, Borax Consolidated Limited, London, 1965.

[8] F.V. Tooley, Handbook of Glass Manufacture, vols. I & II, Ogden Publishing Co., New York, NY, 1960.

[9] N. Abd El-Shafi, M.M. Morsi, Optical absorption and infrared studies of some silicate glasses containing titanium, J. Mater. Sci. 33 (1997) 5185–5189.

[10] H. Keppler, Crystal field spectra and geochemistry of transition metal ions in silicate melts and glasses, Am. Miner. 77 (1992) 62–75.

[11] L.B. Glebov, E.N. Boulo, Absorption of iron and water in the $Na_2O$–CaO–MgO–$SiO_2$ glasses. II. Selection of intrinsic, ferric, and ferrous spectra in the visible and UV regions, J. Non-Cryst. Solids 242 (1998) 49–62.

[12] B. Hannoyer, M. Lenglet, J. Dürr, R. Cortes, Spectroscopic evidence of octahedral iron (III) in soda-lime-silicate glasses, J. Non-Cryst. Solids 151 (1992) 209–316.

[13] L.E. Bausa, F. Jaque, J. Garcia Sole, A. Duran, Photoluminescence of $Ti^{3+}$ in $P_2O_5$–$Na_2O$–$Al_2O_3$ glass, J. Mater. Sci. 23 (1988) 1921–1922.

[14] H. Wen, P.A. Tanner, Optical properties of 3d transition metal ion-doped sodium borosilicate glass, J. Alloys Compd. 625 (2015) 328–335.

[15] Y.G. Choi, K.H. Kim, Y.S. Han, J. Heo, Oxidation state and local coordination of chromium dopant in soda-lime-silicate and calcium-aluminate glasses, Chem. Phys. Lett. 329 (2000) 370–376.

[16] C. Koepke, K. Wisniewski, M. Grinberg, Excited state spectroscopy of chromium ions in various valence states in glasses, J. Alloys Compd. 341 (2002) 19–27.

[17] I.E.C. Machado, L. Prado, L. Gomes, J.M. Prison, J.R. Martinelli, Optical properties of manganese in barium phosphate glasses, J. Non-Cryst. Solids 348 (2004) 113–117.

[18] G. Lakshminarayana, S. Buddhudu, Spectral analysis of $Mn^{2+}$, $Co^{2+}$ and $Ni^{2+}$: $B_2O_3$–ZnO–PbO glasses, Spectrochimica Acta Part A: Molecular and Biomolecular Spectroscopy on ScienceDirect" Spectrochimica Acta A, 63 (2006) 295–304.

[19] Y. Gandhi, N. Krishna Mohan, N. Veeraiah, Role of nickel ion coordination on spectroscopic and dielectric properties of $ZnF_2$–$As_2O_3$–$TeO_2$:NiO glass system, J. Non-Cryst. Solids 357 (2011) 1193–1202.

[20] Y. Ti, F. Qiu, Y. Cao, L. Jia, W. Qin, J. Zheng, G. Farrell, Photoluminescence of copper ion exchange BK7 glass planar waveguides, J. Mater. Sci. 43 (2008) 7073–7078.

[21] W.A. Weyl, Coloured Glasses, Society of Glass Technology, Sheffield, 1951.

[22] F. Almutawa, R. Vandal, S.Q. Wang, H.W. Lim, Current status of photoprotection by window glass, automobile glass, window films, and sunglasses, Photodermatol. Photoimmunol. Photomed. 29 (2013) 65–72.

[23] J.A. Dobrowolski, G.E. Marsh, D.G. Charbonneau, J. Eng, P.D. Josephy, Colored filter glasses: an intercomparison of glasses made by different manufacturers, Appl. Opt. 16 (1977) 1491–1512.

[24] SCHOTT's Optical Filter Glass Catalog, Schott North America, Inc, Duryea, PA, 2009. *www.us.schott.com/advanced_optics*.

[25] D.A. Rayan, Y.H. Elbashar, M.M. Rashad, A. El-Korashy, Optical spectroscopic analysis of cupric oxide doped barium phosphate glass for band pass absorption filter, J. Non-Cryst. Solids 382 (2013) 52–56.

[26] H. Elhaes, M. Attallah, Y. Elbashar, M. El-Okr, M. Ibrahim, Application of $Cu_2O$-doped phosphate glasses for bandpass filter, Physica B 449 (2014) 251–254.

# Functionalization of glasses by 4f rare-earth metal ion doping

3

## Chapter Outline



## 3.1 Introduction

The 4f rare-earth (RE) metal ion-doped glasses exhibit unique functional properties for a wide range of applications ranging from photonics and optoelectronics to various devices of energy, environment, biosciences, medical sciences, sensors, and telecommunications. Many technological progresses would not be possible without their involvement. Some of the specific functional uses of the 4f RE metal ion-doped glasses are as laser glass, radiation-resistant glass, wavelength-standard glass, colored glasses, etc. In addition, there are enormous opportunities for development of new functional glasses doping with 4f RE metal ion. The responsible property of the 4f RE metal ions for their unique functionalities is their incompletely filled 4f orbital ($f^{1-14}$). The atomic number, symbol, name, valences, and outer shell electronic configuration of the 4f rare-earth metal elements (excluding lutetium, Lu) are shown in Fig. 3.1 for understanding this fact.

**Functional Glasses and Glass-Ceramics. http://dx.doi.org/10.1016/B978-0-12-805056-9.00003-9**

**4f rare-earth metal elements**

| | | | | | | | | | | | | |
|---|---|---|---|---|---|---|---|---|---|---|---|---|
| 58<br>Ce<br>Cerium<br>+3, +4<br>$4f^15d^16s^2$ | 59<br>Pr<br>Praseodymium<br>+3<br>$4f^36s^2$ | 60<br>Nd<br>Neodymium<br>+3<br>$4f^46s^2$ | 61<br>Pm<br>Promethium<br>+3<br>$4f^56s^2$ | 62<br>Sm<br>Samarium<br>+2, +3<br>$4f^66s^2$ | 63<br>Eu<br>Europium<br>+2, +3<br>$4f^76s^2$ | 64<br>Gd<br>Gadolinium<br>+3<br>$4f^75d^16s^2$ | 65<br>Tb<br>Terbium<br>+3<br>$4f^85d^16s^2$ | 66<br>Dy<br>Dysprosium<br>+3<br>$4f^{10}6s^2$ | 67<br>Ho<br>Holmium<br>+3<br>$4f^{11}6s^2$ | 68<br>Er<br>Erbium<br>+3<br>$4f^{12}6s^2$ | 69<br>Tm<br>Thulium<br>+3<br>$4f^{13}6s^2$ | 70<br>Yb<br>Ytterbium<br>+2, +3<br>$4f^{14}6s^2$ |

63 ← Atomic number
Eu ← Symbol
Europium ← Name
+2, +3 ← Valences
$4f^76s^2$ ← Outer electronic configuration

**Fig. 3.1** Atomic number, symbol, name, valences, and outer shell electronic configuration of the 4f rare-earth metal elements.

However, the functions (position of the absorption and photoluminescence bands) of the 4f RE metal ions in glasses depend on several factors such as

**(i)** chemical composition and component of the glass,
**(ii)** concentration and combination of 4f RE metal ion in the glass,
**(iii)** glass melting atmosphere (e.g., oxidizing, reducing, and inert),
**(iv)** glass melting temperature and duration,
**(v)** sources of raw materials and impurity content,
**(vi)** volume of glass melted (prepared) per melt, and
**(vii)** glass melting container (e.g., refractory or noble metal crucible).

Therefore, for successful functionalization of glasses by 4f RE metal ion doping, it is necessary to control and consider all of the above factors. As a consequence of this fact, there are a huge number of publications in this domain.

Consequently, an attempt was made using the Web of Science (version 5.17) as described in Chapter 1 to provide a global publication trend and quantitative growth in the area of 4f rare-earth metal ion-doped glasses and glass-ceramics. Here, year-wise data were extracted for the last 25 years (1990–2014) by searching, using the topic keyword *"glass"* followed by refining the document type as *"article"* published in the journals and meetings. It was further refined using the term *"rare-earth."* This plot is shown in Fig. 3.2A. It gives us a clear publication trend and quantitative growth of published papers on rare-earth metal ion containing glasses and glass-ceramics during the last 25 years (1990–2014). To find out the total number of published papers on individual rare-earth metal ion containing glasses and glass-ceramics during the last 25 years (1990–2014), the total publication of 112,633 articles was further refined by term of *"individual 4f rare-earth metal"* (e.g., cerium, praseodymium, and neodymium). This plot is illustrated in Fig. 3.2B. It gives us a clear quantitative idea of paper published on individual 4f rare-earth metal ion containing glasses and glass-ceramics as well as a comparison among them. The decreasing order of publication trend of the individual 4f rare-earth metal ion-doped glasses and glass-ceramics is as follows:

$$\mathrm{Er} \gg \mathrm{Nd} > \mathrm{Eu} > \mathrm{Ce} > \mathrm{Yb} \gg \mathrm{Sm} > \mathrm{Tm} > \mathrm{Gd} \approx \mathrm{Pr} > \mathrm{Tb} > \mathrm{Dy} > \mathrm{Ho}$$

It is seen that during the last 25 years (1990–2014), among the 112,633 published articles, the highest number (1236) of papers was published using erbium (Er), the second highest (549) is with neodymium (Nd), the third position (516) is with europium (Eu), and the lowest position (129 only) is with holmium (Ho). This trend of publications also gives us the idea about the importance of doping the glasses with 4f RE metal ions.

It is necessary to know the underlying theories to understand the activities of these ions in glasses particularly with respect to their functionality, particularly with respect to absorption and photoluminescence phenomena. For this reason, the necessary theories are briefly described in the next section.

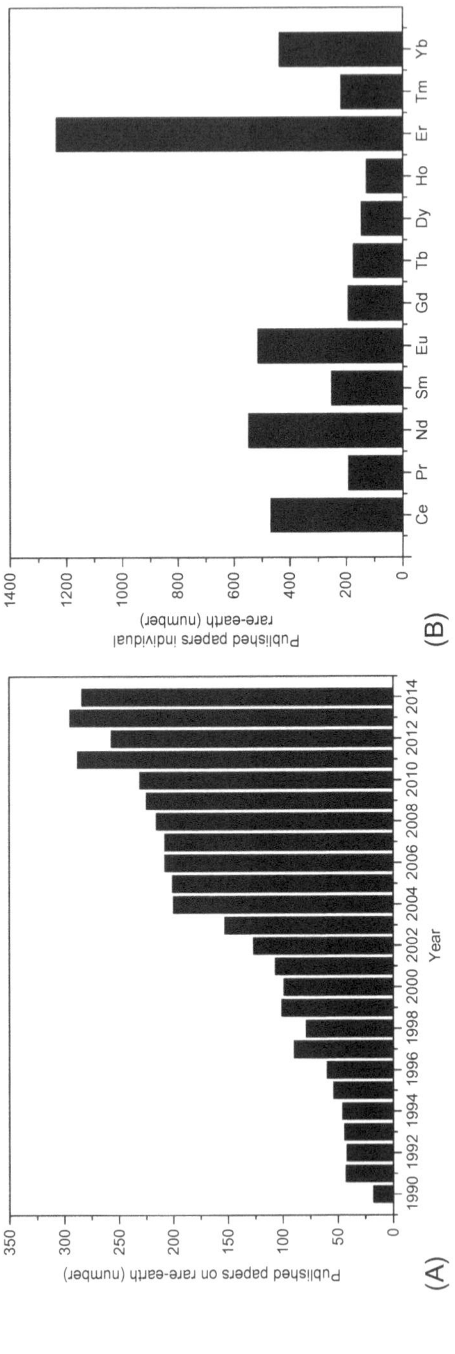


**Fig. 3.2** (A) Year-wise published papers on 4f rare-earth metal ion containing glasses and glass-ceramics during the last 25 years (1990–2014). (B) Total number of published papers on individual 4f rare-earth metal ion during the last 25 years (1990–2014).

## 3.2 Theoretical considerations

The energy levels, $E_i$, for the transitions of the optical absorption and emission bands of the rare-earth metal complex ions in glasses can be determined by solving the following Schrödinger equation:

$$H\psi_i = E_i\psi_i \tag{3.1}$$

where $H$ is a Hamiltonian, which includes all types of interactions of the valence electrons of the rare-earth ion, and $\psi_i$ is the eigenfunctions of the center. Under the field of the ligands of the rare-earth complex ion, the Hamiltonian, $H$, can be written as:

$$H = H_{\mathrm{FI}} + H_{\mathrm{LF}} \tag{3.2}$$

where $H_{\mathrm{FI}}$ is the free-ion Hamiltonian and $H_{\mathrm{LF}}$ is the ligand field Hamiltonian. According to the quantum mechanical perturbation theory, the free-ion Hamiltonian, $H_{\mathrm{FI}}$, can be written as:

$$H_{\mathrm{FI}} = H_0 + H_{\mathrm{ee}} + H_{\mathrm{so}} \tag{3.3}$$

where $H_0$ is the central field Hamiltonian, $H_{\mathrm{ee}}$ represents any perturbation due to Coulombic interactions among the outer (valence) electrons, and $H_{\mathrm{so}}$ symbolizes for the spin-orbit interaction summed over these electrons.

The partial energy-level diagrams of trivalent rare-earth ions (from $Ce^{3+}$ to $Eu^{3+}$) with main visible emission transitions detected in silicate glasses are shown in Fig. 3.3 [1]. From Eq. (3.3), it is seen that the energy levels of transitions of rare-earth ions depend on $H_{\mathrm{ee}}$ and $H_{\mathrm{so}}$. The general spectroscopic notation for the term symbol of the transition energy-levels in rare-earth ions is $^{2S+1}\mathrm{L}_J$. Here, the letter L corresponds to the total orbital angular momentum of the ion obtained by combining the orbital angular momenta of the individual electrons in the ion according to the Clebsch-Gordan series (i.e., L = S, P, D, F, G, H, and I). The left superscript, $2S+1$, is the number of possible orientations of the total spin of the ion where $S$ is the total spin of the ion. The right subscript, $J$, stands for the total angular momentum of the ion and is determined using the Russell-Saunders coupling scheme. The examples of the spectroscopic notation for the term symbol are $^4\mathrm{I}_{13/2}$, $^2\mathrm{F}_{7/2}$, etc. The absorption peak positions and their transition energy-level assignments of trivalent 4f rare-earth ions in silicate glass and aqueous solution are listed in Table 3.1 [2–7]. The energy levels of rare-earth ions in glass are similar to those of free ions or ions in aqueous solution. However, some shifting occurs caused by the so-called *nephelauxetic effect* (literally means *charge cloud expansion*) depending on the covalent or ionic character of the glass matrix. Recently, Som and Karmakar [8] have described this phenomenon elaborately considering different glass matrices with various 4f rare-earth ions. The mathematical expressions for evaluation of the characteristics of absorption and photoluminescence spectra have been described as follows based on the references [9–12].

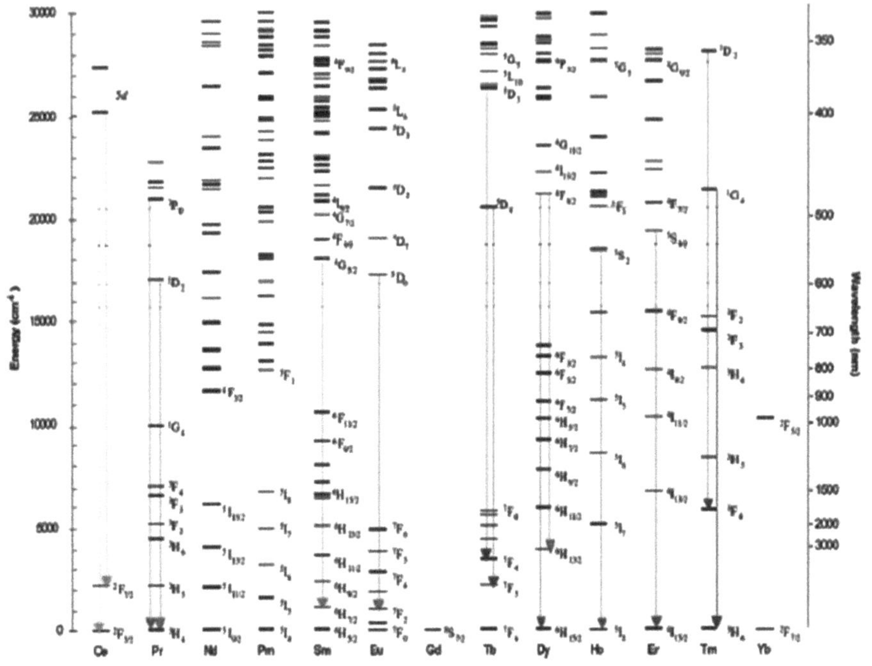


**Fig. 3.3** Partial energy-level diagrams of trivalent rare-earth ions ($Ce^{3+}$ to $Eu^{3+}$) with main visible emission transitions detected in silicate glasses.
Courtesy: M.J. Dejneka, A. Streltsov, S. Pal, A.G. Frutos, C.L. Powell, K. Yost, P.K. Yuen, U. Müller, J. Lahiri, Rare earth-doped glass microbarcodes, PNAS 100 (2003) 389–393.

### 3.2.1 Spectral analysis

The free-ion Hamiltonian, $H_{\mathrm{FI}}$, can be expressed by the following relation:

$$\begin{aligned} H_{\mathrm{FI}} = {} & E_{\mathrm{AVG}} + \sum_k F^k f_k + \xi_{\mathrm{SO}} A_{\mathrm{SO}} + \alpha L(L+1) + \beta G(G_2) \\ & + \gamma G(R_7) + \sum_i T^i t_i + \sum_k P^k P_k + \sum_j M^j m_j \end{aligned} \tag{3.4}$$

where $i$, $k$, and $j$ are the integers that depend on the number of electrons in the f-orbital of the rare-earth ion, that is, their values are different for different rare-earth ions. It contains the central field parameter ($E_{\mathrm{AVG}}$), two-body electrostatic interaction parameter ($F^k$), spin-orbit interaction parameter ($\xi_{\mathrm{SO}}$), two-body configuration interaction parameters ($\alpha$, $\beta$, $\gamma$), three-body configuration interaction parameter ($T^i$), electrostatically correlated spin-orbit interaction parameter ($P^k$), and spin-other orbit interaction parameter ($M^j$). Using the standard least-square method, the fitting is carried out between the experimental and calculated absorption spectra. The integrity of fitting

**Table 3.1 Absorption peak positions and their transition energy-level assignments of trivalent 4f rare-earth ions in silicate glass and aqueous solution**

| Trivalent 4f rare-earth ion | Ground state | Transition energy level | Absorption peak position | |
|---|---|---|---|---|
| | | | Frequency ($cm^{-1}$) | Wavelength (nm) |
| Cerium ion, $Ce^{3+}$ ($4f^1$), in silicate glass [2,3] | $^2F_{5/2}\rightarrow$ | $^2D$ | 31,900 | 313 |
| Praseodymium ion, $Pr^{3+}$ ($4f^2$), in aqueous solution [4] | $^3H_4\rightarrow$ | $^3H_5$ | 2360 | 4237 |
| | | $^3H_6$ | 4500 | 2222 |
| | | $^3F_2$ | 5200 | 1923 |
| | | $^3F_3$ | 6500 | 1538 |
| | | $^3F_4$ | 6950 | 1439 |
| | | $^1G_4$ | 9900 | 1010 |
| | | $^1D_2$ | 16,840 | 594 |
| | | $^3P_0$ | 20,750 | 482 |
| | | $^3P_1$ | 21,300 | 469 |
| | | $^1I_6$ | 21,500 | 465 |
| | | $^3P_2$ | 22,520 | 444 |
| | | $^1S_0$ | 46,900 | 213 |
| Neodymium ion, $Nd^{3+}$ ($4f^3$), in aqueous solution [4] | $^4I_{9/2}\rightarrow$ | $^4I_{11/2}$ | 2028 | 4931 |
| | | $^4I_{13/2}$ | 4050 | 2469 |
| | | $^4I_{15/2}$ | 6050 | 1653 |
| | | $^4F_{3/2}$ | 11,460 | 873 |
| | | $^4F_{5/2}$ | 12,480 | 801 |
| | | $^2H_{9/2}$ | 12,590 | 794 |
| | | $^4F_{7/2}$, $^4S_{3/2}$ | 13,500 | 741 |
| | | $^4F_{9/2}$ | 14,700 | 680 |
| | | $^2H_{11/2}$ | 15,870 | 630 |
| | | $^4G_{5/2}$ | 17,300 | 578 |
| | | $^2G_{7/2}$ | 17,460 | 573 |
| | | $^4G_{7/2}$ | 19,160 | 522 |
| | | $^4G_{9/2}$ | 19,550 | 512 |
| | | $^2K_{15/2}$ | 21,000 | 476 |
| | | $^2G_{9/2}$, $(^2D, {}^2P)_{3/2}$ | 21,300 | 469 |
| | | $^4G_{11/2}$ | 21,650 | 462 |
| | | $^2P_{1/2}$ | 23,250 | 430 |
| | | $^2D_{5/2}$ | 23,900 | 418 |
| | | $(^2P, {}^2D)_{3/2}$ | 26,300 | 380 |
| | | $^4D_{3/2}$ | 28,300 | 353 |
| | | $^4D_{5/2}$ | 28,500 | 351 |
| | | $^4D_{1/2}$ | 28,850 | 347 |
| | | $^4D_{7/2}$ | 30,500 | 328 |
| | | $^2D_{3/2}$ | 33,400 | 299 |
| | | $^2D_{5/2}$ | 34,450 | 290 |

*Continued*

Table 3.1 **Continued**

| Trivalent 4f rare-earth ion | Ground state | Transition energy level | Absorption peak position | |
|---|---|---|---|---|
| | | | Frequency ($cm^{-1}$) | Wavelength (nm) |
| | | $^2F_{5/2}$ | 38,500 | 260 |
| | | $^2F_{1/2}$ | 39,950 | 250 |
| | | $^2G_{9/2}$ | 47,700 | 210 |
| | | $^2G_{7/2}$ | 48,600 | 206 |
| Samarium ion, $Sm^{3+}$ ($4f^5$), in aqueous solution [4] | $^6H_{5/2}\rightarrow$ | $^6H_{7/2}$ | 1080 | 9259 |
| | | $^6H_{9/2}$ | 2286 | 4374 |
| | | $^6H_{11/2}$ | 3608 | 2772 |
| | | $^6H_{13/2}$ | 5000 | 2000 |
| | | $^6F_{1/2}$ | 6400 | 1563 |
| | | $^6F_{3/2}$ | 6630 | 1508 |
| | | $^6F_{5/2}$ | 7100 | 1408 |
| | | $^6F_{7/2}$ | 8000 | 1250 |
| | | $^6F_{9/2}$ | 9200 | 1087 |
| | | $^6F_{11/2}$ | 10,500 | 952 |
| | | $^4G_{5/2}$ | 17,900 | 559 |
| | | $^4F_{3/2}$ | 18,900 | 529 |
| | | $^4G_{7/2}$ | 20,050 | 499 |
| | | $^4M_{15/2}$ | 20,800 | 481 |
| | | $^4I_{11/2}$ | 21,100 | 474 |
| | | $^4I_{13/2}$ | 21,600 | 463 |
| | | $^4F_{5/2}$ | 22,200 | 450 |
| | | $^4G_{9/2}$ | 22,700 | 441 |
| | | $(^6P,^4P)_{5/2}$ | 24,050 | 416 |
| | | $^4L_{13/2}$ | 24,570 | 407 |
| | | $^6P_{3/2}$ | 24,950 | 401 |
| | | $^4L_{15/2}$ | 25,650 | 390 |
| | | $^6P_{7/2}$ | 26,750 | 374 |
| | | $^4D_{3/2}$, $(^4D, ^6P)_{5/2}$ | 27,700 | 361 |
| | | $^4H_{7/2}$ | 28,250 | 354 |
| | | $^4H_{9/2}$, $^4D_{7/2}$ | 29,100 | 344 |
| | | $^4G_{9/2}$ | 30,200 | 331 |
| | | $^4P_{3/2}$ | 31,550 | 317 |
| | | $^4P_{5/2}$ | 32,800 | 305 |
| | | $^2F_{5/2}$, $^4K_{13/2}$ | 33,800 | 296 |
| | | $(^4I, ^4F)_{9/2}$ | 34,600 | 289 |
| | | $^4I_{11/2}$ | 36,000 | 278 |
| | | $^4F_{3/2}$, $^4I_{13/2}$ | 36,700 | 272 |
| | | $(^2H, ^4I)_{9/2}$ | 37,600 | 266 |
| | | $(^2D, ^2F)_{3/2}$ | 38,300 | 261 |
| | | $^2H_{11/2}$ | 39,250 | 255 |
| | | $^2G_{9/2}$ | 42,000 | 238 |

**Table 3.1 Continued**

| Trivalent 4f rare-earth ion | Ground state | Transition energy level | Absorption peak position | |
|---|---|---|---|---|
| | | | Frequency ($cm^{-1}$) | Wavelength (nm) |
| | | $^4G_{7/2}$ | 42,900 | 233 |
| | | $(^4H, ^4G)_{9/2}$ | 43,100 | 232 |
| | | $(^4G, ^3H, ^4H)_{11/2}$ | 43,800 | 228 |
| | | $^4H_{13/2}$, $^4H_{11/2}$, $(^4P, ^2D)_{3/2}$ | 46,500 | 215 |
| | | $^2H_{11/2}$ | 47,400 | 211 |
| Europium ion, $Eu^{3+}$ ($4f^6$), in aqueous solution [5] | $^7F_0 \rightarrow$ | $^7F_1$ | 360 | 27,778 |
| | | $^7F_2$ | 1020 | 9804 |
| | | $^7F_3$ | 1887 | 5299 |
| | | $^7F_4$ | 2865 | 3490 |
| | | $^7F_5$ | 3908 | 2559 |
| | | $^7F_6$ | 4980 | 2008 |
| | | $^5D_0$ | 17,277 | 579 |
| | | $^5D_1$ | 19,028 | 526 |
| | | $^5D_2$ | 21,519 | 465 |
| | | $^5D_3$ | 24,408 | 410 |
| | | $^5L_6$ | 25,400 | 394 |
| | | $^5G_2$ | 26,300 | 380 |
| | | $^5G_4$ | 26,620 | 376 |
| | | $^5G_6$ | 26,700 | 375 |
| | | $^5D_4$ | 27,670 | 361 |
| | | $^5H_4$ | 31,250 | 320 |
| | | $^5H_6$ | 31,520 | 317 |
| | | $^5F_2$ | 33,190 | 301 |
| | | $^5F_4$ | 33,590 | 298 |
| | | $(^5I, ^5H)_6$ | 35,030 | 285 |
| | | $^5K_5$ | 36,205 | 276 |
| | | $^5K_6$ | 37,440 | 267 |
| | | $(^3K, ^3I)_6$ | 39,060 | 256 |
| | | $^5G_4$ | 39,890 | 251 |
| | | $^5G_6$ | 41,370 | 242 |
| Gadolinium ion, $Gd^{3+}$ ($4f^7$), in aqueous solution [6] | $^8S_{7/2} \rightarrow$ | $^6P_{7/2}$ | 32,200 | 311 |
| | | $^6P_{5/2}$ | 32,780 | 305 |
| | | $^6P_{3/2}$ | 33,350 | 300 |
| | | $^6I_{7/2}$ | 35,930 | 278 |
| | | $^6I_{9/2}$ | 36,270 | 276 |
| | | $^6I_{17/2}$ | 36,340 | 275 |
| | | $^6I_{11/2}$ | 36,560 | 274 |
| | | $^6I_{13/2}$ | 36,660 | 273 |
| | | $^6I_{15/2}$ | 36,710 | 272 |

*Continued*

Table 3.1 **Continued**

| Trivalent 4f rare-earth ion | Ground state | Transition energy level | Absorption peak position | |
|---|---|---|---|---|
| | | | Frequency ($cm^{-1}$) | Wavelength (nm) |
| | | $^6D_{9/2}$ | 39,720 | 252 |
| | | $^6D_{1/2}$ | 40,560 | 247 |
| | | $^6D_{7/2}$ | 40,700 | 246 |
| | | $^6D_{3/2}$ | 40,850 | 245 |
| | | $^6D_{5/2}$ | 41,000 | 244 |
| | | $(^6G, ^6F)_{7/2}$ | 49,270 | 203 |
| | | $^6G_{9/2}$, $^6G_{11/2}$, $^6G_{5/2}$ | 49,680 | 201 |
| | | $^6G_{13/2}$ | 51,413 | 195 |
| Terbium ion, $Tb^{3+}$ ($4f^8$), in aqueous solution [7] | $^7F_6 \rightarrow$ | $^7F_5$ | 2100 | 4762 |
| | | $^7F_4$ | 3356 | 2980 |
| | | $^7F_3$ | 4400 | 2273 |
| | | $^7F_2$ | 5038 | 1985 |
| | | $^7F_1$ | 5440 | 1838 |
| | | $^7F_0$ | 5700 | 1754 |
| | | $^5D_4$ | 20,500 | 488 |
| | | $^5G_6$ | 26,500 | 377 |
| | | $^5L_{10}$ | 27,100 | 369 |
| | | $^5G_5$ | 27,800 | 360 |
| | | $^5L_9$ | 28,400 | 352 |
| | | $^5L_8$ | 29,300 | 341 |
| | | $^5L_7$ | 29,450 | 340 |
| | | $^5D_1$ | 30,650 | 326 |
| | | $^5H_7$ | 31,600 | 316 |
| | | $^5H_6$ | 33,000 | 303 |
| | | $^5H_5$ | 33,900 | 295 |
| | | $^5F_5$ | 34,900 | 287 |
| | | $^5I_8$ | 35,200 | 284 |
| | | $^5I_7$ | 36,700 | 272 |
| | | $^5I_6$ | 37,760 | 265 |
| | | $^5K_9$ | 39,100 | 256 |
| Dysprosium ion, $Dy^{3+}$ ($4f^9$), in aqueous solution [4] | $^6H_{15/2} \rightarrow$ | $^6H_{13/2}$ | 3517 | 2843 |
| | | $^6H_{11/2}$ | 5850 | 1709 |
| | | $^6H_{9/2}$, $^6F_{11/2}$ | 7700 | 1299 |
| | | $^6F_{9/2}$, $^6H_{7/2}$ | 9100 | 1099 |
| | | $^6H_{5/2}$ | 10,200 | 980 |
| | | $^6F_{7/2}$ | 11,000 | 909 |
| | | $^6F_{5/2}$ | 12,400 | 806 |
| | | $^6F_{3/2}$ | 13,250 | 755 |
| | | $^4F_{9/2}$ | 21,100 | 474 |
| | | $^4I_{15/2}$ | 22,100 | 452 |

**Table 3.1 Continued**

| Trivalent 4f rare-earth ion | Ground state | Transition energy level | Absorption peak position | |
|---|---|---|---|---|
| | | | Frequency ($cm^{-1}$) | Wavelength (nm) |
| | | $^4G_{11/2}$ | 23,400 | 427 |
| | | $^4F_{7/2}$, $^4I_{13/2}$ | 25,800 | 388 |
| | | $^4K_{17/2}$ | 26,400 | 379 |
| | | $^4M_{19/2}$, ($^4P$, $^4D)_{3/2}$, $^6P_{5/2}$ | 27,400 | 365 |
| | | $^6P_{7/2}$ | 28,550 | 350 |
| | | ($^4F$, $^4D)_{5/2}$ | 29,600 | 338 |
| | | $^6P_{3/2}$ | 30,800 | 325 |
| | | $^4H_{13/2}$ | 33,500 | 299 |
| | | $^4D_{7/2}$ | 33,900 | 295 |
| | | $^4G_{11/2}$, $^4L_{17/2}$ | 34,900 | 287 |
| | | ($^4G$, $^4H)_{7/2}$ | 35,900 | 279 |
| | | ($^4G$, $^4P)_{5/2}$ | 36,550 | 274 |
| | | $^4F_{3/2}$ | 37,900 | 264 |
| | | ($^4G$, $^4P)_{5/2}$ | 38,900 | 257 |
| | | ($^4P$, $^4F)_{3/2}$ | 39,100 | 256 |
| | | ($^4F$, $^2G)_{9/2}$ | 41,050 | 244 |
| | | $^4I_{15/2}$ | 41,700 | 240 |
| Holmium ion, $Ho^{3+}$ ($4f^{10}$), in aqueous solution [4] | $^5I_8 \rightarrow$ | $^5I_7$ | 5130 | 1949 |
| | | $^5I_6$ | 8580 | 1166 |
| | | $^5I_5$ | 11,120 | 899 |
| | | $^5I_4$ | 13,300 | 752 |
| | | $^5F_5$ | 15,500 | 645 |
| | | $^5S_2$, $^5F_4$ | 18,500 | 541 |
| | | $^5F_2$ | 20,600 | 485 |
| | | $^5F_2$ | 21,100 | 474 |
| | | $^3K_8$ | 21,370 | 468 |
| | | $^5G_6$ | 22,100 | 452 |
| | | ($^5G$, $^3G)_5$ | 23,950 | 418 |
| | | $^5G_4$ | 25,800 | 388 |
| | | $^3K_7$ | 26,200 | 382 |
| | | ($^5G$, $^3H)_5$, $^3H_6$ | 27,700 | 361 |
| | | ($^5F$, $^3F$, $^5G)_2$ | 28,250 | 354 |
| | | $^5G_3$ | 28,800 | 347 |
| | | $^3L_9$ | 29,000 | 345 |
| | | ($^5F$, $^3H$, $^3G)_4$, $^3K_6$ | 30,000 | 333 |
| | | $^5G_2$ | 30,900 | 324 |

*Continued*

Table 3.1 **Continued**

| Trivalent 4f rare-earth ion | Ground state | Transition energy level | Absorption peak position: Frequency ($cm^{-1}$) | Absorption peak position: Wavelength (nm) |
|---|---|---|---|---|
| | | $^3D_3$ | 33,200 | 301 |
| | | $^3M_{10}$, $^3L_8$ | 34,200 | 292 |
| | | $(^5G, ^5D, ^3G)_4$ | 34,800 | 287 |
| | | $(^3F, ^3G)_3$ | 35,200 | 284 |
| | | $(^3H, ^5D, ^1G)_4$ | 36,000 | 278 |
| | | $^1L_8$ | 36,500 | 274 |
| | | $(^3H, ^3G)_5$ | 36,800 | 272 |
| | | $^3L_7$ | 38,000 | 263 |
| | | $^3I_7$, $^3F_4$ | 38,500 | 260 |
| | | $^3M_9$ | 39,500 | 253 |
| | | $^5D_3$ | 40,000 | 250 |
| | | $(^3F, ^5D)_4$ | 41,550 | 241 |
| | | $^5D_2$ | 41,900 | 239 |
| | | $(^3F, ^5D)_3$ | 42,850 | 233 |
| | | $(^3H, ^1I)_6$ | 45,700 | 219 |
| | | $(^3H, ^3G, ^3I)_5$ | 48,200 | 207 |
| Erbium ion, $Er^{3+}$ ($4f^{11}$), in aqueous solution [4] | $^4I_{15/2}\rightarrow$ | $^4I_{13/2}$ | 6600 | 1515 |
| | | $^4I_{11/2}$ | 10,250 | 976 |
| | | $^4I_{9/2}$ | 12,400 | 806 |
| | | $^4F_{9/2}$ | 15,250 | 656 |
| | | $^4S_{3/2}$ | 18,350 | 545 |
| | | $^2H_{11/2}$ | 19,150 | 522 |
| | | $^4F_{7/2}$ | 20,450 | 489 |
| | | $^4F_{5/2}$ | 22,100 | 452 |
| | | $^4F_{3/2}$ | 22,500 | 444 |
| | | $(^2G, ^4F, ^2H)_{9/2}$ | 24,550 | 407 |
| | | $^4G_{11/2}$ | 26,400 | 379 |
| | | $^4G_{9/2}$ | 27,400 | 365 |
| | | $^2G_{7/2}$ | 28,000 | 357 |
| | | $(^2P, ^2D, ^4F)_{3/2}$ | 31,600 | 316 |
| | | $^2K_{13/2}$ | 33,200 | 301 |
| | | $^4G_{5/2}$ | 33,400 | 299 |
| | | $^4G_{7/2}$ | 34,050 | 294 |
| | | $^2D_{5/2}$ | 34,850 | 287 |
| | | $(^2H, ^2G)_{9/2}$ | 36,550 | 274 |
| | | $^2D_{5/2}$ | 38,600 | 259 |
| | | $^4D_{7/2}$ | 39,200 | 255 |

**Table 3.1 Continued**

| Trivalent 4f rare-earth ion | Ground state | Transition energy level | Absorption peak position | |
|---|---|---|---|---|
| | | | Frequency ($cm^{-1}$) | Wavelength (nm) |
| | | $^2I_{11/2}$ | 41,150 | 243 |
| | | $^2L_{17/2}$ | 41,650 | 240 |
| | | $^4D_{3/2}$ | 42,300 | 236 |
| | | $^2I_{13/2}$ | 43,550 | 230 |
| | | $(^2D, ^4D)_{5/2}$ | 49,000 | 204 |
| Thulium ion, $Tm^{3+}$ ($4f^{12}$), in aqueous solution [4] | $^3H_6 \rightarrow$ | $^3F_4$ | 5900 | 1695 |
| | | $^3H_5$ | 8400 | 1190 |
| | | $^3H_4$ | 12,700 | 787 |
| | | $^3F_3$ | 14,500 | 690 |
| | | $^3F_2$ | 15,100 | 662 |
| | | $^1G_4$ | 21,350 | 468 |
| | | $^1D_2$ | 28,000 | 357 |
| | | $^1I_6$ | 34,900 | 287 |
| | | $^3P_0$ | 35,500 | 282 |
| | | $^3P_1$ | 36,400 | 275 |
| | | $^3P_2$ | 38,250 | 261 |
| Ytterbium ion, $Yb^{3+}$ ($4f^{13}$), in silicate glass [2,3] | $^2F_{7/2} \rightarrow$ | $^2F_{5/2}$ | 10,310 | 970 |

is tested, evaluating the root mean square (RMS) deviation ($\sigma_{rms}$) using the following relation:

$$\sigma_{rms} = \sqrt{\frac{\left(E_i^{exp} - E_i^{cal}\right)^2}{N}} \tag{3.5}$$

where $E_i^{exp}$ and $E_i^{cal}$ are the experimental and calculated energies for the *i*th levels, respectively. The *N* stands for the total number of energy levels included in the fitting procedure.

### 3.2.2 Experimental oscillator strength

The experimental oscillator strength ($f_{exp}$) is calculated using the following expression:

$$f_{exp} = 4.32 \times 10^{-9} \int_{v1}^{v_2} \in(v)dv \tag{3.6}$$

where $\in(\nu)$ is the molar absorption coefficient at the energy level, $\nu$ (in $cm^{-1}$) and $d\nu$ is the full width at half maximum (FWHM) of the absorption band. The molar absorption coefficient, $\in(\nu)$, is estimated from the following relation obtained using the Lambert-Beer law:

$$\in(v) = \frac{1}{Cl}\log\left(\frac{I_0}{I}\right) \tag{3.7}$$

where $C$, $l$, and $\log(I_0/I)$ are the concentration of rare-earth ion in mol $L^{-1}$, optical path length (i.e., thickness of the sample), and optical density (OD), respectively.

### 3.2.3 Judd-Ofelt (J-O) parameters

The theoretically calculated oscillator strength ($f_{cal}$) of the transition between an initial $J$ manifold ($S$, $L$) and a final $J^1$ manifold ($S^1$, $L^1$) is estimated using the following Judd-Ofelt (J-O) expression:

$$f_{exp} = f_{cal}\left[(S,L)J;(S^1,L^1)J^1\right] = \frac{8\pi^2 mcv}{3h(2J+1)}\left[\frac{(n^2+2)^2}{9n}S_{ed} + nS_{md}\right] \tag{3.8}$$

It includes the terms of mass of electron ($m$), Planck's constant ($h$), refractive index ($n$), mean energy of the transition ($\nu$ in $cm^{-1}$), and total angular momentum of the initial state ($J$). The line strength of the electric dipole transition ($S_{ed}$) is expressed by the following relation:

$$S_{ed}\left[(S,L)J;(S^1,L^1)J^1\right] = \sum_{t=2,4,6} \Omega_t | < (S,L)J||U^t||(S^1,L^1)J^1 > |^2 \tag{3.9}$$

where $\Omega_t$ ($t=2$, 4, 6) stand for the J-O intensity parameters the rare-earth ions in a specific glass matrix and $||U^t||$ is the doubly reduced matrix elements of rank "$t$." The line strength of the magnetic dipole transition ($S_{md}$) is expressed by the following relation:

$$S_{md}\left[(S,L)J;(S^1,L^1)J^1\right] = \frac{h^2}{16\pi^2 m^2 c^2} | < (S,L)J||(L+2S)||(S^1,L^1)J^1 > |^2 \tag{3.10}$$

### 3.2.4 Radiative features

Different radiative features such as radiative transition probability ($A_R$), radiative lifetime ($\tau_R$), fluorescence branching ratio ($\beta_R$), stimulated emission cross section ($\sigma_e$), and effective emission linewidth ($\Delta\lambda_{eff}$) can be calculated using the J-O parameters

($\Omega_t$). The radiative transition probability ($A_R$) for an emission from an initial $J$ manifold ($S, L$) and a final $J^1$ manifold ($S^1, L^1$) is estimated using the following expression:

$$A_R\left[(S,L)J;(S^1,L^1)J^1\right] = \frac{64\pi^4 e^2}{3h(2J+1)\lambda^3}\left[\frac{n(n^2+2)^2}{9}S_{ed} + n^3 S_{md}\right] \tag{3.11}$$

where $[n(n^2+2)^2/9]$ is the Lorentz local field correction factor for the line strength of the electric dipole transition ($S_{ed}$). The total radiative emission probability, $A_T[(S,L)J]$, of an excited state is the sum of the $A_R[(S,L)J$ and $[(S^1,L^1)J^1]$ terms calculated overall final states $(S^1,L^1)J^1$. It can be expressed by the following relation:

$$A_T[(S,L)J] = \sum_{(S^1,L^1)J^1} A_R\left[(S,L)J;(S^1,L^1)J^1\right] \tag{3.12}$$

The radiative lifetime ($\tau_R$) of the emission state $(S,L)J$ is expressed by the following relation:

$$\tau_R = \frac{1}{A_T[(S,L)J]} \tag{3.13}$$

The fluorescence branching ratio ($\beta_R$) is expressed by the following relation:

$$\beta_R = \frac{A_R[(S,L)J;(S^1,L^1)J^1]}{A_T[(S,L)J]} \tag{3.14}$$

The stimulated emission cross section ($\sigma_e$) between the $(S, L)J$ and $(S^1, L^1)J^1$ emission transition level can be written as:

$$\sigma_e\left[(S,L)J;(S^1,L^1)J^1\right] = \frac{\lambda_p^4}{8\pi c n^2 \Delta\lambda_{eff}} A_R\left[(S,L)J;(S^1,L^1)J^1\right] \tag{3.15}$$

where $\lambda_p$ is the peak wavelength of the fluorescence band. The effective emission line width ($\Delta\lambda_{eff}$) is calculated by integrating the intensity of the emission band line shapes and dividing it by the intensity at the peak wavelength ($\lambda_p$), which can be correlated by the following relation:

$$\Delta\lambda_{eff} = \frac{\int I d\lambda}{I_p} \tag{3.16}$$

## 3.3 Processing and functional properties

It is pointed out earlier that the characteristics of the absorption and photoluminescence spectra depend on several factors, for example, chemical composition of glass, concentration of 4f RE metal ion, and glass melting atmosphere. Hence, the absorption and photoluminescence spectra of all the RE ions (from $Ce^{3+}$ to $Yb^{3+}$, from $4f^1$ to $4f^{13}$) in different glasses are described in the following section to understand the influence of these issues. Here, promethium (Pm) has been excluded because generally, it is not used owing to its radioactive character.

## 3.3.1 Absorption and photoluminescence spectra of 4f rare-earth metal ions in different oxide glasses

### 3.3.1.1 Cerium

Fig. 3.4A shows the absorption spectra of $Ce^{3+}$($4f^1$ configuration) and also of $Ce^{4+}$($4f^0$ configuration) ions in different glasses (composition in mol.%) such as (a) $20Li_2O$–$20MgO$–$60SiO_2$, (b) $20BaO$–$80B_2O_3$, and (c) $50CaO$–$50P_2O_5$ [13]. It displays a broad band in the wavelength range of 200–350 nm with five peaks. All the five peaks are assigned to the $4f^1 \rightarrow 5d^1$ ($^2F_{5/2} \rightarrow {}^2D$) transition of the $Ce^{3+}$ ion. It indicates the 5d excited-state ($^2D$) is fully splitted. In sodium trisilicate glass, the single band is peaking around 313 nm (Table 3.1) [2]. It is also seen from the spectra of Fig. 3.4A that the spectral profiles are different from one another. The spectral profiles of cerium in these glasses have pointed out that the charge transfer transition band of $Ce^{4+}$ ion due to the $Ce^{3+}$—O—$Ce^{4+}$ linkage overlaps with that of the $^2F_{5/2} \rightarrow {}^2D$ transition of the $Ce^{3+}$ ion. For this reason, it is difficult to study the redox reaction of cerium ions in these glasses from their absorption spectra. Fig. 3.4B illustrates the fluorescence spectra of $Ce^{3+}$ ions in $xBaO$–$(100-x)B_2O_3$ ($x=20$, 30, and 40) glasses (composition in mol.%) under excitation at 266 nm [13].

The fluorescence band is peaking around 325 nm. Its intensity is decreased and the peak position is shifted toward longer wavelengths with increasing $B_2O_3$ content. This band is attributed to the $5d^1 \rightarrow 4f^1$ ($^2D \rightarrow {}^2F_{5/2}$) transition of the $Ce^{3+}$ ion.

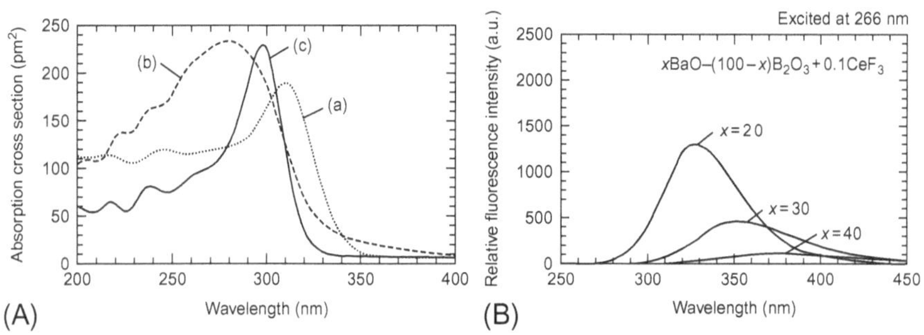


**Fig. 3.4** (A) Absorption spectra of $Ce^{3+}$ (and $Ce^{4+}$) ions in different glasses (composition in mol.%): (a) $20Li_2O$–$20MgO$–$60SiO_2$, (b) $20BaO$–$80B_2O_3$, and (c) $50CaO$–$50P_2O_5$. (B) Fluorescence spectra of $Ce^{3+}$ ions in $x$BaO–$(100-x)B_2O_3$ ($x=20$, 30, and 40) glasses under excitation at 266 nm.

Reproduced with permission from T. Murata, M. Sato, H. Yoshida, K. Morinaga, Compositional dependence of ultraviolet fluorescence intensity of $Ce^{3+}$ in silicate, borate, and phosphate glasses, J. Non-Cryst. Solids, 351 (2005) 312–316, Copyright © Elsevier.

### 3.3.1.2 Praseodymium

The absorption spectrum of $Pr^{3+}$($4f^2$ configuration) doped $20Li_2O–20CaO–60SiO_2$ (mol.%) glass is shown in Fig. 3.5A [14]. Total six absorption bands have been identified in the visible to near infrared (NIR) region and assigned to the transitions from the ground state $^3H_4$. Other low intensity bands could also be identified with the help of the transition levels assigned in the Table 3.1 for the aqueous solution of $Pr^{3+}$ ion.

The luminescence spectra of $Pr^{3+}$-doped $SrO–2B_2O_3$ glass (excitation at 450 nm) recorded at temperatures of 293 and 85 K are depicted in Fig. 3.5B [15]. It is seen that there is an intense broad complex emission band around 600 nm at both the temperature of 293 and 85 K, which is assigned to the $^3P_0 \rightarrow {}^3F_2$, $^3H_6$, and $^1D_2 \rightarrow {}^3H_4$ transitions. The weak emission bands peaking around 450, 690, and 800 nm are corresponding to the $^3P_0 \rightarrow {}^3H_4$, $^3P_0 \rightarrow {}^3F_3$, and $^3P_0 \rightarrow {}^3F_4$ transitions, respectively. The bandwidth and position of the emission bands do not change noticeably with decreasing temperature down to 85 K from 293 K, an evidence of inhomogeneous broadening of spectral lines, which indicates that the disorder of the local structures of the $Pr^{3+}$ ions is less in this borate glass system.

### 3.3.1.3 Neodymium

The absorption spectrum of $Nd^{3+}$($4f^3$ configuration) doped $20Li_2O–20CaO–60SiO_2$ (mol.%) glass is shown in Fig. 3.6A [14]. Total eight absorption bands have been identified in the ultraviolet (UV) to near-infrared (NIR) region and assigned to the transitions from the ground state $^4I_{9/2}$. Other low and high intensity bands could also be

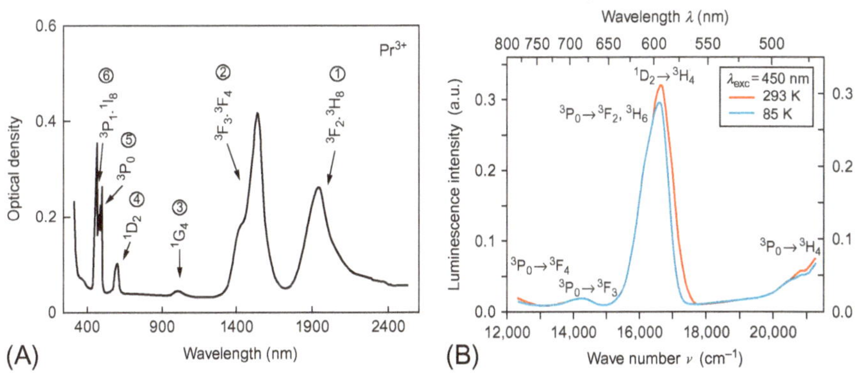


**Fig. 3.5** (A) Absorption spectrum of $Pr^{3+}$-doped $20Li_2O–20CaO–60SiO_2$ glass. (B) Luminescence spectra of $Pr^{3+}$-doped $SrO–2B_2O_3$ glass (excitation at 450 **nm**) at temperatures of 293 and 85 K.

(A) Reproduced with permission from H. Takebe, Y. Nageno, K. Morinaga, Compositional dependence of Judd-Ofelt parameters in silicate, borate, and phosphate glasses, J. Am. Ceram. Soc. 78 (1995) 1161–1168, Copyright © The American Ceramic Society. (B) Courtesy: B. Padlyak, M. Grinberg, B. Kuklinski, Y. Oseledchik, O. Smyrnov, D. Kudryavtcev, A. Prosvirnin, Synthesis and optical spectroscopy of the Eu- and Pr-doped glasses with $SrO–2B_2O_3$ composition, Opt. Appl. XL (2010) 413–426.

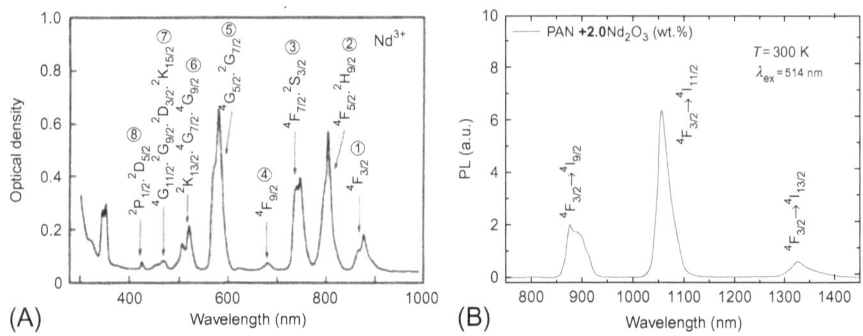


**Fig. 3.6** (A) Absorption spectrum of $Nd^{3+}$-doped $20Li_2O$–20CaO–$60SiO_2$ glass. (B) Photoluminescence (PL) spectrum of $Nd^{3+}$ ion in phosphate glass under excitation at 514 nm.
(A) Reproduced with permission from H. Takebe, Y. Nageno, K. Morinaga, Compositional dependence of Judd-Ofelt parameters in silicate, borate, and phosphate glasses, J. Am. Ceram. Soc. 78 (1995) 1161–1168, Copyright © The American Ceramic Society. (B) Reproduced with permission from V.M. Martins, D.N. Messias, N.O. Dantas, A.M. Neto, Concentration dependent fluorescence quantum efficiency of neodymium doped phosphate glass matrix, J. Lumin. 130 (2010) 2491–2494, Copyright © Elsevier.

identified with the help of the transition levels assigned in the Table 3.1 for the aqueous solution of $Nd^{3+}$ ion.

The room temperature (300 K) photoluminescence (PL) spectrum of $Nd^{3+}$ ion in sodium aluminophosphate, $40P_2O_5$–$20Al_2O_3$–$40Na_2O$ (PAN) glass (mol.%) under excitation at 514 nm ($Ar^+$ laser line) is shown in Fig. 3.6B [16]. Three emission bands peaking around 881, 1061, and 1333 nm are attributed to the $^4F_{3/2} \rightarrow {}^4I_{9/2}$, $^4F_{3/2} \rightarrow {}^4I_{11/2}$, and $^4F_{3/2} \rightarrow {}^4I_{13/2}$ transitions, respectively. Hence, all the important emissions start from the common level of $^4F_{3/2}$ (see Fig. 3.3 for the transitions). This is essential to obtain high quantum efficiency for different functional applications.

### *3.3.1.4 Samarium*

The absorption spectrum of $Sm^{3+}$($4f^5$ configuration) doped $20Li_2O$–20CaO–$60SiO_2$ (mol.%) glass is shown in Fig. 3.7A [14]. Total 11 absorption bands have been identified in the ultraviolet (UV) to near-infrared (NIR) region and assigned to the transitions from the ground state $^6H_{5/2}$. Other low intensity bands could also be identified with the help of the transition levels assigned in the Table 3.1 for the aqueous solution of $Sm^{3+}$ ion.

The room temperature photoluminescence spectrum of $Sm^{3+}$ ion in antimony oxide glass, $15K_2O$–$15B_2O_3$–$70Sb_2O_3$ (mol.%) under excitation at 403 nm is shown in Fig. 3.7B [17]. Three emission bands peaking around 566 (green, weak), 602 (orange, medium), and 652 (red, strong) nm are originated from the $^4G_{5/2} \rightarrow {}^6H_{5/2}$, $^4G_{5/2} \rightarrow {}^6H_{7/2}$, and $^4G_{5/2} \rightarrow {}^6H_{9/2}$ transitions, respectively.

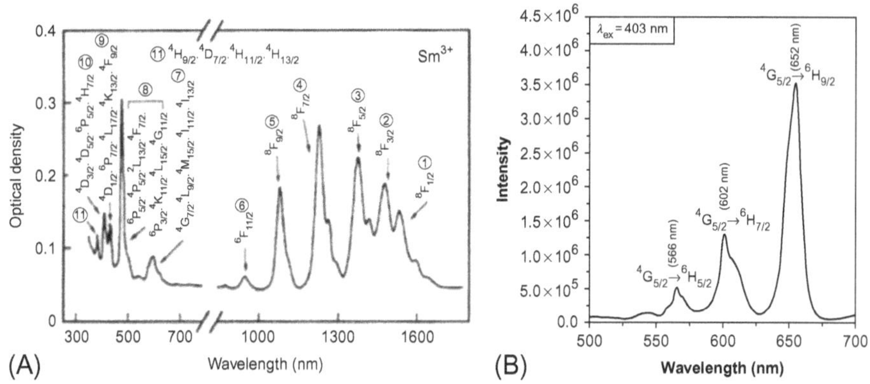


**Fig. 3.7** (A) Absorption spectrum of $Sm^{3+}$-doped $20Li_2O$–$20CaO$–$60SiO_2$ glass. (B) Photoluminescence spectrum of $Sm^{3+}$ ion in antimony oxide glass under excitation at 403 nm.
(A) Reproduced with permission from H. Takebe, Y. Nageno, K. Morinaga, Compositional dependence of Judd-Ofelt parameters in silicate, borate, and phosphate glasses, J. Am. Ceram. Soc. 78 (1995) 1161–1168, Copyright © The American Ceramic Society. (B) Reproduced with permission from T. Som, B. Karmakar, Infrared-to-red upconversion luminescence in samarium-doped antimony glasses, J. Lumin. 128 (2008) 1989–1996, Copyright © Elsevier.

### *3.3.1.4 Europium*

The absorption spectra of different $30BaO$–$30B_2O_3$–$xAl_2O_3$–$(39-x)P_2O_5$–$1Eu_2O_3$ ($x=0$, 1, 2, 3, 4, and 5) glasses (compositions are in mol.%) doped with a fixed concentration of $Eu_2O_3$ (1 mol.%) are shown in Fig. 3.8 [18]. It clearly discloses the influence of $Al^{3+}$ ions on the absorption spectra of $Eu^{3+}$($4f^6$ configuration, ground state $^7F_0$) ion in this borophosphate glass matrix. The absorption bands have been assigned with their respective transition in the figure.

The emission spectra of these samples under excitation at 393 nm are illustrated in Fig. 3.9. Five emission bands peaking around 575, 590, 610, 653, and 703 nm are attributed to the $^5D_0 \rightarrow {}^7F_0$, $^5D_0 \rightarrow {}^7F_1$, $^5D_0 \rightarrow {}^7F_2$, $^5D_0 \rightarrow {}^7F_3$, and $^5D_0 \rightarrow {}^7F_4$ transitions, respectively (see Fig. 3.10A and B). The highest intensity of all the emission bands is obtained in the spectrum of the glass sample $A_3Eu$ with the composition of $30BaO$–$30B_2O_3$–$3Al_2O_336P_2O_5$–$1Eu_2O_3$. The quantum efficiency is also found to be highest in this sample. Hence, in this glass system, 3 mol.% $Al_2O_3$ is the optimum concentration for declustering of $Eu^{3+}$ ions. Infrared spectral studies have disclosed that the $Al^{3+}$ ions largely participate in the octahedral positions, which decluster the $Eu^{3+}$ ions to a greater extent and make possible to increase the luminescence efficiency by minimizing the luminescence quenching as a result of cross relaxation.

The fluorescence excitation and emission spectra of $Eu^{2+}$($4f^7$ configuration, ground state $^8S_{7/2}$) ion in a fluoroaluminophosphate (FP) glass of composition (mol.%) $10Sr(PO_3)_2$–$10MgF_2$–$30CaF_2$–$15SrF_2$–$35AlF_3$ and a borosilicate (Duran) glass of composition (mol.%) $83SiO_2$–$4Na_2O$–$12B_2O_3$–$1Al_2O_3$ are shown in

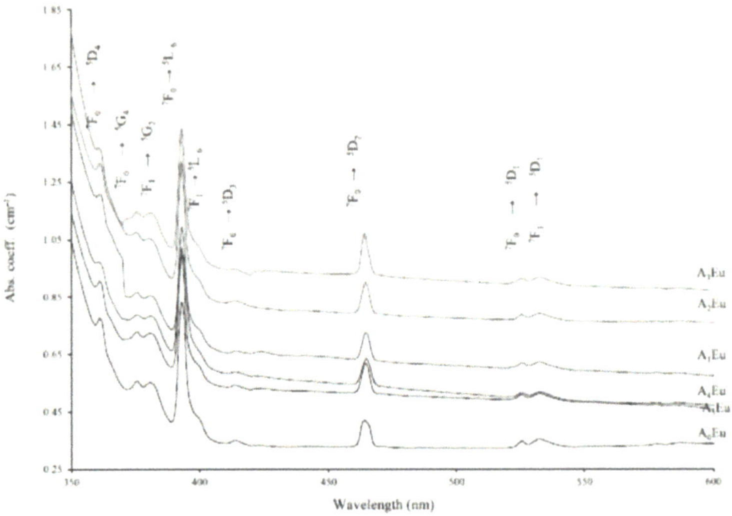


**Fig. 3.8** Absorption spectra of different BaO–$B_2O_3$–$Al_2O_3$–$P_2O_5$ glasses doped with a fixed concentration of $Eu^{3+}$ ion.
Reproduced with permission from T. Kalpana, M. G. Brik, V. Sudarsan, P. Naresh, V. Ravi Kumar, I. V. Kityk, N. Veeraiah, Influence of $Al^{3+}$ ions on luminescence efficiency of $Eu^{3+}$ ions in barium boro-phosphate glasses, J. Non-Cryst. Solids, 419 (2015) 75–81, Copyright © Elsevier.

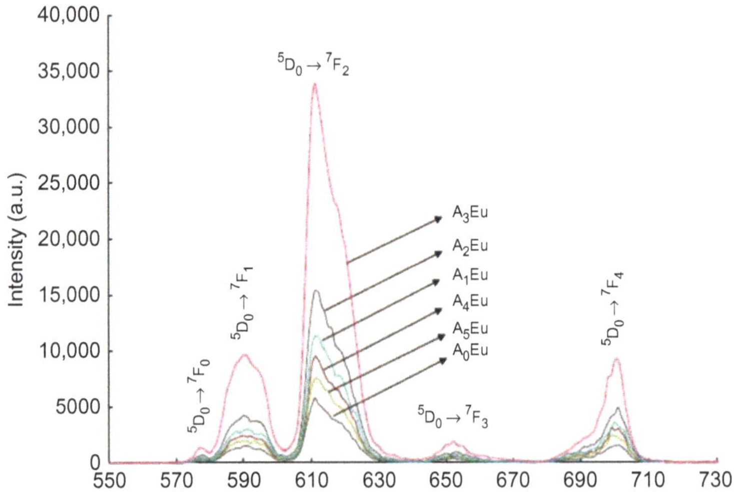


**Fig. 3.9** Emission spectra of different BaO–$B_2O_3$–$Al_2O_3$–$P_2O_5$ glasses doped with a fixed concentration of $Eu^{3+}$ ion under excitation at 393 nm.
Reproduced with permission from T. Kalpana, M. G. Brik, V. Sudarsan, P. Naresh, V. Ravi Kumar, I. V. Kityk, N. Veeraiah, Influence of $Al^{3+}$ ions on luminescence efficiency of $Eu^{3+}$ ions in barium boro-phosphate glasses, J. Non-Cryst. Solids, 419 (2015) 75–81, Copyright © Elsevier.

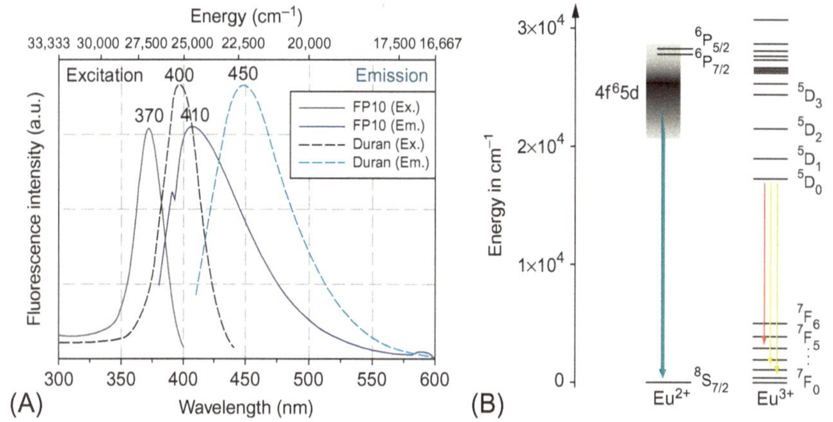


**Fig. 3.10** (A) Fluorescence excitation and emission spectra of $Eu^{2+}$ ion in fluoroaluminophosphate (FP) and borosilicate (Duran) glasses. (B) Energy-level diagram of $Eu^{2+}$ (left) and $Eu^{3+}$ (right) ions.
Reproduced with permission from A. Herrmann, S. Fibkar, D. Ehrt, Time-resolved fluorescence measurements on $Eu^{3+}$- and $Eu^{2+}$-doped glasses, J. Non-Cryst. Solids, 355 (2009) 2093–2101, Copyright © Elsevier.

Fig. 3.10A [19]. The $Eu^{2+}$ ion exhibits a broad blue fluorescence band around 410 nm in the fluoroaluminophosphate (FP10) glass whereas around 450 nm in the borosilicate glass with FWHM about 40 nm. It is observed that its intensity and position is largely dependent on glass composition, concentration of $Eu^{2+}$ ion, and excitation wavelength for the reason that the 5d electron is not shielded by the surrounding. This fluorescence band is attributed to the d-f transition as shown by the transition of $4f^6 5d \rightarrow 4f^7$ in Fig. 3.10B [19].

### *3.3.1.5 Gadolinium*

The absorption spectra of $Gd^{3+}$($4f^7$ configuration, ground state $^8S_{7/2}$) doped $Na_2O{\cdot}3SiO_2$ glasses are illustrated in Fig. 3.11A [20]. The trivalent gadolinium ion, $Gd^{3+}$ has the half-filled 4f orbital ($4f^7$); therefore, it is expected that the intensity of the absorption band due to f-f transitions in the spectrum would obviously be very weak. All of these bands are observed in the very high energy part (UV region, 195–311 nm) of the spectrum (see Fig. 3.11A and Table 3.1). The strongest absorption band around 275 nm is assigned to the $^8S_{7/2} \rightarrow {}^6I_{11/2}$ transition.

The room temperature (25°C) emission spectra of alumina films doped with various concentrations of $Gd^{3+}$ ion (heat-treated at 800°C and excited at 275 nm) are shown in Fig. 3.11B [21]. It may be noted that the emission spectra of $Gd^{3+}$-doped glasses are very rarely available in the literature. For this reason, the emission phenomenon of $Gd^{3+}$ ion is explained, taking the example of $Gd^{3+}$ ion-doped alumina films [21]. The single emission band peaking around 313 nm is assigned to the

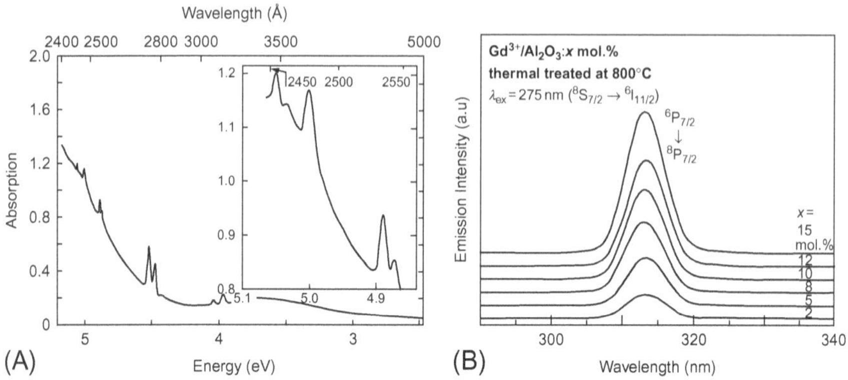


**Fig. 3.11** (A) Absorption spectra of $Gd^{3+}$-doped $Na_2O.3SiO_2$ glasses. (B) Room temperature (25°C) emission spectra of alumina films doped with various concentrations of $Gd^{3+}$ ion (heat-treated at 800°C and excited at 275 nm).
(A) Courtesy: H.L. Smith, A.J. Cohen, Absorption spectra of cations in alkali-silicate glasses of high ultra-violet transmission, Phys. Chem. Glass. 4 (1963) 173–187. (B) Reproduced with permission from T. Ishizaka, Y. Kurokawa, Optical properties of rare-earth ion ($Gd^{3+}$, $Ho^{3+}$, $Pr^{3+}$, $Sm^{3+}$, $Dy^{3+}$ and $Tm^{3+}$)-doped alumina films prepared by sol-gel method, J. Lumin. 92 (2001) 57–63, Copyright © Elsevier.

emission transition of $^6P_{7/2} \rightarrow {}^8S_{7/2}$. It is observed that the intensity and FWHM of the emission band amplify with increase in concentration of $Gd^{3+}$ ion.

### 3.3.1.6 Terbium

The absorption spectrum of $Tb^{3+}$($4f^8$ configuration) doped $20Li_2O$–$20CaO$–$60SiO_2$ (mol.%) glass is shown in Fig. 3.12A [14]. Total seven absorption bands have been identified in the ultraviolet (UV) to near-infrared (NIR) region and assigned to the transitions from the ground state $^7F_6$. Other low intensity bands could also be identified with the help of the transition levels assigned in the Table 3.1 for the aqueous solution of $Tb^{3+}$ ion.

The room temperature photoluminescence spectra of $Tb^{3+}$ ion-doped zinc phosphate glasses of chemical compositions (mol.%) $60P_2O_5$–$15ZnO$–$5Al_2O_3$–$10BaO$–$10PbO$–$xTb_2O_3$ ($x=1$, 2, 3, 4, and 5 wt.%) under excitation at 376 nm are shown in Fig. 3.12B [22]. The emission band located around 540 nm is assigned to the $^5D_4 \rightarrow {}^7F_5$ transition. The emission bands peaking around 412, 434, 455, and 470 nm are assigned to $^5D_3 \rightarrow {}^7F_{5,4,3,2}$ transitions, respectively, and those peaking around 486, 582, and 618 nm are attributed to $^5D_4 \rightarrow {}^7F_{6,4,3}$ transitions, respectively. It is seen that the emission intensities of the bands originated from the $^5D_3$ energy-level decrease with increase in concentration of $Tb^{3+}$ ion, whereas that of those originated from the $^5D_4$ level remain almost unchanged.

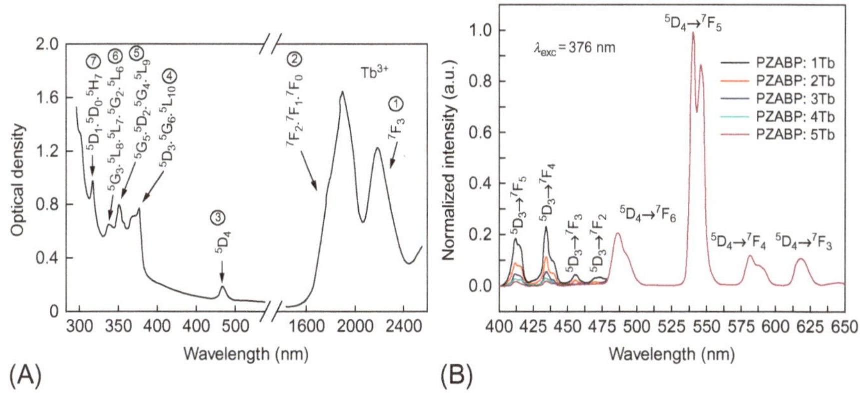


**Fig. 3.12** (A) Absorption spectrum of $Tb^{3+}$-doped $20Li_2O$–$20CaO$–$60SiO_2$ glass. (B) Photoluminescence spectra of $Tb^{3+}$ ion in phosphate glass under excitation at 376 nm. (A) Reproduced with permission from H. Takebe, Y. Nageno, K. Morinaga, Compositional dependence of Judd-Ofelt parameters in silicate, borate, and phosphate glasses, J. Am. Ceram. Soc. 78 (1995) 1161–1168, Copyright © The American Ceramic Society. (B) Reproduced with permission from C.R. Kesavulu, A.C.A. Silva, M.R. Dousti, N.O. Dantas, A.S.S. deCamargo, T. Catunda, Concentration effect on the spectroscopic behaviour of $Tb^{3+}$ ions in zinc phosphate glasses, J. Lumin. 165 (2015) 77–84, Copyright © Elsevier.

### *3.3.1.7 Dysprosium*

The absorption spectrum of $Dy^{3+}$($4f^9$ configuration) doped $20Li_2O$–$20CaO$–$60SiO_2$ (mol.%) glass is shown in Fig. 3.13A [14]. Total seven absorption bands have been identified in the ultraviolet (UV) to near-infrared (NIR) region and assigned to the transitions from the ground state $^6H_{15/2}$. Other low and high intensity bands could also be identified with the help of the transition levels assigned in the Table 3.1 for the aqueous solution of $Dy^{3+}$ ion.

The room temperature photoluminescence spectra of $Dy^{3+}$ ion in alkaline earth titanium phosphate glasses of composition (mol.%) $24(NaPO_3)_6$–$30KH_2PO_4$–$25TiO_2$–$20RCl_2$–$1Dy_2O_3$ (R = Mg, Ca, and Sr) under excitation at 386 nm are shown in Fig. 3.13B [23]. All these glasses exhibit two strong and two very weak emission bands in the visible region. Two strong emission bands located at 483 and 574 nm are assigned to the $^4F_{9/2} \rightarrow {}^6H_{15/2}$ and $^4F_{9/2} \rightarrow {}^6H_{13/2}$ transitions, respectively. Other two weak emission bands peaking at 455 and 664 nm are ascribed to the $^4I_{15/2} \rightarrow {}^6H_{15/2}$ and $^4F_{9/2} \rightarrow {}^6H_{11/2}$ transitions, respectively. The weak emission band of the $^4I_{15/2} \rightarrow {}^6H_{15/2}$ transition has appeared due to thermal population of the $^4I_{15/2}$ level at the expense of the $^4F_{9/2}$ level as result of thermal equilibrium since their energy difference is 1000 $cm^{-1}$ only (see Table 3.1).

### *3.3.1.8 Holmium*

The absorption spectrum of $Ho^{3+}$($4f^{10}$ configuration) doped $20Li_2O$–$20CaO$–$60SiO_2$ (mol.%) glass is shown in Fig. 3.14A [14]. Total nine absorption bands have been identified in the ultraviolet (UV) to near-infrared (NIR) region and assigned to the

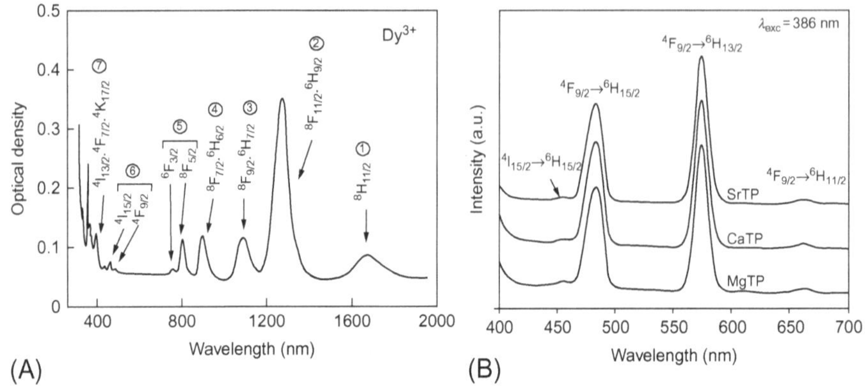


**Fig. 3.13** (A) Absorption spectrum of $Dy^{3+}$-doped $20Li_2O$–20CaO–$60SiO_2$ glass. (B) Photoluminescence spectra of $Dy^{3+}$ ion in phosphate glass under excitation at 386 nm.
(A) Reproduced with permission from H. Takebe, Y. Nageno, K. Morinaga, Compositional dependence of Judd-Ofelt parameters in silicate, borate, and phosphate glasses, J. Am. Ceram. Soc. 78 (1995) 1161–1168, Copyright © The American Ceramic Society. (B) Reproduced with permission from D.V.R. Murthy, B.C. Jamalaiah, A. Mohan Babu, T. Sasikala, L. Rama Moorthy, The luminescence properties of $Dy^{3+}$-doped alkaline earth titanium phosphate glasses, Opt. Mater. 32 (2010) 1112–1116, Copyright © Elsevier.

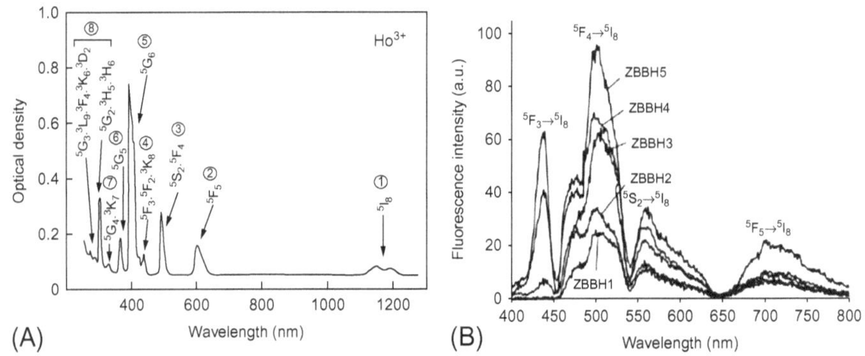


**Fig. 3.14** (A) Absorption spectrum of $Ho^{3+}$-doped $20Li_2O$–20CaO–$60SiO_2$ glass. (B) Visible emission spectra with different concentrations of $Ho^{3+}$ ion in zinc bismuth borate glass under excitation at 405 nm.
(A) Reproduced with permission from H. Takebe, Y. Nageno, K. Morinaga, Compositional dependence of Judd-Ofelt parameters in silicate, borate, and phosphate glasses, J. Am. Ceram. Soc. 78 (1995) 1161–1168, Copyright © The American Ceramic Society. (B) Courtesy: M. Lather, P. Aghamkar, S. Sanghi, Optical absorption and fluorescence spectral analysis of $Ho^{3+}$ ion in zinc bismuth borate glasses. J. Integr. Sci. Technol. 3 (2015) 28–33.

transitions from the ground state $^5I_8$. Other low and high intensity bands could also be identified with the help of the transition levels assigned in the Table 3.1 for the aqueous solution of $Ho^{3+}$ ion.

The room temperature visible emission spectra with different concentrations of $Ho^{3+}$ ion in zinc bismuth borate glass under excitation at 405 nm are shown in Fig. 3.14B [24]. It displays four emission bands peaking around 436, 501, 561, and 698 nm, which are assigned to the $^5F_3 \rightarrow {}^5I_8$, $^5F_4 \rightarrow {}^5I_8$, $^5S_2 \rightarrow {}^5I_8$, and $^5F_5 \rightarrow {}^5I_8$ transitions, respectively. The intensity of the band at 501 nm is found to be the highest due to transition of $^5F_4 \rightarrow {}^5I_8$, which indicates that the population of the $^5F_4$ level is maximum compared with those of others. This is as a result of the energy-level structure of $Ho^{3+}$ ion.

The room temperature near-infrared (NIR) absorption (black) $\sigma_a(\lambda)$ and emission (red) $\sigma_e(\lambda)$ cross sections, calculated by McCumber theory, of $Ho^{3+}$-doped tellurite glass of composition (mol.%) $77TeO_2$–$4.5Bi_2O_3$–$5.5ZnO$–$10.5Li_2O$–$1.5Nb_2O_5$–$1Ho_2O_3$ are shown in Fig. 3.15 [25]. The absorption and emission bands peaking around 1160 and 1950 nm have been assigned to the $^5I_8 \leftrightarrow {}^5I_6$ and $^5I_8 \leftrightarrow {}^5I_7$ transitions, respectively. Considering the gain coefficient calculation and exhibited other lasing properties, it is concluded that the abovesaid glass composition is the best one among all the studied compositions and suitable for different photonic applications.

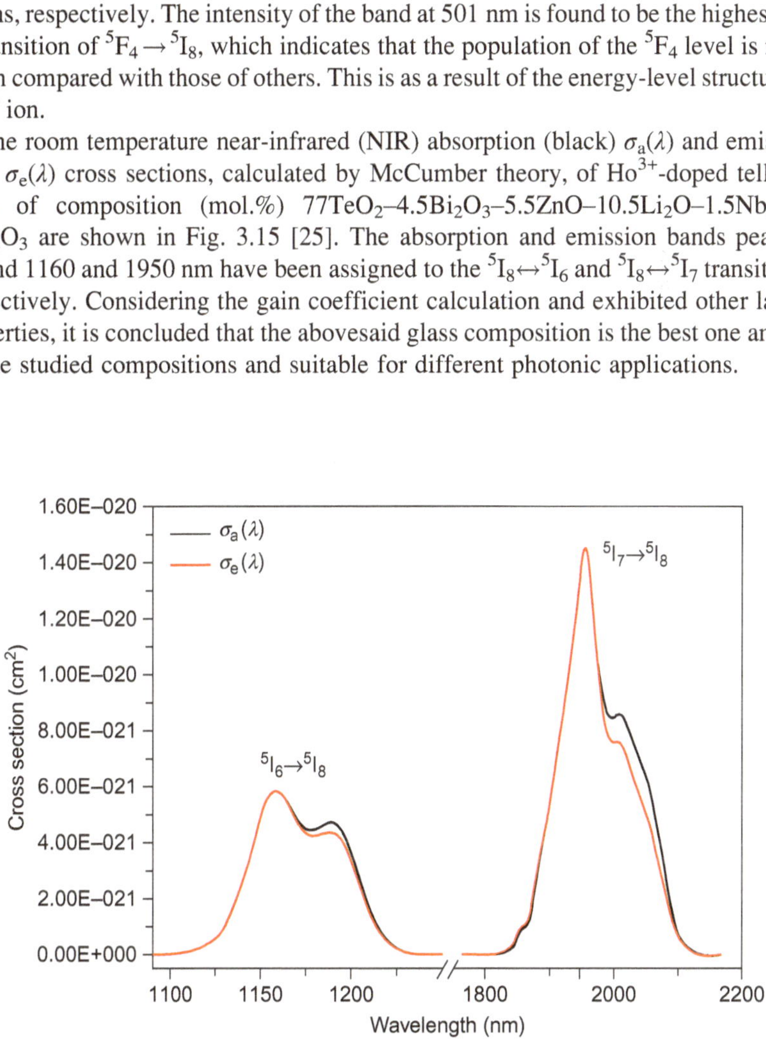


**Fig. 3.15** Near-infrared (NIR) absorption *(black)* and emission *(red)* cross section of $Ho^{3+}$-doped tellurite glass.

Reproduced with permission from M. Seshadri, L.C. Barbosa, M. Radha, Study on structural, optical and gain properties of 1, 2 and 2.0 μm emission transitions in $Ho^{3+}$ doped tellurite glasses, J. Non-Cryst. Solids 406 (2014) 62–72, Copyright © Elsevier.

### 3.3.1.9 Erbium

The absorption spectrum of $Er^{3+}$($4f^{11}$ configuration) doped $20Li_2O$–$20CaO$–$60SiO_2$ (mol.%) glass is shown in Fig. 3.16A [14]. Total nine absorption bands have been identified in the ultraviolet (UV) to near-infrared (NIR) region and assigned to the transitions from the ground state $^4I_{15/2}$. Other low intensity bands could also be identified with the help of the transition levels assigned in the Table 3.1 for the aqueous solution of $Er^{3+}$ ion.

The room temperature visible photoluminescence spectrum of the $Er^{3+}$-doped antimony oxide glass of composition (mol.%) $15K_2O$–$15B_2O_3$–$70Sb_2O_3$ under excitation at 377 nm is shown in Fig. 3.16B [26]. Two emission bands located around 528 and 549 nm are assigned to the $^2H_{11/2} \rightarrow {}^4I_{15/2}$ and $^4S_{3/2} \rightarrow {}^4I_{15/2}$ transitions, respectively. Fig. 3.17A illustrates the visible photoluminescence spectrum of $Er^{3+}$-doped abovesaid glass composition under excitation at 407 nm [26]. Two emission bands are also obtained on changing the excitation wavelength, which is located around 549 and 658 nm due to the $^4S_{3/2} \rightarrow {}^4I_{11/2}$ and $^4F_{9/2} \rightarrow {}^4I_{15/2}$ transitions, respectively. This observation clearly evidences the influence of alteration of excitation wavelength.

The room temperature near-infrared (NIR) absorption (black), $\sigma_a(\lambda)$ and emission (red), $\sigma_e(\lambda)$ cross sections, calculated by McCumber theory, of $Er^{3+}$-doped phosphate glass of composition (mol.%) $61.5P_2O_5$–$10K_2O$–$15.5SrO$–$12Al_2O_3$–$1Er_2O_3$ are shown in Fig. 3.17B [27]. Both the absorption and emission bands peaking around 1535 nm have been assigned to the $^4I_{15/2} \leftrightarrow {}^4I_{13/2}$ transitions.

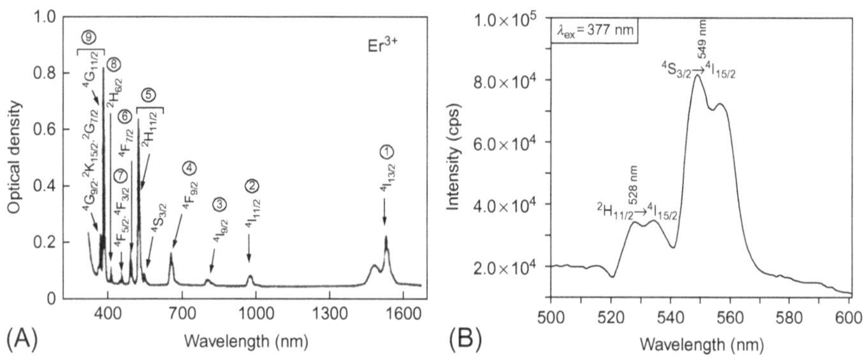


**Fig. 3.16** (A) Absorption spectrum of $Er^{3+}$-doped $20Li_2O$–$20CaO$–$60SiO_2$ glass. (B) Visible photoluminescence spectrum of $Er^{3+}$-doped antimony oxide glass under excitation at 377 nm. (A) Reproduced with permission from H. Takebe, Y. Nageno, K. Morinaga, Compositional dependence of Judd-Ofelt parameters in silicate, borate, and phosphate glasses, J. Am. Ceram. Soc. 78 (1995) 1161–1168, Copyright © The American Ceramic Society. (B) Reproduced with permission from T. Som, B. Karmakar, Efficient green and red fluorescence upconversion in erbium doped new low phonon antimony glasses, Opt. Mater. 31 (2009) 609–618, Fig. 6, Copyright © Elsevier.

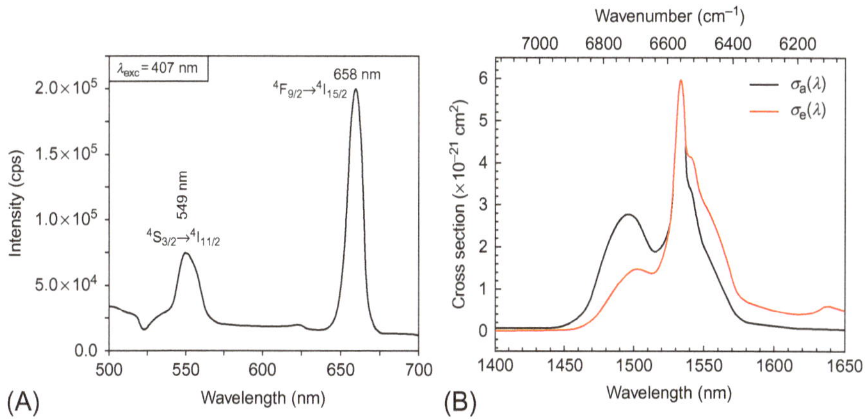


**Fig. 3.17** (A) Visible photoluminescence spectrum of $Er^{3+}$-doped antimony oxide glass under excitation at 407 nm. (B) Near-infrared (NIR) absorption *(black)* and emission *(red)* spectra of $Er^{3+}$-doped phosphate glass.
(A) Reproduced with permission from T. Som, B. Karmakar, Efficient green and red fluorescence upconversion in erbium doped new low phonon antimony glasses, Opt. Mater. 31 (2009) 609–618, Fig. 6, Copyright © Elsevier. (B) Reproduced with permission from K. Linganna, M. Rathaiah, N. Vijaya, Ch. Basavapoornima, C.K. Jayasankar, S. Ju, W.-T. Han, V. Venkatramu, 1.53 μm luminescence properties of $Er^{3+}$-doped K-Sr-Al phosphate glasses, Ceram. Int. 41 (2015) 5765–5771, Copyright © Elsevier.

### *3.3.1.10 Thulium*

The absorption spectrum of $Tm^{3+}$($4f^{12}$ configuration) doped $20Li_2O$–$20CaO$–$60SiO_2$ (mol.%) glass is shown in Fig. 3.18A [14]. Total six absorption bands have been identified in the ultraviolet (UV) to near-infrared (NIR) region and assigned to the transitions from the ground state $^3H_6$. Other low intensity bands could also be identified with the help of the transition levels assigned in the Table 3.1 for the aqueous solution of $Tm^{3+}$ ion.

Effect of glass composition on visible photoluminescence spectra of $Tm^{3+}$-doped tellurite glasses, composition (molar ratio) $(1-x)TeO_2$–$xWO_3$ ($x=0.15$, 0.25, and 0.30) + 1.0 mol.% $Tm_2O_3$, under excitation at 457.9 nm(laser light) is shown in Fig. 3.18B [28]. Three emission bands peaking around 475, 650, and 800 nm have been assigned to the $^1G_4 \rightarrow {}^3H_6$, $^1G_4 \rightarrow {}^3H_4$, and $^1G_4 \rightarrow {}^3H_5$ transitions, respectively. It is seen that the intensity of the band around 475 nm gradually increases with increase in $WO_3$ content, whereas that of the bands around 650 and 800 nm initially decreases then increases. Therefore, all the sites of the $Tm^{3+}$ ion do not behave equally with respect to $WO_3$ content, that is, the glasses are structurally different.

The effects of spectral measurement temperature (300 and 9 K) on the visible photoluminescence bandwidth, band shape, and band position of $Tm^{3+}$-doped above tellurite glasses under excitation at 457.9 nm due to (A)$^1G_4 \rightarrow {}^3H_6$ and (B) $^1G_4 \rightarrow {}^3H_5$ transitions are illustrated in Fig. 3.19A and B, respectively [28]. The full width at half maximum

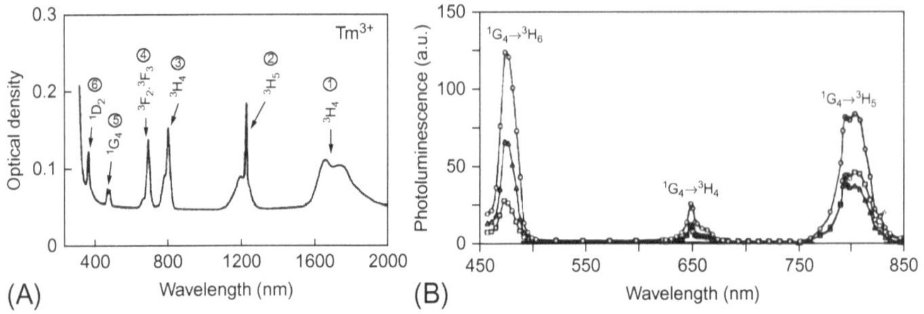


**Fig. 3.18** (A) Absorption spectrum of $Tm^{3+}$-doped $20Li_2O$–$20CaO$–$60SiO_2$ glass. (B) Effect of glass composition on visible photoluminescence spectra of $Tm^{3+}$-doped tellurite glasses under excitation at 457.6 nm (□, 0.15 mol; △, 0.25 mol; and ∘, 0.30 mol $WO_3$ content).
(A) Reproduced with permission from H. Takebe, Y. Nageno, K. Morinaga, Compositional dependence of Judd-Ofelt parameters in silicate, borate, and phosphate glasses, J. Am. Ceram. Soc. 78 (1995) 1161–1168, Copyright © The American Ceramic Society. (B) Reproduced with permission from G. Özen, A. Aydinli, S. Cenk, A. Sennaroğlu, Effect of composition on the spontaneous emission probabilities, stimulated emission cross-sections and local environment of $Tm^{3+}$ in $TeO_2$-$WO_3$ glass, J. Lumin. 101 (2003) 293–306, Copyright © Elsevier.

(FWHM) of the emission band around 800 nm due to ${}^1G_4 \rightarrow {}^3H_5$ transition broadens at room temperature (300 K) despite the fact that its peak position does not change. On the other hand, the FWHM of the band around 475 nm attributable to the ${}^1G_4 \rightarrow {}^3H_6$ transition originating from the same level of ${}^1G_4$ is found to be temperature-independent, but its peak position shifts to shorter wavelength at room temperature. This shift is due to the temperature effect on the ground state ${}^3H_6$ level of $Tm^{3+}$ ion.

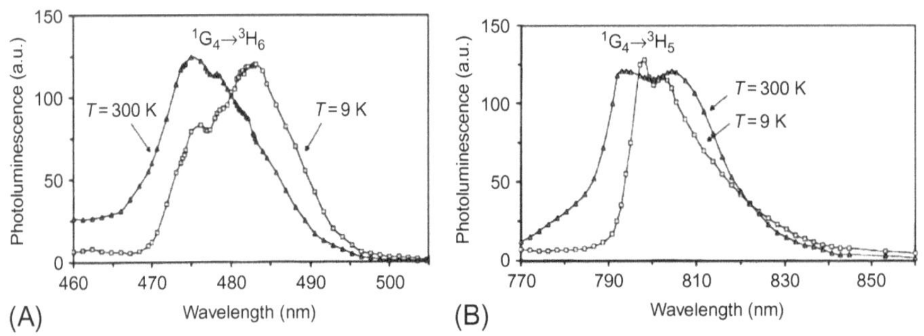


**Fig. 3.19** (A) Effect of temperature on visible photoluminescence spectra of $Tm^{3+}$-doped tellurite glasses under excitation at 457.9 nm due to ${}^1G_4 \rightarrow {}^3H_6$ transition. (B) Effect of temperature on visible photoluminescence spectra of $Tm^{3+}$-doped tellurite glasses under excitation at 457.9 nm due to ${}^1G_4 \rightarrow {}^3H_5$ transition.
Reproduced with permission from G. Özen, A. Aydinli, S. Cenk, A. Sennaroğlu, Effect of composition on the spontaneous emission probabilities, stimulated emission cross-sections and local environment of $Tm^{3+}$ in $TeO_2$–$WO_3$ glass, J. Lumin. 101 (2003) 293–306, Copyright © Elsevier.

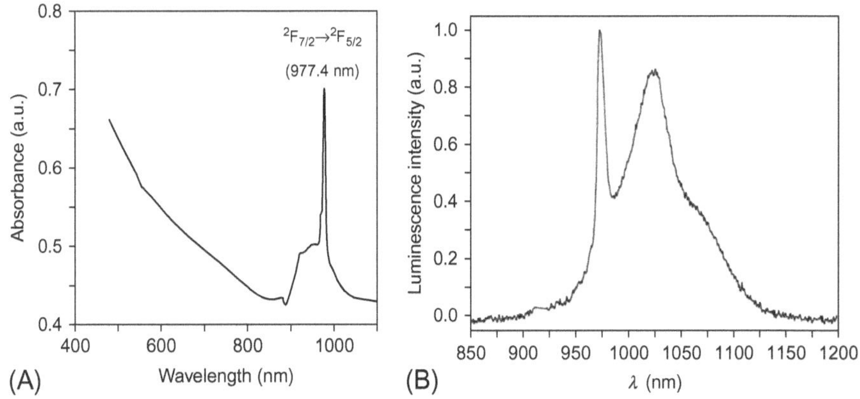


**Fig. 3.20** (A) Absorption spectrum of $Yb^{3+}$-doped antimony oxide glass. (B) Luminescence spectrum of $Yb^{3+}$-doped aluminosilicate glass.
(A) Reproduced with permission from T. Som, B. Karmakar, Nephelauxetic effect of low phonon antimony oxide glass in absorption and photoluminescence of rare-earth ions, Spectrochim. Acta A 79 (2011) 1766–1782, Copyright © Elsevier. (B) Reproduced with permission from B. Schaudel, P. Goldner, M. Prassas, F. Auzel, Cooperative luminescence as a probe of clustering in $Yb^{3+}$ doped glasses, J. Alloys Compd. 300–301 (2000) 443–449, Copyright © Elsevier.

#### 3.3.1.11 Ytterbium

The absorption spectrum of $Yb^{3+}$($4f^{13}$ configuration, ground state $^2F_{7/2}$) doped $15K_2O$–$15B_2O_3$–$70Sb_2O_3$ (mol.%) glass is shown in Fig. 3.20A [8]. Only one broad absorption band is observed in the near-infrared (NIR) region with a very strong sharp peak at 977.4 nm. It is assigned to the $^2F_{7/2}\rightarrow ^2F_{5/2}$ transition. In sodium trisilicate glass, the single band is peaking around 970 nm (Table 3.1) [2,3]. Hence, it is clear that the band position is strongly dependent on the glass components.

The near-infrared (NIR) luminescence band, corresponding to the $^2F_{5/2}\rightarrow ^2F_{7/2}$ transition, of 5 wt.% $Yb^{3+}$ doped aluminosilicate glass under excitation at 975 nm (laser diode) is shown in Fig. 3.20B [29]. Three emission peaks are detected at 975, 1025, and 1065 nm, which are attributed to the transitions from the excited-state multiplets to the ground state multiplets.

## 3.4 Functional applications

Many types of functional glasses were created by doping glasses with 4f rare-earth metal ions. Some of these functional glasses are discussed in this section.

### 3.4.1 Laser glass

From the foregoing discussion, it is clear that glasses on doping with 4f rare-earth metal ions the resultant glasses exhibit unique photoluminescence properties due to f-f transition. They cover almost the entire spectrum, starting from ultraviolet (UV, e.g., $Ce^{3+}$ and $Gd^{3+}$)-to-visible (e.g., $Pr^{3+}$, $Sm^{3+}$, $Eu^{3+}$, $Tb^{3+}$, and $Dy^{3+}$) and near-infrared (NIR, e.g., $Nd^{3+}$, $Er^{3+}$, $Ho^{3+}$, $Tm^{3+}$, and $Yb^{3+}$)regions. The first solid-state laser was demonstrated in 1961 by Snitzer [30] in $Nd^{3+}$-doped barium crown glass of composition (wt.%) $59SiO_2$–25BaO–$15K_2O$–$1Sb_2O_3$+(0.13–2.0) $Nd_2O_3$. However, all of the rare-earth-doped glasses do not exhibit laser (originated as an acronym for *light amplification by stimulated emission of radiation*) property. Only those 4f rare-earth metal ion-doped glasses could be used as laser glasses, which show evidences of "*population inversion*" at the metastable state with high lifetime. This is schematically explained in Fig. 3.21 with the energy-level diagrams of operation of optically pumped (A) three-level and (B) four-level laser systems. The dominant process for a stimulated emission is larger population of the excited state (e.g., $N_2$) than that of the lower or ground state (e.g., $N_1$), that is, for a medium to produce laser light, there must be a "*population inversion*" where $N_2 > N_1$ and $\Delta N = N_2 - N_1 =$ positive.

$Er^{3+}$- and $Nd^{3+}$-doped laser glasses are the examples of the four-level and three-level laser systems, respectively. Two conditions must be satisfied for a three-level laser system as shown in Fig. 3.21A. Firstly, as the ground state level is highly populated, therefore, more than half of the ground level ($E_1$) of the rare-earth ions must be in the excited-state level of metastable state ($E_2$), that is, $N_2 > N_1 = N_1/2$. Secondly, the lifetime ($\tau_{21}$) of the excited-state level of metastable state ($E_2$) should be larger than the relaxation time ($\tau_{32}$) of the $E_3 \rightarrow E_2$ transition, that is, $\tau_{21} \gg \tau_{32}$. Similarly, in the case of four-level laser system as shown in Fig. 3.21B, the following conditions should be satisfied, $\Delta N = N_3 - N_2 = N_3 > 0$ and $\tau_{32} \gg \tau_{43}$.

There are several distinct advantages for the use of rare-earth-doped glasses as laser materials over crystalline counterparts [31–34]:

**(i)** Glass can be formed in high optical quality in various shapes and sizes.
**(ii)** The concentration of laser active rare-earth ions can be made high (>15 wt.%).
**(iii)** Different important laser, optical, thermal, mechanical, and chemical properties such as radiative lifetime, refractive index, coefficient of thermal expansion, hardness, and chemical durability can be altered through changes in the glass component and composition.

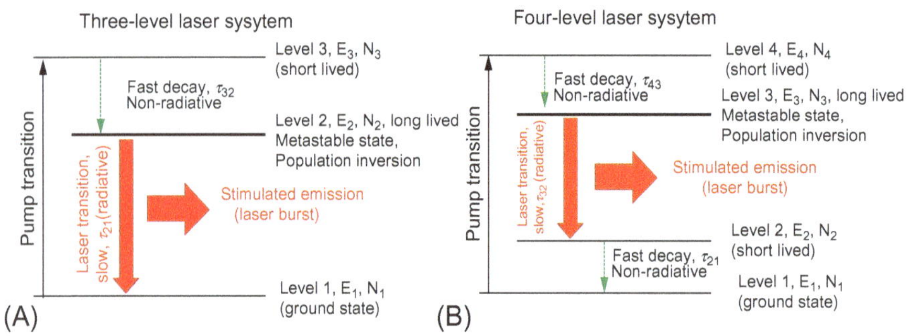


**Fig. 3.21** Schematic representation of energy-level diagram for explanation of operation of optically pumped (A) three-level and (B) four-level laser systems.

**(iv)** Glass can be produced with excellent homogeneity and uniform distribution of rare-earth ions.
**(v)** The inhomogeneously broadened emission bands in glasses make them suitable for use in Q-switched (in contrast, crystals are suitable for continuous wave, cw operation due to their sharp emission line) and amplification applications.
**(vi)** Glasses are most suitable for high energy pulsed operation because of their large size, flexibility in controlling properties, and broadened fluorescent line.

The 4f rare-earth metal ions that have been reported to exhibit laser action in glass are recorded in Table 3.2. It is noticed that $Gd^{3+}$ is the only rare-earth ion that displays lasing action in the UV region (312 nm) [2,3], whereas the other ions (e.g., $Nd^{3+}$, $Ho^{3+}$, $Er^{3+}$, $Tm^{3+}$, and $Yb^{3+}$) exhibit lasing action in the NIR region (920–2020 nm) [2,3,30,32]. It is also observed that the lasing action of $Er^{3+}$ ions in the range of 1500–1600 nm is enhanced through the process of sensitization by codoping with $Yb^{3+}$ (single ion) [35], $Yb^{3+}$–$Cr^{3+}$(double ions) [36,37], and $Yb^{3+}$–$Cr^{3+}$–$Ce^{3+}$(triple ions) [38]. The lasing of 4f rare-earth metal ions in glass has been demonstrated not only in a single type but also in a wide range of glass

Table 3.2 **Laser glasses created by 4f rare-earth metal ion doping**

| 4f rare-earth metal ion | Glass system | Transition | Laser wavelength (nm) | Reference |
|---|---|---|---|---|
| $Nd^{3+}$ | Silicate and orate | $^4F_{3/2} \rightarrow {}^4I_{9/2}$ | 920–921 | [2,3] |
| | Silicate, borate, borosilicate, phosphate, fluorophosphate, germanate, tellurite, fluoride, and chloride | $^4F_{3/2} \rightarrow {}^4I_{11/2}$ | 1047–1080 | [2,3,30,32] |
| | Silicate, borate, phosphate, and fluoride | $^4F_{3/2} \rightarrow {}^4I_{13/2}$ | 1320–1370 | [2,3,32] |
| $Gd^{3+}$ | Silicate | $^6P_{7/2} \rightarrow {}^8S_{7/2}$ | 312 | [2,3] |
| $Ho^{3+}$ | Silicate and fluoride | $^5I_7 \rightarrow {}^5I_8$ | 1950–2080 | [2,3,32] |
| $Er^{3+}$ | Silicate, phosphate, and fluoride | $^4I_{13/2} \rightarrow {}^4I_{15/2}$ | 1530–1550 | [2,3,32] |
| $Tm^{3+}$ | Silicate and fluoride | $^3H_4 \rightarrow {}^3H_6$ | 1850–2020 | [3] |
| $Yb^{3+}$ | Silicate, borate, and fluoride | $^2F_{5/2} \rightarrow {}^2F_{7/2}$ | 1010–1060 | [2,3,32] |
| $Er^{3+}$-$Yb^{3+}$ | Fluorophosphate | $^4I_{13/2} \rightarrow {}^4I_{15/2}$ | 1500 | [35] |
| $Er^{3+}$-$Yb^{3+}$-$Cr^{3+}$ | Fluoroaluminate | $^4I_{13/2} \rightarrow {}^4I_{15/2}$ | 1600 | [36,37] |
| $Er^{3+}$-$Yb^{3+}$-$Cr^{3+}$-$Ce^{3+}$ | Phosphate | $^4I_{13/2} \rightarrow {}^4I_{15/2}$ | 1540 | [38] |

systems, for example, silicate, borate, borosilicate, phosphate, fluorophosphate, germanate, tellurite, fluoride, and chloride.

Laser glasses are used for various purposes, for example, high power, ultra-short pulse, laser range finding, medical, and biophotonics applications. Two most important commercial laser glasses are (i) neodymium-doped phosphate laser glass for high energy applications and (ii) erbium-ytterbium-chromium-cerium-doped eye-safe phosphate laser glass. Various aspects of these two types of laser glasses are briefly described as follows. In these two cases, phosphate glasses are selected as host matrix as a result of advantages such as high solubility (>15 wt.%) of rare-earth ions, large stimulated emission cross section, medium phonon energy, good thermo-optical properties, and lower nonlinear refractive index.

**(i)** Neodymium doped phosphate laser glass for high energy applications

Neodymium-doped phosphate laser glass is widely used in high power generation through the process of inertial confinement fusion (ICF), laser material processing, range finder, industrial, and scientific applications. As a result of multipurpose uses of this laser glass, virtually all the leading optical firms are producing it. For example, Schott, Germany, produces LG-750, LG-760, and LG-770 [39]; Hoya Corporation, the United States, produces LHG-80, LHG-8, and LHG-5 [40]; Kigre, Inc., the United States, produces Q88 and Q98 [41]; Shanghai Institute of Optics and Fine Mechanics (SIOM), China, produces N21 and N31 [42]. A comparison of different properties of some of these commercial Nd-phosphate laser glasses is presented in Table 3.3. It is seen that the differences of properties are small and have minor impact for most of the applications. Some property values may differ within the range of 5%–10% because of measurement variations between laboratories, new measurement techniques, and small variation in chemical compositions. Nevertheless, the concentration level of $Nd_2O_3$ can create a small variation in many properties but big changes in laser properties.

The representative chemical compositions (mol.%) of LG-770 and LHG-8 are (58–62) $P_2O_5$–(6–10)$Al_2O_3$–(20–25)$K_2O$–(5–10)MgO–(2.0–3.5)$Nd_2O_3$–2(other) with O/P = 3 and (55–60)$P_2O_5$–(8–12)$Al_2O_3$–(13–17)$K_2O$–(10–15)BaO–(2.0–3.5)$Nd_2O_3$–2(other) with O/P = 3, respectively [43,44]. The manufacturing process of laser glass is very important because of many properties are related with the production techniques. Such properties are fluorescence lifetime, optical loss at lasing wavelength (1054 nm), optical quality (should be free from bubble, seeds, and cords), platinum inclusion, hydroxyl (OH) group content, and residual stress. Very high-purity metaphosphate raw materials containing a trace amount (<10 ppm) of transition metal ions (with very low amount of copper and iron to avoid their absorption at 1054 nm) and <0.1 wt.% chemically or physically absorbed or adsorbed water [42,43]. There are two melting technologies of Nd-phosphate laser glass: discontinuous glass melting and continuous glass melting [42,43]. These two methods are briefly described as follows.

*Discontinuous glass melting*: In the first step, the premelting is done in a refractory crucible made of high-purity silica ($SiO_2$) in the temperature range of 1100–1250°C depending on the glass composition. A gas (e.g., oxygen) may be bubbled through the melt for expulsion of unwanted volatile products and to set the redox state. In the second step, the glass (cullet) obtained in the first step is remelted in a platinum or platinum-lined refractory crucible attached with a bottom-pouring flow-casting facility (platinum downpipe) under stirring. These are shown in Fig. 3.22A [43]. The major purposes of remelting are to remove bubbles, dissolve platinum inclusions, and homogenize the molten glass to obtain defect-free (free

**Table 3.3 Comparison of different properties of commercial Nd-phosphate laser glasses**

| Glass properties | LG-770, Schott, Germany [39] | LHG-8, Hoya, Japan [40] | Q-88, Kigre, Inc., the United States [41] | N-31, SIOM, China [42] |
|---|---|---|---|---|
| ***Neodymium laser properties*** | | | | |
| $Nd_2O_3$ concentration (wt.%) | 2.0 | 2.0 | – | – |
| $Nd^{3+}$ ion concentration ($10^{20}$ ions $cm^{-3}$) | – | 3.03 | 2.9 | 3.0 |
| Emission peak, $\lambda$ (nm) | 1053 | 1053 | 1054 | 1054 |
| Emission band width, $\Delta\lambda_{em}$ (nm) | 25.4 | 26.5 | 21.9 | 20.1 |
| Radiative lifetime, $\tau_{Rad}$ (μs) | 350 | 365 | 326 | 320 |
| Emission cross section, $\sigma_{em}$ ($10^{-20}$ $cm^2$) | 3.9 | 3.6 | 4.0 | 3.9 |
| ***Optical properties*** | | | | |
| Refractive index | | | | |
| $n_d$ (at 587.3 nm) | 1.5086 | 1.5296 | 1.5449 | 1.5406 |
| $n_{1053nm}$ (at 1053 nm) | 1.4996 | 1.5201 | 1.5363 | 1.5341 |
| Abbe number, $\nu_d$ | 68.4 | 66.5 | 64.8 | 64.9 |
| Nonlinear refractive index at 1053 nm, $n_2$ ($10^{-13}$ esu) | 1.02 | 1.12 | 1.14 | 1.15 |
| Temperature coefficient of refractive index, $dn/dT$ ($10^{-6}$ $K^{-1}$) | −4.7 | −5.3 | −0.5 | −4.2 |
| ***Physical property*** | | | | |
| Density (g $cm^{-3}$) | 2.585 | 2.83 | 2.71 | 2.86 |
| ***Thermal properties*** | | | | |
| Glass transition temperature (°C) | 461 | 485 | 367 | 455 |
| Coefficient of thermal expansion, $\alpha_{20\text{–}300°C}$ ($10^{-7}$ °$C^{-1}$) | 134 | 127 | 104[a] | 127 |
| Thermal conductivity at 25°C, $\kappa_{25°C}$ (W $m^{-1}$ $K^{-1}$) | 0.57 | 0.58 | 0.84 | – |
| ***Mechanical properties*** | | | | |
| Hardness (GPa) | 3.58 | 3.43 | – | – |
| Fracture toughness, $K_{1C}$ (MPa $m^{1/2}$) | 0.48 | 0.51 | – | – |
| Young's modulus, $E$ (GPa) | 47.29 | 50.1 | 70.1 | – |
| Poisson's ratio, $\nu$ | 0.253 | 0.26 | 0.24 | 0.27 |

[a]20–40°C.

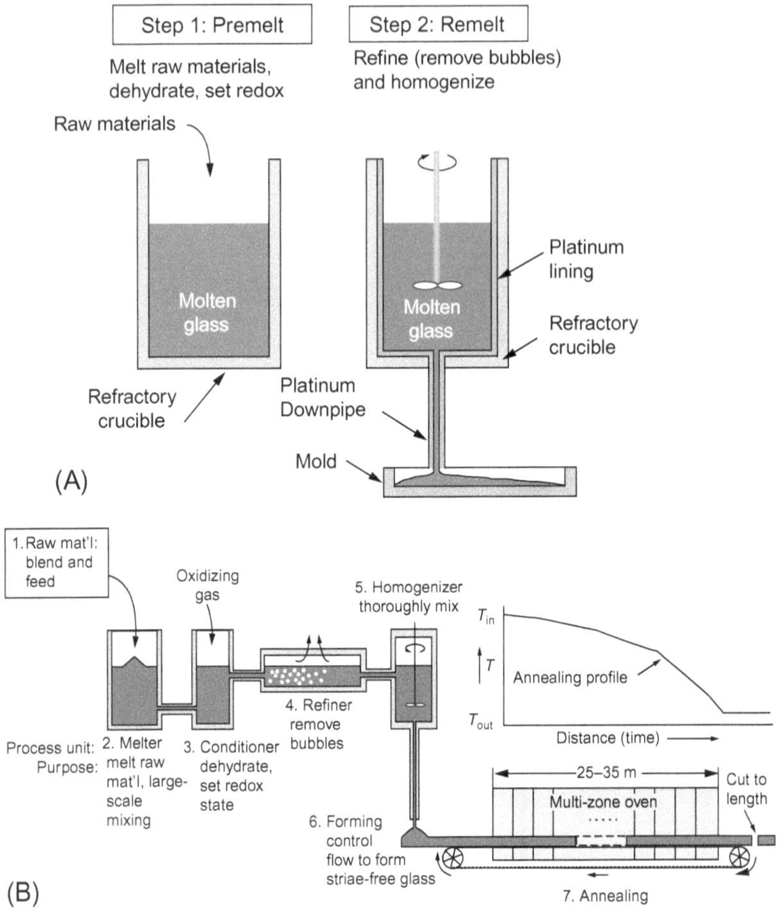


**Fig. 3.22** (A) Schematic representation of laser glass melting by a discontinuous process. (B) Schematic representation of laser glass melting by a continuous process.
Reproduced with permission from J.H. Campbell, T.I. Suratwala, C.B. Thorsness, J.S. Hayden, A.J. Thorne, J.M. Cimino, A.J. Marker III, K. Takeuchi, M. Smolley, G.F. Ficini-Dorn, Continuous melting of phosphate laser glasses, J. Non-Cryst. Solids 263 & 264 (2000) 342–357, Copyright © Elsevier.

from bubbles, seeds, and striae) high optical quality glass suitable for laser application. In the second step of glass melting, the elimination of platinum inclusion is carried out by bubbling oxidizing gas (e.g., oxygen, chlorine, or their combination) through the process of platinum dissolution (in reality it is an ionization process) as per the following reactions:

$$Pt^{o} + \frac{n}{4}O_2 \rightarrow Pt^{n+} + \frac{n}{2}O^{2-} \tag{3.17}$$

$$Pt^{o} + \frac{n}{2}Cl_2 \rightarrow Pt^{n+} + nCl^{-} \tag{3.18}$$

This gas bubbling through the molten glass, in addition to the abovesaid reactions also eliminates the hydroxyl (−OH) groups from the glass matrix. In the case of nonreactive oxygen, the −OH group removal is achieved through the following condensation reaction as a result of expulsion of water molecule ($H_2O$):

$$2(P-OH) \rightarrow H_2O\uparrow + P-O-P \tag{3.19}$$

Reactive chlorine gas is more effective for −OH group removal as it reacts with the resultant water molecule according to the following reaction and catalyzes the elimination process:

$$2H_2O + 2Cl_2 \rightarrow 4HCl\uparrow + O_2\uparrow \tag{3.20}$$

After these treatments, the glass is cast into suitable mold, coarse and fine annealed for removal of any residual thermal stresses in the glass.

*Continuous glass melting*: In continuous laser glass melting, there are seven interconnected processing stations such as (1) raw material blending and feeding, (2) melting, (3) conditioning, (4) refining, (5) homogenization, (6) continuous strip forming, and (7) annealing. These are shown in Fig. 3.22B with the purpose of each station [43]. All the processing steps as described in the discontinuous laser glass melting are carried out at the respective work stations.

The photograph of Nd-laser glasses showing the evolution of glass composition type (silicate → phosphate) and manufacturing technology (discontinuous → continuous glass melting) is shown in Fig. 3.23A [44]. It is seen that up to around 1980 $Nd^{3+}$-doped silicate laser glass was produced by the discontinuous glass melting technique, then after $Nd^{3+}$-doped phosphate laser glass was produced by the discontinuous glass melting technique followed by development of continuous laser glass melting technique around 1990. Fig. 3.23B shows the schematic representation of xenon flashlamp and laser glass setup for obtaining lasing action, flashlamp pumping, partial energy-level diagram showing excitation, relaxation, and stimulated emission at 1054 nm in Nd-phosphate laser glass, matching of absorption spectrum of Nd-phosphate laser glass and spectrum of xenon flashlamp output [45]. It is seen that there is a good spectral overlap of the output of the xenon flashlamp with the $Nd^{3+}$ absorption bands (that is, pumping). The laser emission at 1054 nm is due to the $^3F_{3/2} \rightarrow {}^4I_{11/2}$ transition. The same laser emission is also obtained by pumping at 808 nm with a laser diode. Various properties of different commercial Nd-phosphate laser glasses are already provided in Table 3.3. All these glasses are used for terawatt (TW, $10^{12}$ W) to petawatt (PW, $10^{15}$ W) power generation [45]. Schematic exemplification of inertial confinement fusion using high-power laser beam of Nd-phosphate laser glass is shown in Fig. 3.24. The well-known nuclear fusion reaction is as follows:

$$^2D_1 + {}^3T_1 \rightarrow {}^4He_2 + {}^1n_0 + 17.6\,MeV \tag{3.21}$$

where D, T, He, and n are the symbols of deuterium, tritium, helium, and neutron, respectively. It is seen that a huge amount of thermal energy is obtained from this fusion reaction and it could be used for the generation of electric power.

**(ii)** Erbium-ytterbium-chromium-cerium-doped eye-safe phosphate laser glass

The erbium-ytterbium-chromium-cerium-doped phosphate laser glass has the "*eye-safe*" laser wavelength of 1533 nm. Here, ytterbium and chromium ions function as sensitizer and cerium as radiation-resistant; these facts will be discussed later. This laser

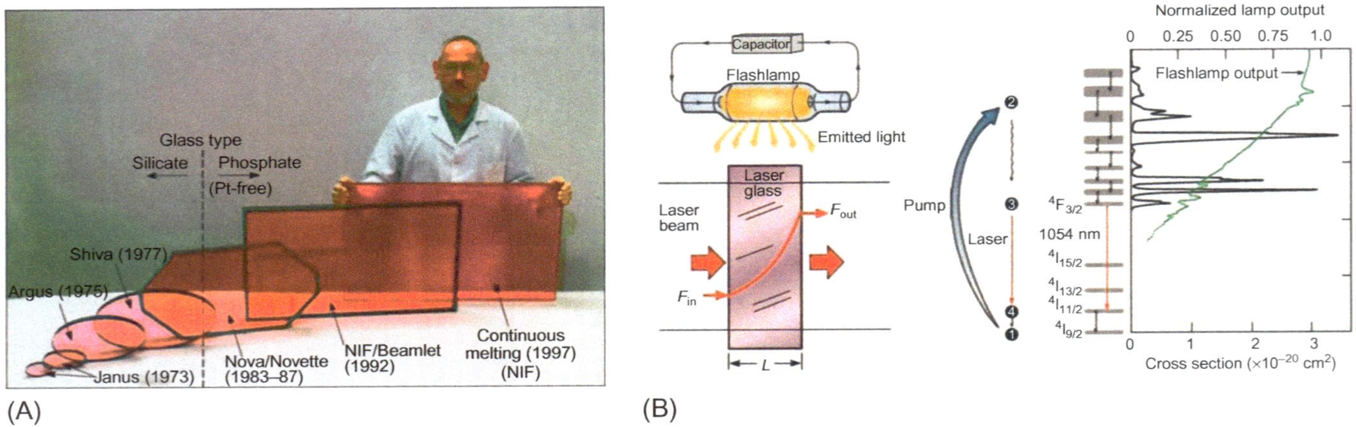


**Fig. 3.23** (A) Photograph of Nd-laser glasses showing the evolution of glass composition type (silicate → phosphate) and manufacturing technology (discontinuous → continuous glass melting) carried out at the Lawrence Livermore National Laboratory (LLNL), United States. (B) (Left) Schematic representation of xenon flashlamp and laser glass setup for obtaining lasing action. (Right) Flashlamp pumping; partial energy-level diagram showing excitation, relaxation, and stimulated emission at 1054 nm in Nd-phosphate laser glass, matching of absorption spectrum of Nd-phosphate laser glass and spectrum of xenon flashlamp output.
(A) Reproduced with permission from J.S. Hayden, Overcoming technical challenges and moving into the future with laser glass, Int. J. Appl. Glass Sci. 6 (2015) 19–25, Copyright © The American Ceramic Society. (B) Reproduced with permission from J.H. Campbell, J.S. Hayden, A. Marker, High-power solid-state lasers: a laser glass perspective, Intl. J. Appl. Glass Sci. 2 (2011) 3–29, Copyright © The American Ceramic Society.

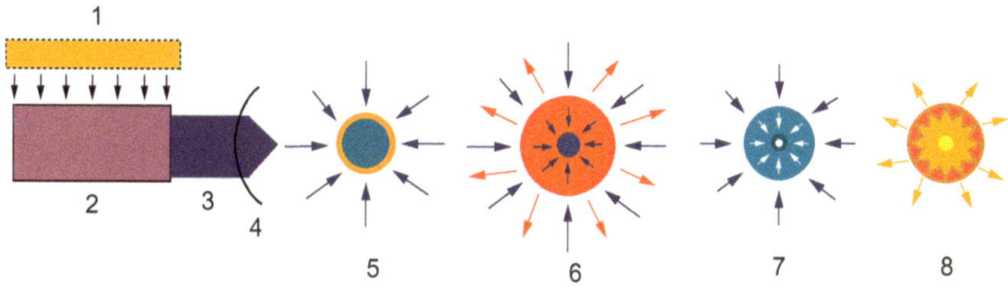


**Fig. 3.24** Schematic exemplification of inertial confinement fusion using laser beam of Nd-phosphate laser glass. (1) Flashlamp (*yellow*, *black arrows* represent radiation of flashlamp); (2) Nd-phosphate laser glass *(violet-pink)*; (3) laser beam *(blue)*; (4) target chamber; (5) laser beam *(blue)* rapidly heat the surface of the fusion target forming a surrounding plasma envelope *(yellow)*; (6) fuel is compressed by the rocket-like blowoff of the hot surface material (*orange arrows* represent blowoff ); (7) during the final part of the capsule implosion, the fuel core reaches 20 times the density of lead and ignites at 100,000,000°C (*white arrows* represent inwardly transported thermal energy); (8) thermonuclear burn spreads rapidly through the compressed fuel, yielding many times the input energy (*yellow arrows* represent release of giant thermal energy).

glass exhibits lasing action under pumping either with flash lamp or laser diode. It is used for range finding (time of flight), medical (skin rejuvenation), and biophotonics. Schott, Germany, produces this laser glass under the commercial code LG-940. Some of the properties of this glass are listed in Table 3.4 [46].

Liu et al. [47] have reported the spectral and laser properties of $Er^{3+}$–$Yb^{3+}$-codoped phosphate glasses in the $P_2O_5$–$Na_2O$–SrO–$Al_2O_3$–$Yb_2O_3$–$Er_2O_3$ system. Evaluating the energy-transfer efficiency and spectral parameters, the optimum concentrations of $Yb_2O_3$ and $Er_2O_3$ were found to be 6 mol.% $Yb_2O_3$ and 0.2–0.4 mol.% $Er_2O_3$ for laser-diode-pumped microchip laser application. OH groups in glass greatly affect the fluorescent intensity and lifetime of this glass, and its absorption coefficient at 3000 $cm^{-1}$ should be <1 $cm^{-1}$ for laser applications. Pumping with a laser diode at 974 nm, CW laser emission centered at 1530 nm with slope efficiency of 10.6% and maximum output of 43 mW was achieved in a 2 mm-thick glass sample at room temperature. Philipps et al. [35] have reported the $Er^{3+}$ emission and lasing properties at 1500 nm in an $Er^{3+}$–$Yb^{3+}$-codoped fluorophosphates glass containing 80% fluorides of the $Sr(PO_3)_2$–$AlF_3$–$CaF_2$–$SrF_2$–$MgF_2$ system. The results of laser and tuning studies indicate that this glass is a good candidate for applications in broadband amplification and as gain material for ultra-short laser pulse sources.

Heumann et al. [36] and Ledig et al. [37] have studied spectroscopic and laser properties of $Er^{3+}$ ion in $Cr^{3+}$–$Yb^{3+}$–$Er^{3+}$ triply doped fluorophosphate (also called fluoride phosphate and fluoroaluminate) glasses having the most preferred composition of 20 mol.% $Sr(PO_3)_2$ or $Ba(PO_3)_2$ and 80 mol.% of ($AlF_3$–$CaF_2$–$SrF_2$–$MgF_2$). $Cr^{3+}$ and $Yb^{3+}$ ions are used to doubly sensitize the emission of $Er^{3+}$ ion around 1550 nm as a result of $^4I_{13/2} \rightarrow {}^4I_{15/2}$ emission transition. Energy-level diagrams of $Cr^{3+}$, $Yb^{3+}$, and $Er^{3+}$ in fluorophosphate glass with energy-transfer channel of $Cr^{3+} \rightarrow Yb^{3+} \rightarrow Er^{3+}$ under krypton laser pumping at 647.1 nm ($Cr^{3+}$ ion has a broad absorption band around 450 nm) and laser transition of $Er^{3+}$ around 1550 nm are shown in Fig. 3.25. However, the emission could be tuned between 1536 and 1596 nm. Fig. 3.26 (A) shows the fluorescence spectra of $Cr^{3+}$ in (spectrum (a)) Cr-doped fluorophosphates glass, and $Yb^{3+}$ in (spectrum (b)) Cr–Yb-codoped fluorophosphates glass and (c) Cr–Yb–Er triply doped fluorophosphates glass. The singly doped (a) $Cr^{3+}$ glass shows a broad fluorescence band peaking around 820 nm. This band

**Table 3.4** **Properties of commercial "eye-safe" Er-Yb-Cr-Ce-phosphate laser glass (LG-940, Schott) [46]**

| Glass properties | Corresponding value | Glass properties | Corresponding value |
|---|---|---|---|
| ***Erbium laser properties*** | | ***Physical property*** | |
| Emission peak, $\lambda$ (nm) | 1532.5 | Density (g cm$^{-3}$) | 3.04 |
| Linewidth, FWHM (nm) | 25.3 | | |
| Radiative lifetime, $\tau_{Rad}$ (ms) | 11.1 | ***Thermal properties*** | |
| Fluorescence lifetime (ms) | 9.4 | Glass transition temperature (°C) | 456 |
| Emission cross section, $\sigma_{em}$ ($10^{-21}$ cm$^2$) | 7.1 | Coefficient of thermal expansion, $\alpha_{20\text{–}300°C}$ ($10^{-7}$ °C$^{-1}$) | 119.6 |
| | | Thermal conductivity at 25°C, $\kappa_{25°C}$ (W m$^{-1}$ K$^{-1}$) | 0.51 |
| ***Optical properties*** | | Thermal conductivity at 90°C, $\kappa_{90°C}$ (W m$^{-1}$ K$^{-1}$) | 0.61 |
| Refractive index | | | |
| $n_d$ (at 587.3 nm) | 1.533 | ***Mechanical properties*** | |
| $n_{1540nm}$ (at 1540 nm) | 1.522 | Knoop hardness, $HK_{0.1/20}$ | 380 |
| Abbe number, $\nu_d$ | 62.2 | Fracture toughness, $K_{1C}$ (MPa m$^{1/2}$) | 0.7 |
| Nonlinear refractive index, $n_2$ ($10^{-13}$ esu) | 1.3 | Young's modulus, $E$ (GPa) | 57.6 |
| Temperature coefficient of refractive index at 1540 nm, $dn/dT$ ($10^{-6}$K$^{-1}$) | −3.6 | Poisson's ratio, $\nu$ | 0.26 |

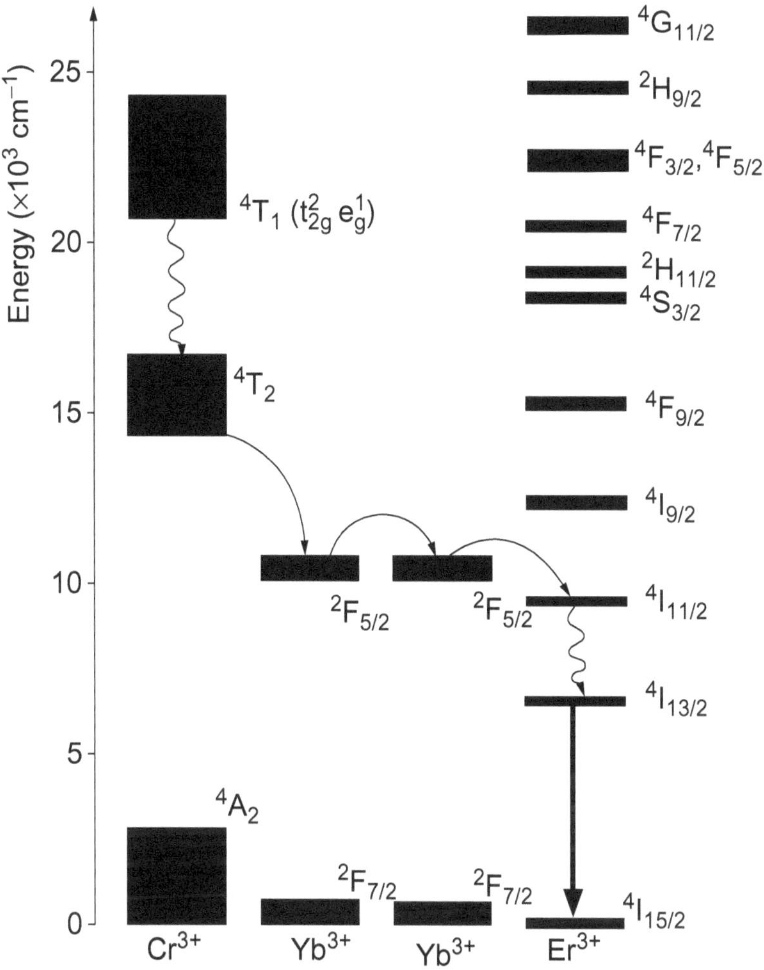


**Fig. 3.25** Energy-level diagrams of $Cr^{3+}$, $Yb^{3+}$, and $Er^{3+}$ in fluorophosphate glass with energy-transfer channel of $Cr^{3+} \rightarrow Yb^{3+} \rightarrow Er^{3+}$ and laser transition of $Er^{3+}$ around 1550 nm. Reproduced from M. Ledig,E. Heumann, D. Ehrt, W. Seeber, Spectroscopic and laser properties of $Cr^{3+}$: $Yb^{3+}$: $Er^{3+}$: fluoride phosphate glass, Opt. Quant. Electron. 22 (1990) S107–S122.

completely disappears by codoping with $Yb^{3+}$ ion (b). This fact clearly indicates an effective energy-transfer from $Cr^{3+}$ ion to $Yb^{3+}$ ion in the glass. Further doping of this glass with $Er^{3+}$ ion leads to significant decrease of intensity of the fluorescence band of $Yb^{3+}$ ion (spectrum (c)). This has happened due to the second step energy-transfer from $Yb^{3+}$ ion to $Er^{3+}$ ion. Emission spectrum of $Er^{3+}$ due to $^4I_{13/2} \rightarrow {}^4I_{15/2}$ transition in Cr–Yb–Er triply doped fluorophosphates glass is shown in Fig. 3.26B. Fig. 3.26C shows the output as a function of input power around 1600 nm emission under cw operation at room temperature of Cr–Yb–Er triply doped fluorophosphates glass. A threshold of 80 mW of pump power was obtained with 0.6% power slope efficiency. This points out the high losses of pump energy and is attributed to the poor quality of the experimental glass.

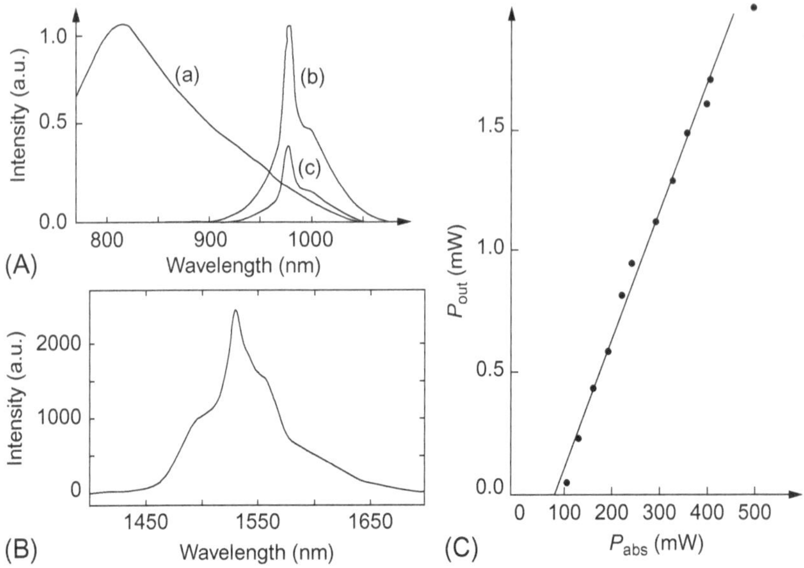


**Fig. 3.26** (A) Fluorescence of $Cr^{3+}$ in (a) Cr-doped fluorophosphates glass, $Yb^{3+}$ in (b) Cr–Yb-codoped fluorophosphates glass, and (c) Cr–Yb–Er triply doped fluorophosphates glass. (B) Emission spectrum of $Er^{3+}$ due to $^4I_{13/2} \rightarrow {^4I_{15/2}}$ transition in Cr–Yb–Er triply doped fluorophosphates glass. (C) Output as a function of input power at 1600 nm emission under cw operation at room temperature of Cr–Yb–Er triply doped fluorophosphates glass. Reproduced with permission from E. Heumann, M. Ledig, D. Ehrt, W. Seeber, E. W. Duczynski, H.-J. v.d. Heide, G. Huber, Cw laser action of $Er^{3+}$ in double sensitized fluoroaluminate glass at room temperature, Appl. Phys. Lett. 52 (1988) 255–256, Copyright © AIP.

Myers [38] has patented the glass compositions of the erbium-ytterbium-chromium-cerium-doped phosphate laser glass for the *"eye-safe"* laser application. The most preferred compositional (mol.%) ranges are (55–60)$P_2O_5$–(15–20)$R_2O$–(10–15)MO–(2–5)$Al_2O_3$–(0.2–0.7)$Nb_2O_5$–(0.1–0.3)$Sb_2O_3$–(1–4)$CeO_2$–(0.03–0.1)$Er_2O_3$–(4–17)$Yb_2O_3$–(0.04–0.06)$Cr_2O_3$, where R = Li, K, and M = Ca, Sr. Cerium prevents solarization, which is a dominant issue known to limit the pump power conversion efficiency. This is because of the fact that $Ce^{3+}$ is an efficient antisolarant and inhibits the color center formation from the UV light part of the pump lamp. Byshevskaya-Konopko et al. [48] have disclosed the role of cerium in cerium-chromium-ytterbium-erbium phosphate laser glass. They have studied the influences of UV and gamma-ray radiations on the lasing properties of the glass. They have observed that doping the glass with cerium oxide improves its radiation resistance particularly with respect to the gamma rays. This aspect will be discussed in details later.

## 3.4.2 Luminescent glasses for white light emitting diodes (WLEDs)

White light emitting diodes (WLEDs) as the next-generation solid-state lighting have many important economic and technological advantages such as efficiency in power and environmental friendliness. Their potential applications in diverse technological areas are electronic display, automobile headlight, backlight, device indicator, illumination, etc. In comparison with the conventional phosphors used in development of WLEDs, the use of rare-earth-doped glasses have several benefits, for example, better optical and mechanical properties, lower production cost and the manufacturing technique is easy, and finally, the assemblage procedure is completely free from photo-decomposable epoxy resin. In recent times, luminescent glasses for WLEDs have engrossed much attention as a result of these advantages [49–56].

Consequently, various white light emitting glasses doped by different 4f rare-earth metal ions are reported, which are listed in Table 3.5 [49–56]. It is seen that the white light can be obtained in glasses by doping with single rare-earth ion (e.g., $Dy^{3+}$ and $Ce^{3+}$) [49,50], double rare-earth ions (e.g., $Ce^{3+}$–$Eu^{3+}$, $Ce^{3+}$–$Tb^{3+}$, and $Tb^{3+}$–$Sm^{3+}$) [51–53], triple rare-earth (e.g., $Eu^{3+}$–$Tb^{3+}$–$Tm^{3+}$) [54,55] and also by the combination of 4f rare-earth and 3d transition metal ions [56].

Vijaya et al. [49] have reported the white light photoluminescence in $Dy^{3+}$-doped fluorophosphate glasses of composition (mol.%) $41P_2O_5$–$17K_2O$–$8Al_2O_3$–$(24-x)$ $ZnF_2$–10LiF–$xDy_2O_3$ ($x=0.05$, 0.1, 0.5, 1.0, 2.0, and 4.0). Emission spectra of these glasses doped with different concentrations of $Dy^{3+}$ ion under excitation at 385 nm is shown in Fig. 3.27A. They exhibit four emission bands peaking around 457, 486, 577, and 666 nm, which are assigned to the ${}^4I_{13/2} \rightarrow {}^6H_{13/2}$, ${}^4F_{9/2} \rightarrow {}^6H_{15/2}$, ${}^4F_{9/2} \rightarrow {}^6H_{13/2}$, and ${}^4F_{9/2} \rightarrow {}^6H_{11/2}$ transitions, respectively. The blue band at 486 nm and the yellow band at 577 nm are strong, whereas other two bands at 457 and 666 nm are very weak. The strong yellow (577 nm) and blue (486 nm) luminescence are very essential for displaying white color emission. The intensities of the yellow and blue emission bands increase with increase in the concentration of $Dy^{3+}$ ion up to 1.0 mol.% and then decrease with their constant intensity ratio. Fig. 3.27B demonstrates the CIE 1931 chromaticity diagram showing the color coordinates ($x$,$y$) of the fluorophosphate glasses doped with different concentrations of $Dy^{3+}$ ion. The chromaticity coordinates are found to vary in the ranges of $x=0.294$–0.350 and $y=0.312$–0.367. The ideal white light chromaticity coordinates are $x=0.333$ and $y=0.333$. Therefore, some $Dy^{3+}$ ion-doped fluorophosphates glasses emit bright white light, which are found to be suitable for use in development of WLEDs. Andrade et al [50] have reported the white light emission in a Ce-doped low-silica calcium aluminosilicate (LSCAS) glass of composition (wt.%) 41.5-$Al_2O_3$–47.4CaO–7.0$SiO_2$–2.1MgO–2.0$CeO_2$. The major broad emission bands are peaking at 475 and 540 nm. The chromaticity coordinates are found to vary in the ranges of $x=0.20$–0.40 and $y=0.28$–0.49.

Rocha et al. [51] have reported the white light emission in a $Ce^{3+}$–$Eu^{2+,\,3+}$-codoped low-silica calcium aluminosilicate (LSCAS) glass of composition (wt.%) 41.5-$Al_2O_3$–47.4CaO–7.0$SiO_2$–2.1MgO–2.0$CeO_2$+0.5$Eu_2O_3$. The single very broad emission band was peaked at 550 nm. The chromaticity coordinates are found to vary in the ranges of $x=0.29$–0.41 and $y=0.27$–0.51. Wei et al. [52] have demonstrated the blue-white-green tunable luminescence for WLEDs of $Ce^{3+}$–$Tb^{3+}$-codoped silicate glasses of composition (wt.%) 73.2$SiO_2$–15.3$Na_2CO_3$–6MgO–2ZnO–3.5$CaCO_3$–1$CeO_2$–$xTbF_3$ ($x=0$, 1, 2, 3, 5, and 7). The major emission bands are peaking around 400 and 542 nm. The chromaticity coordinates are found to vary in the ranges of $x=0.25$–0.35 and $y=0.24$–0.45. Fig. 3.27C shows the photographs exhibiting the color of photoluminescence in GCe1 (blue), GCeTb2 (white), and GCeTb7 (yellowish

**Table 3.5 White light emitting glasses by 4f rare-earth metal ion doping**

| RE ion | Glass | Glass composition | Chromaticity coordinates[a] | Major emission band (nm) | Reference |
|---|---|---|---|---|---|
| $Dy^{3+}$ | Fluorophosphate | $41P_2O_5$–$17K_2O$–$8Al_2O_3$–$(24-x)$ $ZnF_2$–10LiF–$xDy_2O_3$ ($x=0.05$, 0.1, 0.5, 1.0, 2.0, and 4.0) (mol.%) | $x=0.294$–0.350 and $y=0.312$–0.367 | 486, 577 | [49] |
| $Ce^{3+}$ | Aluminosilicate | 41.5-$Al_2O_3$–47.4CaO–7.0$SiO_2$–2.1MgO–2.0$CeO_2$ (wt.%) | $x=0.20$–0.40 and $y=0.28$–0.49 | 475, 540 | [50] |
| $Ce^{3+}$–$Eu^{2+,\ 3+}$ | Aluminosilicate | 41.5-$Al_2O_3$–47.4CaO–7.0$SiO_2$–2.1MgO–2.0$CeO_2$+0.5$Eu_2O_3$ (wt.%) | $x=0.29$–0.41 and $y=0.27$–0.51 | 550 (very broad) | [51] |
| $Ce^{3+}$–$Tb^{3+}$ | Silicate | 73.2$SiO_2$–15.3$Na_2CO_3$–6MgO–2ZnO–3.5$CaCO_3$–1$CeO_2$–$xTbF_3$ ($x=0$, 1, 2, 3, 5, and 7) (wt.%) | $x=0.25$–0.35 and $y=0.24$–0.45 | 400, 542 | [52] |
| $Tb^{3+}$–$Sm^{3+}$ | Silicate and borosilicate | 50$SiO_2$–37CaO–13BaO–$xB_2O_3$ 0.125$Tb_4O_7$–$ySm_2O_3$ ($x=0$, 10; $y=0.25$, 0.50, 0.75, and 1.0) (mol.%) | – | 543, 601 | [53] |
| $Ce^{3+}$–$Tb^{3+}$–$Eu^{3+}$ | Silicate, borosilicate, and phosphate | $xP_2O_5$–$ySiO_2$–$z$SrO–37CaO–13BaO–$pB_2O_3$–$qCe_2O_3$–$rTb_2O_3$–$mEu_2O_3$ ($x=0$, 60; $y=45$, 50; $z=0$, 20; $p=5$, 10; $q=0.25$, 1.0; $r=0.125$, 0.5; $m=0.5$, 0.75, 1.0) (mol.%) | – | 488, 542, 611 | [54] |
| $Eu^{3+}$–$Tb^{3+}$–$Tm^{3+}$ | Aluminoborate | CaAlB = (18–40)CaO–(17–23)$Al_2O_3$–(29–38)$B_2O_3$ (wt. %). $Ca_{0.931-x}AlB:Eu^{3+}_{0.038},Tb^{3+}_{0.031}$, $Tm^{3+}{}_x$($x=0$–0.011 mol) and $Ca_{0.959-y}AlB$: $Tb^{3+}_{0.031}$, $Tm^{3+}_{0.0094}$, $Eu^{3+}{}_y$($y=0$–0.025 mol) | $x=0.32$ and $y=0.34$ | 454, 544, 615 | [55] |
| $(Cu^+)_2$–$Eu^{3+}$–$Ce^{3+}$ | Silicate | 73.2$SiO_2$–15.3$Na_2CO_3$–6MgO–2ZnO–3.5$CaCO_3$ (wt.%) +2CuO, (1–2)$EuF_3$, and 0.4$CeO_2$ | $x=0.336$ and $y=0.346$ | 525, 612 | [56] |

[a]Ideal white light chromaticity coordinates, $x=0.333$ and $y=0.333$.

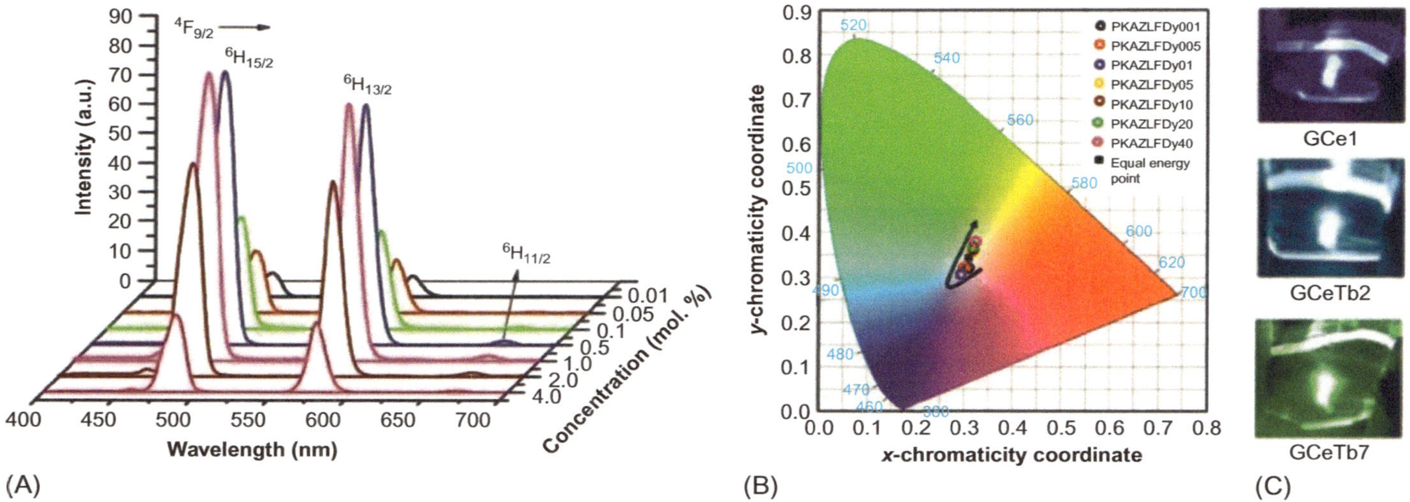


**Fig. 3.27** (A) Emission spectra of fluorophosphates glasses doped with different concentrations of $Dy^{3+}$ ion (excitation at 385 nm). (B) CIE 1931 chromaticity diagram showing the color coordinates $(x,y)$ of fluorophosphates glasses doped with different concentrations of $Dy^{3+}$ ion.
(C) Photographs showing the color of photoluminescence in GCe1 *(blue)*, GCeTb2 *(white)*, and GCeTb7 *(yellowish green)* glass samples excited at 340 nm in dark.
(A) Reproduced with permission from N. Vijaya, K. U. Kumar, C.K. Jayasankar, $Dy^{3+}$-doped zinc fluorophosphate glasses for white luminescence applications, Spectrochim. Acta A 113 (2013) 145–153, Copyright © Elsevier. (B) Reproduced with permission from R.F. Wei, H. Zhang, F. Li, H. Guo, Blue-white-green tunable luminescence of $Ce^{3+}$, $Tb^{3+}$ co-doped sodium silicate glasses for white LEDs, J. Am. Ceram. Soc. 95 (2012) 34–36, Copyright © The American Ceramic Society.

green) glass samples excited at 340 nm in dark. The sample GCeTb2 with 2 wt.% $TbF_3$ emits very bright white light. Liang et al. [53] have reported white light emission in $Tb^{3+}$–$Sm^{3+}$-codoped silicate and borosilicate glasses. Here the major emission bands are at 543 and 601 nm.

Zhu et al. [54] have demonstrated the white light emission in $Ce^{3+}$–$Tb^{3+}$–$Eu^{3+}$ triply doped silicate, borosilicate, and phosphate glasses. The major emission bands are at 488, 542, and 611 nm. Whereas Zhao et al. [55] have demonstrated the white light emission in $Eu^{3+}$–$Tb^{3+}$–$Tm^{3+}$ triply doped aluminoborate glasses excited by UV light. The major emission bands are at 454, 544, and 615 nm. The chromaticity coordinates are found to be $x=0.32$ and $y=0.34$. Guo et al. [56] have described the tunable white light luminescence in $(Cu^+)_2$–$Eu^{3+}$–$Ce^{3+}$ triply doped sodium silicate glasses with the emission bands at 525 and 612 nm. The chromaticity coordinates of $x=0.336$ and $y=0.346$ for perfect white light are also achieved.

### 3.4.3 Radiation-resistant glasses

The electromagnetic radiations such as ultraviolet (UV) light, X- and gamma (γ) rays are regarded as the ionizing radiations as a result of their higher energies. The energy for one photon varies in the ranges of 3.1–124 eV for UV lights, $1.24 \times 10^2$–$1.24 \times 10^5$ eV for X-rays, and $1.24 \times 10^5$–$1.24 \times 10^7$ eV for γ-rays. These radiations interacting with glasses produce defect centers, for example, hole ($h^+$) and electron ($e^-$) in the glass matrix. These defect centers are called the *"color centers,"* which absorb light in a spectral region (in particular UV and visible regions) where the glasses do not absorb light and exhibit any coloration. This phenomenon limits the use of glasses in several technological employments. Solarization effect is defined as a color change of the glass under prolonged exposure in sunlight. This is usually observed in window glasses. Actually, the UV part, as indicated above, of the sunlight creates the color and transmission changes. Radiation-shielding glasses that are used in *"hot cells"* (nuclear radiation containment chambers) should have perfect combination of high γ-ray shielding capability and resistance against ionizing irradiation (high long-term light transmittance and discharge resistance). Similarly, the cases of glasses, which are used for shielding of radiations emit from X-ray installations of clinical applications. The $Ce^{3+}$ ion (in reality is a combination of $Ce^{3+}$ and $Ce^{4+}$ ions) is one of the important members of the 4f rare-earth metal ion family as a very efficient inhibitor of color center formation. Consequently, glasses are doped with cerium dioxide ($CeO_2$) for preventing browning of the glasses as a result of high energy radiation-induced ionization to make them radiation-resistant glasses [57–60].

Recently, Fu et al. [58] have reported the detailed studies on radiation-induced color centers in cerium-doped and cerium-free in two multicomponent silicate glasses (K509 and K9). Fig. 3.28A shows the comparison of photographs showing the influence of doses of gamma-ray radiation on extent of color center generation in cerium (Ce)-free (left side) and cerium (Ce)-doped (right side) silicate glasses. The evolution of absorption spectra of cerium (Ce)-doped silicate glasses with doses of gamma-ray radiation is illustrated in Fig. 3.28B. The cerium dioxide ($CeO_2$) produces both the $Ce^{3+}$ and $Ce^{4+}$

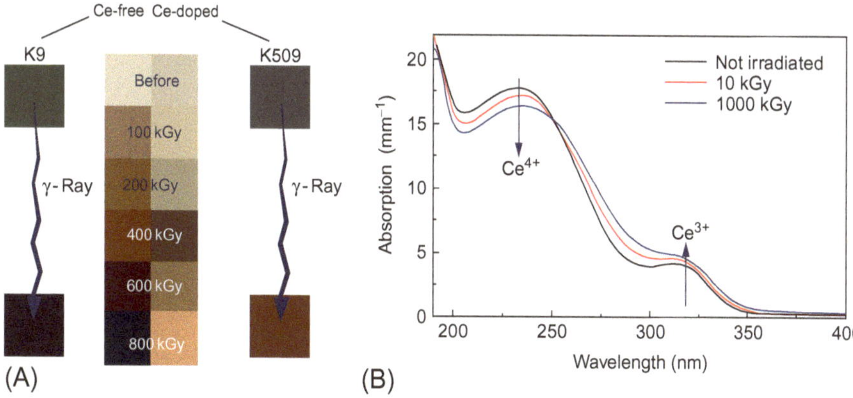


**Fig. 3.28** (A) Comparison of photographs showing the influence of doses of gamma-ray radiation on extent of color center generation in cerium (Ce) free (left side) and cerium (Ce) doped (right side) silicate glasses. (B) Evolution of absorption spectra of cerium (Ce) doped silicate glasses with doses of gamma-ray radiation.
Reproduced with permission from X. Fu,L. Song, J. Li, Radiation induced color centers in cerium-doped and cerium-free multicomponent silicate glasses, J. Rare Earth. 32(2014) 1037–1042, Copyright © Elsevier.

ions in the glass during its melting at high temperature as a result of reduction-oxidation (redox) equilibrium. The absorption band peaking at 328 nm is assigned to 4f → 5d transition of $Ce^{3+}$ ion and the broad band peaking at 240 nm to the charge transfer between $Ce^{4+}$ ion and the surrounding oxygen, that is, the $Ce^{4+}$—O bonds. It is seen that intensity of absorption band at 240 nm due to $Ce^{4+}$ ion decreases while that of $Ce^{3+}$ ion increases with the progress of γ-ray irradiation. This results in the value of the induced spectra (difference absorption) at 240 negative whereas that at 328 nm positive. In this way, both the $Ce^{3+}$ and $Ce^{4+}$ ions stabilize the glasses against high energy radiations (i.e., UV, X-, and γ-rays). In such stabilized glasses, the $Ce^{3+}$ and $Ce^{4+}$ ions of the glass react with both of the electron ($e^-$) and hole ($h^+$) centers in equal numbers, which results in no effect on its change of transmission. These are shown below with the following chemical reactions [61]:

$$\text{Glass} + \gamma - \text{ray} \rightarrow \text{glass} + \text{electron}\,(e^-) + \text{hole}\,(h^+) \tag{3.22}$$

$$Ce^{4+} + \text{electron}\,(e^-) \rightarrow Ce^{3+} \tag{3.23}$$

$$Ce^{3+} + \text{hole}\,(h^+) \rightarrow Ce^{3++} \tag{3.24}$$

### 3.4.4 Wavelength-standard glass

Holmium oxide ($Ho_2O_3$)-doped glass is being used as a wavelength-standard for a period of more than five decades [62,63]. This is widely used for calibration of

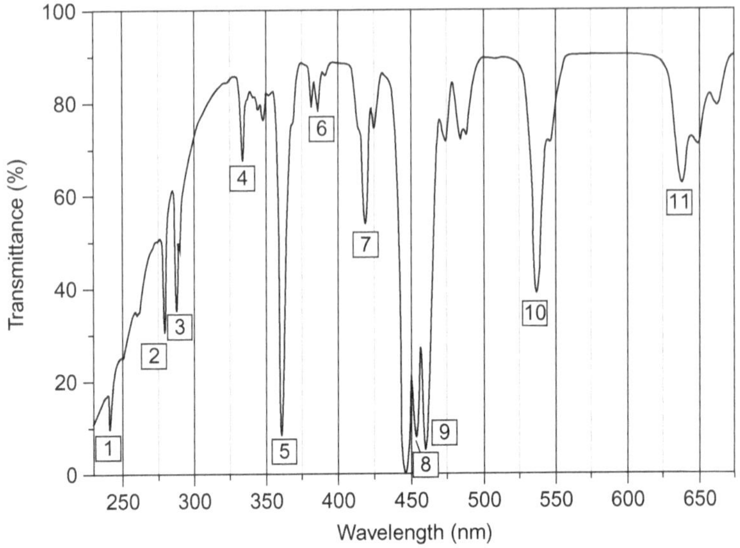


**Fig. 3.29** Transmission spectrum of holmium oxide glass showing eleven National Institute of Standards and Technology (NIST) certified bands.
Courtesy: D. W. Allen, Holmium oxide glass wavelength standards, J. Res. Natl. Inst. Stand. Technol. 112 (2007) 303–306.

ultraviolet (UV) visible near-infrared (NIR) spectrophotometers. This glass was developed by Corning Glass Works, Corning, New York, the United States, as glass filter designated as Corning CS 3-138 (Glass No. 3130) [62,63]. The typical thickness of the glass is 3 mm and the absorption band positions were decided by agreement of several laboratories using different instruments. Eleven absorption bands were selected based on their sharp and symmetrical appearance. Transmission spectrum of the holmium oxide glass showing 11 National Institute of Standards and Technology (NIST), formerly known as the National Bureau of Standards (NBS), certified bands is depicted in Fig. 3.29 [64]. The NIST certified band positions of holmium

Table 3.6 **National Institute of Standards and Technology (NIST) certified band position of holmium oxide glass wavelength-standard [63]**

| Band no. | Wavelength, ±0.2 (nm) | Band no. | Wavelength, ±0.2 (nm) | Band no. | Wavelength, ±0.2 (nm) |
|---|---|---|---|---|---|
| 1 | 241.5 | 5 | 360.8 | 9 | 459.9 |
| 2 | 279.3 | 6 | 385.8 | 10 | 536.4 |
| 3 | 287.6 | 7 | 418.5 | 11 | 637.5 |
| 4 | 333.8 | 8 | 453.4 | | |

oxide glass wavelength-standard are also listed in Table 3.6 [64]. The uncertainty of the certified values is ±0.2 nm. The holmium oxide glass has high mechanical strength and chemical durability. Practically, there is no variation in its properties within the normal range of temperature and humidity.

### 3.4.5 Gamma-ray dosimeter glass

Glass dosimeters are widely used in diversified areas, for instance, personnel and environment monitoring, as battlefield personnel dosimetry, nuclear disaster as accidental dosimetry, and high level radiation and high temperature dosimetry. These are mainly radiophotoluminescence (RPL) dosimeters based on silver-activated phosphate glasses [65]. The X- or γ-ray dose is measured using these RPL glass dosimeters from the correlation between the absorbed dose and RPL intensity. Pedroza et al. [66] have reported the detection of γ-radiation dose using fluorescence intensity measurement of rare-earth (e.g., $Eu^{3+}$) ion-doped silica glass prepared by the sol-gel technique. Photoluminescence spectra of $Eu^{3+}$-doped silica xerogel (A) before and after (B) gamma-ray irradiation are illustrated in Fig. 3.30A [66]. It exhibits two emission bands peaking at 5900 and 6120 Å (590 and 612 nm), which are corresponding to magnetic dipole (MD) $^5D_0 \rightarrow {}^7F_1$ and electric dipole (ED) $^5D_0 \rightarrow {}^7F_2$ transitions, respectively. It is seen that the intensity of the magnetic dipole (MD) $^5D_0 \rightarrow {}^7F_1$ transition enhanced considerably (up to 900%) after γ-ray irradiation compared with that of the electric dipole (ED) $^5D_0 \rightarrow {}^7F_2$ transition. The enhancement is explained as a decrease of nonradiative decay pathway, where these pathways, mainly related to water molecule ($H_2O$) or hydroxyl (O—H) groups, are eliminated by the gamma-ray irradiation.

The intensity of the emission band due to magnetic dipole (MD) $^5D_0 \rightarrow {}^7F_1$ transition at 590 nm is found to be a strong dependence on the gamma radiation dose. Plot

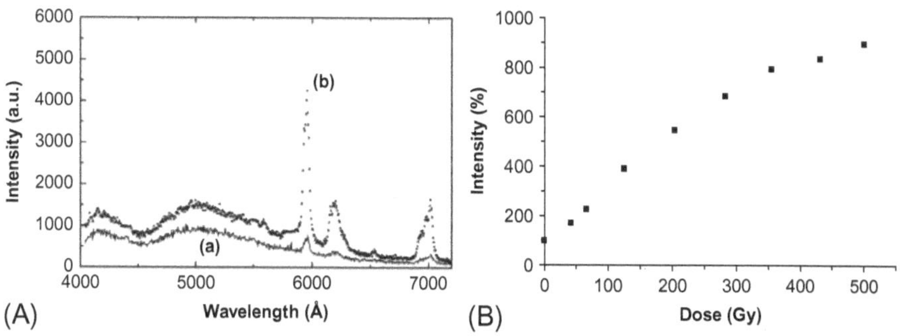


**Fig. 3.30** (A) Photoluminescence spectra of $Eu^{3+}$-doped silica xerogel (a) before and after (b) gamma-ray irradiation. (B) Plot of intensity at 590 nm of $^7D_0 \rightarrow {}^7F_1$ emission transition as a function of dose of gamma-ray irradiation.
Courtesy: G. Pedroza, W.M. de Azevedo, H.J. Khoury, E.F. da Silva Jr., Gamma ray detection using sol-gel glass doped with lanthanide ions, Appl. Radiat. Isotop. 56 (2002) 563–566.

of intensity at 590 nm as a function of dose of gamma-ray irradiation is shown in Fig. 3.30B. It is seen that there is a good correlation of the emission intensity for $\gamma$-ray irradiation doses up to 400 Gy. This relation, therefore, suggests the use of $Eu^{3+}$ ion-doped sol-gel silica glass as a dosimeter glass for measurement of dose of gamma radiation.

### 3.4.6 Colored glasses

The color of glass is the absorption or emission of light by the active ions of its constituents. In Chapter 2, we have seen that the colored glasses have various important applications. There, it is observed that those colored glasses are produced by doping with 3d transition metal ions. Similarly, a variety of colored glasses could also be created by doping with suitable 4f rare-earth metal ions. Absorption and emitted colors of different 4f rare-earth ion-doped glasses are listed in Table 3.7 [1,2,67]. Some ions produce intense-colored glasses (e.g., $Nd^{3+}$), whereas some ions create light-colored glasses (e.g., $Ce^{3+}$, $Er^{3+}$, and $Ho^{3+}$). Compared with 3d transition metal ions, the 4f rare-earth metal ions generally generate light-colored glasses.

**Table 3.7 Absorption and emitted colors of 4f rare-earth ion-doped glasses [1,2,67]**

| 4f configuration | Rare-earth ion | Absorption color | Emitted color on ultraviolet excitation at 360 nm |
|---|---|---|---|
| $4f^0$ | $La^{3+}$ | Colorless | |
| | $Ce^{4+}$ | Light yellow | |
| $4f^1$ | $Ce^{3+}$ | Light yellow | Blue |
| $4f^2$ | $Pr^{3+}$ | Light Green | |
| $4f^4$ | $Nd^{3+}$ | Violet-pink | |
| $4f^5$ | $Sm^{3+}$ | Colorless | Orange |
| $4f^6$ | $Sm^{2+}$ | Green | |
| | $Eu^{3+}$ | Colorless | Orange-red |
| $4f^7$ | $Eu^{2+}$ | Brown | |
| | $Gd^{3+}$ | Colorless | |
| $4f^8$ | $Tb^{3+}$ | Colorless | Blue-green |
| $4f^9$ | $Dy^{3+}$ | Colorless | Yellow |
| $4f^{10}$ | $Dy^{2+}$ | Brown | |
| | $Ho^{3+}$ | Light Yellow | |
| $4f^{11}$ | $Er^{3+}$ | Light pink | |
| $4f^{12}$ | $Tm^{3+}$ | Colorless | Deep blue |
| $4f^{13}$ | $Tm^{2+}$ | Colorless | |
| | $Yb^{3+}$ | Colorless | |
| $4f^{14}$ | $Lu^{3+}$ | Colorless | |

## 3.5 Summary

Here, functionalizations of different oxide glasses by doping with 4f rare-earth metal ions are described. Initially, the publication trend in this area along with total number of published papers on individual rare-earth ion during the last 25 years (1990–2014) is presented. This is followed by a brief description of relevant theories, partial energy level diagram, and absorption peak positions with their transition energy-level assignments of trivalent 4f rare-earth ions in silicate glass and aqueous solution. Absorption and photoluminescence spectra of 12 rare-earth ions (e.g., Ce, Pr, Nd, Sm, Eu, Gd, Tb, Dy, Ho, Er, Tm, and Yb) in different oxide glasses (e.g., silicate, borosilicate, borate, phosphate, fluorophosphates, and antimony oxide) are described in details with spectral evidences and interpretations. Functional applications of these glasses are illustrated elaborately with reference to their uses as laser glass, luminescent glasses for white light emitting diodes (WLEDS), radiation-resistant glasses, wavelength standard glass, gamma-ray dosimeter glass, and colored glasses.

## 3.6 Future outlook

It is seen that the major applications of rare-earth-doped glasses is as glass lasers. The $Nd^{3+}$-doped terawatt ($10^{12}$ W) and petawatt ($10^{15}$ W) laser glasses necessitate operating with higher repetition rates to obtain greater output energy. Hence, the major objectives of future laser glass development will be to improve its thermal shock resistance and other laser, optical, and mechanical properties [45]. It is also necessary to develop new laser glass compositions for the next-generation laser glass of very high energy of exawatt ($10^{18}$ W) ranges [44]. All the laser glasses developed up to now are operating in the infrared wavelengths (see Table 3.2). It is, therefore, an open area to develop new laser glass compositions, which can operate in the visible wavelengths (380–780 nm). Jha et al. [68] have enlightened in this direction with reference to the $Sm^{3+}$- and $Tm^{3+}$-doped glasses. It is witnessed that the rare-earth-doped oxide glasses are used in many functional applications. Consequently, it is expected that further exploration of the fundamental aspects of these glasses will overcome their present drawbacks leading to greatly extended possibilities for novel functional applications.

## Exercises

**3.1** Describe the outer shell electronic configurations and oxidation states of all rare-earth elements.

**3.2** State five factors on which the functional properties of 4f rare-earth-doped glasses depend.

**3.3** Name the rare-earth element on which the highest numbers of papers on 4f rare-earth metal ion-doped glass were published during the period of 1990–2014. What is the number? Explain the reason.

**3.4** What are the mathematical expressions that are required to evaluate the (i) spectral analysis, (ii) oscillator strength, (iii) Judd-Ofelt (J-O) parameters, and (iv) different radiative features of 4f rare-earth-doped glasses?

**3.5** What do you mean by charge transfer transition and band? Explain with an example of 4f rare-earth metal ion-doped glass.
**3.6** What are the differences between the emission spectra of $Eu^{3+}$ ion and $Eu^{2+}$ ion-doped oxide glasses? Explain with spectral profiles, partial energy-level diagrams, and emission transitions.
**3.7** What are the major functional applications of 4f rare-earth metal ion-doped glasses? Give specific examples.
**3.8** There is only one ultraviolet (UV) laser glass. What is that? Describe its different features.
**3.9** Why is Nd-phosphate laser glass so important? Cite its most important use with explanation.
**3.10** What are the different uses of this laser glass? Explain its laser emission mechanism with spectral profile, partial energy-level diagram, and emission transition.
**3.11** What are the different uses of erbium-ytterbium-chromium-cerium-doped phosphate laser glass? Why is it called "*eye-safe*" laser? Explain relevant energy-transfer mechanisms in this laser glass with schematic energy-level diagram.
**3.12** What are the luminescent glasses that are used for white light emitting diodes (WLEDs)? Cite examples of glass with (i) one, (ii) two, and (iii) three 4f rare-earth metal ion(s).
**3.13** Which ion of the 4f rare-earth metal ion family is found to be the most efficient for making glass radiation-resistant? Explain the mechanism of making glass radiation-resistant with schematic spectral profile and relevant chemical reactions.
**3.14** Which glass is used as wavelength-standard and used for calibration of spectrophotometers?
**3.15** Which 4f rare-earth metal ion and glass are used as gamma-ray dosimeter? Which emission transition exhibit this gamma-ray sensor property? Will the melt-route-derived glasses demonstrate gamma-ray sensor property? Explain the reasons.
**3.16** Write the name of various colored glasses that can be produced doping with different 4f rare-earth metal ions.
**3.17** Describe the future outlook of 4f rare-earth metal ion-doped oxide glasses.

# References


[1] M.J. Dejneka, A. Streltsov, S. Pal, A.G. Frutos, C.L. Powell, K. Yost, P.K. Yuen, U. Müller, J. Lahiri, Rare earth-doped glass microbarcodes, PNAS 100 (2003) 389–393.
[2] K. Pátek, J.G. Edwards (Ed.), Glass Lasers, Butterworth & Co (Publishers) Ltd., London, 1970.
[3] G. Fuxi, Optical and Spectroscopic Properties of Glass, Springer-Verlag, Berlin, 1992.
[4] W.T. Carnall, P.R. Fields, K. Rajnak, Electronic energy levels in the trivalent lanthanide aquo ions. I. $Pr^{3+}$, $Nd^{3+}$, $Pm^{3+}$, $Sm^{3+}$, $Dy^{3+}$, $Ho^{3+}$, $Er^{3+}$, and $Tm^{3+}$, J. Chem. Phys. 49 (1968) 4424–4442.
[5] W.T. Carnall, P.R. Fields, K. Rajnak, Electronic energy levels of the trivalent lanthanide aquo ions. IV. $Eu^{3+}$, J. Chem. Phys. 49 (1968) 4450–4455.
[6] W.T. Carnall, P.R. Fields, K. Rajnak, Electronic energy levels of the trivalent lanthanide aquo ions. II. $Gd^{3+}$, J. Chem. Phys. 49 (1968) 4443–4446.
[7] W.T. Carnall, P.R. Fields, K. Rajnak, Electronic energy levels of the trivalent lanthanide aquo ions. III. $Tb^{3+}$, J. Chem. Phys. 49 (1968) 4447–4449.
[8] T. Som, B. Karmakar, Nephelauxetic effect of low phonon antimony oxide glass in absorption and photoluminescence of rare-earth ions, Spectrochim. Acta A 79 (2011) 1766–1782.

[9] B.C. Jamalaiah, J. Suresh Kumar, T. Suhasini, K. Jang, H.S. Lee, H. Choi, L. Rama Moorthy, Visible luminescence characteristics of $Dy^{3+}$-doped LBTAF glasses, J. Alloys Compd. 474 (2009) 382–387.
[10] B.R. Judd, Optical absorption intensities of rare earth ions, Phys. Rev. B 127 (1962) 750–761.
[11] G.S. Ofelt, Intensities of crystal spectra of rare-earth ions, J. Chem. Phys. 37 (1962) 511–520.
[12] W.T. Carnall, P.R. Fields, K. Rajnak, Spectral intensities of the trivalent lanthanides and actinides in solution. II. $Pm^{3+}$, $Sm^{3+}$, $Eu^{3+}$, $Gd^{3+}$, $Tb^{3+}$, $Dy^{3+}$, and $Ho^{3+}$, J. Chem. Phys. 49 (1968) 4412–4423.
[13] T. Murata, M. Sato, H. Yoshida, K. Morinaga, Compositional dependence of ultraviolet fluorescence intensity of $Ce^{3+}$ in silicate, borate, and phosphate glasses, J. Non-Cryst. Solids 351 (2005) 312–316.
[14] H. Takebe, Y. Nageno, K. Morinaga, Compositional dependence of Judd-Ofelt parameters in silicate, borate, and phosphate glasses, J. Am. Ceram. Soc. 78 (1995) 1161–1168.
[15] B. Padlyak, M. Grinberg, B. Kuklinski, Y. Oseledchik, O. Smyrnov, D. Kudryavtcev, A. Prosvirnin, Synthesis and optical spectroscopy of the Eu- and Pr-doped glasses with $SrO\text{-}2B_2O_3$ composition, Opt. Appl. XL (2010) 413–426.
[16] V.M. Martins, D.N. Messias, N.O. Dantas, A.M. Neto, Concentration dependent fluorescence quantum efficiency of neodymium doped phosphate glass matrix, J. Lumin. 130 (2010) 2491–2494.
[17] T. Som, B. Karmakar, Infrared-to-red upconversion luminescence in samarium-doped antimony glasses, J. Lumin. 128 (2008) 1989–1996.
[18] T. Kalpana, M.G. Brik, V. Sudarsan, P. Naresh, V. Ravi Kumar, I.V. Kityk, N. Veeraiah, Influence of $Al^{3+}$ ions on luminescence efficiency of $Eu^{3+}$ ions in barium boro-phosphate glasses, J. Non-Cryst. Solids 419 (2015) 75–81.
[19] A. Herrmann, S. Fibkar, D. Ehrt, Time-resolved fluorescence measurements on $Eu^{3+}$- and $Eu^{2+}$-doped glasses, J. Non-Cryst. Solids 355 (2009) 2093–2101.
[20] H.L. Smith, A.J. Cohen, Absorption spectra of cations in alkali-silicate glasses of high ultra-violet transmission, Phys. Chem. Glasses 4 (1963) 173–187.
[21] T. Ishizaka, Y. Kurokawa, Optical properties of rare-earth ion ($Gd^{3+}$, $Ho^{3+}$, $Pr^{3+}$, $Sm^{3+}$, $Dy^{3+}$ and $Tm^{3+}$)-doped alumina films prepared by sol-gel method, J. Lumin. 92 (2001) 57–63.
[22] C.R. Kesavulu, A.C.A. Silva, M.R. Dousti, N.O. Dantas, A.S.S. deCamargo, T. Catunda, Concentration effect on the spectroscopic behaviour of $Tb^{3+}$ ions in zinc phosphate glasses, J. Lumin. 165 (2015) 77–84.
[23] D.V.R. Murthy, B.C. Jamalaiah, A. Mohan Babu, T. Sasikala, L. Rama Moorthy, The luminescence properties of $Dy^{3+}$-doped alkaline earth titanium phosphate glasses, Opt. Mater. 32 (2010) 1112–1116.
[24] M. Lather, P. Aghamkar, S. Sanghi, Optical absorption and fluorescence spectral analysis of $Ho^{3+}$ ion in zinc bismuth borate glasses, J. Integr. Sci. Technol. 3 (2015) 28–33.
[25] M. Seshadri, L.C. Barbosa, M. Radha, Study on structural, optical and gain properties of 1, 2 and 2.0 μm emission transitions in $Ho^{3+}$ doped tellurite glasses, J. Non-Cryst. Solids 406 (2014) 62–72.
[26] T. Som, B. Karmakar, Efficient green and red fluorescence upconversion in erbium doped new low phonon antimony glasses, Opt. Mater. 31 (2009) 609–618. Fig. 6.
[27] K. Linganna, M. Rathaiah, N. Vijaya, Ch. Basavapoornima, C.K. Jayasankar, S. Ju, W.-T. Han, V. Venkatramu, 1.53 μm luminescence properties of $Er^{3+}$-doped K-Sr-Al phosphate glasses, Ceram. Int. 41 (2015) 5765–5771.

[28] G. Özen, A. Aydinli, S. Cenk, A. Sennaroğlu, Effect of composition on the spontaneous emission probabilities, stimulated emission cross-sections and local environment of $Tm^{3+}$ in $TeO_2$-$WO_3$ glass, J. Lumin. 101 (2003) 293–306.
[29] B. Schaudel, P. Goldner, M. Prassas, F. Auzel, Cooperative luminescence as a probe of clustering in $Yb^{3+}$ doped glasses, J. Alloys Compd. 300–301 (2000) 443–449.
[30] E. Snitzer, Optical maser action of $Nd^{3+}$ in a barium crown glass, Phys. Rev. Lett. 7 (1961) 444–446.
[31] T.C. MacAvoy, M.L. Charters, R.D. Maurer, Development of corning code 0580 laser glass, SCP Solid State Technol. 8 (1965) 23–31.
[32] E. Snitzer, Glass lasers, Appl. Opt. 5 (1966) 1487–1499.
[33] E. Snitzer, Lasers and glass technology, Am. Ceram. Soc. Bull. 52 (1973) 516–525.
[34] Y. Asahara, Glass and optical materials for lasers, J. Ceram. Soc. Jpn. 99 (1991) 903–911.
[35] J.F. Philipps, T. Töpfer, H. Ebendorff-Heidepriem, D. Ehrt, R. Sauerbrey, Spectroscopic and lasing properties of $Er^{3+}$:$Yb^{3+}$-doped fluoride phosphate glasses, Appl. Phys. B 72 (2001) 399–405.
[36] E. Heumann, M. Ledig, D. Ehrt, W. Seeber, E.W. Duczynski, H.J. v. d. Heide, G. Huber, Cw laser action of $Er^{3+}$ in double sensitized fluoroaluminate glass at room temperature, Appl. Phys. Lett. 52 (1988) 255–256.
[37] M. Ledig, E. Heumann, D. Ehrt, W. Seeber, Spectroscopic and laser properties of $Cr^{3+}$ : $Yb^{3+}$ : $Er^{3+}$ :fluoride phosphate glass, Opt. Quant. Electron. 22 (1990) S107–S122.
[38] J.D. Myers, Erbium laser glass composition, US patent 4,962,067, 1990.
[39] LG-770 Phosphate Laser Glass, Schott AG, Mainz, Germany. www.schott.com/advanced_optics.
[40] Hoya Corporation USA, Optics Division, Santa Clara, USA. www.hoyaoptics.com/contact.index.htm.
[41] Kigre, Inc., Marshland Road, Hilton Head, SC, USA. www.kigre.com/glass.html.
[42] J.-P. Tang, L.-L. Hu, S.-B. Chen, B. Wang, Y.-S. Jiang, D.-B. He, J.-Z. Zhang, S.-G. Li, J.-J. Hu, Y.-C. Xu, Properties of phosphate laser glass by continuous melting, Acta Photonica Sin. 37 (1) (2008) 248–251.
[43] J.H. Campbell, T.I. Suratwala, C.B. Thorsness, J.S. Hayden, A.J. Thorne, J.M. Cimino, A. J. Marker III, K. Takeuchi, M. Smolley, G.F. Ficini-Dorn, Continuous melting of phosphate laser glasses, J. Non-Cryst. Solids 263 & 264 (2000) 342–357.
[44] J.S. Hayden, Overcoming technical challenges and moving into the future with laser glass, Int. J. Appl. Glass Sci. 6 (2015) 19–25.
[45] J.H. Campbell, J.S. Hayden, A. Marker, High-power solid-state lasers: a laser glass perspective, Int. J. Appl. Glass Sci. 2 (2011) 3–29.
[46] LG-940 'Eye-Safe' Laser Glass, Schott AG, Mainz, Germany. www.schott.com/advanced_optics.
[47] Z. Liu, C. Qi, S. Dai, Y. Jiang, L. Hu, Spectra and laser properties of $Er^{3+}$, $Yb^{3+}$:phosphate glasses, Opt. Mater. 21 (2003) 789–794.
[48] L.O. Byshevskaya-Konopko, A.A. Izyneev, Y.S. Pavlov, P.I. Sadovskii, On the role of cerium in chromium-ytterbium-erbium phosphate glass, Quantum. Electron. 30 (2000) 767–770.
[49] N. Vijaya, K.U. Kumar, C.K. Jayasankar, $Dy^{3+}$-doped zinc fluorophosphate glasses for white luminescence applications, Spectrochim. Acta A 113 (2013) 145–153.
[50] L.H.C. Andrade, S.M. Lima, M.L. Baesso, A. Novatski, J.H. Rohling, Y. Guyot, G. Boulon, Tunable light emission and similarities with garnet structure of Ce-doped LSCAS glass for white-light devices, J. Alloys Compd. 510 (2012) 54–59.

[51] A.C.P. Rocha, L.H.C. Andrade, M. Lima, A.M. Farias, A.C. Bento, M.L. Baesso, Y. Guyot, G. Boulon, Tunable color temperature of $Ce^{3+}/Eu^{2+,\ 3+}$ codoped low silica aluminosilicate glasses for white Lighting, Opt. Exp. 20 (2012) 10034–10041.
[52] R.F. Wei, H. Zhang, F. Li, H. Guo, Blue-white-green tunable luminescence of $Ce^{3+}$, $Tb^{3+}$ co-doped sodium silicate glasses for white LEDs, J. Am. Ceram. Soc. 95 (2012) 34–36.
[53] X. Liang, Y. Yang, C. Zhu, S. Yuan, G. Chen, A. Pring, F. Xia, Luminescence properties of $Tb^{3+}$–$Sm^{3+}$ codoped glasses for white light emitting diodes, Appl. Phys. Lett. 91 (2007) 091104 (3 pages).
[54] C. Zhu, Y. Yang, X. Liang, S. Yuan, G. Chen, Rare earth ions doped full-color luminescence glasses for white LED, J. Lumin. 126 (2007) 707–710.
[55] C. Zhao, J. Cai, R. Li, S. Tie, X. Wan, J. Shen, White light emission from $Eu^{3+}/Tb^{3+}/Tm^{3+}$ triply-doped aluminoborate glass excited by UV light, J. Non-Cryst. Solids 358 (2012) 604–608.
[56] H. Guo, R.F. Wei, X.Y. Liu, Tunable white luminescence and energy transfer in $(Cu^{+})_2$, $Eu^{3+}$ codoped sodium silicate glasses, Opt. Lett. 37 (2012) 1670–1672.
[57] S. Baccaro, A. Cemmi, I.D. Sarcina, F. Menchini, Gamma rays effects on the optical properties of cerium-doped glasses, Int. J. Appl. Glass Sci. 6 (2015) 295–301.
[58] X. Fu, L. Song, J. Li, Radiation induced color centers in cerium-doped and cerium-free multicomponent silicate glasses, J. Rare Earth 32 (2014) 1037–1042.
[59] J. Du, J. Wu, L. Zhao, L. Song, The influence of calcium oxide on the $^{60}Co$-γ ray irradiation resistance of Ce-doped (alumino-) borosilicate glass, Radiat. Effects Defects Solids 169 (2014) 57–60.
[60] S. Baccaro, A. Cecilia, G. Chen, Y. Du, L. Nencini, S. Wang, Optical transmittance and irradiation resistance of rare-earth ($Ce^{3+}$, $Tb^{3+}$, $Pr^{3+}$) doped heavy germanate glasses, Radiat. Effects Defects Solids 158 (2003) 451–456.
[61] B. Karmakar, S. Mandal, Glass development and production at CSIR-CGCRI for optical applications: some success stories, Sci. Cult. 8 (2015) 325–334.
[62] J.M. Vandenbelt, Holmium filter for checking the wavelength scale of recording spectrophotometers, J. Opt. Soc. Am. 51 (1961) 802–803.
[63] J. McNeirney, W. Slavin, A wavelength standard for ultraviolet-visible-near infrared spectrophotometry, Appl. Opt. 1 (1962) 365–367.
[64] D.W. Allen, Holmium oxide glass wavelength standards, J. Res. Natl. Inst. Stand. Technol. 112 (2007) 303–306.
[65] B. Karmakar, Functionality of reversible glass nanocomposites and their applications, in: B. Karmakar, K. Rademann, A.L. Stepanov (Eds.), Glass Nanocomposites: Synthesis, Properties and Applications, Elsevier, Amsterdam, 2016, pp. 265–278 (Chapter 11).
[66] G. Pedroza, W.M. de Azevedo, H.J. Khoury, E.F. da Silva Jr., Gamma ray detection using sol-gel glass doped with lanthanide ions, Appl. Radiat. Isot. 56 (2002) 563–566.
[67] C.R. Bamford, Control of Colours and Generation in Glass, Elsevier-North Holland Publications, Amsterdam, 1977.
[68] A. Jha, S. Shen, L. Huang, B. Richards, J. Lousteau, Rare-earth doped glass waveguides for visible, near-IR and mid-IR lasers and amplifiers, J. Mater. Sci. Mater. Electron. 18 (2007) S315–S320.

# Functional glass-ceramics

**4**

## Chapter Outline



## 4.1 Introduction

Glasses of appropriate composition can produce one or more crystal phases (regarded as ceramics) on thermal or any other suitable processing action. These crystal phases remain embedded in the residual glass matrix. For this reason, glass-ceramics are defined as *nano- and/or micro-grained polycrystalline solids consisting of embedded crystal phase in residual glass matrix and prepared by controlled crystallization (devitrification or ceramization) of precursor glasses*. Only some specific glasses (not all) with suitable compositions can generate glass-ceramics by the well-controlled heat treatment program following the progressions of nucleation and crystal growth within the glass matrix. These glass-ceramics possess several unique properties that cannot be achieved in any other materials.

Functional glass-ceramics are those glass-ceramics that exclusively function (accomplish purposes) based on their unique properties. These unique properties may be of any category such as optical, physical, mechanical, electric, electronic, and optoelectronic. Some examples of such unique property-correlated applications are as follows. Transluscent and/or tinted low expansion glass-ceramics of the system $Li_2O$–$Al_2O_3$–$SiO_2$ ($TiO_2$, $ZrO_2$, and $P_2O_5$ as nucleants) containing beta-quartz solid solution (β-quartz ss) and/or beta-spodumene solid solution (β-spodumene ss) crystal phase(s) are used as cook oven and work table tops. Transparent varieties of these glass-ceramics are used as astronomical mirror blanks and in laser devices. Machinable glass-ceramics of the system $K_2O$–$MgO$–$Al_2O_3$–$SiO_2$–F containing mica (fluorophlogopite) crystal phase are used as high temperature hermetic seals and in ultrahigh vacuum equipment. Bioactive

Functional Glasses and Glass-Ceramics. http://dx.doi.org/10.1016/B978-0-12-805056-9.00004-0

glass-ceramics of the system $CaO–MgO–SiO_2–P_2O_5–CaF_2$ containing apatite and wollastonite crystal phases are used for replacement of bone tissue. All these examples disclose the importance of functional glass-ceramics in advanced science, technology, and medicine. Various aspects of functional glass-ceramics have been scholarly described in the following recent reviews.

Yasumori [1] has described the preparation of different functional glass-ceramics using stable immiscibility in glasses. Zanotto [2–4], based on his long 36-year research experience on fundamentals of glass crystallization resulting in glass-ceramics, their impressive range of properties, commercial products, and exciting potential applications, has anticipated the bright future for glass-ceramics. Pinckney and Beall [5] have reviewed the microstructural evolution in a wide range of silicate glass-ceramics. In recent times, Rüssel [6] has very nicely reviewed the processing, properties, and applications of various types of glass-ceramics obtained by means of glass crystallization. Pannhorst [7] has reported about the recent developments of commercial applications of low expansion glass-ceramics. Sakamoto and Yamamoto [8] have reviewed the history and typical achievements in the development of glass-ceramics on the basis of the engineering principles of crystallization in glass and its applications. Jain [9] has reviewed the processing, properties, and applications of transparent ferroelectric glass-ceramics in an ample range of glass systems. Komatsu and Honma [10] have reviewed the synthesis and characterization of several transparent glass-ceramics with optical active nanocrystals. Teng et al. [11] have reviewed the photonic glass-ceramics embedded with nanocrystals for applications in optical amplifiers, tunable solid-state lasers, luminescent solar concentrators, upconversion luminescence devices, etc. Chen et al. [12] have reviewed the progresses in transparent glass-ceramics used in white light emitting diodes (WLEDs). Zhou et al. [13] have reviewed the quantum cutting phenomenon in luminescent glasses and glass-ceramics.

An attempt was made using the Web of Science (Version 5.17) as described in Chapter 1 to provide a global publication trend and quantitative intensification in the area of glass-ceramics. Here, year-wise data were extracted for the last 25 years (1990–2014) by searching, using the topic keyword "glass" followed by refining the document type as "article" published in the journals and meetings. It was further refined using the term "glass-ceramics." This plot is shown in Fig. 4.1. It gives us a clear publication trend and quantitative growth of published papers on glass-ceramics during the last 25 years (1990–2014).

A classification and hierarchical relationships among different functional glass-ceramics are represented in Fig. 4.2. All these topics of functional glass-ceramics will be described in the following sections. Before that, synthesis of glass-ceramics is described as follows.

## 4.2 Synthesis of glass-ceramics

Glass-ceramics are prepared from precursor glasses of appropriate compositions by a well-controlled heat treatment program following the progressions of nucleation and crystal growth phenomena within the glass matrix. A schematic representation of a

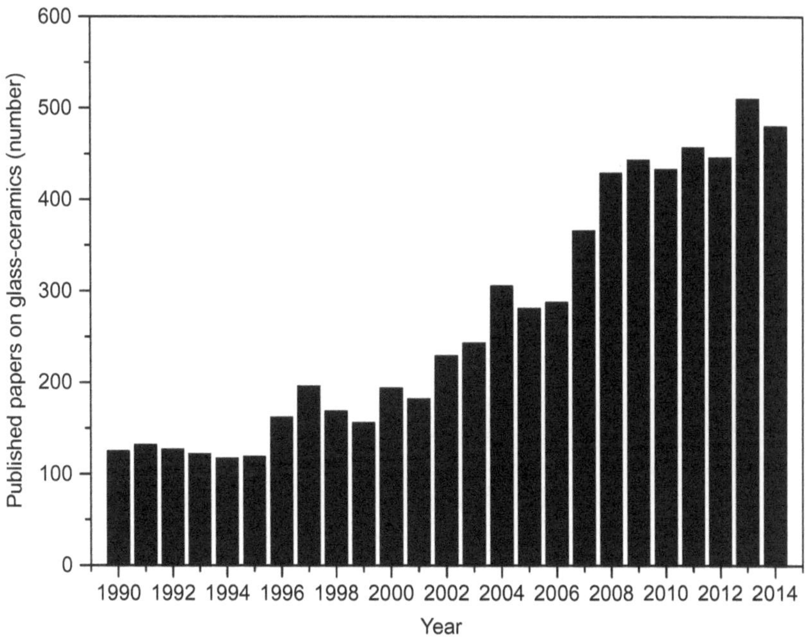


**Fig. 4.1** Year-wise published papers on glass-ceramics during the last 25 years (1990–2014).

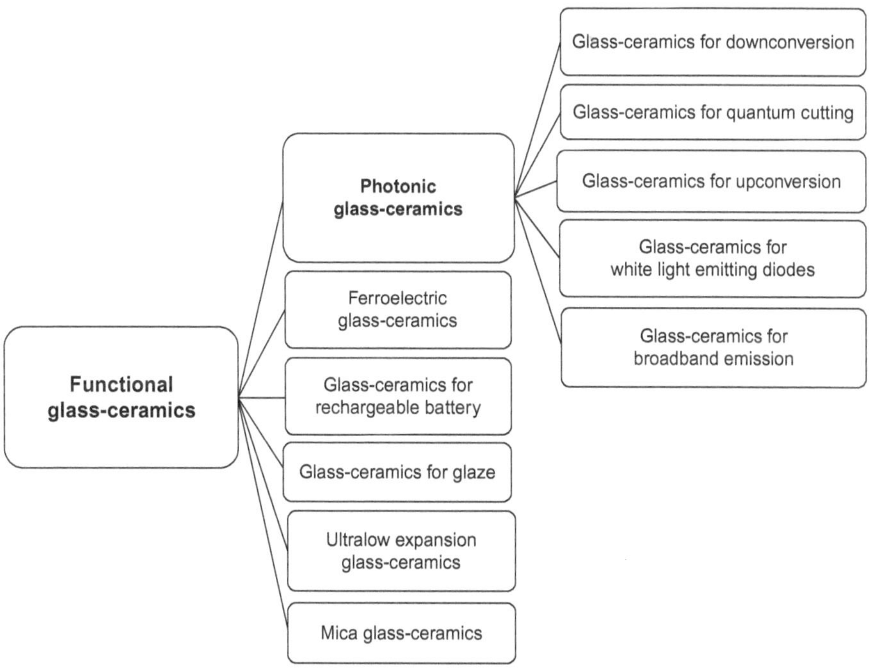


**Fig. 4.2** Classification and hierarchical relationships among different functional glass-ceramics.

typical temperature schedule for the preparation of precursor glass and thereafter glass-ceramics by the nucleation and crystallization processes is shown in Fig. 4.3A. The two-step (nucleation and crystallization) procedure is convenient for obtaining the glass-ceramics, when there is a small overlap of rates of nucleation ($I$) and crystal growth ($U$) curves, that is, the temperatures of maximum nucleation rate ($T_n$) and crystal growth ($T_c$) are far apart. This fact is schematically illustrated in Fig. 4.3B. The rates of nucleation ($I$) and crystal growth ($U$) are given by the following expressions:

$$I = \left(n_v kT / 3\pi\alpha_0^3\eta\right) \exp\left(-W^*/kT\right) \tag{4.1}$$

$$U = \left(kT/3\pi\alpha_0^2\eta\right)\left[1 - \exp\left(\Delta G/kT\right)\right] \tag{4.2}$$

where $n_v$ is the number of formula unit of the crystal per unit volume of the melt, $k$ is the Boltzmann constant, $T$ is the absolute temperature in Kelvin, $\alpha_0$ is the average distance of atomic jump (approximately equal to the interatomic separation distance), $\eta$ is the melt viscosity, $W^*$ is the thermodynamic barrier of nucleation, and $\Delta G$ is the thermodynamic barrier of crystal growth. Eqs. (4.1), (4.2) are used to find out the correlations of the rates of nucleation ($I$) and crystal growth ($U$) with the temperature ($T$) of a glass-forming melt.

The optimum temperatures for these thermal treatments must be experimentally determined by using the differential thermal analyzer (DTA) or differential scanning calorimeter (DSC) curve. A schematic illustration of a typical DTA or DSC curve of a crystallizable glass is depicted in Fig. 4.4A for guidance, where $T_g$, $T_x$, $T_p$, and $T_m$ are the temperatures of the glass transition, onset of crystallization exotherm, exothermic peak of crystallization, and endothermic peak of first melting, respectively.

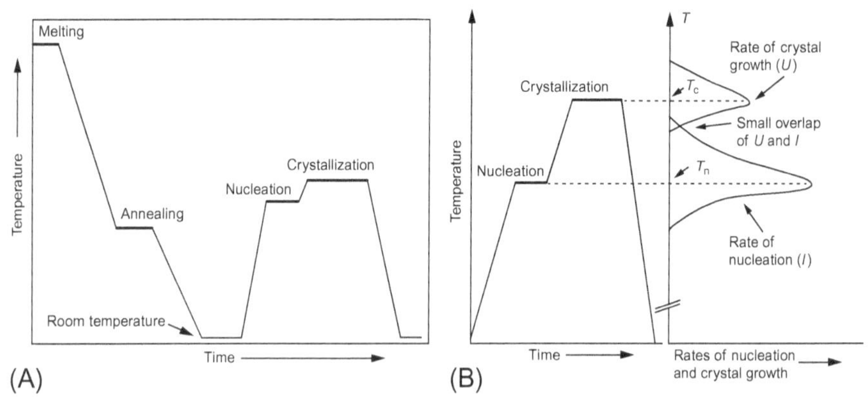


**Fig. 4.3** (A) Schematic representation of temperature schedule for preparation of glass and glass-ceramics. (B) (Left) Temperature schedule for preparation of glass-ceramics by a two-step heat treatment protocol and (right) small overlap of temperature dependence rates of nucleation ($I$) and crystal growth ($U$) ($T$ = temperature, $T_n$ = temperature of maximum nucleation rate, and $T_c$ = temperature of maximum crystallization rate).

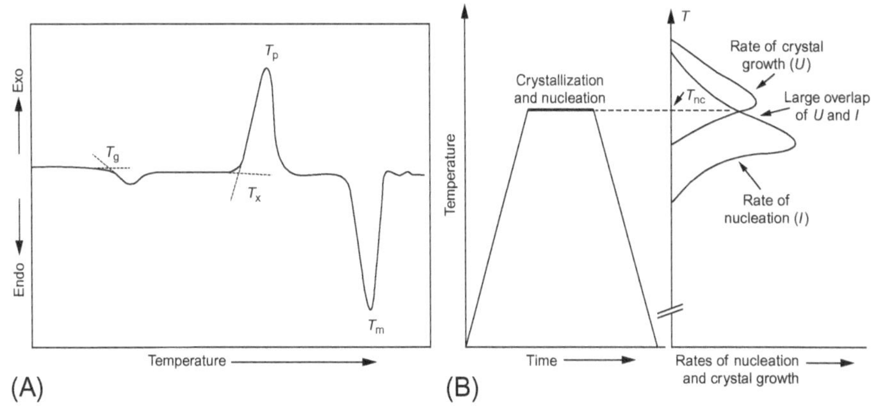


**Fig. 4.4** (A) Schematic illustration of a typical DTA or DSC curve of a crystallizable glass ($T_g$ = glass transition temperature, $T_x$ = onset of crystallization exotherm, $T_p$ = exotherm peak of crystallization, and $T_m$ = endothermic peak of first melting). (B) (Left) Temperature schedule for preparation of glass-ceramics by a single-step heat treatment protocol and (right) large overlap of temperature dependence rates of nucleation ($I$) and crystal growth ($U$) ($T$ = temperature and $T_{nc}$ = single temperature of maximum nucleation and crystallization).

The optimum temperature for nucleation is in a range between $T_g$ and $T_x$ whereas that of crystal growth is in a range between $T_x$ and $T_p$. Whereas the single-step procedure is appropriate for obtaining the glass-ceramics when there is a large overlap of the rates of nucleation ($I$) and crystal growth ($U$) curves, the temperatures of maximum nucleation rate ($T_n$) and crystal growth ($T_c$) are very close. This fact is schematically illustrated in Fig. 4.4B. Under this situation, both the nucleation and crystal growth are carried out at a single temperature, $T_{nc}$. In this case, the difference between the $T_g$ and $T_p$ of the DTA or DSC curve is very small. The most important precondition for obtaining good and fruitful glass-ceramics is the internal or bulk crystallization of the precursor glasses. To accomplish this, the mother glass composition is selected judiciously comprising of nucleating agents such as metallic (e.g., Ag, Au, Cu, and Pt) or nonmetallic (e.g., $TiO_2$, $ZrO_2$, $TiO_2$–$ZrO_2$, $P_2O_5$, $CaF_2$). Further in-depth information on glass-ceramic processes can be found in the books of McMillan [14] and Höland and Beall [15].

## 4.3 Processing, properties, and applications of functional glass-ceramics

### 4.3.1 Photonic glass-ceramics

Glass-ceramics doped with 3d transition and 4f rare-earth metal ions have great potential as a photonic medium. The practical advances of these glass-ceramics have significant roles in the next-generation technologies. Hence, the recent progress of processing, properties, and applications of this type of functional glass-ceramics is described in this section under the following subheadings:

#### 4.3.1.1 Glass-ceramics for downconversion

The term "downconversion" describes optical processes that convert high energy input photons to a lower-energy output photon. It has a number of diversified application areas such as compact solid-state lasers, lightings, displays, high-density optical storage, optical amplification, solar cell, light emitting diodes, optoelectronic devices, biomedical imaging, and other photonic devices. The major advantage of glass-ceramics over their counterpart precursor glasses is that they exhibit enhanced downconversion photoluminescence as a result of partitioning of active 3d transition or 4f rare-earth metal ions into the crystallized low phonon crystal phases. Low phonon energy of crystal suppresses the nonradiative decay processes. However, the enhanced downconversion photoluminescence property of glass-ceramics is largely dependent on several factors such as nature of active dopant ion, glass system, crystal phases, and processing parameters (e.g., crystallization temperature and time). For these reasons, some of these factors and properties of several downconversion glass-ceramics are summarized in Table 4.1 [16–27]. The glass-ceramics may be singly doped or doped by multiple active ions. It is seen that when these glass-ceramics are excited at higher energies (lower wavelengths), they exhibit enhanced downconverted photoluminescence at lower energies (higher wavelengths) of visible and/or near-infrared (NIR) regions. Some of these examples are described below.

Lakshminarayana et al. [16] have reported the enhanced photoluminescence emission properties of $Sm^{3+}$-, $Dy^{3+}$-, and $Tm^{3+}$-doped transparent oxyfluoride silicate glass-ceramics containing low phonon (400 $cm^{-1}$) $CaF_2$ nanocrystals. For detailed information, see Table 4.1. Fig. 4.5 illustrates the emission spectra of (A) 1 mol.% $Sm^{3+}$ (excitation at 402 nm) and (B) 1 mol.% $Dy^{3+}$ (excitation at 386 nm) doped glasses and glass-ceramics [16]. Energy-level diagrams for observed emission transitions are also shown in the respective insets.

Liu et al. [20] have reported the dual valence Eu-doped transparent glass-ceramics containing $LuPO_4$ nanocrystals. For comprehensive information, see Table 4.1. XRD patterns, TEM image, SAED pattern, excitation and emission spectra of precursor glass (PG), and glass-ceramics (GC) heat-treated at different temperatures are shown in Fig. 4.6 [20]. The prominent Stark splitting, low-forced electric dipole $^5D_0 \rightarrow {}^7F_2$ transition, Fig. 4.6B (b), and long decay lifetimes of $Eu^{3+}$ emission for glass-ceramic disclose the incorporation of $Eu^{3+}$ into $LuPO_4$ nanocrystals. The $Eu^{2+}$-related excitation ($\lambda_{em} = 430$ nm) and emission ($\lambda_{ex} = 310$ nm) spectra of PG and glass-ceramics are shown in Fig. 4.6B (c) and (d), respectively. The broad excitation spectrum in the range 250–400 nm, which matches well with emission of UV-LED chips, is ascribed to $4f^7 \rightarrow 4f^65d$ transition of $Eu^{2+}$. A very intense blue emission in the range 380–550 nm is attributed to $4f^65d \rightarrow 4f^7$ transition of $Eu^{2+}$ ion. It is evidenced that the intensity of $Eu^{2+}$ emission increases remarkably with increase in heat treatment temperature. A parallel change is also observed in the excitation spectra. The enhancement of $Eu^{2+}$ emission is caused by the increase in $Eu^{2+}$ quantity as a result of reduction of $Eu^{3+}$ to $Eu^{2+}$ to an increasing extent. When the raw materials were melted in a corundum ($\alpha$-$Al_2O_3$) crucible, the fluorine was evaporated from the melt, leading to the reduction of $Eu^{3+}$ ions according to the following reaction:

$$2F^- + 2Eu^{3+} = F_2\uparrow + 2Eu^{2+} \quad (4.3)$$

**Table 4.1 Dopant ions, glass system, crystal phases, and properties of glass-ceramics for downconversion**

| Rare-earth ion(s) | Glass composition (mol.%) | Major crystal phase in glass-ceramics | Excitation | | Downconverted emission band | | Ref. |
|---|---|---|---|---|---|---|---|
| | | | nm | $cm^{-1}$ | nm | $cm^{-1}$ | |
| $Sm^{3+}$ | $45SiO_2$–$20Al_2O_3$–$10CaO$–$24CaF_2$–$1SmF_3$ | $CaF_2$ | 402 | 24,876 | 562<br>598<br>645 | 17,794<br>16,722<br>15,504 | [16] |
| $Dy^{3+}$ | $45SiO_2$–$20Al_2O_3$–$10CaO$–$24CaF_2$–$1DyF_3$ | $CaF_2$ | 386 | 25,907 | 481<br>575 | 20,790<br>17,391 | [16] |
| $Tm^{3+}$ | $45SiO_2$–$20Al_2O_3$–$10CaO$–$24CaF_2$–$1TmF_3$ | $CaF_2$ | 467 | 21,413 | 650<br>795 | 15,385<br>12,579 | [16] |
| $Eu^{3+} \rightarrow Eu^{2+}$ | $44SiO_2$–$28Al_2O_3$–$17NaF$–$11YF_3$–$xEuF_3$ ($x=0.1$, 0.2, 0.5, and 0.8) | $YF_3$ | 394 | 25,381 | 433 | 23,095 | [17] |
| $Eu^{3+} \rightarrow Eu^{2+}$ | $50SiO_2$–$22Al_2O_3$–$20SrF_2$–$6NaF$–$2EuF_3$ | $SrF_2$ | 394 | 25,381 | 440 | 22,727 | [18] |
| $Eu^{3+} \rightarrow Eu^{2+}$ | $68SiO_2$–$15BaF_2$–$13K_2O$–$2.75\ La_2O_3$–$1.0Sb_2O_3$–$0.25Eu_2O_3$ | $BaF_2$ | 300 | 33,333 | 390 | 25,641 | [19] |
| $Eu^{3+} \rightarrow Eu^{2+}$ | $43SiO_2$–$22Al_2O_3$–$18Na_2O$–$3NaF$–$10LuF_3$–$4P_2O_5$–$1EuF_3$ | $LuPO_4$ | 310 | 32,258 | 430 | 23,256 | [20] |
| $Eu^{3+} \rightarrow Eu^{2+}$ | $48SiO_2$–$24Li_2O$–$16ZnO$–$8Al_2O_3$–$3K_2O$–$1P_2O_5+0.2Eu_2O_3$ | $Li_2ZnSiO_4$ | 350 | 28,571 | 450 | 22,222 | [21] |
| $Eu^{3+} \rightarrow Eu^{2+}$ | $60SiO_2$–$40BaO+0.06Eu_2O_3$ | $Ba_2Si_3O_8$ | 390 | 25,641 | 518 | 19,305 | [22] |
| $(Eu^{3+} \rightarrow Eu^{2+})$–$Gd^{3+}$ | $44SiO_2$–$28Al_2O_3$–$17NaF$–$11GdF_3$ | $GdF_3$ | 274 | 36,496 | 420 | 23,810 | [23] |

*Continued*

**Table 4.1 Continued**

| Rare-earth ion(s) | Glass composition (mol.%) | Major crystal phase in glass-ceramics | Excitation | | Downconverted emission band | | Ref. |
|---|---|---|---|---|---|---|---|
| | | | nm | $cm^{-1}$ | nm | $cm^{-1}$ | |
| $Pr^{3+}$–$Yb^{3+}$ | $40SiO_2$–$30Al_2O_3$–$18Na_2O$–$12LaF_3+0.5PrF_3+(0–5)YbF_3$ | $LaF_3$ | 442 | 22,624 | 525<br>600<br>645<br>715<br>975<br>1025 | 19,048<br>16,667<br>15,504<br>13,986<br>10,256<br>9756 | [24] |
| $Dy^{3+}$–$Tb^{3+}$ | $95SiO_2$–$5LaF_3+0.1Dy^{3+}+(0.1–0.3)Tb^{3+}$ | $LaF_3$ | 390 | 25,641 | 542<br>584<br>621 | 18,450<br>17,123<br>16,103 | [25] |
| $Ho^{3+}$–$Yb^{3+}$ | $60SiO_2$–$16Na_2O$–$12B_2O_3$–$8BaF_2$–$4YbF_3+0.5HoF_3$ | $Ba_2YbO_7$ | 980 | 10,204 | 2050 | 4878 | [26] |
| $Ce^{3+}$–$Er^{3+}$–$Yb^{3+}$ | $50GeO_2$–$40PbO$–$10PbF_2+0.5ErF_3+1YbF_3+0.5CeF_3$ | $\beta$-$PbF_2$ | 975 | 10,256 | 1540 | 6494 | [27] |

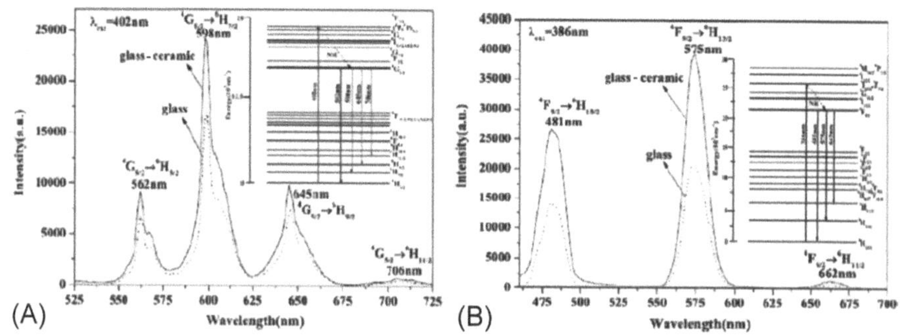


**Fig. 4.5** Emission spectra of (A) 1 mol.% $Sm^{3+}$ (excitation at 402 nm) and (B) 1 mol.% $Dy^{3+}$ (excitation at 386 nm) doped glasses and glass-ceramics. Energy-level diagrams for observed emission transitions are shown in insets.
Reproduced with permission from G. Lakshminarayana, R. Yang, M. Mao, J. Qiu, I.V. Kityk, Photoluminescence of $Sm^{3+}$, $Dy^{3+}$, and $Tm^{3+}$-doped transparent glass ceramics containing $CaF_2$ nanocrystals, J. Non-Cryst. Solids 355 (2009) 2668–2673, Copyright © Elsevier.

Additionally, when glasses were heat-treated in air for crystallization, a larger amount of fluorine might evaporate from the glasses, and then, more $Eu^{2+}$ ions were produced, resulting in enhancement of $Eu^{2+}$ emission after crystallization. The matching broad excitation band in the range 250–400 nm with emission of UV-LED chips reveals the use of this glass-ceramics in producing WLEDs.

Chen et al. [23] have reported structure and optical spectroscopy of Eu-doped and $GdF_3$ nanocrystal containing glass-ceramics of composition $44SiO_2$–$28Al_2O_3$–17NaF–$11GdF_3$ (mol.%). Fig. 4.7A and B exemplify the emission spectra (excited at 274 nm) of as-quenched precursor glass and glass-ceramic (GC670) sample and TEM image of glass-ceramic (GC670) sample, respectively [23]. The TEM image and the selected area electron diffraction (SAED) pattern demonstrate 20–30 nm $GdF_3$ nanocrystals in the glass-ceramic sample (GC670) after heating at 670°C. The emission spectra of as-quenched glass and glass-ceramics (GC670) exhibit the characteristic emissions of $Eu^{2+}$ and $Eu^{3+}$ ions, indicating the existence of the energy transfer from $Gd^{3+}$ to $Eu^{2+}$ and $Eu^{3+}$ ions. By means of 274 nm excitation, $Gd^{3+}$ ions were excited to $^6I_J$ states and then decayed nonradiatively to the $^6P_J$ states of lower energy. This is due to the fact that the $Gd^{3+}$:$^6P_J$, $Eu^{2+}$:5d, and $Eu^{3+}$:$^5H_J$ states are energetically close to each other and the energy transfer from $Gd^{3+}$ to $Eu^{2+}$ or $Eu^{3+}$ is feasible. Subsequently, the $Eu^{2+}$ ions are excited to the 5d state from which the blue emission happens, and the $Eu^{3+}$ ions are excited to the $^5H_J$ states and then decay nonradiatively to the $^5D_J$ states from which the sharp $f \rightarrow f$ emission transition occur. It is also seen that the $Eu^{3+}$ ions mainly partitioned into the $GdF_3$ nanocrystals, whereas the $Eu^{2+}$ ions resided in the glassy phase of the glass-ceramics heat-treated around 670°C.

Xu et al. [24] have reported the efficient near-infrared downconversion in $Pr^{3+}$–$Yb^{3+}$ codoped glasses and glass-ceramics containing $LaF_3$ nanocrystals in a glass of composition (mol.%) $40SiO_2$–$30Al_2O_3$–$18Na_2O$–$12LaF_3$. For wide-ranging information, see Table 4.1. Fig. 4.8 illustrates the (A) XRD patterns and (B) visible and

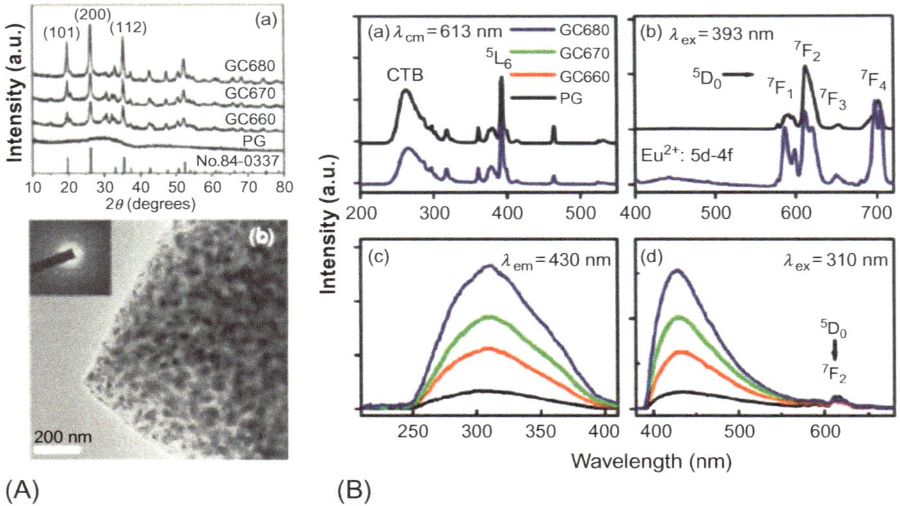


**Fig. 4.6** (A) (a) XRD patterns of precursor glass (PG) and glass-ceramics heat-treated at different temperature. (b) TEM image of glass-ceramic sample GC680. (B) (a) Excitation spectra ($\lambda_{em} = 613$ nm), (b) emission spectra ($\lambda_{ex} = 393$ nm) of PG and GC680, (c) excitation spectra ($\lambda_{em} = 430$ nm), and (d) emission spectra ($\lambda_{ex} = 310$ nm) of PG and glass-ceramics.
Reproduced with permission from X.Y. Liu, Y. Wei, H.Guo, Luminescence of Eu-doped transparent $LuPO_4$ glass-ceramics, J. Am. Ceram. Soc. 96 (2013) 369–371, Copyright © The American Ceramic Society.

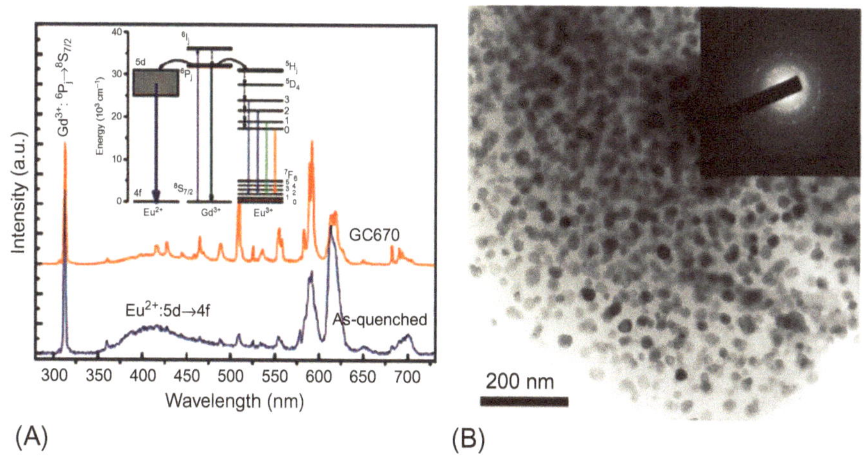


**Fig. 4.7** (A) Emission spectra of as-quenched precursor glass and glass-ceramic (GC670) sample excited at 274 nm; inset illustrates the energy-level diagram showing energy transfer from $Gd^{3+}$ to $Eu^{2+}$ and $Eu^{3+}$ ions. (B) TEM image of glass-ceramic (GC670) sample; inset shows the selected area electron diffraction (SAED) pattern.
Reproduced with permission from D. Chen, Y. Wang, Y. Yu, P. Huang, Structure and optical spectroscopy of Eu-doped glass ceramics containing $GdF_3$ nanocrystals, J. Phys. Chem. C 112 (2008) 18943–18947, Copyright © ACS.

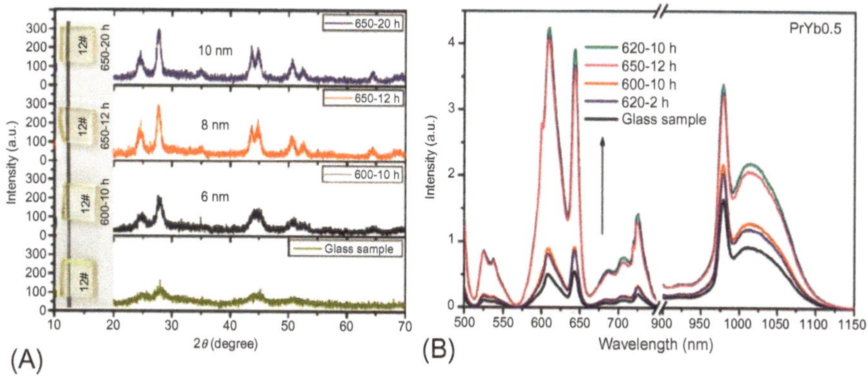


**Fig. 4.8** (A) XRD patterns and (B) visible and near-infrared (NIR) emission spectra of $Pr^{3+}$–$Yb^{3+}$ (0.5 mol.% each) codoped $40SiO_2$–$30Al_2O_3$–$18Na_2O$–$12LaF_3$ (mol.%) glasses under different heat treatment conditions.
Reproduced with permission from Y. Xu, X. Zhang, S. Dai, B. Fan, H. Ma, J. Adam, J. Ren, G. Chen, Efficient near-infrared down-conversion in $Pr^{3+}$–$Yb^{3+}$ codoped glasses and glass ceramics containing $LaF_3$ nanocrystals, J. Phys. Chem. C 115 (2011) 13056–13062, Copyright © ACS.

near-infrared (NIR) emission spectra (excited at 442 nm) of $Pr^{3+}$–$Yb^{3+}$ (0.5 mol.% each) codoped glasses under different heat treatment conditions. Following the Scherrer equation, the size of the $LaF_3$ crystals varies from 6 to 12 nm in the transparent glass-ceramics. The hexagonal $LaF_3$ crystals are precipitated more easily from the glass

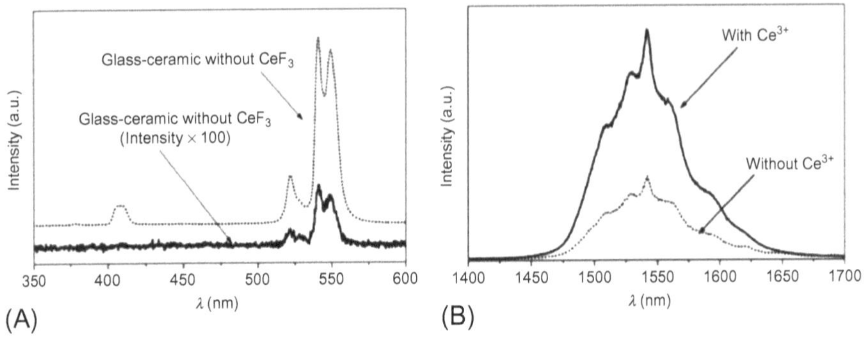


**Fig. 4.9** (A) Upconversion ($\lambda_{exc}=920$ nm) and (B) downconversion emission ($\lambda_{exc}=975$ nm) spectra of $50GeO_2$–40PbO–$10PbF_2+0.5ErF_3+1YbF_3$ (mol.%) glass-ceramics with *(solid line)* and without *(dotted line)* $0.5CeF_3$.
Reproduced with permission from G. Dantelle, M. Mortier, D. Vivien, G. Patriarche, Effect of $CeF_3$ addition on the nucleation and up-conversion luminescence in transparent oxyfluoride glass-ceramics, Chem. Mater. 17 (2005) 2216–2222, Copyright © ACS.

matrix as a consequence of the hexagonal phase that requires smaller interstitial space for its dense packing structure. Therefore, the $Pr^{3+}$ ion ($PrF_3$ has the same hexagonal crystal lattice) predominantly enters into the $LaF_3$ crystals compared with the $Yb^{3+}$ ion that has dissimilar crystal lattice ($YbF_3$ has the orthorhombic crystal structure). For this reason, $Pr^{3+}$ ion exhibits enhanced photoluminescence in the visible region, whereas the energy transfer efficiency between $Pr^{3+}$ and $Yb^{3+}$ ions cannot be enhanced.

Dantelle et al. [27] have reported the effect of $CeF_3$ addition on up- and down-conversion luminescence of $Er^{3+}$ ion in $Er^{3+}$–$Yb^{3+}$ codoped transparent oxyfluoride glass-ceramics of composition (mol.%) $50GeO_2$–40PbO–$10PbF_2+0.5ErF_3+1YbF_3$. For more information, see Table 4.1. Fig. 4.9 shows the (A) upconversion ($\lambda_{exc}=920$ nm) and (B) downconversion emission ($\lambda_{exc}=975$ nm) spectra of glass-ceramics with (solid line) and without (dotted line) $0.5CeF_3$. It is seen that the addition of $CeF_3$ dramatically reduces the upconversion emission intensity around 545 nm, whereas the downconversion emission intensity around 1.54 μm increases about three times higher than that of the glass-ceramic without $CeF_3$. This happened due to the cross relaxation occurring between $Er^{3+}$ and $Ce^{3+}$ ions. It gives the impression that the synthesis of glass-ceramics tridoped with $ErF_3$, $YbF_3$, and $CeF_3$ is very promising for phonic applications around 1.54 μm.

#### 4.3.1.2 Glass-ceramics for quantum cutting

In 1957, Dexter [28] first proposed that under suitable conditions, the quantum efficiency above 100% could be achieved by creating multiple photons through "cutting" a single photon of higher energy into two or more photons of lower energy. This process is known as "quantum cutting." In fact, this is a distinctive case of downconversion process. The quantum cutting can be grasped by photon cascade emission from a single rare-earth (RE) ion (e.g., $Pr^{3+}$ or $Gd^{3+}$) or downconversion via energy transfer (ET) through various combinations of rare-earth (RE) ions (e.g., $Pr^{3+}$–$Yb^{3+}$, $Tb^{3+}$–$Yb^{3+}$, $Ho^{3+}$–$Yb^{3+}$, and $Tm^{3+}$–$Yb^{3+}$). Various quantum cutting (downconversion) mechanisms are shown using two different rare-earth ions (RE1 and RE2) illustrated in Fig. 4.10. Fig. 4.10A shows the quantum cutting on a single rare-earth ion (RE1) by sequential emission of two photons, and Fig. 4.10B–D depicts the quantum cutting as a result of energy transfer (ET) from rare-earth ion 1 (RE1) to rare-earth ion 2 (RE2), indicated by ET1 and ET2. Fig. 4.10E shows quantum cutting by means of cooperative energy transfer (CET) from RE1 to RE2. In 1999, Wegh et al. [29] were the first to demonstrate successfully the quantum cutting downconversion in $LiGdF_4:Eu^{3+}$ phosphor with quantum efficiency close to 200% in the orange or red spectral region. In recent times, the importance of this type of glass-ceramics has gained much attention for their promising applications in plasma display panels (PDPs), mercury-free fluorescent lighting technologies, and silicon-based solar cell technologies. Table 4.2 summarizes the rare-earth ion combinations, glass systems, crystal phases, excitation wavelengths, and quantum efficiencies of various glass-ceramics established for quantum cutting downconversion process [30–39]. Some are described as follows.

Ye et al. [30] have reported the spectral modification and quantum cutting in $RE^{3+}$–$Yb^{3+}$ (RE = Pr, Tm, and Tb) codoped transparent glass-ceramics containing $CaF_2$ nanocrystals in the glass composition (mol.%) $60SiO_2$–$20Al_2O_3$–$20CaF_2$+$0.4REF_3$–+$1YbF_3$. Details are given in Table 4.2. Fig. 4.11A (left side) shows the excitation spectra of $Pr^{3+}$ 605 nm emission monitored in 0.4 mol.% $Pr^{3+}$ single-ion doped glass-ceramics (black spectrum) and $Yb^{3+}$ 980 nm emission monitored in 0.4 mol.% $Pr^{3+}$–1.0 mol.% $Yb^{3+}$ codoped glass-ceramics (red spectrum) and (right side) emission spectra of 0.4 mol.% $Pr^{3+}$ and 0.4 mol.% $Pr^{3+}$–1.0 mol.% $Yb^{3+}$ under excitation at 482 nm (green spectrum and blue spectrum, respectively).

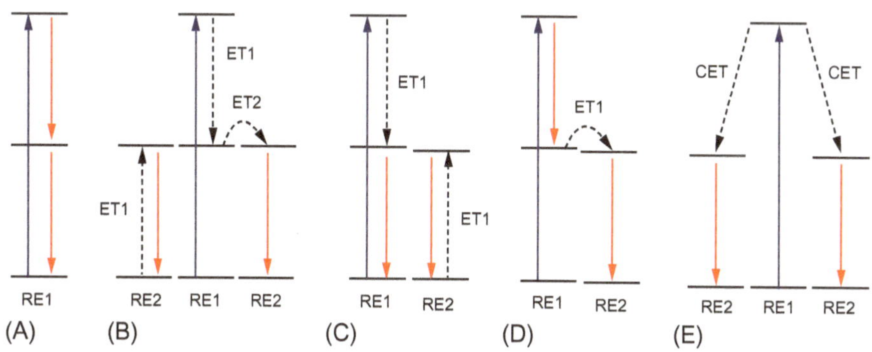


**Fig. 4.10** Various quantum cutting (downconversion) mechanisms are shown using two different rare-earth ions. (A) Quantum cutting on a single rare-earth ion (RE1) by sequential emission of two photons. (B–D) Quantum cutting as a result of energy transfer (ET) from rare-earth ion 1 (RE1) to rare-earth ion 2 (RE2), indicated by ET1 and ET2. (E) Quantum cutting by means of cooperative energy transfer (CET) from RE1 to RE2.

**Table 4.2 Rare-earth ion combinations, glass systems, crystal phases, excitation wavelengths, and quantum efficiencies of various glass-ceramics established for quantum cutting downconversion**

| Rare-earth ion combination | Glass system/glass composition (mol.%) | Major crystal phase in glass-ceramics | Excitation (nm) | Maximum quantum efficiency (%) | Ref. |
|---|---|---|---|---|---|
| $Pr^{3+}$–$Yb^{3+}$ | $60SiO_2$–$20Al_2O_3$–$20CaF_2+0.4PrF_3+1YbF_3$ | $CaF_2$ | 482 | 184 | [30] |
| $Pr^{3+}$–$Yb^{3+}$ | $55SiO_2$–$5Na_2O$–$20Al_2O_3$–$20CaF_2+0.5Pr^{3+}+xYb^{3+}$ ($x=0$, 2, 4, 8, 16) | $CaF_2$ | 482 | 191 | [31] |
| $Pr^{3+}$–$Yb^{3+}$ | $SiO_2$–$Al_2O_3$–NaF–$YF_3+0.1PrF_3+xYbF_3$ ($x=0$, 0.2, 0.5, 1.0, 1.5) | β-$YF_3$ | 482 | 200 | [32] |
| $Tb^{3+}$–$Yb^{3+}$ | $60SiO_2$–$20Al_2O_3$–$20CaF_2+0.4TbF_3+4YbF_3$ | $CaF_2$ | 484 | 152 | [30] |
| $Tb^{3+}$–$Yb^{3+}$ | $23Li_2O$–$(16-x)LaF_3$–$6Al_2O_3$–$55SiO_2$–$1TbF_3$–$xYbF_3$ ($x=1$, 2, 4, 8) | $LaF_3$ | 488 | – | [33] |
| $Tb^{3+}$–$Yb^{3+}$ | $(44-x)SiO_2$–$28Al_2O_3$–17NaF–$11TbF_3$–$xYbF_3$ ($x=0$, 0.25, 0.5, 1.0, 1.5, 3.0) | $TbF_3$ | 485 | – | [34] |
| $Tb^{3+}$–$Yb^{3+}$ | $58SiO_2$–$15B_2O_3$–$15Na_2O$–$12BaF_2$–$xYbF_3$–$0.5TbF_3$ ($x=0$, 1, 2, 4, 6) | $Ba_{2+1.5x}Yb_{1-x}F_7$ | 485 | – | [35] |
| $Ho^{3+}$–$Yb^{3+}$ | $SiO_2$–$Al_2O_3$–NaF–$YF_3+0.1HoF_3+1.0YbF_3$ | $YF_3$ | 488 | 159 | [36] |
| $Ho^{3+}$–$Yb^{3+}$ | $75TeO_2$–20ZnO–$5Na_2O+0.5Ho_2O_3+(0$–$20)Yb_2O_3$ | $TeO_2$ (cubic) | 360 | 166 | [37] |
| $Tm^{3+}$–$Yb^{3+}$ | $60SiO_2$–$20Al_2O_3$–$20CaF_2+0.4TmF_3+4YbF_3$ | $CaF_2$ | 468 | 183 | [30] |
| $Tm^{3+}$–$Yb^{3+}$ | $45SiO_2$–$12Na_2O$–$23Al_2O_3$–$20LaF_3$–$0.5TmF_3$–$xYbF_3$ ($x=0$, 4, 8) | $LaF_3$ | 468 | 162 | [38] |
| $Tm^{3+}$–$Yb^{3+}$ | $45SiO_2$–$25Al_2O_3$–$10Na_2CO_3$–$18CaF_2$–$2CaCO_3+0.5Tm^{3+}+1Yb^{3+}$ | $CaF_2$ (cubic) | 468 | 163 | [39] |

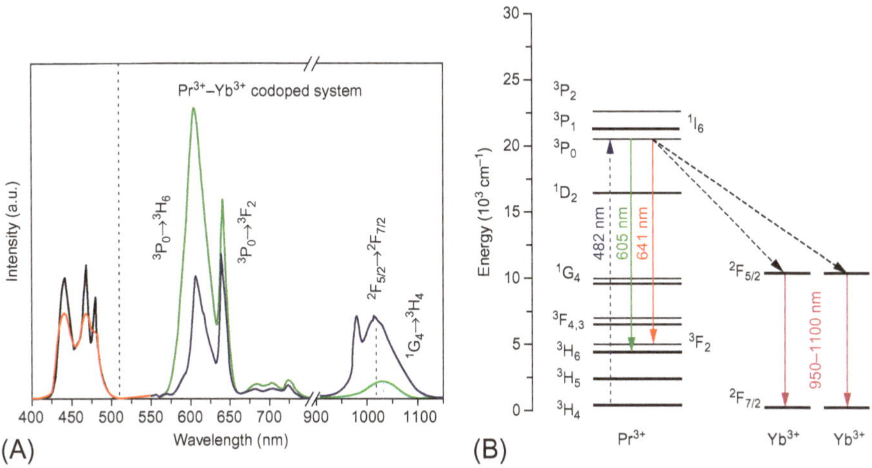


**Fig. 4.11** (A) Left hand side, excitation spectra of $Pr^{3+}$ 605 nm emission monitored in 0.4 mol.% $Pr^{3+}$ single-ion-doped glass-ceramics *(black spectrum)* and $Yb^{3+}$ 980 nm emission monitored in 0.4 mol.% $Pr^{3+}$, 1.0 mol.% $Yb^{3+}$, codoped glass-ceramics *(red spectrum)*. Right hand side, emission spectra of 0.4 mol.% $Pr^{3+}$ and 0.4 mol.% $Pr^{3+}$, 1.0 mol.% $Yb^{3+}$ under excitation at 482 nm (*green spectrum* and *blue spectrum*, respectively). (B) Schematic energy-level diagram of $Pr^{3+}$ and $Yb^{3+}$ showing the transition responsible for cooperative energy transfer.
Reproduced with permission from S. Ye, B, Zhu, Y.X. Zhuang, X.F. Liu, J. Luo, L. Wang, J.R. Qiu, Spectral modification and quantum cutting in $RE^{3+}$–$Yb^{3+}$ (R = Pr, Tm, and Tb) codoped transparent glass-ceramics containing $CaF_2$ nanocrystals, IOP Conf. Ser. Mater. Sci. Eng. 1 (2009) 012008 (7 pages), Copyright © IOP Science.

It is seen from the excitation and emission spectra as shown in Fig. 4.11A that the excitation bands peaking at 482, 468, and 441 nm are assigned to the $^3H_4 \rightarrow {}^3P_j$ ($j = 0, 1,$ and 2) transitions of $Pr^{3+}$ ion. These bands were identified by monitoring both the $^3P_0 \rightarrow {}^3H_6$ transition of $Pr^{3+}$ at 605 nm and the $^2F_{5/2} \rightarrow {}^2F_{7/2}$ transition of $Yb^{3+}$ ion at 980 nm, which points toward the existence of energy transfer (ET) from $Pr^{3+}$ ion to $Yb^{3+}$ ion. In addition, in the $Pr^{3+}$–$Yb^{3+}$ codoped glass-ceramics, the observation of $Yb^{3+}$ infrared emission and the decrease of $^3P_0 \rightarrow {}^3H_6$ and $^3P_0 \rightarrow {}^3F_2$ transition emission intensities of $Pr^{3+}$ under 482 nm excitation demonstrates the existence of energy transfer (ET). Incidentally, the weak infrared emission at 1030 nm in $Pr^{3+}$ single-doped glass-ceramics originated from $Pr^{3+}$ $^1G_4 \rightarrow {}^3H_4$ transition. As it was proved experimentally that the second-order process is responsible for the upconversion luminescence in the $Yb^{3+}$–$Pr^{3+}$ codoped system, the cooperative downconversion may also happen, and two infrared photons emitted by $Yb^{3+}$ could be obtained through on visible photon absorbed by $Pr^{3+}$ ion. Fig. 4.11B displays the schematic energy-level diagram of $Pr^{3+}$ and $Yb^{3+}$ showing the transition responsible for cooperative energy transfer.

Yu et al. [39] have reported the three-photon near-infrared quantum cutting in $Tm^{3+}$-doped transparent oxyfluoride glass-ceramics yielding cubic $CaF_2$ nanocrystals in the composition (mol.%) $45SiO_2$–$25Al_2O_3$–$10Na_2CO_3$–$18CaF_2$–$2CaCO_3$ + $0.5Tm^{3+}$ + $1Yb^{3+}$. Details are given in Table 4.2. Fig. 4.12A shows the near-infrared (NIR) photoluminescence (PL) spectra of glass-ceramic doped with $Tm^{3+}$ (GC:$Tm^{3+}$) and

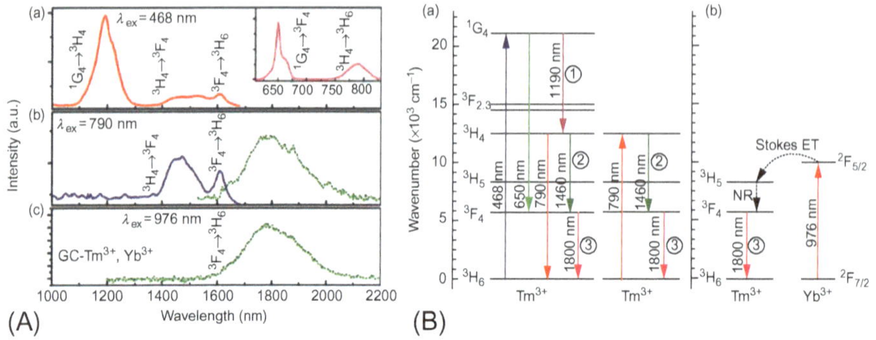


**Fig. 4.12** (A) Near-infrared (NIR) photoluminescence spectra of glass-ceramic doped with $Tm^{3+}$ (GC:$Tm^{3+}$) and $Tm^{3+}$–$Yb^{3+}$ codoped (GC:$Tm^{3+}$-$Yb^{3+}$). *Solid-line spectra* were recorded under xenon lamp excitation at 468 nm (a) and (b). The inset of (a) shows visible photoluminescence spectrum of GC:$Tm^{3+}$ under xenon lamp excitation at 468 nm. *Dotted-line spectra* were recorded under laser diode excitation at 808 nm (b) and 976 nm (c). (B) Schematic energy-level diagram of three-step sequential three-photon NIR-QC (a) via 468 and 790 nm directly exciting $Tm^{3+}$ in GC:$Tm^{3+}$, respectively, and (b) with 976 nm laser diode excitation of $Yb^{3+}$ to excite $Tm^{3+}$ through a Stokes energy transfer (ET) in GC:$Tm^{3+}$-$Yb^{3+}$.
Reproduced with permission from D. C. Yu, J.P. Zhang, Q.J. Chen, W.J. Zhang, Z.M. Yang, Q.Y. Zhang, Three-photon near-infrared quantum cutting in $Tm^{3+}$-doped transparent oxyfluoride glass ceramics, Appl. Phys. Lett. 101 (2012) 171108 (4 pages), Copyright © AIP.

$Tm^{3+}$–$Yb^{3+}$ codoped (GC:$Tm^{3+}$–$Yb^{3+}$). Solid-line spectra were recorded under xenon lamp excitation at 468 nm (A) and (B). The inset of (A) shows visible photoluminescence spectrum of GC:$Tm^{3+}$ under xenon lamp excitation at 468 nm. Dotted-line spectra were recorded under laser diode excitation at 808 nm (B) and 976 nm (C).

On excitation at 468 nm (Fig. 4.12A (a)), the GC:$Tm^{3+}$ sample not only yields typical visible emissions at 650 nm ($^1G_4 \rightarrow {}^3F_4$) and 790 nm ($^3H_4 \rightarrow {}^3H_6$) of $Tm^{3+}$ (the inset of Fig. 4.12A (a)) but also produces NIR PL bands peaking at 1190, 1460, and 1612 nm, which easily assigned to the electronic transitions of $^1G_4 \rightarrow {}^3H_4$, $^3H_4 \rightarrow {}^3F_4$, and $^3F_4 \rightarrow {}^3H_6$ of $Tm^{3+}$ ion, respectively. In order to investigating the NIR fluorescence in detail, additional NIR PL spectra of GC:$Tm^{3+}$ and GC:$Tm^{3+}$–$Yb^{3+}$ were recorded under excitation at 790, 808, and 976 nm by laser diodes (LDs), respectively. As GC:$Tm^{3+}$ is excited at 790 nm (Fig. 4.12A (b)), the 1460 and 1612 nm NIR PL peaks can be obtained due to the transitions of $^3H_4 \rightarrow {}^3F_4$ and $^3F_4 \rightarrow {}^3H_6$ of $Tm^{3+}$, respectively, while, as shown in Fig. 4.12A (c), the NIR transition of $^3F_4 \rightarrow {}^3H_6$ around 1800 nm can be observed from $Tm^{3+}$ ions excited efficiently via an effective Stokes energy transfer (ET) from $Yb^{3+}$ sensitizer ions pumped at 976 nm by LD. For comparison, PL spectra were recorded also with a PbSe photoconductor with optimal spectral response in this wavelength region. These are shown in Fig. 4.12A (b) and (c).

Fig. 4.12B shows the schematic energy-level diagram of three-step sequential three-photon NIR-QC (a) via 468 and 790 nm directly exciting $Tm^{3+}$ in GC:$Tm^{3+}$, respectively, and (b) with 976 nm laser diode excitation of $Yb^{3+}$ to excite $Tm^{3+}$ through a Stokes energy transfer (ET) in GC:$Tm^{3+}$–$Yb^{3+}$.

### 4.3.1.3 Glass-ceramics for upconversion

The "upconversion" process is just reverse of the "downconversion" process. The term "upconversion" defines the nonlinear anti-Stokes optical processes that convert two or more low energy photons to a higher energy photon. It was discovered by Auzel in the 1960s [40]. The upconversion process is used in many advanced technological applications such as compact solid-state lasers, silicon-based solar cells, lightings, displays, temperature sensors, quantum counter detectors, biomedical imaging, and security. Different mechanisms of upconversion process are illustrated in Fig. 4.13. In excited state absorption (ESA) upconversion mechanism, two low energy photons excite an ion into a higher energy level, and a single high energy photon is emitted when the ion transitions to the ground state, whereas in energy transfer upconversion (ETU) mechanism, an energy transfer between two rare-earth ions is terminating in the release of high energy photons. In photon avalanche (PA) upconversion mechanism, a combination of ESA and ETU mechanisms causes terminating in the release of some high energy photons. All the ions of d- and f-block elements do not exhibit upconversion process. Some examples of ions of the d-block elements which exhibit upconversion process in solids are $Ti^{2+}$ ($3d^2$), $Cr^{3+}$($3d^3$), $Ni^{2+}$($3d^8$), $Mn^{2+}$($3d^5$), $Mo^{3+}$($4d^3$), $Re^{4+}$($5d^3$), and $Os^{4+}$($5d^4$). Examples of the ions of the f-block elements are $Pr^{3+}$($4f^2$), $Nd^{3+}$($4f^3$), $Gd^{3+}$($4f^7$), $Dy^{3+}$($4f^9$), $Ho^{3+}$ ($4f^{10}$), $Er^{3+}$($4f^{11}$), $Tm^{3+}$($4f^{12}$), $Tm^{2+}$($4f^{13}$), $U^{4+}$($5f^2$), and $U^{3+}$($5f^5$) [40]. Upconversion emission processes of these ions or their combinations in glass-ceramics are largely dependent on glass chemical composition, processing parameters, crystal phases, excitation wavelength, etc. For these reasons, rare-earth ions, glass compositions, crystal phases, excitation wavelengths, and upconverted emission band peak positions of various glass-ceramics established for upconversion are summarized in Table 4.3 [41–82]. Some of these examples are briefly described as follows.

Chen et al. [51] have reported the influences of $Er^{3+}$ content on structure and upconversion emission of oxyfluoride glass-ceramics containing $CaF_2$ nanocrystals in the glass composition (mol.%) $45SiO_2–25Al_2O_3–5CaO–10NaF–15CaF_2+xErF_3$

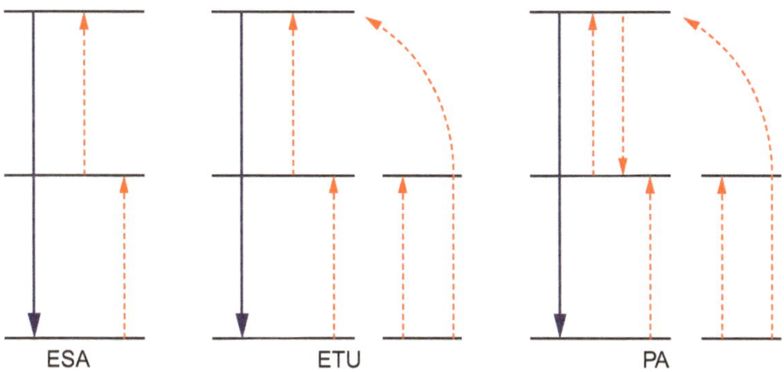


**Fig. 4.13** Different mechanisms of upconversion (*ESA*, excited state absorption; *ETU*, energy transfer upconversion; *PA*, photon avalanche).

**Table 4.3 Rare-earth ions, glass compositions, crystal phases, excitation wavelengths, and upconverted emission band peak positions of various glass-ceramics established for upconversion**

| Rare-earth ion(s) | Glass system/glass composition (mol.%) | Major crystal phase in glass-ceramics | Excitation | | Main upconverted emission band | | Ref. |
|---|---|---|---|---|---|---|---|
| | | | nm | $cm^{-1}$ | nm | $cm^{-1}$ | |
| $Pr^{3+}$ | $50.52SiO_2$–$11.09Na_2O$–$29.26Al_2O_3$–$9.09LaF_3$+(0.1–0.5)$PrF_3$ | $LaF_3$ | 591.6 | 16,903 | 479.1 | 20,872 | [41] |
| $Nd^{3+}$ | $30SiO_2$–$15Al_2O_3$–$29CdF_2$–$22PbF_2$–$3YF_3$–$1NdF_3$ | β-$PbF_2$ | 806.4 | 12,008 | 589.5 | 16,964 | [42] |
| $Nd^{3+}$ | $44SiO_2$–$28Al_2O_3$–17LiF–$11YF_3$+$0.1NdF_3$ | β-YF3 | 796 | 12,563 | 382 | 26,178 | [43] |
| $Ho^{3+}$ | $60PbGeO_3$–$10PbF_2$–$30CdF_2$+(0.1–1.0)$Ho^{3+}$ | β-$PbF_2$ | 980 | 10,204 | 650 | 15,385 | [44] |
| $Er^{3+}$ | $41.2SiO_2$–$29.4Al_2O_3$–$17.6Na_2CO_3$–$11.8LaF_3$–$xErF_3$ ($x$=0.05, 0.1, 0.5, 1, 4) | $LaF_3$ | 976 | 10,246 | 540 | 18,519 | [45] |
| $Er^{3+}$ | $51SiO_2$–$20Al_2O_3$–$15Na_2O$–$14LaF_3$–$0.07ErF_3$ | $LaF_3$ | 980 | 10,204 | 540 | 18,519 | [46] |
| $Er^{3+}$ | $95SiO_2$–$5LaF_3$+$0.1Er^{3+}$ (sol-gel) | $LaF_3$ | 980 | 10,204 | 540 | 18,519 | [47] |
| $Er^{3+}$ | $63SiO_2$–$15B_2O_3$–$16Na_2O$–$5.5LaF_3$–$0.5ErF_3$ | LaOF | 980 | 10,204 | 540 | 18,519 | [48] |
| $Er^{3+}$ | $60SiO_2$–$17ZnF_2$–$20BaF_2$–$3ErF_3$ | $BaF_2$ | 980 | 10,204 | 651 | 15,361 | [49] |
| $Er^{3+}$ | $xErF_3$–$5BaF_2$–$95SiO_2$ ($x$=0, 0.5, 1.0) (sol-gel) | $BaF_2$ | 980 | 10,204 | 545 | 18,349 | [50] |
| $Er^{3+}$ | $45SiO_2$–$25Al_2O_3$–5CaO–10NaF–$15CaF_2$+$xErF_3$ ($x$=0, 0.1, 0.5, 1, 2) | $CaF_2$ | 980 | 10,204 | 550 | 18,182 | [51] |
| $Er^{3+}$ | $45SiO_2$–$25Al_2O_3$–$5CaCO_3$–10NaF–$15CaF_2$–$0.5ErF_3$ | $CaF_2$ | 980 | 10,204 | 660 | 15,152 | [52] |
| $Er^{3+}$ | $50SiO_2$–$20Al_2O_3$–$30CaF_2$–$5ErF_3$ | $CaF_2$ | 980 | 10,204 | 544 | 18,382 | [53] |
| $Er^{3+}$ | 97.5($90SiO_2$–$10SrF_2$)–$2.5Al_2O_3$–$0.5ErF_3$ (sol-gel) | $SrF_2$ | 976 | 10,246 | 540 | 18,519 | [54] |
| $Er^{3+}$ | 36PbO–$9PbF_2$–$45B_2O_3$–$6Al_2O_3$–$3WO_3$–$1Er_2O_3$ | $PbF_2$ | 980 | 10,204 | 546 | 18,315 | [55] |
| $Er^{3+}$ | $40SiO_2$–$25Al_2O_3$–$18NaCO_3$–$10YF_3$–7NaF +$xErF_3$ ($x$=0.05, 0.2, 0.5, 1.0, 2.0) | $NaYF_4$ | 980 | 10,204 | 660 | 15,152 | [56] |

| | | | | | | | |
|---|---|---|---|---|---|---|---|
| $Er^{3+}$ | $66SiO_2$–$8B_2O_3$–$10Na_2O$–8NaF–$8GdF_3+xErF_3$ ($x=0.1, 0.2, 0.4, 1.0, 1.5, 2.0$) | β-$NaGdF_4$ | 1530 | 6536 | 540, 653 | 18,519 15,314 | [57] |
| $Er^{3+}$ | $40SiO_2$–$25Al_2O_3$–$15BaCO_3$–$10YF_3$–$10BaF_2$–$xErF_3$ ($x=0.5, 1.0, 2.0, 3.0$) | $BaYF_5$ | 980 | 10,204 | 545 | 18,349 | [58] |
| $Er^{3+}$ | $45SiO_2$–$15Al_2O_3$–$12Na_2O$–$21BaF_2$–$7La_2O_3$ $+0.5ErF_3$ | $Ba_2LaF_7$ | 980 | 10,204 | 545 | 18,349 | [59] |
| $Er^{3+}$ | $71TeO_2$–18ZnO–$7ZnF_2$–$3YF_3$–$1ErF_3$ | $Y_2Te_6O_{15}$ or $Er_2Te_5O_{13}$ | 970 | 10,309 | 546, 660 | 18,315 15,152 | [60] |
| $Er^{3+}$ | $30SiO_2$–$15Al_2O_3$–$29CdF_2$–$22PbF_2$–$1.5YF_3$–$2.5ErF_3$ | $PbF_2$–$CdF_2$ | 975 | 10,256 | 545 | 18,349 | [61] |
| $Tb^{3+}$–$Yb^{3+}$ | $PbGeO_3$–$PbF_2$–$CdF_2+0.4Yb^{3+}+0.2Tb^{3+}$ | $PbF_2$ | 975 | 10,256 | 545 | 18,349 | [62] |
| $Tb^{3+}$–$Yb^{3+}$ | $40SiO_2$–$19Na_2CO_3$–$25Al_2O_3$–$12LaF_3$–$1TbF_3$–$3YbF_3$ | $LaF_3$ | 980 | 10,204 | 543 | 18,416 | [63] |
| $Tb^{3+}$–$Yb^{3+}$ | $(44-x)SiO_2$–$28Al_2O_3$–17NaF–$11TbF_3$–$xYbF_3$ ($x=0, 0.25, 0.5, 1.0, 1.5, 3.0$) | $TbF_3$ | 976 | 10,246 | 543 | 18,416 | [64] |
| $Tb^{3+}$–$Yb^{3+}$ | $45SiO_2$–$10Na_2CO_3$–$15Al_2O_3$–$4CaCO_3$–18NaF–$8LuF_3$–$0.5YbF_3$–$0.1TbF_3$ | $NaLuF_4$ | 980 | 10,204 | 476 | 21,008 | [65] |
| $Ho^{3+}$–$Yb^{3+}$ | $PbGeO_3$–$PbF_2$–$CdF_2+0.1Yb^{3+}+0.2Ho^{3+}$ | $PbF_2$ | 975 | 10,256 | 545 | 18,349 | [62] |
| $Ho^{3+}$–$Yb^{3+}$ | $40SiO_2$–$25Al_2O_3$–$15BaCO_3$–$10YF_3$–$10BaF_2$–$0.5HoF_3$–$xYbF_3$ ($x=0,1, 2, 3$) | $BaYF_5$ | 976 | 10,246 | 545 | 18,349 | [66] |
| $Ho^{3+}$–$Yb^{3+}$ | $30SiO_2$–$15AlO_{1.5}$–$28PbF_2$–$22CdF_2$–$(4.9-x)$ $GdF_3 0.1HoF_3$–$xYbF_3$ ($x=0,0.1, 0.2, 0.5, 1, 2, 3, 4$) | $Pb_xCd_{1-x}F_2$ | 980 | 10,204 | 545 | 18,349 | [67] |
| $Ho^{3+}$–$Yb^{3+}$ | $54SiO_2$–$10K_2CO_3$–$15Al_2O_3$–15KF–$6YbF_3$–$0.1HoF_3$ | $KYb_2F_7$ | 980 | 10,204 | 545 | 18,349 | [68] |
| $Ho^{3+}$–$Yb^{3+}$ | $45SiO_2$–$10Na_2CO_3$–$5Al_2O_3$–$4CaCO_3$–18NaF–$8YbF_3$–$0.05Ho_2O_3$ | $NaYbF_4$ | 980 | 10,204 | 545 | 18,349 | [69] |
| $Er^{3+}$–$Yb^{3+}$ | $0.1ErF_3$–$0.1YbF_3$–$5LaF_3$–$94.8SiO_2$ (sol-gel) | $LaF_3$ | 973 | 10,277 | 540, 660 | 18,519 15,152 | [70] |

*Continued*

Table 4.3 **Continued**

| Rare-earth ion(s) | Glass system/glass composition (mol.%) | Major crystal phase in glass-ceramics | Excitation | | Main upconverted emission band | | Ref. |
|---|---|---|---|---|---|---|---|
| | | | nm | $cm^{-1}$ | nm | $cm^{-1}$ | |
| $Er^{3+}$–$Yb^{3+}$ | $60SiO_2$–$20ZnF_2$–$20BaF_2$–$5YbF_3$–$0.5ErF_3$ | $BaF_2$ | 980 | 10,204 | 651 | 15,361 | [71] |
| $Er^{3+}$–$Yb^{3+}$ | $45SiO_2$–$20Al_2O_3$–10CaO–(20–22)$CaF_2$–(2–5) $YbF_3$–(0.5–1.0)$ErF_3$ | $CaF_2$ | 970 | 10,309 | 660 | 15,152 | [72] |
| $Er^{3+}$–$Yb^{3+}$ | $45SiO_2$–$25Al_2O_3$–$5CaCO_3$–10NaF–$15CaF_2$–$0.5ErF_3$–$xYbF_3$ ($x=0$, 0.5) | $CaF_2$ | 980 | 10,204 | 655 | 15,267 | [73] |
| $Er^{3+}$–$Yb^{3+}$ | $44SiO_2$–$28Al_2O_3$–17NaF–$11YF_3$–$xErF_3$–$yYbF_3$ ($x=0$, 0.1, 0.5; $y=0$, 0.1, 0.2, 0.4, 0.8, 1.0) | $YF_3$ | 980 | 10,204 | 545 | 18,349 | [74] |
| $Er^{3+}$–$Yb^{3+}$ | $64SiO_2$–$14.5B_2O_3$–$9.5Na_2O$–6NaF–$6YF_3$–$0.5ErF_3$–$1YbF_3$ | $NaYF_4$ | 980 | 10,204 | 538 | 18,587 | [75] |
| $Er^{3+}$–$Yb^{3+}$ | $40NaPO_3$–$30YF_3$–$20BaF_2$–$10CaF_2$+4 wt.% $YbF_3$+0.5, 1, 2, 4, 6 wt.% $ErF_3$ | $NaYF_4$, $CaF_2$, $(CaF_2)_{0.75}$–$(YF_3)_{0.25}$ | 975 | 10,256 | 542 | 18,450 | [76] |
| $Er^{3+}$–$Yb^{3+}$ | $30SiO_2$–$12P_2O_5$–20CaO–$20Al_2O_3$–$18CaF_2$–$0.2Er_2O_3$–$1Yb_2O_3$ | $Ca_5(PO_4)_3F$ | 980 | 10,204 | 547, 667 | 18,282 14,993 | [77] |
| $Tm^{3+}$–$Yb^{3+}$ | $90SiO_2$–$10YF_3$+$1.5Yb^{3+}$+$0.1Tm^{3+}$ (sol-gel) | $YF_3$ | 980 | 10,204 | 348, 807 | 28,736 12,392 | [78] |
| $Tm^{3+}$–$Yb^{3+}$ | $50SiO_2$–$10Al_2O_3$–$20ZnF_2$–$20SrF_2$+$4YbF_3$+$0.05TmF_3$ | $SrF_2$ | 980 | 10,204 | 479 | 20,877 | [79] |
| $Tm^{3+}$–$Yb^{3+}$ | $PbF_2$–$GeO_2$–$3Al_2O_3$+$0.05Tm_2O_3$+$4Yb_2O_3$ ($PbF_2/GeO_2=1.5$–2.0) | $PbF_2$ | 1000 | 10,000 | 478.6 | 20,894 | [80] |
| $Tm^{3+}$–$Yb^{3+}$ | $30SiO_2$–$50PbF_2$–$15AlO_{1.5}$–$(4.9-x)AlF_3$–$0.1TmF_3$–$xYbF_3$ ($x=0.5$, 1.0, 2.0) | β-$PbF_2$ | 980 | 10,204 | 476 | 21,008 | [81] |
| $Tm^{3+}$–$Er^{3+}$–$Yb^{3+}$ | $40SiO_2$–$10AlO_{1.5}$–$45PbF_2$–$5AlF_3$+$0.1Er^{3+}$+$0.1Tm^{3+}$+$2Yb^{3+}$ | β-$PbF_2$ | 970 | 10,309 | 480 | 20,833 | [82] |

($x=0$, 0.1, 0.5, 1.0, and 2.0). For details, see Table 4.3. Fig. 4.14A shows the upconversion luminescence spectra ($\lambda_{exc}=980$ nm) of glass-ceramics with different concentration of $Er^{3+}$ ion. It is seen that for precursor glasses, no appreciable upconversion luminescence was observed, whereas for all the glass-ceramics, the upconversion emission bands are centered around 550 (green) and 660 (red) nm. These bands are attributed to $^2H_{11/2}$, $^4S_{3/2} \rightarrow {}^4I1_{5/2}$, and $^4F_{9/2} \rightarrow {}^4I_{15/2}$ transitions. The intensity ratio of red to green emission is 7–8 times for the 0.1 mol.% $Er^{3+}$-doped samples. It increases to about 20 times at $Er^{3+}$ concentration of 0.5 mol.% or more. Nevertheless, both the intensities of red and green emission decreased dramatically when the $Er^{3+}$ content reaches up to 2 mol.%. This fact confirmed the incorporation of $Er^{3+}$ ions into low phonon $CaF_2$ nanocrystals during crystallization. The enhanced upconversion is explained based on low phonon energy (300–400 $cm^{-1}$) of $CaF_2$ nanocrystal. Partial energy-level diagram depicting mechanism for the upconversion emission is shown in Fig. 4.14B.

Wei et al. [65] have reported intense blue upconversion in novel transparent $Tb^{3+}$–$Yb^{3+}$ codoped glass-ceramics containing $NaLuF_4$ nanocrystals in the glass composition (mol.%) $45SiO_2$–$10Na_2CO_3$–$15Al_2O_3$–$4CaCO_3$–18NaF–$8LuF_3$–$0.5YbF_3$–$0.1TbF_3$. Details are given in Table 4.3. The upconversion spectra ($\lambda_{exc}=980$ nm) of (a) $Tb^{3+}$ and $Yb^{3+}$ codoped and (b) $Yb^{3+}$ single-doped glass and glass-ceramics are illustrated in Fig. 4.15A. The emission bands peaking at 381, 415, 437, 543, 585, and 623 nm are attributed to $^5D_3 \rightarrow {}^7F_6$ ($A_1$), $^5D_3 \rightarrow {}^7F_5$($A_2$), $^5D_3 \rightarrow {}^7F_4$ ($A_3$), $^5D_4 \rightarrow {}^7F_5$ ($B_1$), $^5D_4 \rightarrow {}^7F_4$ ($B_2$), and $^5D_4 \rightarrow {}^7F_3$($B_3$) transitions of $Tb^{3+}$, respectively. The strongest blue emission at 476 nm is credited to the cooperative luminescence (CL) of $Yb^{3+}$ ion pairs. The additional emissions around 525 ($E_1$) and 650 nm ($E_2$) cannot be assigned to $Tb^{3+}$ ions. To settle the origin of $E_1$ and $E_2$ emissions, upconversion spectra of $Yb^{3+}$ single-doped Y-PG, Y-GC650, and Y-GC660 are depicted in Fig. 4.15A (b). Except for the CL of $Yb^{3+}$ ion pairs, extra emissions around 525 ($E_1$), 550 ($E_3$), and 650 nm

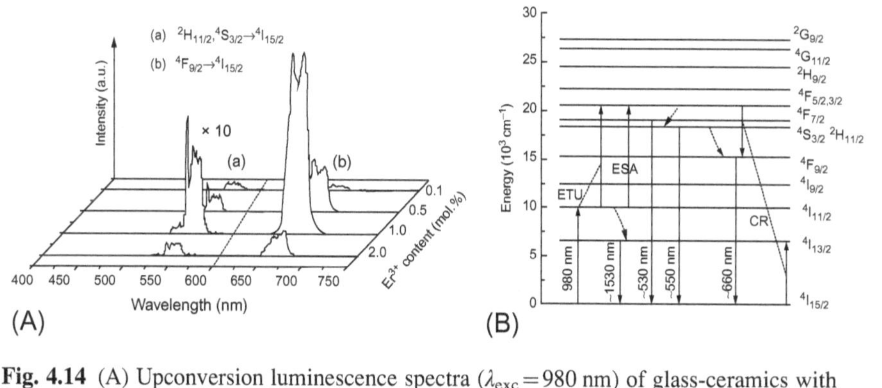


**Fig. 4.14** (A) Upconversion luminescence spectra ($\lambda_{exc}=980$ nm) of glass-ceramics with different concentration of $Er^{3+}$ ion. (B) Partial energy-level diagram depicting mechanism for same upconversion.

Reproduced with permission from D. Chen, Y. Wang, Y. Yu, E. Ma, F. Bao, Z. Hu, Y. Cheng, Influences of $Er^{3+}$ content on structure and upconversion emission of oxyfluoride glass ceramics containing $CaF_2$ nanocrystals, Mater. Chem. Phys. 95 (2006) 264–269, Copyright © Elsevier.

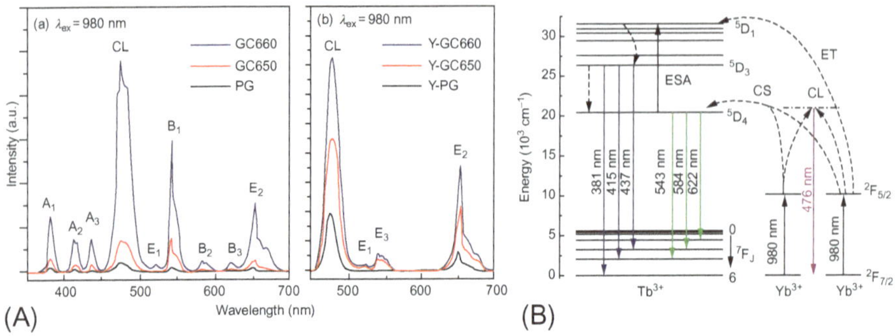

**Fig. 4.15** (A) Upconversion spectra ($\lambda_{exc} = 980$ nm) of (a) $Tb^{3+}$ and $Yb^{3+}$ codoped and (b) $Yb^{3+}$ single-doped glass and glass-ceramics. (B) Partial energy-level diagram depicting mechanism for same upconversions.
Reproduced with permission from Y. Wei, X. Liu, X. Chi, R. Wei, H. Guo, Intense upconversion in novel transparent $NaLuF_4$:$Tb^{3+}$, $Yb^{3+}$ glass–ceramics, J. Alloys Compd. 578 (2013) 385–388, Copyright © Elsevier.

($E_2$) are detected as well. Such additional emissions can be attributed to $^2H_{11/2} \rightarrow {}^4I_{15/2}$, $^4S_{3/2} \rightarrow {}^4I_{15/2}$, and $^4F_{9/2} \rightarrow {}^4I_{15/2}$ transitions of trace $Er^{3+}$ impurities admixed with $YbF_3$. Due to the extremely limited content of trace $Er^{3+}$, $Er^{3+}$ impurities can hardly affect the upconversion process of $Tb^{3+}$. Moreover, no upconversion luminescence can be detected in $Tb^{3+}$ single-doped precursor glasses and glass-ceramics, indicating the upconverted $Tb^{3+}$ emissions are produced by the effective ET from $Yb^{3+}$ to $Tb^{3+}$. Partial energy-level diagram depicting mechanism of these upconversions is given in Fig. 4.15B.

Of late, Wei et al. [68] have described the elaboration, structure, and intense green upconversion of $Ho^{3+}$ in transparent glass-ceramics containing $KYb_2F_7$ nanocrystals in the glass composition (mol.%) $54SiO_2$–$10K_2CO_3$–$15Al_2O_3$–15KF–$6YbF_3$–$0.1HoF_3$. Details are given in Table 4.3. Fig. 4.16 represents the (A) XRD patterns

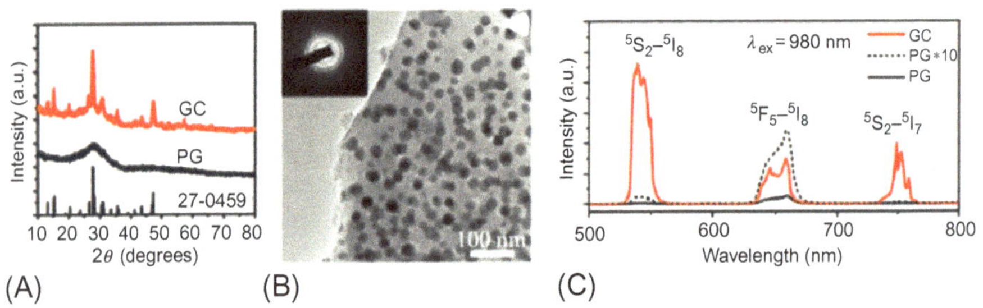

**Fig. 4.16** (A) XRD patterns of precursor glass (PG) and its glass-ceramics (GC), (B) TEM image of GC, inset of SAED patterns, and (C) upconversion spectra ($\lambda_{exc} = 980$ nm) of PG and GC.
Reproduced with permission from Y. Wei, H.M. Yang, X.M. Li, L.J. Wang, H. Guo, Elaboration, structure, and intense upconversion in transparent KYb2F7:$Ho^{3+}$ glass-ceramics, J. Am. Ceram. Soc. 97 (2014) 2012–2015, Copyright © The American Ceramic Society.

of precursor glass (PG) and its glass-ceramics (GC), (B) TEM image of GC, inset of SAED patterns, and (C) upconversion spectra ($\lambda_{exc} = 980$ nm) of PG and GC.

The mean crystallite size ($D$) of $KYb_2F_7$ crystals was determined from the XRD pattern using the following Scherrer formula [83]:

$$D = k\ \lambda/\beta \cos\theta \tag{4.4}$$

where $k = 0.89$, $\lambda = 0.154056$ nm (wavelength of $CuK_{\alpha 1}$ radiation), $\theta$ and $\beta$ are the Bragg angle and corrected full width at half maximum (FWHM) of diffraction peak. The determined mean crystallite size of $KYb_2F_7$ was found to be 18 nm. The TEM image and SAED patterns (Fig. 4.16B) disclose the homogeneously grown spherical $KYb_2F_7$ nanocrystals of ~22 nm, which is close to the size determined by Scherrer formula. The upconversion emission bands, as shown in Fig. 4.16C, positioned at 545 (green), 660 (red), and 750 nm are corresponding to $^5S_2 \rightarrow {}^5I_8$, $^5F_5 \rightarrow {}^5I_8$, and $^5S_2 \rightarrow {}^5I_7$ transitions of $Ho^{3+}$, respectively. Only one dominant and weak red emission band is observed for precursor glass (PG). Comparing with that of PG, it becomes distinct that the green upconversion intensity of $Ho^{3+}$ ion in glass-ceramics (GC) is greatly enhanced and increased by 180–190 times, whereas the red emission intensity increases by only five times. Moreover, all the emission bands in GC exhibit Starks splitting. All these changes point toward incorporation of $Ho^{3+}$ into the $KYb_2F_7$ nanocrystals.

Duan et al. [82] have described the multicolor upconversions of $Er^{3+}$–$Tm^{3+}$–$Yb^{3+}$ triply doped oxyfluoride glass-ceramics containing β-$PbF_2$ crystals in the glass composition (mol.%) $40SiO_2$–$10AlO_{1.5}$–$45PbF_2$–$5AlF_3$ + $0.1Er^{3+}$ + $0.1Tm^{3+}$ + $2Yb^{3+}$. Details are provided in Table 4.3. Fig. 4.17A illustrates the upconversion spectra ($\lambda_{exc} = 970$ nm) of these $Er^{3+}$–$Tm^{3+}$–$Yb^{3+}$ triply doped oxyfluoride glass-ceramics with varied dopant concentrations. It is seen that with increase in concentration of

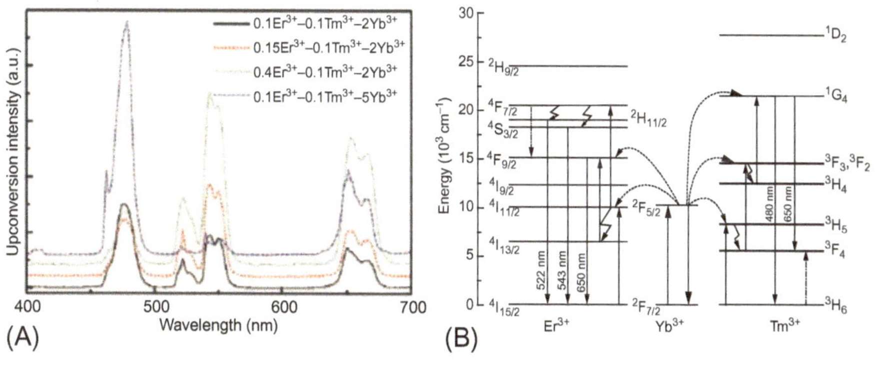


**Fig. 4.17** (A) Upconversion spectra ($\lambda_{exc} = 970$ nm) of $Er^{3+}$–$Tm^{3+}$–$Yb^{3+}$ triply doped oxyfluoride glass-ceramics with varied dopant concentrations. (B) Partial energy-level diagram depicting mechanism for same upconversions.

Reproduced with permission from Z. Duan, J. Zhang, W. Xiang, H. Sun. L. Hu, Multicolor upconversion of $Er^{3+}/Tm^{3+}/Yb^{3+}$ doped oxyfluoride glass ceramics, Mater. Lett. 61 (2007) 2200–2203, Copyright © Elsevier.

$Er^{3+}$ ion, the upconversion intensity of green and red increase greatly, while the blue upconversion decreased in nominal magnitude. With the purpose of comparison of their intensity changes, the integrated upconversion intensity of red ($I_R$), green ($I_G$), and blue ($I_B$) was evaluated. In $0.1Er^{3+}$–$0.1Tm^{3+}$–$2Yb^{3+}$-doped sample, $I_R$:$I_G$:$I_B$ is calculated to be 1:1.23:1.59. Once the concentration of $Er^{3+}$ increases to 0.15 mol.%, the $I_R$:$I_G$:$I_B$ changes to 1:1.25:1.05. Even though $Er^{3+}$ increases to 0.4 mol.%, the $I_R$:$I_G$:$I_B$ changes to 1:1.18:0.30. It is reasonable to conclude that the $I_R$:$I_G$ ratio changes unnoticeably, while $I_R$:$I_B$ and/or $I_G$:$I_B$ increases considerably with increasing $Er^{3+}$. Consistent with the upconversion mechanism, as shown in Fig. 4.17B, it is conceivable to explain this observation. With increasing $Er^{3+}$, more $Er^{3+}$ ions are involved in upconversion emission. Instead, the probability of energy transfer between $Yb^{3+}$ and $Er^{3+}$ increases with rise in concentration of $Er^{3+}$. Consequently, red and green upconversion of $Er^{3+}$ is enhanced significantly.

#### *4.3.1.4 Glass-ceramics for WLEDs*

The white light (WL) can be generated by employing many methods. Nevertheless, the phosphor method has gained renewed attention after receiving adjudge for the Nobel Prize in Physics for 2014 in favor of the invention of blue light emitting diodes. The white light producing phosphor method consists of a combination of a short-wavelength (300–500 nm) light emitting diode (LED such as blue, violet, or UV) and a coating of yellow light emitting diode. The blue, violet, or UV photons generated in the LED either passes through the phosphor layer without modification or are converted into yellow photons in the phosphor layer. The outcoming blue and yellow photons combine together to produce white light as the resultant. This new lighting technology of the 21st century has several advantages over other existing technologies of the 19th (incandescent lamp) and 20th (fluorescent lamp) centuries. WLED lamps emit bright white light; they are long lasting and energy efficient. The lighting efficiency (known as "efficacy" and expressed as lumens per watt, lm $W^{-1}$) of a WLED is over 300 lm $W^{-1}$ compared with 16 and 70 lm $W^{-1}$ for incandescent and fluorescent lamps, respectively. WLEDs contribute significantly for earning carbon credit as they require about one-fourth of world electricity consumption used for lighting purposes. Moreover, they remarkably reduce materials consumption (save Earth's resources) as the lifetime of WLEDs is up to 100,000 h, compared with 1000 and 10,000 h for incandescent and fluorescent lamps, respectively. It is seen that the most important ingredient of this lighting technology is the short-wavelength (blue, violet, or UV) light emitting diode (LED). Akasaki et al. [84], Amano et al. [85], and Nakamura et al. [86–88] have disclosed the invention of blue light emitting diodes (LEDs) using GaN, InGaN, and their associated semiconductors. This invention has outstandingly contributed to create white and other lights in an entirely new manner to the benefit of mankind, and the world got the lighting technology of the 21st century. Consequently, Isamu Akasaki, Hiroshi Amano, and Shuji Nakamura were rewarded with the Nobel Prize in Physics in 2014 by the Royal Swedish Academy of Sciences quoting "for the invention of efficient blue light emitting diodes which has enabled bright and energy-saving white light sources."

In the phosphor method of WLEDs, the phosphor layer is applied by mixing phosphor powders with organic resins in the form of a paste. However, these organic resins undergo deterioration by the action of high energy short-wavelength lights of LEDs due to photodecomposition resulting in diminishing the efficacy of white light emitting devices with time. On the other hand, if glass-ceramics of suitable thickness (thin or thick according to performance) are used, the problem of photodecomposition can be easily eliminated, as a consequence of glass-ceramics made up of undecomposable and weather-resistant inorganic materials. Of late, Chen et al. [12] have reviewed the current advances in transparent glass-ceramic phosphors for WLEDs. A schematic representation of the design of WLEDs constructed using glass-ceramics is illustrated in Fig. 4.18.

Rare-earth ions, glass compositions, crystal phases, excitation wavelengths, emitted band peak wavelengths, and chromaticity coordinates of some glass-ceramics established for WLEDs are summarized in Table 4.4. It is seen that the WLEDs can be obtained employing either the downconversion process or the upconversion process. For this reason, these glass-ceramics are divided into these two categories in this table. It is also seen that the properties, particularly with respect to the white light emission, are strongly dependent on the processing parameters. To explain this fact, some examples are described as follows.

Wang et al. [89] have described a new fabrication technique of phosphor (Ce:YAG) in glass (PiG) by tape-casting method and the chromaticity property in the combination of $PbO–B_2O_3–SiO_2–ZnO + x$Ce:YAG ($x = 6, 10, 12, 20$, and 30 wt.%). Details are provided in Table 4.4. Photograph of a WLED lamp fabricated with 20 wt.% Ce:YAG phosphor (TCPG-20 sample) in glass (left) and its lamp in operation (right) are shown

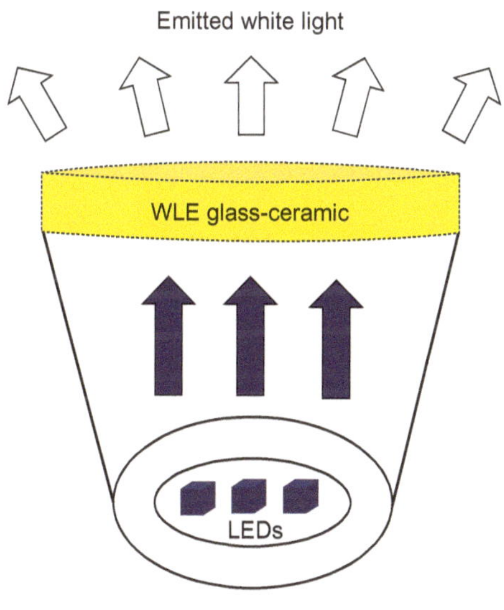


**Fig. 4.18** Schematic representation of the design of white light emitting diodes constructed using glass-ceramics.

**Table 4.4 Rare-earth ions, glass compositions, crystal phases, excitation wavelengths, emitted band peak wavelengths, and chromaticity coordinates of some glass-ceramics established for white light emitting diodes (WLEDs)**

| Rare-earth ion | Glass system/glass composition (mol.%) | Major crystal phase in glass-ceramics | Excitation (nm) | Emitted band (nm) | Chromaticity coordinates[a] | Ref. |
|---|---|---|---|---|---|---|
| *Glass-ceramics for downconverted WLEDs* | | | | | | |
| $Ce^{3+}$ | $PbO–B_2O_3–SiO_2–ZnO+x$Ce:YAG ($x=6$, 10, 12, 20, 30 wt.%) | $Y_3Al_5O_{12}$ (YAG) | 458 (blue LED chip) | 450, 555 | $x=0.169–0.328$, $y=0.067–0.352$ | [89] |
| $Ce^{3+}$ | $Sb_2O_3–ZnO–K_2O–B_2O_3+3$ wt.% Ce:YAG | $Y_3Al_5O_{12}$ (YAG) | 465 (blue LED chip) | 450, 555 | $x=0.342$, $y=0.339$ | [90] |
| $Ce^{3+}$ | $50Al_2O_3–29.85Y_2O_3–20SiO_2–0.15Ce_2O_3$ | $Y_3Al_5O_{12}$ (YAG) | 465 | 530 | $x=0.32–0.45$, $y=0.32–0.45$ | [91] |
| $Eu^{3+}$ | $44SiO_2–28Al_2O_3–17NaF–11LaF_3+(0.005–0.5)$ $Eu^{3+}$ | $LaF_3$ | 394 | 410, 425, 460, 510, 555, 590, 620 | $x=0.30–0.36$, $y=0.24–0.33$ | [92] |
| $Eu^{3+}/Eu^{2+}$ | $40SiO_2–25Al_2O_3–15BaCO_3–10BaF_2–10YF_3–0.25Eu_2O_3$ | $BaAl_2Si_2O_8$ | 380 | 475, 585, 615, 695 | $x=0.333$, $y=0.307$ | [93] |
| $Eu^{3+}/Eu^{2+}$ | $43SiO_2–22Al_2O_3–18Na_2CO_3–3NaF–10YF_3–4P_2O_5–0.5Eu_2O_3$ | $YPO_4$ | 365 | 440, 590, 620 | – | [94] |
| $Eu^{2+}$ | $60(0.99CaO–0.01EuO_{1.5})–40SiO_2$ | $\beta$-$Ca_2SiO_4$, $Ca_3Si_2O_7$ | 378 | 490, 610 | $x=0.31–0.43$, $y=0.36–0.47$ | [95,96] |
| $Eu^{2+}$ | $42SiO_2–xCaO–(38-x)SrO–20MgO–0.2EuO_{1.5}$ ($x=0$, 10, 20, 30, 38) | $M_2MgSi_2O_7$ (M = Sr, Ca) | 378 | 470–530 | $x=0.15–0.30$, $y=0.25–0.60$ | [97] |
| $Dy^{3+}$ | $44SiO_2–28Al_2O_3–17NaF–11GdF_3+(0.1–2.5)$ $DyF_3$ | $GdF_3$ | 350 | 480, 571 | $x=0.334–0.355$, $y=0.391–0.403$ | [98] |

| | | | | | | |
|---|---|---|---|---|---|---|
| $Dy^{3+}$ | $30SiO_2$–$15Al_2O_3$–$29CdF_2$–$22PbF_2$–$(4-x)YF_3$–$xDyF_3$ ($x$=0.01, 0.1, 1.0) | β-$PbF_2$ | 451 | 480, 572 | $x$=0.29–0.31, $y$=0.34–0.36 | [99] |
| $Dy^{3+}$ | $47SiO_2$–$23Al_2O_3$–$19Na_2O$–$(9-x)YF_3$–$2B_2O_3$–$xDyF_3$ ($x$=0.8, 1.6) | $NaAlSiO_4$, $NaY_9Si_6O_{26}$ | 350 | 460, 555 | $x$=0.32, $y$=0.38 | [100] |
| $Dy^{3+}$–$Ce^{3+}$ | $50SiO_2$–$15Al_2O_3$–$(32-x)$ $LaF_3$–$3DyF_3$–$xCeF_3$ ($x$=3, 6, 9, 12) | $LaF_3$ | 320 | 480, 575 | $x$=0.300–0.450, $y$=0.300–0.438 | [101] |
| $Dy^{3+}$–$Eu^{2+}$ | $45SiO_2$–$25Al_2O_3$–$10CaO$–$20CaF_2$–(0.3–0.9) $DyF_3$–$0.03EuF_3$ | $CaF_2$ | 385 | 480, 573 | $x$=0.296, $y$=0.311 | [102] |
| $Dy^{3+}$–$Tm^{3+}$ | $50SiO_2$–$20B_2O_3$–$18Na_2O$–$8SrF_2$–$4GdF_3+xDyF_3+yTmF_3$ ($x$=0.05, 0.1, 0.2, 0.3, 0.4; $y$=0.15, 0.3, 0.45, 0.6, 0.75) | $GdSr_2F_7$ | 354 | 450, 480, 571 | $x$=0.2–0.4, $y$=0.3–0.4 | [103] |
| $Eu^{3+}$/$Eu^{2+}$–$Tb^{3+}$ | $43SiO_2$–$22Al_2O_3$–$8Na_2CO_3$–$3NaF$–$10GdF_3$–$4P_2O_5+1Eu_2O_3+(1.0–2.5)Tb_4O_7$ | $GdPO_4$ | 350 | 450, 545, 590, 620 | $x$=0.2–0.3, $y$=0.2–0.3 | [104] |
| $Eu^{3+}$/$Eu^{2+}$–$Tb^{3+}$ | $40.2SiO_2$–$13.4Al_2O_3$–$7La_2O_3$–$10.7Na_2O$–$26.7SrF_2$–$1EuF_3$–$1TbF_3$ | $SrLaF_5$ | 405 | 485, 545, 590, 620 | $x$=0.36–0.38, $y$=0.34–0.35 | [105] |
| $Tm^{3+}$–$Tb^{3+}$–$Mn^{2+}$ | $80P_2O_5$–$15Li_2O$–$5Sb_2O_3$+0.2 wt.% $Tm_2O_3$+0.3 wt.% $Tb_2O_3$+0.5 wt.% $MnO_2$ | – | 360 | 450, 550, 630 | $x$=0.324, $y$=0.335 | [106] |
| $Dy^{3+}$–$Tb^{3+}$–$Eu^{3+}$ | $50SiO_2$–$20B_2O_3$–$18Na_2O$–$8SrF_2$–$4GdF_3+0.3DyF_3+xTbF_3+yEuF_3$ ($x=y$=0.15, 0.30, 0.45, 0.60, 0.75) | $GdSr_2F_7$ | 382 | 475, 575, 620 | $x$=0.32–0.52, $y$=0.32–0.52 | [107] |
| $Tm^{3+}$–$Tb^{3+}$–$Eu^{3+}$ | $70P_2O_5$–$15Li_2O$–$15Sb_2O_3+0.1Tm_2O_3+0.1Tb_2O_3+xEu_2O_3$ ($x$=0.05, 0.07, 0.10, 0.20 wt.%) | – | 365 | 450, 485, 545, 590, 615 | $x$=0.339, $y$=0.341 | [108] |
| $Eu^{2+}$–$Tb^{3+}$–$Sm^{3+}$ | $45SiO_2$–$15Al_2O_3$–$7La_2O_3$–$12Na_2O$–$5BaF_2$–$0.3EuF_2$–$2.7TbF_3$–$3SmF_3$ | $Ba_2LaF_7$ | 405 | 489, 542, 562, 598, 646 | $x$=0.33, $y$=0.38 | [109] |

*Continued*

**Table 4.4 Continued**

| Rare-earth ion | Glass system/glass composition (mol.%) | Major crystal phase in glass-ceramics | Excitation (nm) | Emitted band (nm) | Chromaticity coordinates[a] | Ref. |
|---|---|---|---|---|---|---|
| ***Glass-ceramics for upconverted WLEDs*** | | | | | | |
| $Er^{3+}$–$Tm^{3+}$–$Yb^{3+}$ | $52SiO_2$–$20Al_2O_3$–$13Na_2O$–$5BaO$–$10BaF_2+1Yb_2O_3+0.1Er_2O_3+0.1Tm_2O_3$ | $BaF_2$ | 980 | 476, 543, 656 | $x=0.278$, $y=0.358$ | [110] |
| $Ho^{3+}$–$Tm^{3+}$–$Yb^{3+}$ | $44SiO_2$–$28Al_2O_3$–$7NaF$–$11YF_3+0.02Ho^{3+}+0.1Tm^{3+}+0.5Yb^{3+}$ | β-$YF_3$ | 976 | 452, 478, 545, 648, 790 | $x=0.351$, $y=0.306$ | [111] |
| $Yb^{3+}$–$Er^{3+}$–$Tm^{3+}$ | $89.5SiO_2$–$10PbF_2$–$0.3Yb^{3+}$–$0.1Er^{3+}$–$0.1Tm^{3+}$ (sol-gel) | $PbF_2$ | 980 | 480, 540, 660 | $x=0.36$–$0.42$, $y=0.42$–$0.43$ | [112] |
| $Yb^{3+}$–$Ho^{3+}$–$Tm^{3+}$ | $89.5SiO_2$–$10PbF_2$–$0.3Yb^{3+}$–$0.1Ho^{3+}$–$0.1Tm^{3+}$ (sol-gel) | $PbF_2$ | 980 | 480, 540, 660 | $x=0.33$–$0.42$, $y=0.35$–$0.40$ | [112] |
| $Yb^{3+}$–$Ho^{3+}$–$Tm^{3+}$ | $90SiO_2$–$5LaF_3+0.3Yb^{3+}+0.1Ho^{3+}+0.1Tm^{3+}$ (sol-gel) | $LaF_3$ | 980 | 475, 530, 640 | $x=0.27$–$0.46$, $y=0.32$–$0.45$ | [113] |
| $Yb^{3+}$–$Er^{3+}$–$Tm^{3+}$ | $90SiO_2$–$10YF_3+0.3Yb^{3+}+(0.025$–$0.1)Er^{3+}+0.1Tm^{3+}$ (sol-gel) | $YF_3$ | 980 | 480, 540, 650 | $x=0.24$–$0.42$, $y=0.22$–$0.42$ | [114] |
| $Nd^{3+}$–$Tm^{3+}$–$Ho^{3+}$–$Yb^{3+}$ | $44SiO_2$–$28Al_2O_3$–$17NaF$–$11YF_3+0.1NdF_3+0.05HoF_3+0.5YbF_3+xTmF_3$ ($x=0.1$–$1.0$) | β-$YF_3$ | 796 | 451, 478, 544, 640 | $x=0.25$, $y=0.25$ | [115] |

[a] Ideal white light chromaticity coordinates, $x=0.333$ and $y=0.333$.

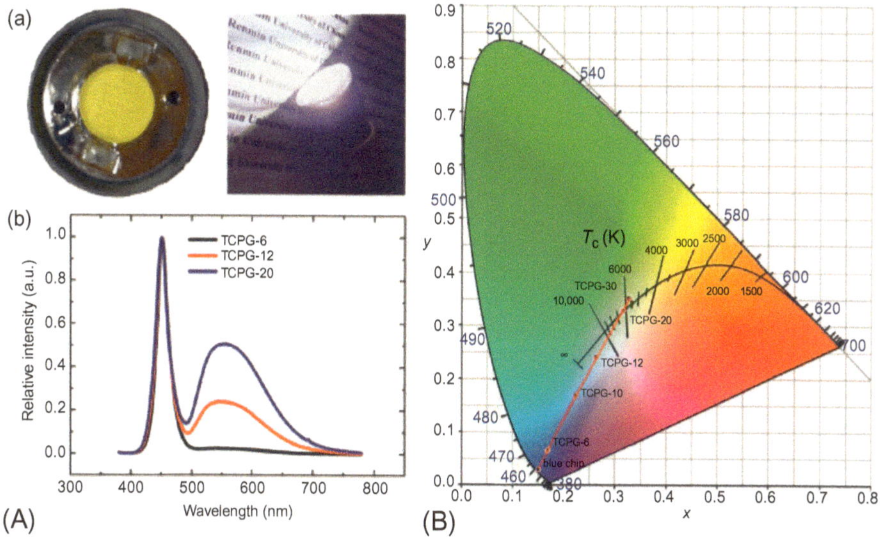


**Fig. 4.19** (A) (a) Photograph of a WLED lamp fabricated with TCPG-20 sample (left) and the lamp in operation (right) and (b) normalized spectra of different TCPG samples. (B) CIE chromaticity diagram of TCPG-based WLEDs with different concentration of phosphors. Reproduced with permission from F. Wang, Y. Lin, H. Shi, W. Wang, Z. Deng, J. Chen, X. Yuan, Y. Cao, Introduction on the fabrication technique of phosphor in glass by tape-casting and investigation on the chromaticity property, Opt. Express 22 (2014) A1355–A1362, Copyright © OSA

in Fig. 4.19A (a). Normalized spectra ($\lambda_{exc}=458$ nm) of different TCPG samples are depicted in Fig. 4.19A (b). The CIE chromaticity diagram of TCPG-based WLEDs with different concentration of phosphors is shown in Fig. 4.19B. It is seen that the luminescence intensity in the yellow portion rises with increase in concentration of the phosphor. As a result, the chromaticity parameters ($x$ and $y$) of these obtained lights change. Therefore, by logical combination of phosphor and glass contents in the tape-cast samples, the WLEDs with chromaticity coordinates close to standard white light ($x=0.33$ and $y=0.33$) can be achieved.

Ye et al. [102] have described white light emission under UV light excitation ($\lambda_{exc}=385$ nm) in $Eu^{2+}/Dy^{3+}$ codoped $CaF_2$ crystal containing glass-ceramics in the glass composition (mol.%) $45SiO_2$–$25Al_2O_3$–10CaO–$20CaF_2$–(0.3–0.9)$DyF_3$–0.03$EuF_3$. Details are provided in Table 4.4. Emission spectra ($\lambda_{exc}=385$ nm) of glass and glass-ceramics obtained after heat treating at different temperatures are illustrated in Fig. 4.20A. It is found that the emission spectra of the glass and glass-ceramics of the sample $SACD_{0.03}E_{0.9}$ consist of a broad emission band of $Eu^{2+}$ ions and three emission bands of $Dy^{3+}$ ions. The broadband centered at 425 nm attributed to the 5d → 4f transition of $Eu^{2+}$ ions. Three sharp emission bands centered at 480, 573, and 661 nm are assigned to $^4F_{9/2} \rightarrow {}^6H_{15/2}$, $^4F_{9/2} \rightarrow {}^6H_{13/2}$, and $^4F_{9/2} \rightarrow {}^6H_{11/2}$ transitions of $Dy^{3+}$ ions, respectively. The emission spectra demonstrate the increase of emission intensities of all the bands due to incorporation of both the $Eu^{2+}$ and $Dy^{3+}$ ions into $CaF_2$ crystals. Partial energy-level diagram displaying the energy transfer (ET) mechanisms for

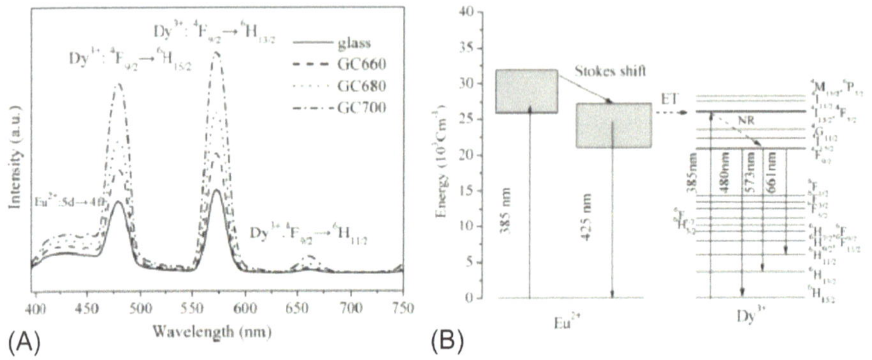

**Fig. 4.20** (A) Emission spectra ($\lambda_{exc}=385$ nm) of glass and glass-ceramics obtained after heat treating at different temperatures. (B) Partial energy-level diagram displaying the energy transfer (ET) mechanisms for exhibiting the same emissions.
Reproduced with permission from R. Ye, Z. Cui, Y. Hua, D. Deng, S. Zhao, C. Li, S. Xu, $Eu^{2+}$/$Dy^{3+}$ co-doped white light emission glass ceramics under UV light excitation, J. Non-Cryst. Solids 357 (2011) 2282–2285, Copyright © Elsevier.

exhibiting the same emissions is shown in Fig. 4.20B. The demonstrated evidences signify that the chromaticity coordinates of these glass-ceramics can be controlled by changing the concentration of the $Eu^{2+}$ and $Dy^{3+}$ ions in the samples.

Song et al. [106] have reported the enhanced white light emission in $Tm^{3+}$–$Tb^{3+}$–$Mn^{2+}$-tridoped phosphate glass-ceramic of composition (mol.% for glass) $80P_2O_5$–$15Li_2O$–$5Sb_2O_3$+0.2 wt.% $Tm_2O_3$+0.3 wt.% $Tb_2O_3$+0.5 wt.% $MnO_2$. Details are provided in Table 4.4. Emission spectra ($\lambda_{exc}=360$ nm) of these glass and glass-ceramics are shown in Fig. 4.21A. Inset shows emission of lights by the same glass

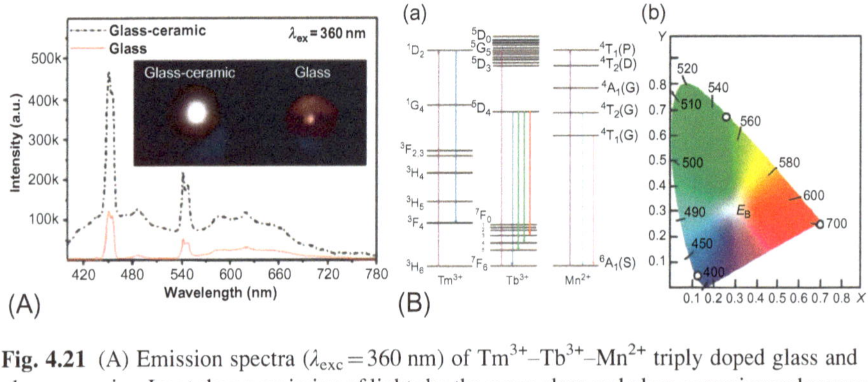

**Fig. 4.21** (A) Emission spectra ($\lambda_{exc}=360$ nm) of $Tm^{3+}$–$Tb^{3+}$–$Mn^{2+}$ triply doped glass and glass-ceramics. Inset shows emission of lights by the same glass and glass-ceramics under same excitation. (B) (a) Partial energy-level diagram displaying the mechanisms for exhibiting the same emissions. (b) CIE chromaticity diagram depicting the position of glass (*B*) and glass-ceramics (*E*) under same excitation.
Reproduced with permission from F. Song, C. Ming, L. An, Q. Wang, Y. Yu, H. Yu, T. Sun, J. Tian, $Tm^{3+}$/$Tb^{3+}$/$Mn^{2+}$ tri-doped phosphate glass ceramic for enhanced white-light-emitting material, Mater. Lett. 65 (2011) 3140–3142, Copyright © Elsevier.

and glass-ceramics under same excitation. Partial energy-level diagram displaying the mechanisms for exhibiting the same emissions is illustrated in Fig. 4.21B (a). The CIE chromaticity diagram depicting the position of glass (*B*) and glass-ceramics (*E*) under same excitation is shown in Fig. 4.21B (b). The 450 nm blue emission due to the ${}^1D_2 \rightarrow {}^3F_4$ transition of $Tm^{3+}$ and the population of the ${}^1D_2$ level is the result of ground-state absorption (GSA) according to ${}^3H_6 + h\nu \rightarrow {}^1D_2$. The 484, 548, 580, and 620 nm emissions are ascribed to the transitions of ${}^5D_4 \rightarrow {}^7F_{6,5,4,3}$ of $Tb^{3+}$. The population processes of ${}^5D_4$ level happened by GSA2 according to ${}^7F_6 + h\nu \rightarrow {}^5G_5$, the $Tb^{3+}$ ions in the ground state are excited to ${}^5G_5$ level, and then, the ions are relaxed to ${}^5D_4$ level by a nonradiative transition. The wide 550–700 nm emission band comes from the ${}^4T_1(G) \rightarrow {}^6A_1(S)$ transition of $Mn^{2+}$ ions. The population of ${}^4T_1(P)$ level of $Mn^{2+}$ in the ground state is pumped to ${}^4T_1(P)$ level via the absorption transition of ${}^6A_1(S) + h\nu \rightarrow {}^4T_1(P)$ and then relaxed to ${}^4T_1(G)$ state by nonradiative transition. The CIE chromaticity coordinates of the emissions from the glass and glass-ceramic are $x = 0.351$ and $y = 0.282$ and $x = 0.324$ and $y = 0.335$, respectively. The coordinates are close to the standard equal energy white light illumination ($x = 0.333$ and $y = 0.333$). The emission color of the glass and glass-ceramic is white, and the emission intensity of the glass-ceramic is about 20 times stronger than that of the glass.

Lin et al. [115] have reported that $Nd^{3+}$-sensitized upconversion (UC) white light emission of $Tm^{3+}/Ho^{3+}$ bridged by $Yb^{3+}$ in $\beta$-$YF_3$ nanocrystals embedded transparent glass-ceramics in the glass composition (mol.%) $44SiO_2$–$28Al_2O_3$–$17NaF$–$11YF_3 + 0.1NdF_3 + 0.05HoF_3 + 0.5YbF_3 + xTmF_3$ ($x = 0.1$–$1.0$). Details are given in Table 4.4. Normalized upconversion emission spectra ($\lambda_{exc} = 796$ nm) of $0.1Nd^{3+}$–$0.5Yb^{3+}$–$0.5Ho^{3+}$–$xTm^{3+}$ quadruply doped glass-ceramics are shown in Fig. 4.22A. Inset depicts change of red to green intensity ratio as a function of concentration ratio

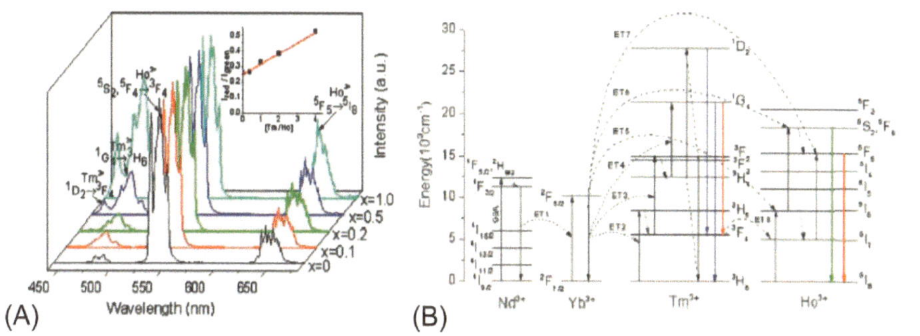


**Fig. 4.22** (A) Normalized upconversion emission spectra ($\lambda_{exc} = 796$ nm) of $0.1Nd^{3+}$–$0.5Yb^{3+}$–$0.5Ho^{3+}$–$xTm^{3+}$ quadruply doped glass-ceramics. Inset depicts change of *red* to *green* intensity ratio as a function of concentration ratio of $Tm^{3+}$ to $Ho^{3+}$. (B) Partial energy-level diagram displaying the mechanisms for exhibiting the same emissions. *Solid arrows* represent for absorption or emission transitions and *dashed* and *curved arrows* represent for nonradiative relaxations (NR) and energy transfer (ET) between the rare-earth ions, respectively. Reproduced with permission from H. Lin, D. Chen, Y. Yu, Z. Shan, P. Huang, Y. Wang, J. Yuan, $Nd^{3+}$-sensitized upconversion white light emission of $Tm^{3+}/Ho^{3+}$ bridged by $Yb^{3+}$ in $\beta$-$YF_3$ nanocrystals embedded transparent glass ceramics, J. Appl. Phys. 107 (2010) 103511 (4 pages), Copyright © AIP.

of $Tm^{3+}$ to $Ho^{3+}$. The blue emission bands peaking at 451 and 478 nm are attributed to the $^1D_2 \rightarrow ^3F_4$ and $^1G_4 \rightarrow ^3H_6$ transitions of $Tm^{3+}$, respectively. The green emission band centered at 544 nm is due to the $^5S_2$ and $^5F_4 \rightarrow ^5I_8$ transition of $Ho^{3+}$ ion, while the red band at 640 nm originates from the $^5F_5 \rightarrow ^5I_8$ transition of $Ho^{3+}$ ion. In the UC emission spectra of the $Nd^{3+}$–$Tm^{3+}$ or $Nd^{3+}$–$Ho^{3+}$ codoped glass-ceramics under excitation at 796 nm, no $Tm^{3+}$ UC signals are noticed, while the UC emission intensity of $Ho^{3+}$ is very feeble, only about 1/20 of that in the $Nd^{3+}$–$Yb^{3+}$–$Tm^{3+}$–$Ho^{3+}$ quadruply doped sample. These evidences support the low probability of energy transfer (ET) from $Nd^{3+}$ to $Tm^{3+}$ or $Nd^{3+}$ to $Ho^{3+}$. It is also found that the $Ho^{3+}$ could not be excited by the light of 796 nm. Hence, doping with $Yb^{3+}$ ion is found to be essential to obtain intense visible UC emissions in the glass-ceramics. Partial energy-level diagram displaying the mechanisms for exhibiting the same emissions is depicted in Fig. 4.22B. Solid arrows represent for absorption or emission transitions, and dashed and curved arrows represent for nonradiative relaxations (NR) and energy transfer (ET) between the rare-earth ions, respectively.

#### *4.3.1.5 Glass-ceramics for broadband emission*

Broadband amplifiers and tunable lasers are the key devices for development of dense wavelength division multiplexing (DWDM) network systems. Thus, there is an unrelenting demand to increase the transmission capacity of the DWDM system that becomes vital as a result of rapid development of the optical telecommunication. In recent times, expanding the gain bandwidths of the fiber amplifiers and the laser sources to achieve more efficient DWDM transmission network with higher capacity and faster bit rate has become a solution and attractive object for the future development of the optical communication. In order to further develop new amplifiers or lasers with much broader bandwidth, it is essential to explore new luminescent materials that can cover the 1200–1700 nm regions. Broadband infrared emission is also necessitated in several other areas such as optical sensors, tunable solid-state lasers, luminescent solar concentrators, and surgery.

Many attempts have been made on broadening and flattening of gain spectra of rare-earth doped fiber amplifiers (RDFAs) and fiber Raman amplifiers (FRAs) such as hybrids of different RDFAs, and multiwavelength-pumping FRAs have been developed. However, these amplifiers have some drawbacks such as the gain bandwidths of RDFAs are limited to around 100 nm due to the narrow f-f transitions of rare-earth metal ions and FRAs have complex system structure and require high power utilization. Therefore, there is an urgent need to find out new candidate materials for all these applications. Among the possible materials, transparent glass-ceramics doped with transition metal ions, combination of rare-earth metal ions (use of different rare-earth codopants with overlapping emission bands and similar lifetimes), or combination of both transition metal and rare-earth metal ions has attracted much attention in view of the fact that glass-ceramics have the combined advantages of both crystalline and glass materials.

Of late, Teng et al. [11] have reviewed the recent research findings on the design and control of optical properties of singly and doubly doped glass-ceramics intended for use to achieve broadband emission. Dopants, glass compositions, crystal phases, excitation wavelength, emitted band range, FWHM, and transitions of dopants in

Table 4.5 **Dopants, glass compositions, crystal phases, excitation wavelength, emitted band range, FWHM, and transitions of dopants in glass-ceramics established for broadband emission**

| Dopant ion(s) | Glass system/glass composition (mol.%) | Major crystal phase in glass-ceramics | Excitation (nm) | Emitted band range (nm) | Full width at half maximum (FWHM) (nm) | Transition | Ref. |
|---|---|---|---|---|---|---|---|
| $Ni^{2+}$ | $50SiO_2$–$16.7Al_2O_3$–$8.3Ga_2O_3$–$6.7MgO$–$8.3TiO_2+0.30NiO$ | $MgAl_2O_4$ | 980 | 1050–1600 | 240 | $^3T_{2g}(^3F) \rightarrow {}^3A_{2g}(^3F)$ | [116] |
| $Ni^{2+}$ | $Li_2O$–$Ga_2O_3$–$SiO_2+0.1NiO$ | $LiGa_5O_8$ | 976 | 1100–1600 | 300 | $^3T_{2g}(^3F) \rightarrow {}^3A_{2g}(^3F)$ | [117] |
| $Ni^{2+}$ | $3.4Na_2O$–$3.3K_2O$–$13.3Ga_2O_3$–$80.0SiO_2+0.1NiO$ | β-$Ga_2O_3$ | 980 | 1050–1600 | 250 | $^3T_{2g}(^3F) \rightarrow {}^3A_{2g}(^3F)$ | [118] |
| $Ni^{2+}$ | $58SiO_2$–$21ZnO$–$10Al_2O_3$–$5TiO_2$–$3Ga_2O_3$–$3K_2O+0.1$ wt.% NiO | $ZnAl_2O_4$ | 976 | 1100–1600 | 300 | $^3T_{2g}(^3F) \rightarrow {}^3A_{2g}(^3F)$ | [119] |
| $Ni^{2+}$ | $59.4SiO_2$–$2.9AlO_{1.5}$–$7.8ZnO$–$3.8MgO$–$0.9BaO$–$3.9TiO_2$–$1.5ZrO_2+xNiO$ ($x=0.01$, 0.05, 0.1, 0.5, 1.0 wt.%) | $ZnAl_2O_4$ | 980 | 1050–1550 | 300 | $^3T_{2g}(^3F) \rightarrow {}^3A_{2g}(^3F)$ | [120] |
| $Ni^{2+}$ | $66.5SiO_2$–$9.5Ga_2O_3$–$6.5Al_2O_3$–$7.5Na_2O$ $+0.15$ NiO | β-$Ga_2O_3$ | 980 | 1000–1600 | – | $^3T_{2g}(^3F) \rightarrow {}^3A_{2g}(^3F)$ | [121] |
| $Ni^{2+}$ | $30TiO_2$–$30BaO$–$30SiO_2$–$10Al_2O_3+xNiO$ ($x=0$, 0.05, 0.1, 0.2, 0.4, 0.8) | $BaAl_2Ti_6O_{16}$ | 365 | 1000–1600 | 350 | $^3T_{2g}(^3F) \rightarrow {}^3A_{2g}(^3F)$ | [122] |
| $Ni^{2+}$ | $SiO_2$–$Al_2O_3$–$Ga_2O_3$–$Li_2O+0.1$ wt.% NiO | $LiAl_5O_8$ | 980 | 1050–1600 | 250 | $^3T_{2g}(^3F) \rightarrow {}^3A_{2g}(^3F)$ | [123] |
| $Cr^{4+}$ | $SiO_2$–$Al_2O_3$–$ZnO$–$Li_2O$–$K_2O+0.1$ wt.% $Cr_2O_3$ | $Li_2ZnSiO_4$ | 808 | 1000–1550 | 250 | – | [124] |
| $Cr^{4+}$ | (wt.%) $29.8CaO$–$13.5Al_2O_3$–$27.8GeO_2$–$19.7Li_2CO_3$–$0.65\ Cr_2O_3$; (15.6–24.0) $Al_2O_3$–(16.4–25.2)MgO–(21.4–48.9) $SiO_2$–(14.3–22)$K_2O$–(4.8–7.4) $TiO_2+0.25Cr_2O_3$ | $Ca_2GeO_4$, $Mg_2SiO_4$ | 633 | 950–1350 | 240 | $^3T_2 \rightarrow {}^3A_2$ | [125] |

*Continued*

Table 4.5 **Continued**

| Dopant ion(s) | Glass system/glass composition (mol.%) | Major crystal phase in glass-ceramics | Excitation (nm) | Emitted band range (nm) | Full width at half maximum (FWHM) (nm) | Transition | Ref. |
|---|---|---|---|---|---|---|---|
| $Bi^{2+}$, $Bi^{+}$, or $Bi^{0}$ | $59SiO_2$–$19Al_2O_3$–6MgO–$6Li_2O$–$10TiO_2+0.5Bi_2O_3$ | β-$LiAlSiO_4$ | 808 | 1000–1550 | 270 | – | [126] |
| $Bi^{2+}$ or $Bi^{0}$ | $57.3SiO_2$–$26.2AlO_{1.5}$–10.6ZnO–0.9BaO–$4.0TiO_2$–$1.0Bi_2O_3$; $59.4SiO_2$–$22.9AlO_{1.5}$–6.8ZnO–3.8MgO–0.9BaO–$3.9TiO_2$–$1.5ZrO_2$–$1.0Bi_2O_3$; $30Na_2O$–$10Al_2O_3$–$59SiO_2$–$1.0Bi_2O_3$ | $ZnAl_2O_4$ | 808 | 1000–1600 | 250 | – | [127] |
| $Bi^{+}$–$Ni^{2+}$ | $58SiO_2$–21ZnO–$13Al_2O_3$–$5TiO_2$–$3Ga_2O_3+0.1NiO+0.5Bi_2O_3$ | $ZnAl_2O_4$ | 980 | 1100–1500 | 350 | $Bi^{+}$:$^3P_1 \rightarrow {}^3P_0$; $Ni^{2+}$ :$^3T_2(^3F) \rightarrow {}^3A_2(^3F)$ | [128] |
| $Cr^{3+}$–$Ni^{2+}$ | $13Li_2O$–$23Ga_2O_3$–$64SiO_2$–0.1NiO–$0.1Cr_2O_3$ | $LiGa_5O_8$ | 532 | 1100–1650 | 300 | $Cr^{3+}$:$^4T_2 \rightarrow {}^4A_2$; $Ni^{2+}$ :$^3T_2(^3F) \rightarrow {}^3A_2(^3F)$ | [129] |
| $Ni^{2+}$–$Er^{3+}$ | $51SiO_2$–$15Na_2O$–$20Ga_2O_3$–$14LaF_3+0.05ErF_3+0.1NiO$ | $Ga_2O_3$, $LaF_3$ | 980 | 1050–1650 | – | $Ni^{2+}$ :$^3T_2(^3F) \rightarrow {}^3A_2(^3F)$; $Er^{3+}$:$^4I_{13/2} \rightarrow {}^4I_{15/2}$ | [130] |
| $Yb^{3+}$–$Ni^{2+}$ | $(100-y)(13Li_2O$–$23Ga_2O_3$–$64SiO_2 0.1NiO)$–$yYb_2O_3$ ($y=0$, 0.25, 0.5,0.75, 1.0) | $LiGa_5O_8$ | 980 | 1100–1700 | 290 | $Yb^{3+}$:$^2F_{5/2} \rightarrow {}^2F_{7/2}$; $Ni^{2+}$ :$^3T_2(^3F) \rightarrow {}^3A_2(^3F)$ | [131] |
| $Tm^{3+}$–$Er^{3+}$ | $44SiO_2$–$28Al_2O_3$–17NaF–$11YF_3+0.1ErF_3+xTmF_3$ ($x=0.1$, 0.2, 0.4, 0.8, 1.6) | β-$YF_3$ | 792 | 1300–1700 | – | $Tm^{3+}$:$^3H_4 \rightarrow {}^3F_4$; $^3F_4 \rightarrow {}^3H_6$; $Er^{3+}$ :$^4I_{13/2} \rightarrow {}^4I_{15/2}$ | [132] |
| $Yb^{3+}$–$Er^{3+}$–$Tm^{3+}$ | $32SiO_2$–$9AlO_{1.5}$–$31.5CdF_2$–$18.5PbF_2$ – $5.5ZnF_2+1.5YbF_3+1.0ErF_3+1.0TmF_3$ | β-$PbF_2$ | 980 | 1500–1900 | – | $Tm^{3+}$:$^3H_4 \rightarrow {}^3F_4$; $^3F_4 \rightarrow {}^3H_6$; $Er^{3+}$ :$^4I_{13/2} \rightarrow {}^4I_{15/2}$ | [133] |

various glass-ceramics established for broadband emission use are summarized in Table 4.5 [116–133]. It is found from this table that different processing parameters significantly control the broadband emissions and their bandwidths. Some of them are described as follows.

Suzuki et al. [119] have reported the structural and optical properties of $Ni^{2+}$-doped $ZnAl_2O_4$ nanocrystal-grown glass-ceramics in the glass composition (mol.% for glass) of $58SiO_2$–21ZnO–$10Al_2O_3$–$5TiO_2$–$3Ga_2O_3$–$3K_2O$+0.1 wt.% NiO. Details are given in Table 4.5. Alteration of emission spectra ($\lambda_{exc}$=976 nm) of $Ni^{2+}$-doped ZnO–$Al_2O_3$–$SiO_2$ (ZAS) glass-ceramics with heat treatment time is illustrated in Fig. 4.23A. Energy-level diagram of $Ni^{2+}$ ion showing the transition levels in these glass-ceramics is also depicted in Fig. 4.23B. The emission in the 1100–1600 nm wavelength region is attributed to the $^3T_{2g}(^3F) \rightarrow {}^3A_{2g}(^3F)$ transition of $Ni^{2+}$ ions positioned in octahedral site of the $ZnAl_2O_4$ nanocrystals for the reason that the emission induced by 976 nm excitation was not detected from glasses heat-treated for <10 min. The emission intensity of the glasses increases, along with shifting toward lower wavelengths, with increase in heat treatment time. These trends suggest that there are several different kinds of octahedral $Ni^{2+}$ sites of different crystal field strengths. Additionally, octahedral $Ni^{2+}$ ions in low crystal field were formed in the early stage of heat treatment, and then, octahedral $Ni^{2+}$ ions of high crystal field strength were formed with the progress of heat treatment duration.

Dai et al. [126] have described the broadband NIR luminescence of Bi-doped $Li_2O$–$Al_2O_3$–$SiO_2$ (LAS) glass-ceramics yielding β-eucryptite (β-$LiAlSiO_4$) crystals in the glass composition (mol.%) of $59SiO_2$–$19Al_2O_3$–6MgO–$6Li_2O$–$10TiO_2$+$0.5Bi_2O_3$. Details are provided in Table 4.5. Emission spectra ($\lambda_{exc}$=808 nm),

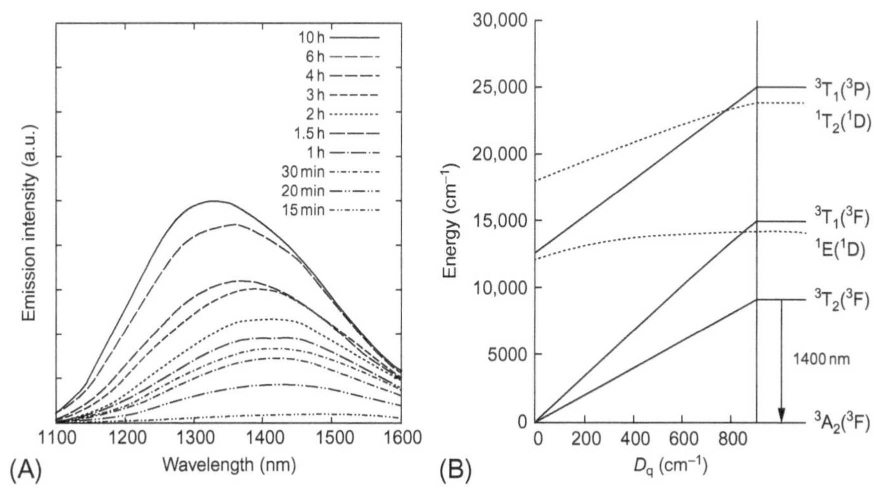


**Fig. 4.23** (A) Alteration of emission spectra of $Ni^{2+}$-doped ZnO–$Al_2O_3$–$SiO_2$ (ZAS) glass-ceramics with heat treatment time. (B) Energy-level diagram of $Ni^{2+}$ ion in ZAS glass-ceramics. Reproduced with permission from T. Suzuki, K. Horibuchi, Y. Ohishi, Structural and optical properties of ZnO–$Al_2O_3$–$SiO_2$ system glass–ceramics containing $Ni^{2+}$-doped nanocrystals, J. Non-Cryst. Solids 351 (2005) 2304–2309, Copyright © Elsevier.

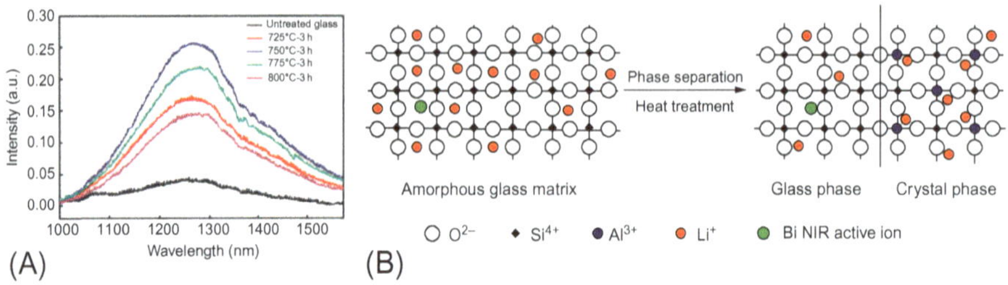


**Fig. 4.24** (A) Emission spectra ($\lambda_{exc} = 808$ nm) of Bi-doped $Li_2O$–$Al_2O_3$–$SiO_2$ (LAS) glass and glass-ceramics heat-treated at different temperatures for 3 h. (B) Mechanisms of phase separation and ion migration caused by heat treatment.
Reproduced with permission from N. Dai, H. Luan, Z. Liu, Y. Sheng, J. Peng, Z. Jiang, H. Li, L. Yang, J. Li, Broadband NIR luminescence of Bi-doped $Li_2O$–$Al_2O_3$–$SiO_2$ glass ceramics, J. Non-Cryst. Solids 358 (2012) 2970–2973, Copyright © Elsevier.

in the 1000–1550 nm range with 270 nm bandwidth, of Bi-doped LAS glass and glass-ceramics heat-treated at different temperatures for 3 h are illustrated in Fig. 4.24A. These NIR emission centers of Bi-doped glass and glass-ceramics are attributed to the lower valence state (e.g., $Bi^{2+}$ and $Bi^{+}$) of Bi ions. Mechanism of phase separation and ion migration caused by heat treatment is depicted in Fig. 4.24B. During phase separation, $Al^{3+}$ ion replaces the $Si^{4+}$ ion in the quartz tetrahedral sites of the glass matrix, and the charge balance is preserved by $Li^{+}$ ion. Since nearly all the $Li^{+}$ ions and partial $Al^{3+}$ ions enter into the crystal phase and replace nearly one-half of $Si^{4+}$ ions; consequently, very few $Li^{+}$ ions and residual $Al^{3+}$ ions still reside in the glassy phase. This phase separation and ion migration result in reduction of optical basicity in the glass phase. According to Duffy and Ingram's theory of optical basicity, the reduced optical basicity of the host glass facilitates the formation of low valence state of Bi [134–136].

Zhang et al. [128] have reported the bandwidth broadening of near-infrared emission through $ZnAl_2O_4$ nanocrystallization in the $Bi^{+}$-$Ni^{2+}$codoped glass in the composition (mol.%) of $58SiO_2$–$21ZnO$–$13Al_2O_3$–$5TiO_2$–$3Ga_2O_3 + 0.1NiO + 0.5Bi_2O_3$. Details are provided in Table 4.5. Emission spectra ($\lambda_{exc} = 980$ nm) of $Bi^{+}$-, $Ni^{2+}$-, and $Bi^{+}$–$Ni^{2+}$-doped glass-ceramics are illustrated in Fig. 4.25A. Singly $Ni^{2+}$-doped glass exhibits no emission in the NIR region, and the $Bi^{+}$ singly doped and $Bi^{+}/Ni^{2+}$ codoped glasses exhibit a single-band NIR peaking at 1100 nm. After heat treatment, the single-band luminescence feature of $Bi^{+}$-doped glass remains similar to the as-prepared glass, whereas another single-band NIR emission peaking at 1270 nm was perceived in $Ni^{2+}$ singly doped $ZnAl_2O_4$ nanocrystallized glass. The 1100 nm emission band of the $Bi^{+}$-doped glass and 1270 nm band of the $Ni^{2+}$-doped sample are assigned to the $^3T_2(F) \rightarrow {}^3A_2(F)$ transition of octahedral $Ni^{2+}$ and $^3P_1 \rightarrow {}^3P_0$ transition of $Bi^{+}$, respectively. Considerably, the codoped nanocrystallized glass shows an ultrabroadband emission peaking at 1160 nm with a FWHM of 350 nm. This is larger than that of the $Ni^{2+}$ singly doped sample and twice the width of $Bi^{+}$ singly doped one (around 160 nm). The observed unusual super-broadband NIR emission in $Bi^{+}/Ni^{2+}$ codoped nanocrystallized glass is mainly created from the overlapping fluorescence of $Bi^{+}$ and $Ni^{2+}$ active centers. This fact points

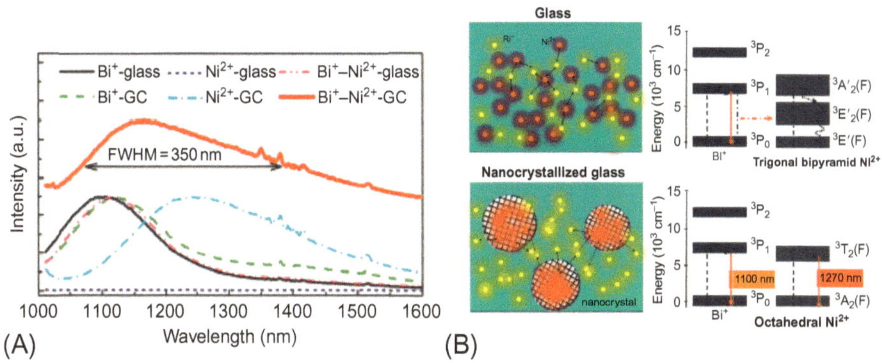


**Fig. 4.25** (A) Emission spectra ($\lambda_{exc}$ = 980 nm) of $Bi^+$-, $Ni^{2+}$-, and $Bi^+$–$Ni^{2+}$-doped glass-ceramics. (B) (Left) Schematic illustration of spatial distribution of $Bi^+$ and $Ni^{2+}$ ions in glass (top) and glass-ceramics (bottom) and (right) partial energy-level diagram of $Bi^+$ and $Ni^{2+}$ ions depicting the mechanisms of energy transfer (ET) and radiative/nonradiative transitions in glass and glass-ceramics. Excitation, nonradiative relaxation, ET, and emission transitions are represented by *dashed*, *dotted*, *dashed-dotted*, and *solid arrows*, respectively.
Reproduced with permission from K. Zhang, S. Zhou, Y. Zhuang, R. Yang, J. Qiu, Bandwidth broadening of near-infrared emission through nanocrystallization in Bi/Ni co-doped glass, Opt. Express 20 (2012) 8675–8680, Copyright © OSA.

toward the preferential incorporation of $Ni^{2+}$ ions into nanocrystals during nanocrystallization leading to an effective isolation of $Ni^{2+}$ from $Bi^+$ active centers. This results in a significant suppression of energy transfer between them. Fig. 4.25B depicts (left) schematic illustration of spatial distribution of $Bi^+$ and $Ni^{2+}$ ions in glass (top) and glass-ceramics (bottom) and (right) partial energy-level diagram of $Bi^+$ and $Ni^{2+}$ ions depicting the mechanisms of energy transfer (ET) and radiative/nonradiative transitions in glass and glass-ceramics. Excitation, nonradiative relaxation, ET, and emission transitions are represented by dashed, dotted, dashed-dotted, and solid arrows, respectively.

Wu et al. [129] have described the enhanced near-infrared emission from $Ni^{2+}$ in $Cr^{3+}$–$Ni^{2+}$ codoped transparent glass-ceramics containing $LiGa_5O_8$ crystals in the glass composition (mol.%) of $13Li_2O$–$23Ga_2O_3$–$64SiO_2$–$0.1NiO$–$0.1Cr_2O_3$. Details are given in Table 4.5. Emission spectra ($\lambda_{exc}$ = 532 nm) of $Cr^{3+}$-, $Ni^{2+}$-, and $Cr^{3+}$–$Ni^{2+}$-doped glass-ceramics are illustrated in Fig. 4.26A. Inset shows red emission ($\lambda_{exc}$ = 532 nm) at 720 nm from $Cr^{3+}$ in $Cr^{3+}$ and $Cr^{3+}$–$Ni^{2+}$-doped glass-ceramics. Fig. 4.26B shows the energy-level diagram of the broadband emission mechanism in $Cr^{3+}$–$Ni^{2+}$ codoped glass-ceramics. Excitation and emission transitions are represented by solid arrows, and nonradiative and energy transfer are represented by dashed arrows. The peaks around 1064 and 1440 nm pointed out by black arrows are the second diffraction peaks of 532 nm excitation and 720 nm emission lights, respectively. Under excitation at 532 nm, $Cr^{3+}$-doped glass-ceramics exhibit a NIR emission peaking at 960 nm with a small hump around 880 nm in the 800–1200 nm wavelength region. The broadband emission at 960 nm is assigned to the $^4T_2 \rightarrow {}^4A_2$ transition of

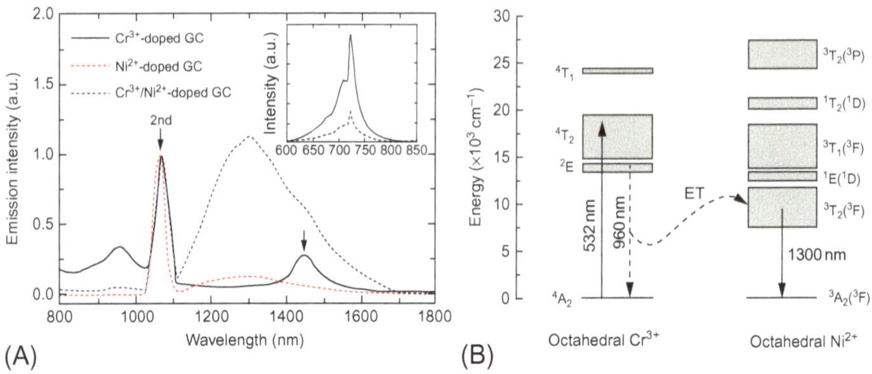


**Fig. 4.26** (A) Emission spectra ($\lambda_{exc}=532$ nm) of $Cr^{3+}$-, $Ni^{2+}$-, and $Cr^{3+}$–$Ni^{2+}$-doped glass-ceramics. Inset shows red emission ($\lambda_{exc}=532$ nm) at 720 nm from $Cr^{3+}$ in $Cr^{3+}$ and $Cr^{3+}$–$Ni^{2+}$-doped glass-ceramics. (B) Energy-level diagram depicts the broadband emission mechanism in $Cr^{3+}$–$Ni^{2+}$ codoped glass-ceramics. Excitation and emission transitions are represented by solid arrows, and nonradiative and energy transfer are represented by *dashed arrows.*
Reproduced with permission from B. Wu, S. Zhou, J. Ruan, Y. Qiao, D. Chen, C. Zhu, J. Qiu, Enhanced near-infrared emission from $Ni^{2+}$ in $Cr^{3+}$/$Ni^{2+}$ codoped transparent glass ceramics, Appl. Phys. Lett. 92 (2008) 151102 (3 pages), Copyright © AIP.

$Cr^{3+}$ ions, and the weak hump around 880 nm is due to the vibronic sideband induced by the $^4T_2$ vibrational transition. For $Ni^{2+}$-doped glass-ceramics, an NIR emission peaking at 1300 nm with the FWHM of nearly 300 nm is observed. This broadband emission is attributed to the $^3T_2(^3F)\rightarrow{}^3A_2(^3F)$ transition of octahedral $Ni^{2+}$ in $LiGa_5O_8$ nanocrystals of the glass-ceramics. In $Cr^{3+}$–$Ni^{2+}$ codoped glass-ceramics, two most ostensible features are noticed. Firstly, the emission intensity of $Cr^{3+}$ at 960 nm is considerably reduced compared with that in $Cr^{3+}$-doped glass-ceramics, and secondly, the emission intensity of $Ni^{2+}$ at 1300 nm is greatly intensified with almost 7.3 times compared with that in $Ni^{2+}$-doped glass-ceramics. This singularity is ascribed to energy transfer (ET) from $Cr^{3+}$ to $Ni^{2+}$ in $LiGa_5O_8$ nanocrystals of the glass-ceramics. Furthermore, an intense red emission at 720 nm in $Cr^{3+}$ single and $Cr^{3+}$–$Ni^{2+}$ codoped glass-ceramics is also obtained, which is attributed to the zero-phonon line of $^2E\rightarrow{}^4A_2$ transition of octahedral $Cr^{3+}$ in the $LiGa_5O_8$ nanocrystals of glass-ceramics.

Zhou et al. [130] have demonstrated the broadband emission in the near-infrared (NIR) range 1050–1650 nm in $Ni^{2+}$–$Er^{3+}$ codoped $Ga_2O_3$ and $LaF_3$ nanocrystals precipitated glass-ceramics in the glass composition (mol.%) of $51SiO_2$–$15Na_2O$–$20Ga_2O_3$–$14LaF_3+0.05ErF_3+0.1NiO$. Details are given in Table 4.5. NIR emission spectra ($\lambda_{exc}=980$ nm) of these glass and glass-ceramics heat-treated at 700°C and 740°C are depicted in Fig. 4.27A. Energy-level diagram illustrating the mechanisms of energy transfer and radiative/nonradiative transitions in the glass and glass-ceramics is also shown in Fig. 4.27B. Excitation, nonradiative relaxation, energy transfer, and emission transitions are represented by dashed, dotted, dashed-dotted, and solid arrows, respectively. It is found that the glass-ceramics heat-treated at 740°C exhibits a distinctive ultrabroadband NIR emission under excitation at 980 nm, which

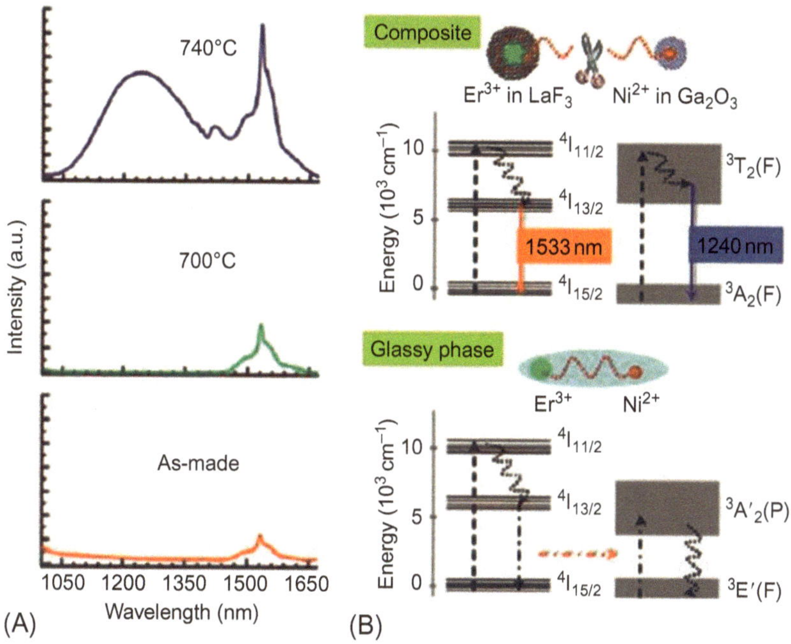


**Fig. 4.27** (A) Near-infrared emission spectra ($\lambda_{exc} = 980$ nm) of $Er^{3+}$–$Ni^{2+}$ codoped glass and glass-ceramics heat-treated at 700°C and 740°C. (B) Energy-level diagram illustrating the mechanisms of energy transfer and radiative/nonradiative transitions in the glass and glass-ceramics. Excitation, nonradiative relaxation, energy transfer, and emission transitions are represented by *dashed*, *dotted*, *dashed-dotted*, and *solid arrows*, respectively.
Reproduced with permission from S. Zhou, N. Jiang, K. Miura, S. Tanabe, M. Shimizu, M. Sakakura, Y. Shimotsuma, M. Nishi, J. Qiu, K. Hirao, Simultaneous tailoring of phase evolution and dopant distribution in the glassy phase for controllable luminescence, J. Am. Chem. Soc.132 (2010) 17945–17952, Copyright © ACS.

covers the entire telecommunication low loss window of 1200–1600 nm. The experimental two emission bands peaking at 1240 and 1533 nm produce from the $^3T_2(^3F) \rightarrow {}^3A_2(^3F)$ transition of $Ni^{2+}$and the $^4I_{13/2} \rightarrow {}^4I_{15/2}$ transition of $Er^{3+}$, respectively. On the other hand, the glass-ceramics heat-treated at low temperature without incorporation of $Ni^{2+}$ and $Er^{3+}$ ions into the $Ga_2O_3$ and $LaF_3$ nanocrystals display only very feeble and narrow emission around 1533 nm as a consequence of predominant energy transfer between the $Ni^{2+}$ and $Er^{3+}$ ions (Fig. 4.27B, bottom).

### 4.3.2 Ferroelectric glass-ceramics

Ferroelectric crystals are those that have polar noncentrosymmetrical crystal structure, and their polarization can be reversed along the polar axis by reversing the polarity of the electric field, that is, they are spontaneously polarized materials with reversible polarization. Therefore, all the ferroelectric materials, by default, are simultaneously pyroelectric and piezoelectric. Some examples of ferroelectric crystals are

$LiNbO_3$, $KNbO_3$, $SrBi_2Nb_2O_9$, $BaTiO_3$, $PbTiO_3$, $Bi_4Ti_3O_{12}$, $BaBi_4Ti_4O_{15}$, $LaBGeO_5$, $Pb_5Ge_3O_{11}$, $KNbSi_2O_7$, $LiTaO_3$, $SrBi_2Ta_2O_9$, $Bi_2WO_6$, $Bi_2VO_{5.5}$, etc. As a result of structural anisotropy, they exhibit several nonlinear optical (NLO) properties, for example, harmonic generation (change of wavelength of light), electro-optical effect (change of refractive index with electric field), and photorefraction (change of refractive index under influence of light). Ferroelectric crystals are used in capacitors, piezoelectric, pyroelectric, and different optical devices. Nevertheless, the use of single crystals for these applications is often limited due to difficulties in obtaining defect-free, homogeneous, and large-size crystals. Moreover, the preparation technique of single crystals is sophisticated and expensive.

In contrast, ferroelectric glass-ceramics, in comparison with single crystals and sintered ceramics, offer several advantages such as the following:

- **(i)** Preparation technique of glass-ceramics is simple, easy, and cost effective.
- **(ii)** Homogeneous, pore-free, and low dielectric loss materials can be obtained.
- **(iii)** Transparency, crystal size, and microstructure can be engineered precisely for particular application by means of controlled heat treatment protocol.
- **(iv)** Properties can be tailored by changing the type of ferroelectric crystal.
- **(v)** High energy density, dielectric constant, and breakdown voltage capacitors can be obtained for power electronics and pulsed power applications.
- **(vi)** Dielectric constant is almost independent of temperature.
- **(vii)** Coefficient of thermal expansion is adjustable to compatible with other multicomponent system.

There is a wide range of applications of ferroelectric glass-ceramics in various technologies, for example, in energy storage devices; optoelectronic devices; second harmonic generators; multilayer capacitors; multilayer inductors; multilayer microwave resonators and filters; hydrophone detectors; pyroelectric detectors; microwave absorbers; frequency-agile microwave electronic devices; microelectromechanical system; electro-optical, nonlinear optical, and integrated optical systems; and dielectric pastes in thick film and ceramic integration technology.

Jain [9] has reviewed various types of transparent ferroelectric glass-ceramics and scholarly described the material systems, devitrification by heat treatment, surface and volume crystallization, alignment of crystallites, tailored structure, effects of crystallite size on ferroelectric response, optical characteristics, and future prospects of these glass-ceramics. Xi [137] has reviewed various preparation techniques and guidelines for optimization of different process parameters to achieve ferroelectric and ferromagnetic microcrystalline glass-ceramics. Crystal phases, glass compositions, processing, and dielectric constants of various ferroelectric glass-ceramics are summarized in Table 4.6 [138–185]. These glass-ceramics are categorized into the ferroelectric crystal families of niobate, titanate, germanate, niobosilicate, tantalate, tungstate, and vanadate. It is found that properties of glass-ceramics for a particular crystal are largely altered with glass composition and processing parameters. Therefore, for a particular functional use, these parameters should be optimized to acquire the required properties. Some examples are described as follows.

Chaliha et al. [143] have reported the structure, dielectric, and optical properties of transparent $Nd^{3+}$-doped and undoped $KNbO_3$ nanocrystal-grown glass-ceramics in

Table 4.6 **Crystal phases, glass compositions, processing, and dielectric constants of various ferroelectric glass-ceramics**

| Major crystal phase in glass-ceramics | Glass system/glass composition (mol.%) | Glass melting (time)/ annealing (time) temperature (°C) | Heat treatment temperature/time (°C $h^{-1}$) | Dielectric constant at 25°C, $\varepsilon_r$ | Ref. |
|---|---|---|---|---|---|
| ***Niobate*** | | | | | |
| $LiNbO_3$ | $(100-x)LiNbO_3$–$xSiO_2$($20 \leq x \leq 35$) | 1300–1350/550(1) | 600, 650, 700, 975/1 | 75–190 | [138] |
| $LiNbO_3$ | $30Li_2O$–$30Nb_2O_5$–$35SiO_2$–$5Al_2O_3$ | 1400(1)/400(1) | 500, 625, 750/2 | – | [139] |
| $LiNbO_3$ | (17.5–35.8)$Li_2O$–(17.0–32.3) $Nb_2O_5$–(29.8–61.0)$SiO_2$–(4.5–7.6) $Al_2O_3$ | 1300–1450(1)/500 (2–3) | 610–900/(0.17–1.0) | 155–185 | [140] |
| $LiNbO_3$ | $50Li_2O$–$40P_2O_5$–$10Nb_2O_5$ | 1200(1)/300 | 450–500/(2–5) | 41–98 | [141] |
| $LiNbO_3$ | $(100-x)LiBO_2$–$xNb_2O_5$ ($5 \leq x \leq 25$) | 1100–1250(1)/300(6) | 500/3 | 13–30 | [142] |
| $KNbO_3$ | $25K_2O$–$25Nb_2O_5$–$50SiO_2$ | 1550(2)/600(1) | 800/(1–100) | 31–33 | [143] |
| $KNbO_3$ | $15K_2O$–$15Nb_2O_5$–$70TeO_2$ | 900-1050(0.67) | 425/1 | 44 | [144] |
| $NaNbO_3$ | $35SiO_2$–$31Nb_2O_5$–$19Na_2O$–$11K_2O$–$2CdO$–$2B_2O_3$ | 1550(2) | 610/(8–142) | – | [145] |
| $Sr^{II}Nb^{IV}O_3$ | $53.75SiO_2$–$18.25K_2O$–$9Bi_2O_3$–$9SrO$–$9Nb_2O_5$–$0.5CeO_2$–$0.5Eu_2O_3$ | 1350(2)/450(2) | 450/10+500/(15–200) | – | [146] |
| $K_{0.5}Na_{0.5}NbO_3$ | $SiO_2$–$Nb_2O_5$–$Na_2O$–$K_2O$ | 1300(1) | 525–575 | – | [147] |
| $NaNbO_3+NaBa_2Nb_5O_{15}$ | $20Na_2O$–$4BaO$–$26Nb_2O_5$–$50SiO_2$ | 1500(2) | 790, 880, 1000/1 | 2.16–21.13 | [148] |
| $NaNbO_3+NaBa_2Nb_5O_{15}$ | $17Na_2O$–$17BaO$–$34Nb_2O_5$–$31SiO_2$–$xLa_2O_3$ ($x=0$–3) | 1500(3)/450 | 1000/3 | 328–754 | [149] |
| $Sr_{0.47}Ba_{0.53}Nb_2O_6$ | $SrO$–$BaO$–$Nb_2O_5$–$SiO_2$ | 1300–1600(6)/600(12) | 850, 865, 900/6 | 20–180 | [150] |

*Continued*

**Table 4.6 Continued**

| Major crystal phase in glass-ceramics | Glass system/glass composition (mol.%) | Glass melting (time)/annealing (time) temperature (°C) | Heat treatment temperature/time (°C h$^{-1}$) | Dielectric constant at 25°C, $\varepsilon_r$ | Ref. |
|---|---|---|---|---|---|
| $Sr_{0.33}Ba_{0.67}Nb_2O_6$ | 10.2PbO–19.4BaO–8.8SrO–1.15$Al_2O_3$–19.55$Nb_2O_5$–5.8$B_2O_3$–17.1$SiO_2$ | 1500(2) | 700–1000/2 | – | [151] |
| $Sr_{0.75}Ba_{0.25}Nb_2O_6$ | 21BaO–21SrO–18$Nb_2O_5$–10$B_2O_3$–30$SiO_2$ + (0–8.8)$CeO_2$ | 1450(1)/600(10) | 800/2 | 39–49 | [152] |
| $SrBi_2Nb_2O_9$ | $(100-x)Li_2B_4O_7$–$x$(SrO–$Bi_2O_3$–$Nb_2O_5$) ($10 \leq x \leq 60$) | 1100(1)/325(6) | 500, 550, 625/6 | 49–86 | [153–155] |
| $SrBi_2Nb_2O_9$ | 16.66SrO–16.66($(1-x)Bi_2O_3$–$xSm_2O_3$)–16.66$Nb_2O_5$–50$Li_2B_4O_7$ ($0 \leq x \leq 0.5$) | 1300(1)/400(6) | 530/3 | 45 | [156] |
| $Bi_3TiNbO_9$ | **(i)** (33–50)$B_2O_3$–(20–40)$Bi_2O_3$–(10–20)$TiO_2$–(10–15)$Nb_2O_5$<br>**(ii)** (30–50)$SiO_2$–(25–40)$Bi_2O_3$–(10–23)$TiO_2$–(8–20)$Nb_2O_5$ | 1100–1250(2) | **(i)** 700/2<br>**(ii)** 860/2 | **(i)** 48–78<br>**(ii)** 82–154 | [157] |
| ***Titanate*** | | | | | |
| $BaTiO_3$ | (33–40)BaO–(27–34)$TiO_2$–11$Al_2O_3$–22$SiO_2$ | 1550(2–3)/650(3) | 850–1000/2 | 75–430 | [158] |
| $SrTiO_3$ | (SrO–$TiO_2$)–(2$SiO_2$–$B_2O_3$)–($(1-y)K_2O \cdot yBaO$) ($y = 0.3$, 0.5, 0.7, 1.0) | 1250–1300(1)/450(3) | 850–900/(3–12) | 27–53 | [159] |
| $PbTiO_3$ | 35PbO–35$TiO_2$–20$SiO_2$–10$B_2O_3$ | 1450–1470(0.33)/400(2) | 455–485/(0–3) + 560–630/1 | 140–392 | [160] |
| $PbTiO_3$ | 39PbO–1BaO–25$TiO_2$–9.8$Al_2O_3$–1$SiO_2$–24.2$B_2O_3$ | 1200(2) | 550–700/4 | 75–490 | [161] |

| | | | | | |
|---|---|---|---|---|---|
| $Pb_xBa_{1-x}TiO_3$ (ss) | **(i)** $29BaTiO_3$–$17PbTiO_3$–$38TiO_2$–$13Al_2O_3$–$3B_2O_3$ **(ii)** $29BaTiO_3$–$17PbTiO_3$–$37TiO_2$–$15B_2O_3$ | 1450(2) | 750–950/(0.33–2.0) | 20–680 | [162] |
| $Ba_xSr_{1-x}TiO_3$ | 29.36BaO–7.34SrO–21.2$SiO_2$–(30.5–31.5)$TiO_2$–(6.6–10.6) $Al_2O_3$+(0–8)$AlF_3$+(0–1)$MnO_2$ | 1550(2–3)/650(3) | 1000–1100/2 | 210–384 | [163] |
| $Ba_{0.7}Sr_{0.3}TiO_3$ | $70BaTiO_3$–$30SrTiO_3$+($B_2O_3$, $BaF_2$, $P_2O_5$) | 1500(2)/600(10) | 550–750/3 +850–1200/2 | 450–520 | [164] |
| $Ba_xSr_{1-x}TiO_3$ | $28.8BaCO_3$–$7.2SrCO_3$–$14Al_2O_3$–$20SiO_2$–$28TiO_2$–$2BaF_2$ | 1550(2–3)/650(3) | **(i)** 850–1100/2 **(ii)** 750–950/2 +1100/2 | 60–510 | [165] |
| $Pb_{1-x}Sr_xTiO_3$ | (($Pb_{1-x}Sr_x$)O.$TiO_2$)–($2SiO_2$·$B_2O_3$)–($K_2O$) ($x$=0, 0.1, 0.2, 0.3, 0.4) | 1050–1250(1)/400(3) | 560–750/3 | – | [166] |
| $Bi_4Ti_3O_{12}$ | (35–43)$Bi_2O_3$–(26–35)$TiO_2$–(25–40)$B_2O_3$ | 1100–1200(2) | 700/2 | – | [167] |
| $Bi_4Ti_3O_{12}$ | $(100-x)$($K_2O$–$2SiO_2$)–$xBi_4Ti_3O_{12}$ ($x$=10, 20, 30, 40, 45, 50) | 1000–1150(2)/465(2) | 650–930/2 | 34–38 | [168] |
| $BaBi_4Ti_4O_{15}$ | $(100-x)$($K_2O$–$2SiO_2$)–$xBaBi_4Ti_4O_{15}$ ($x$=10, 20, 30, 40, 50, 60) | 1150–1200(2)/500(2) | 520/10+610/(1–100) | 28–56 | [169] |
| ***Germanate*** | | | | | |
| $LaBGeO_5$, $PrBGeO_5$ | (0–35)$La_2O_3$–(5–25)$Pr_2O_3$–(20–35)$B_2O_3$–(45–60)$GeO_2$ | 1100–1200 | 870–985 | 60–100 | [170] |
| $Pb_5Ge_3O_{11}$ | $Pb_5Ge_3O_{11}$–$xPbNb_2O_6$ ($x$=0.0, 0.5, 1, 2, 3) | 1000–1150 | 667/48 | 12–13 | [171] |
| $Bi_2GeO_5$ | $60BiO_{1.5}$–$20GeO_2$–$20BO_{1.5}$ | 1000–1050(0.25–0.50) | 475/(4–12) | 5–8 | [172] |

*Continued*

Table 4.6 **Continued**

| Major crystal phase in glass-ceramics | Glass system/glass composition (mol.%) | Glass melting (time)/ annealing (time) temperature (°C) | Heat treatment temperature/time (°C $h^{-1}$) | Dielectric constant at 25°C, $\varepsilon_r$ | Ref. |
|---|---|---|---|---|---|
| ***Niobosilicate*** | | | | | |
| $KNbSi_2O_7$, $K_3Nb_3O_6Si_2O_7$, $K_3Nb_3Si_2O_{12}$, $K_2(NbO)_2Si_4O_{12}$ | $K_2O$–$Nb_2O_5$–$SiO_2$ | 1530–1560(1.5) | 700–1100/0.5–4 | – | [173–175] |
| ***Tantalate*** | | | | | |
| $LiTaO_3$ | 25.53$Li_2O$–21.53$Ta_2O_5$–35.29$SiO_2$–17.65$Al_2O_3$ | 1600(2)/600(4) | 680/(0,1,3,5,7,10) | 19–29 | [176–178] |
| $SrBi_2Ta_2O_9$ | (100 – x)$Li_2B_4O_7$–$x$$SrBi_2Ta_2O_9$ ($0 \leq x \leq 20$) | 1100(0.5)/200(6) | 500/8 | – | [179] |
| $SrBi_2Ta_2O_9$ | 17.5$K_2O$–51.3$SiO_2$–10SrO–10$Bi_2O_3$–10$Ta_2O_5$+0.5 $Dy_2O_3$+0.7$CeO_2$ | 1470(2)/530(2) | 550/2+580/ (0,20,30,50,100) | 16.6–17.6 | [180] |
| ***Tungstate*** | | | | | |
| $Bi_2WO_6$ | $(1-x)Li_2B_4O_7$–$xBi_2WO_6$ ($0 \leq x \leq 0.35$) | 1100(1)/200(6) | 450/6 | 17–32 | [181–184] |
| ***Vanadate*** | | | | | |
| $Bi_2VO_{5.5}$ | $2xBi_2O_3$–$xV_2O_5$–$(100-3x)SrB_4O_7$ ($x$=15, 20, 25) | 1100(1)/300(12) | 410/12 | 15–30 | [185] |

**Fig. 4.28** (A) Dielectric constant of $K_2O$–$Nb_2O_5$–$SiO_2$ (KNS) glass and glass-ceramics as a function of heat treatment time at 800°C. Field emission scanning electron microscopic photomicrographs depicting the morphology of strontium barium niobate crystals heat treatment at (B) 725°C/120 min and (C) 775°C/22 min.
(A) Reproduced with permission from R.S. Chaliha, K. Annapurna, A. Tarafder, V.S. Tiwari, P.K. Gupta, B. Karmakar, Structure, dielectric and optical properties of transparent $Nd^{3+}$: $KNbO_3$ nanocrystalline glass–ceramics, Opt. Mater. 32 (2010) 1202–1209, Copyright © Elsevier.
(B) Reproduced with permission from C.-T. Cheng, M. Lanagan, B. Jones, J.-T. Lin, M.-J. Pan, Crystallization kinetics and phase development of PbO–BaO–SrO–$Nb_2O_5$–$B_2O_3$–$SiO_2$-based glass–ceramics, J. Am. Ceram. Soc., 88 (2005) 3037–3042, Copyright © The American Ceramic Society.

the glass composition (mol.%) of $25K_2O$–$25Nb_2O_5$–$50SiO_2$. Details are given in Table 4.6. Fig. 4.28A shows the change of dielectric constant of these glass-ceramics as a function of heat treatment time (1–100 h) at 800°C. As-quenched $Nd^{3+}$-doped potassium niobium silicate glass has displayed relatively a higher value of dielectric constant ($\varepsilon = 19$) compared with that of the sodalime silicate ($\varepsilon = 7$–$10$) and borosilicate glasses ($\varepsilon = 4.5$–$8$) as a result of high spontaneous polarization because of empty d-orbital ($4d^0$) electronic configuration and high ionic refraction of $Nb^{5+}$ ion ($R_i = 24.5$). The magnitude of dielectric constant rises steeply (from $\varepsilon = 19$ to $\varepsilon = 33$) in heat-treated glass-ceramics, and thereafter, it maintains saturation for any further heat treatment duration. This advocates that on heat treatment, in the initial stages, separation of silica-rich phase and K-Nb-enriched phases occurs and, with the further progress of heat treatment, initial precipitation of $KNbO_3$ having high dielectric constant ($\varepsilon = 137$) and spontaneous polarization, $P_s = 0.41$ C $m^{-2}$, occurs. Thus, accumulation of $K^+$ ions in the phase-separated glass matrix, in the beginning, causes a significant increase of dielectric constant. With further heat treatment time, the heavy $Nb^{5+}$ ion enters into the residual glass matrix in order to preserve the required crystallite composition, which tends to maintain a saturated value of dielectric constant. The variation in crystallite size distributions and also the distribution of the $KNbO_3$ phase in the microstructure are the causes for the differences in the dielectric constant values among the heat-treated samples. Fig. 4.28 illustrates the field emission scanning electron microscopic (FESEM) photomicrographs depicting the change of morphology of strontium barium niobate crystals in the glass-ceramics obtained by heat treatment at (B) 725°C/120 min and (C) 775°C/22 min (see Table 4.6 for details) [151].

Prasad et al. [154] have reported the evolution and characterization of fluorite-like nano-$SrBi_2Nb_2O_9$ (SBN) crystal phase in the SrO–$Bi_2O_3$–$Nb_2O_5$–$Li_2B_4O_7$ glass

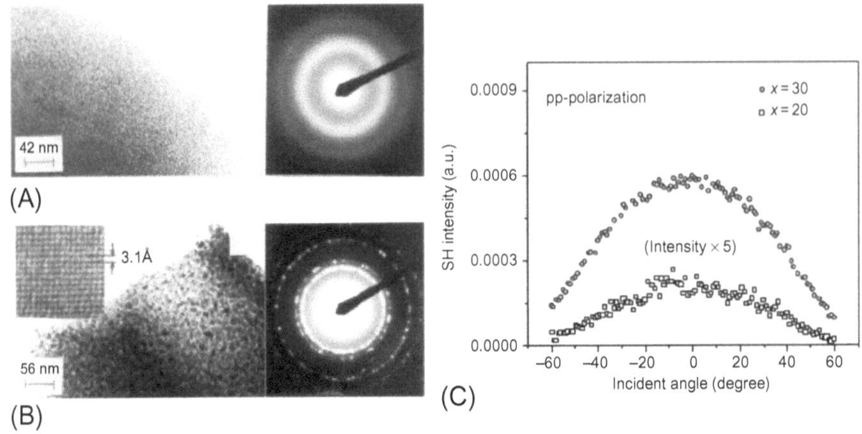


**Fig. 4.29** Transmission electron microscopic images with SAED patterns of (A) glass and (B) glass-ceramics heat-treated at 500°C. Inset of (B) shows the HRTEM lattice image. (C) Second harmonic (SH) intensities of two different compositional glass-ceramics as a function of incident angle.
Reproduced with permission from N.S. Prasad, K.B.R. Varma, Y. Takahashi, Y. Benino, T. Fujiwara, T. Komatsu, Evolution and characterization of fluorite-like nano-$SrBi_2Nb_2O_9$ phase in the $SrO–Bi_2O_3–Nb_2O_5–Li_2B_4O_7$ glass system, J. Solid State Chem. 173 (2003) 209–215, Copyright © Elsevier.

system. Details are given in Table 4.6. Fig. 4.29 shows the transmission electron microscopic (TEM) images with SAED patterns of (A) glass and (B) glass-ceramics heat-treated at 500°C. Inset of (B) shows the HRTEM lattice image. The SAED pattern (Fig. 4.29A) of the as-quenched glass depicts its amorphous nature. The occurrence of virtually sharp spherical rings around the bright central region advocates the presence of local-ordered regions. Fig. 4.29B shows the 12–20 nm spherical crystallites of $SrBi_2Nb_2O_9$ in the glass matrix. The d-spacings obtained from the SAED pattern are in close agreement with that of the fluorite ($CaF_2$) phase with lattice spacing of 3.1 Å as revealed in HRTEM. Fig. 4.29C illustrates the second harmonic (SH) intensities of two different compositional glass-ceramics as a function of incident angle. It is seen that the SH intensity increases with increase in ferroelectric SBN crystal content in the glassy phase. There is also a remarkable variation in SHG intensity with the angle of incidence.

Molla et al. [169] have described the processing and properties of $Eu^{3+}$-doped barium bismuth titanate ($BaBi_4Ti_4O_{15}$) glass-ceramic nanocomposites in the glass composition (mol.%) $(100-x)(K_2O–2SiO_2)–xBaBi_4Ti_4O_{15}$ where $x=10$, 20, 30, 40, 50, and 60. Details are given in Table 4.6. The dielectric constant of these glass-ceramics has found to be increased in the range 28–56 with heat treatment. XRD patterns of $Eu^{3+}$-doped barium bismuth titanate glass-ceramics heat-treated at 610°C for different durations are depicted in Fig. 4.30A. The patterns will match with the data of JCPDS file card 35-0757 for the crystal phase of $BaBi_4Ti_4O_{15}$ (BBiT). Fig. 4.30B shows the (left) photoluminescence excitation ($\lambda_{em}=613$ nm) and (right) emission ($\lambda_{exc}=466$ nm) spectra of $Eu^{3+}$-doped barium bismuth titanate glass-ceramics heat-treated at 520°C/10 h for

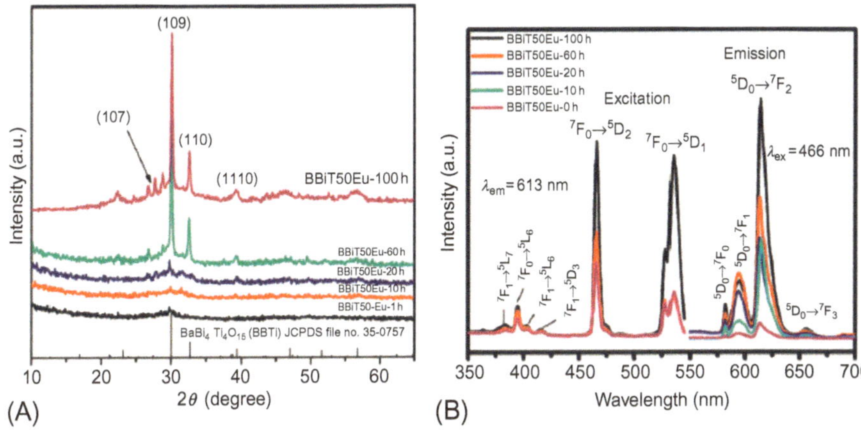


**Fig. 4.30** (A) XRD patterns of $Eu^{3+}$-doped barium bismuth titanate glass-ceramics heat-treated at 610°C for different durations. (B) (Left) Photoluminescence excitation ($\lambda_{em} = 613$ nm) and (right) emission ($\lambda_{exc} = 466$ nm) spectra of $Eu^{3+}$-doped barium bismuth titanate glass-ceramics heat-treated at 520°C/10 h for nucleation and 610°C for different durations for crystal growth. Reproduced with permission from A.R. Molla, A. Tarafder, S. Mukherjee, S.K. Mohanty, B. Karmakar, Processing and properties of $Eu^{3+}$-doped barium bismuth titanate ($BaBi_4Ti_4O_{15}$) glass–ceramic nanocomposites, J. Am. Ceram. Soc. 96 (2013) 2387–2395, Copyright © The American Ceramic Society.

nucleation followed by at 610°C for different durations (0–100 h) for crystal growth. The excitation spectra consist of two intense bands at 396 and 466 nm are assigned to the f-f transitions of $^7F_0 \rightarrow {}^5L_6$ and $^7F_0 \rightarrow {}^5D_2$ levels of $Eu^{3+}$ ions, respectively. Their intensities increase with progress of heat treatment time. The main emission band peaking at 613 nm is attributed to the induced electric dipole (ED) transition of $^5D_0 \rightarrow {}^7F_2$ of $Eu^{3+}$ and is dominant over the emission band near 592 nm, which is ascribed to the $Eu^{3+}$ magnetic dipole (MD) transition of $^5D_0 \rightarrow {}^7F_1$. The emission intensity of the $^5D_0 \rightarrow {}^7F_2$ (ED) transition is enhanced up to 15-fold on heat treatment for 100 h compared with the precursor glass. This fact advocates that the $Eu^{3+}$ ions entered into the low phonon crystal phase of BBiT by replacing $Bi^{3+}$ ions of the same valency and closeness of their ionic radii ($Eu^{3+} = 0.95$ Å, $Bi^{3+} = 1.02$ Å, and $Ti^{4+} = 0.61$ Å). In addition, the generated BBiT crystals are noncentrosymmetrical, and for this reason, the electric dipole (ED) transition of $^5D_0 \rightarrow {}^7F_2$ (ED) is the dominant one in these glass-ceramics. Consequently, the local field asymmetry factor based on the intensity ratio of $I_{ED}$ to $I_{MD}$ is found to be 2.47 for the precursor glass and 5.50, 2.05, 2.13, and 3.99 for the glass-ceramics heat-treated for 10, 20, 60, and 100 h, respectively.

Pengpat and Holland [172] have reported the glass-ceramics containing ferroelectric bismuth germanate ($Bi_2GeO_5$) orthorhombic crystal in the glass composition (mol.%) of $60BiO_{1.5}$–$20GeO_2$–$20BO_{1.5}$. Details are given in Table 4.6.The microstructures of the glass-ceramics have been found to be altered remarkably on changing the crystallization heat treatment temperature. Examples are shown in Fig. 4.31, which illustrates the SEM backscattered electron images of cross section of $BiGeB_6$

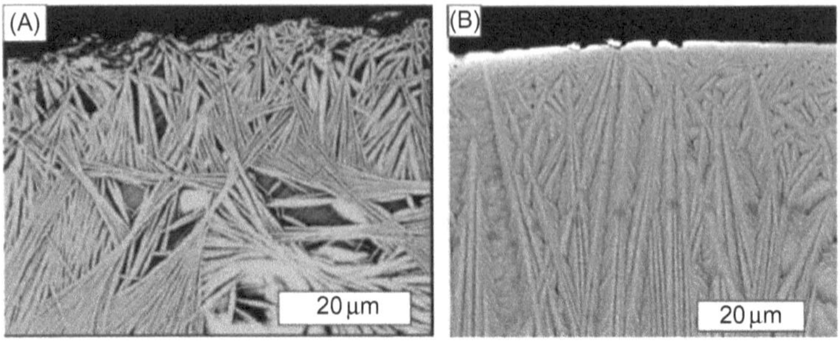


**Fig. 4.31** SEM backscattered electron images of cross section of $BiGeB_6$ glass-ceramics heat-treated at (A) 467°C/4 h and (B) 475°C/4 h.
Reproduced with permission from K. Pengpat, D. Holland, Glass-ceramics containing ferroelectric bismuth germanate ($Bi_2GeO_5$), J. Euro. Ceram. Soc. 23 (2003) 1599–1607, Copyright © Elsevier.

glass-ceramics heat-treated at (A) 467°C/4 h and (B) 475°C/4 h. The sample heat-treated at 467°C exhibits spherulites whereas at 475°C the surface dendritic crystallization. The latter is preferable for enhance pyroelectric or piezoelectric properties. The dielectric and ferroelectric properties of the glass-ceramics demonstrate that they are ferroelectric at room temperature with spontaneous polarization ($P_s$) of 14 μC $cm^{-2}$ and coercive field ($E_r$) of 5.6 kV $cm^{-1}$.

Lately, Tarafder et al. [177] have reported the second harmonic generation (SHG) in ferroelectric $LiTaO_3$ (LT) and $KNbO_3$ (KN) containing bulk nano glass-ceramics in silicate glass systems. Details are provided in Table 4.6. Fig. 4.32 shows the variation of SHG output power with angle of rotation of $3 \times 1 \times 1$ $cm^3$ nano

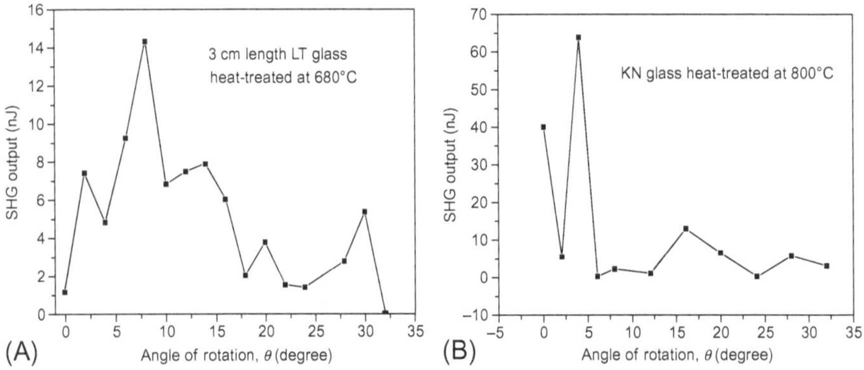


**Fig. 4.32** Variation of second harmonic generation (SHG) output power with angle of rotation of $3 \times 1 \times 1$ $cm^3$ nano glass-ceramic samples of (A) lithium tantalate (LT) heat-treated at 680°C/10 h and (B) potassium niobate (KN) heat-treated at 800°C/10 h.
Reproduced with permission from A. Tarafder, K. Annapurna, R.S. Chaliha, S. Satapathy, V.S. Tiwari, P.K. Gupta, B. Karmakar, Second harmonic generation in ferroelectric $LiTaO_3$ and $KNbO_3$ containing bulk nano glass-ceramics, J. Nonlinear Opt. Phys. Mater. 20 (2011) 49–61, Copyright © World Scientific Publishing Company.

glass-ceramic samples of (A) lithium tantalate (LT) heat-treated at 680°C/10 h and (B) potassium niobate (KN) heat-treated at 800°C/10 h. The SHG power output was found to increase up to 14 and 62.4 nJ with variation of rotation angle of monolithic LT and KN bulk nano glass-ceramics, respectively, due to alignment of ferroelectric domains under applied field.

### 4.3.3 Glass-ceramics for rechargeable battery

At present, rechargeable lithium-ion batteries are the widely used mobile power source, due to their high energy density, for many electronic devices, for example, laptop, cell phone, watch, electric vehicles (EV), electrochromic sensors, and camcorder. Lithium-ion battery is the choice for acceptable storage technologies, which are necessary to allow a transition to renewable energy sources from fossil fuels. Common lithium-ion batteries are extensively used but are limited by availability of materials, price, and safety. Increasing battery size creates the safety issues of lithium-ion batteries more serious due to increasing volume of flammable organic liquid electrolytes. On the other hand, all solid-state rechargeable lithium-ion batteries have drawn much consideration because of its several advantages. Some of these advantages are the following:

**(i)** Replacement of flammable organic liquid electrolyte by nonflammable inorganic solid electrolyte makes it safer.
**(ii)** Design is simple (this is schematically shown in Fig. 4.33).
**(iii)** Possibility of stacking in one package to achieve high voltage reduces device volume for use in large scale battery system.
**(iv)** High durability.
**(v)** Use of large capacity electrode materials.

Glass-ceramics are class of dense materials of great potential for use as cathode material and solid electrolyte of such batteries. Compared with the sintered polycrystalline materials, they can be prepared into desired shape and size more easily. In addition, they possess good chemical stability, which is most important for practical use. Glass system, processing, crystal phase, conductivity, and activation energy of glass-ceramics established for all solid-state rechargeable lithium battery are summarized in Table 4.7 [186–208]. It is seen from this table that both the cathode material and solid electrolyte of this battery can be derived from glass-ceramics. Some of them are described as follows.

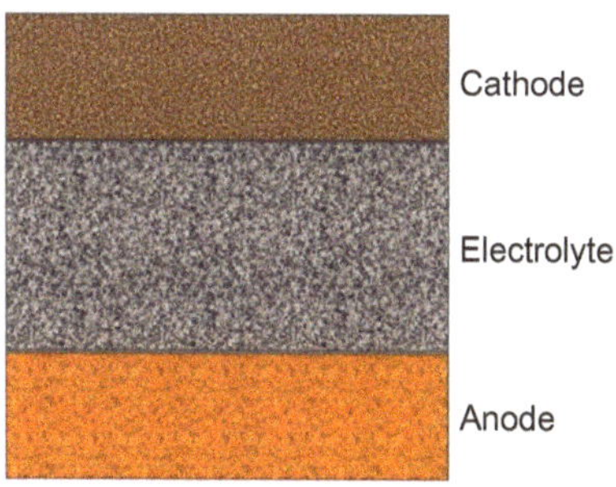


**Fig. 4.33** Schematic representation of construction of an all solid-state lithium-ion battery.

**Table 4.7 Glass system, processing, crystal phase, conductivity, and activation energy of glass-ceramics established for all solid-state rechargeable lithium battery**

| Glass system/composition (mol.%) | Heat treatment temperature/time (°C $h^{-1}$) | Major crystal phase | Conductivity at 25°C (S $cm^{-1}$) | Activation energy (kJ $mol^{-1}$) | Ref. |
|---|---|---|---|---|---|
| ***Cathode material*** | | | | | |
| $26Li_2O$–43FeO–$5Nb_2O_5$–$26P_2O_5$ | YAG laser irradiation | $LiFePO_4$ | $<10^{-9}$ | – | [186] |
| $26Li_2O$–43FeO–$5Nb_2O_5$–$26P_2O_5$ | 560, 580, 700/0.5 | $LiFePO_4$ | – | – | [187] |
| $33.3Li_2O$–33.3FeO–$33.3P_2O_5$+(0–5) $Nb_2O_5$ | 800/0.5 | $LiFePO_4$ | – | – | [188] |
| $Li_2O$–$MnO_2$–$Fe_2O_3$–$P_2O_5$ | 550/1.5 | $LiMn_xFe_{1-x}PO_4$ ($x$=0–0.8) | – | – | [189] |
| $33.3Li_2O$–$33.3V_2O_5$–$33.3P_2O_5$ | 450, 475, 594/1.5 | β-$LiVOPO_4$ | $1.7\times10^{-7}$ | 36.7 | [190] |
| ***Solid electrolyte*** | | | | | |
| $37.5Li_2O$–$(25-x)Fe_2O_3$–$xNb_2O_5$–$37.5P_2O_5$ ($x$=5, 10, 15) | 620–710/1.5 | $Li_3Fe_2(PO_4)_3$ | $3\times10^{-6}$ | 46.3–49.2 | [191] |
| $Li_2O$–$Al_2O_3$–$GeO_2$–$P_2O_5$ | 650–700/12 | $Li_{1+x}Al_xGe_{2-x}(PO_4)_3$ ($x$=0–0.8) | $4.0\times10^{-4}$ | 36.0 | [192] |
| (wt.%) (5.3–5.4)$Li_2O$–(5.6–6.1) $Al_2O_3$–(37.5–41.4)$GeO_2$–(47.6–51.0)$P_2O_5$ | 524/6–48 +571/0.5–1.0 | $Li_{1.5}Al_{0.5}Ge_{1.5}(PO_4)_3$ | $1.3$–$2.3\times10^{-5}$ | 37.6–44.4 | [193] |
| $Li_2O$–$Al_2O_3$–$GeO_2$–$P_2O_5$+(10–20) $Li_2O$ | 800/2 | $Li_{1.5}Al_{0.5}Ge_{1.5}(PO_4)_3$ | $1.9\times10^{-4}$ | 39.6 | [194] |
| $Li_2O$–$Al_2O_3$–$GeO_2$–$P_2O_5$–$xB_2O_3$ ($x$=0.05, 0.1, 0.2, 0.4 wt.%) | 825/5 | $Li_{1.5}Al_{0.5}Ge_{1.5}(PO_4)_3$ | $6.94\times10^{-4}$ | 21.6 | [195] |
| $Li_2O$–$Al_2O_3$–$GeO_2$–$P_2O_5$ | 800, 850, 900 | $Li_{1.5}Al_{0.5}Ge_{1.5}(PO_4)_3$ | $5.21\times10^{-4}$ | 32.8–35.7 | [196] |
| $Li_2O$–$Al_2O_3$–$GeO_2$–$P_2O_5$ | 550–650/12–100 | $Li_{1.4}Al_{0.4}Ge_{1.6}(PO_4)_3$ | $3.8\times10^{-5}$ | 50.2 | [197] |

| | | | | | |
|---|---|---|---|---|---|
| $Li_2O$–$Al_2O_3$–$GeO_2$–$P_2O_5$–0.05$Li_2O$ | 600/2 | $Li_{1.5}Al_{0.5}Ge_{1.5}(PO_4)_3$–$xLi_2O$ ($x$=0–0.20) | $7.25 \times 10^{-4}$ | 29.9 | [198] |
| 19.75$Li_2O$–6.17$Al_2O_3$–37.04$GeO_2$–37.04$P_2O_5$ | 750–850/12–48 | $Li_{1.5}Al_{0.5}Ge_{1.5}(PO_4)_3$–$Li_2O$ | $\sim 10^{-2}$ | 43.4–59.8 | [199] |
| $Li_2O$–$Cr_2O_3$–$GeO_2$–$P_2O_5$ | 800, 850, 900, 950, 1000/12 | $Li_{1.3}Cr_{0.3}Ge_{1.7}(PO_4)_3$ | $5.7 \times 10^{-4}$ | 27.8 | [200] |
| $Li_2O$–$Al_2O_3$–$GeO_2$–$TiO_2$–$P_2O_5$ | 546–688/2 | $Li_{1.4}Al_{0.4}(Ge_{1-x}Ti_x)_{1.6}(PO_4)_3$ ($x$=0–1.0) | $6.21 \times 10^{-4}$ | 30.9 | [201] |
| $Li_2O$–$Al_2O_3$–$GeO_2$–$P_2O_5$ | 750/12 | $Li_{1.4}Al_{0.4}Ge_{1.6}(PO_4)_3$ | $3.99 \times 10^{-4}$ | 29.9 | [202] |
| $Li_2O$–$Al_2O_3$–$TiO_2$–$P_2O_5$ | 950/12 | $Li_{1.3}Al_{0.3}Ti_{1.7}(PO_4)_3$ | $6.53 \times 10^{-4}$ | 30.9 | [202] |
| 14$Li_2O$–9$Al_2O_3$–38$TiO_2$–39$P_2O_5$ | 700, 800, 900, 950, 1000/12 | $Li_{1+x}Al_xTi_{2-x}(PO_4)_3$ | $1.3 \times 10^{-3}$ | 31.5 | [203] |
| $Li_2O$–$Al_2O_3$–$TiO_2$–$P_2O_5$ | 700, 800, 900, 1000/6 | $Li_{1.4}Al_{0.4}Ti_{1.6}(PO_4)_3$ | $1.09 \times 10^{-3}$ | 27.0 | [204] |
| 16.2$Li_2O$–3.8$Al_2O_3$–42.5$TiO_2$–37.5$P_2O_5$ | 700, 800, 900, 1000/0.33 | $Li_{1+x}Al_xTi_{2-x}(PO_4)_3$ | $1.3 \times 10^{-3}$ | 26.1 | [205] |
| $Li_2O$–$Al_2O_3$–$TiO_2$–$P_2O_5$ | 950, 1050, 1100/12 | $Li_{1.275}Al_{0.275}Ti_{1.75}(PO_4)_3$ + $AlPO_4$ | $\sim 1.5 \times 10^{-3}$ | 31.6 | [206] |
| $Li_2O$–$Al_2O_3$–$TiO_2$–$SiO_2$–$P_2O_5$ | 644/12 + 700/24 | $Li_{1.4+y}Al_yTi_{2-y}Si_{0.4}P_{2.6}O_{12}$ | $1.5 \times 10^{-3}$ | 30.0 | [207] |
| $Li_2O$–$Al_2O_3$–$TiO_2$–$SiO_2$–$P_2O_5$ | – | $Li_{1+x+y}Al_xTi_{2-x}Si_yP_{3-y}O_{12}$ | $>10^{-3}$ | – | [208] |

#### 4.3.3.1 Cathode material

Komatsu and his coworkers [186–188] have described the synthesis, properties, and application of glass-ceramics with lithium iron phosphate ($LiFePO_4$) crystals, in the $Li_2O$–FeO–$Nb_2O_5$–$P_2O_5$ glass system, for use as cathode material in lithium-ion rechargeable battery. Details are given in Table 4.7. $LiFePO_4$ has an olivine type crystal structure and is shown in Fig. 4.34 [209]. In this orthorhombic structure, the $Li^+$ ions diffuse through one dimensional channels with high energy barriers to cross between the channels, and the most favorable $Li^+$ ion diffusion path is the chains along the $a$-axis. This process increases its electronic conductivity and inhibits its ionic conductivity [210]. Nagakane et al. [188] have described the synthesis of $LiFePO_4$ glass-ceramics by melting glass in the above system in air and subsequent crystallization of glass powders. A lithium-ion rechargeable battery was fabricated with this glass-ceramic powder, and its performance is compared with that of a battery made using commercial $LiFePO_4$ ceramic powder. Fig. 4.35 shows the discharge capacity (left) and voltage (right) as a function of discharge rate of lithium-ion battery fabricated with glass-ceramics (GC-LFP) and ceramic of $LiFePO_4$. The battery with the glass-ceramic powder demonstrates better high-rate performance than that of the battery of the $LiFePO_4$ ceramic powder. The better performance of the glass-ceramic battery is credited to the existence of the surface amorphous layer. Fig. 4.36A shows the charge and discharge curves for the cells made of $LiMn_xFe_{1-x}PO_4$ glass-ceramics as the cathode and lithium metal as the anode at the charge/discharge rate of $C/10$ in each cell [189]. The charging curve for $LiMn_{0.5}Fe_{0.5}PO_4$ glass-ceramics shows a clear plateau at 3.4 and 4.1 V, which represents the redox potentials of the $Fe^{3+}/Fe^{2+}$ and $Mn^{3+}/Mn^{2+}$ systems, whereas the clear plateau at 4.1 V was not observed with the cathode made of $LiMn_xFe_{1-x}PO_4$ ceramics.

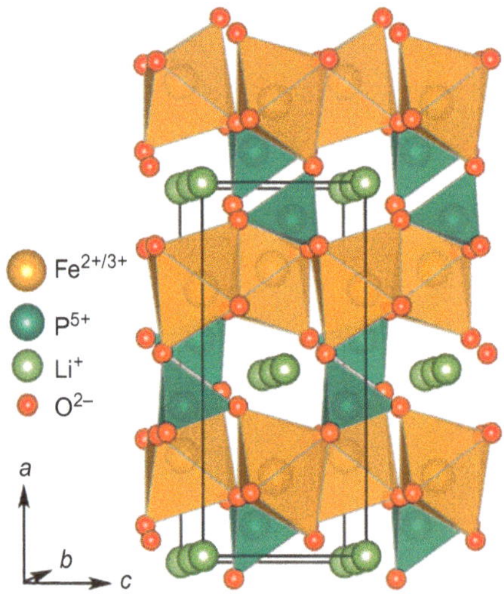


**Fig. 4.34** Olivine type crystal structure of $LiFePO_4$.
Courtesy: https://crystallography365.wordpress.com/2014/04/29/lifepo4-the-unexpected-battery-success-story/ Accessed 10 February 2016.

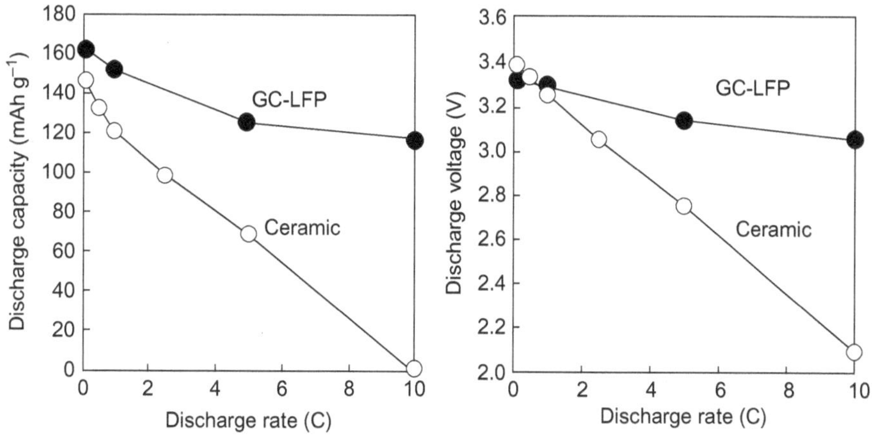


**Fig. 4.35** Discharge capacity (left) and voltage (right) as a function of discharge rate of lithium-ion battery fabricated with glass-ceramics (GC-LFP) and ceramic of $LiFePO_4$. Reproduced with permission from T. Nagakane, H. Yamauchi, K. Yuki, M. Ohji, A. Sakamoto, T. Komatsu, T. Honma, M. Zou, G. Park, T. Sakai, Glass-ceramic $LiFePO_4$ for lithium-ion rechargeable battery, Solid State Ion. 206 (2012) 78–83, Copyright © Elsevier.

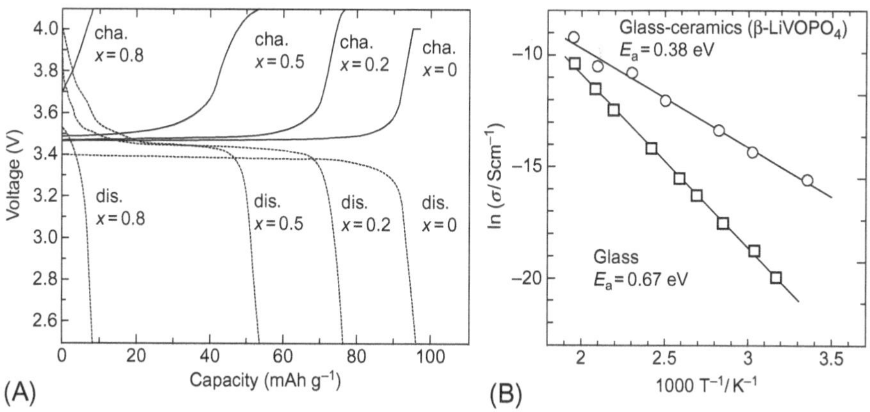


**Fig. 4.36** (A) Charge and discharge curves for the cells made of $LiMn_xFe_{1-x}PO_4$ glass-ceramics as the cathode and lithium metal as the anode at the charge/discharge rate of $C/10$ in each cell. (B) Plot of electric conductivity as a function of temperature of precursor glass and β-$LiVOPO_4$ glass-ceramics (heat-treated at 600°C/12 h) showing the evaluation of activation energy.
(A) Reproduced with permission from T. Honma, K. Nagamine, T. Komatsu, Fabrication of olivine-type $LiMn_xFe_{1-x}PO_4$ crystals via the glass–ceramic route and their lithium ion battery performance, Ceram. Int. 36 (2010) 1137–1141, Copyright © Elsevier. (B) Reproduced with permission from K. Nagamine, T. Honma, T. Komatsu, Selective synthesis of lithium ion-conductive β-$LiVOPO_4$ crystals via glass–ceramic processing, J. Am. Ceram. Soc. 91 (2008) 3920–3925, Copyright © The American Ceramic Society.

Nagamine et al. [190] have studied the selective synthesis of lithium-ion conductive β-$LiVOPO_4$ crystals via glass-ceramic processing in the glass composition (mol.%) of 33.3$Li_2O$–33.3$V_2O_5$–33.3$P_2O_5$. Details are given in Table 4.7. Fig. 4.36B illustrates the plot of electric conductivity as a function of temperature of precursor glass and β-$LiVOPO_4$ glass-ceramics (heat-treated at 600°C/12 h) showing the evaluation of activation energy. The glass-ceramics show the electric conductivity of $1.7 \times 10^{-7}$ S $cm^{-1}$ at room temperature and the activation energy of 0.38 eV (36.7 kJ $mol^{-1}$) in the temperature range of 25–240°C. The corresponding glass has the activation energy of 0.67 eV. It is also observed that the synthesis of β-$LiVOPO_4$ crystals using a conventional powder-sintering method is difficult to obtain comparable performance as the glass-ceramics. Consequently, these authors propose that the glass-ceramic processing is advantageous for fabrication of β-$LiVOPO_4$ crystals for use in lithium-ion rechargeable battery.

### *4.3.3.2 Solid electrolyte*

The majority of the glass-ceramic-based solid electrolytes of all solid-state rechargeable lithium-ion battery contains the derivatives of the $LiGe_2(PO_4)_3$ compound—lithium analogs of Na superionic conductor (NASICON), that is, the lithium super ionic conductor (LISICON).Construction of different derivatives of $LiGe_2(PO_4)_3$ is illustrated in Fig. 4.37. The chemical formula of NASICON is $NaZr_2(PO_4)_3$. In the structure, the $LiGe_2(PO_4)_3$ contains covalent anionic skeletons of $[Ge_2P_3O_{12}]^-$ constituted of $GeO_6$ octahedra and $PO_4$ tetrahedra, which form the 3D interconnected channels, in which two types of interstitial sites are available for the $Li^+$ ions—the conductive ion. The NASICON-type (lithium analog) crystal structure of different derivatives of $LiGe_2(PO_4)_3$ is shown in Fig. 4.38 [211]. Various summarized data about the glass-ceramics established creating the different derivatives of $LiGe_2(PO_4)_3$ for all solid-state rechargeable lithium-ion battery are provided in Table 4.7, and their original research findings are available in Refs. [192–208]. A few are described as follows.

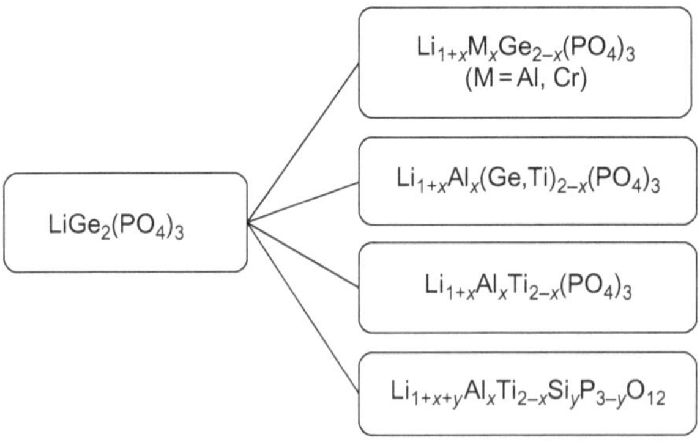


**Fig. 4.37** Construction of different derivatives of $LiGe_2(PO_4)_3$ having NASICON-type (lithium analog) structure.

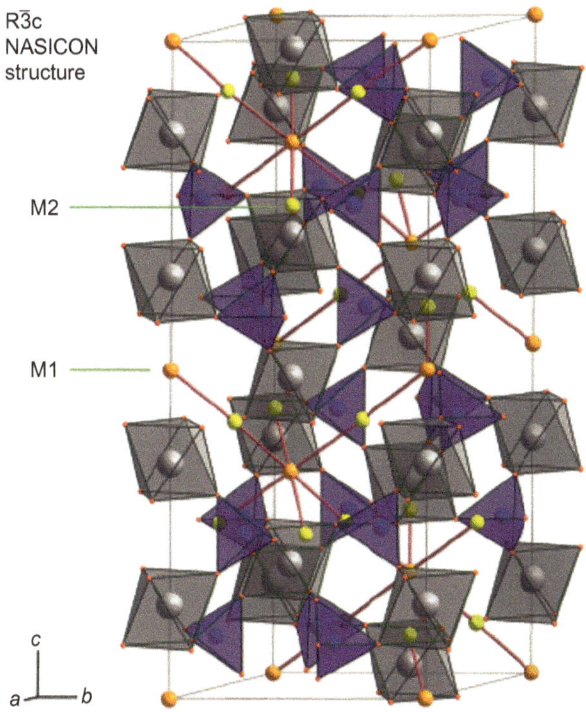


**Fig. 4.38** NASICON-type (lithium analog) crystal structure of different derivatives of $LiGe_2(PO_4)_3$. *Orange* (M1) and *yellow* (M2) spheres represent two different types of lithium-ion sites that are used to diffuse through the crystal lattice. Reproduced with permission from S. Breuer, D. Prutsch, Q. Ma, V. Epp, F. Preishuber-Pflügl, F. Tietz, M. Wilkening, Separating bulk from grain boundary Li ion conductivity in the sol–gel prepared solid electrolyte $Li_{1.5}Al_{0.5}Ti_{1.5}(PO_4)_3$, J. Mater. Chem. A 3 (2015) 21343–21350, Copyright © RSC.

The development of rechargeable lithium-ion batteries has been slowed down by the need of a suitable, high conductivity, and chemically stable electrolyte. Many efforts have been made searching the solids with liquid-like conductivity during the last two decades. In 1997, Fu [192,203] reported the LAGP and LATP glass-ceramics of creating the crystals of $Li_{1+x}Al_x\ Ge_{2-x}\ (PO_4)_3$ and $Li_{1.3}Al_{0.3}Ti_{1.7}(PO_4)_3$ in the $Li_2O–Al_2O_3–GeO_2–P_2O_5$ and $Li_2O–Al_2O_3–TiO_2–P_2O_5$ glass systems, respectively. They exhibited high room temperature conductivities of $4.0\times10^{-4}$ and $6.53\times10^{-4}$S cm$^{-1}$ with activation energies of 36.0 and 30.9 kJ mol$^{-1}$, respectively. In these glass-ceramics, thermodynamically stable crystalline phases precipitate from the precursor glasses during heat treatment crystallization, and the grain boundaries stay behind filled with very thin layers of amorphous phases, which reduces the grain-boundary resistance resulting in high conductivity. Fig. 4.39A shows the electric conductivity at room temperature and activation energy of $Li_{1+x}Al_xGe_{2-x}\ (PO_4)_3$ glass-ceramics as a function of $x$ [192]. The conductivity ($\sigma$) gradually increases by more than fourfold of magnitude with rise in $x$ and arrives at a maximum value of $4.0\times10^{-4}$S cm$^{-1}$ at $x=0.5$ and then decreases. This conductivity value is more than twofold greater than that of sintered ceramics of $Li_{1+x}Al_xGe_{2-x}\ (PO_4)_3$. Similarly, the activation energy ($E_a$) passes through a minimum value of 36.0 kJ mol$^{-1}$ at $x=0.5$ and then increases. These correspondences well with that of conductivity that indicates that the changes in activation energy control the alterations in conductivity. Fig. 4.39B demonstrates the SEM photomicrograph of the fracture surface of the

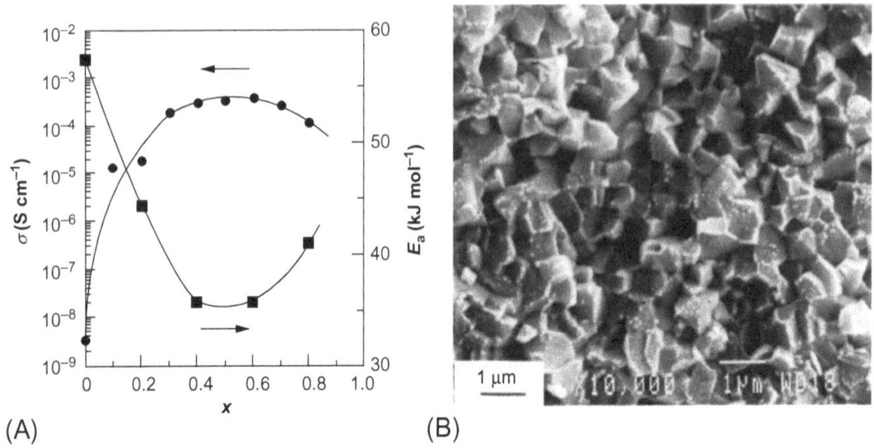


**Fig. 4.39** (A) Electric conductivity at room temperature and activation energy of $Li_{1+x}Al_xGe_{2-x}(PO_4)_3$ glass-ceramics as a function of $x$. (B) SEM photomicrograph of the fracture surface of the $Li_{1.4}Al_{0.4}Ge_{1.6}(PO_4)_3$ glass-ceramics.
Reproduced with permission from J. Fu, Fast $Li^+$ ion conducting glass-ceramics in the system $Li_2O$–$Al_2O_3$–$GeO_2$–$P_2O_5$, Solid State Ionics 104 (1997) 191–194, Copyright © Elsevier.

$Li_{1.4}Al_{0.4}Ge_{1.6}(PO_4)_3$ glass-ceramics [192].The glass-ceramics acquire a very dense microstructure without any cracks or pores with more close contact with the grains resulting in higher mobility of $Li^+$ ions and hence higher conductivity. In contrast, the sintered ceramics contain more cracks or pores with less contact with the grains resulting in lower mobility of $Li^+$ ions and hence lower conductivity.

Narváez-Semanate and Rodrigues [205] have established a very good correlation with the microstructure and ionic conductivity of $Li_{1+x}Al_xTi_{2-x}(PO_4)_3$ NASICON glass-ceramics. Fig. 4.40 shows the SEM photomicrographs of the fracture surface of the $Li_{1+x}Al_xTi_{2-x}(PO_4)_3$ glass-ceramics obtained by single stage heat treatment protocol of the precursor glass at (A) 700°C, (B) 800°C, (C) 900°C, and (D) 1000°C. The insets demonstrate the transparency or opacity of the respective samples. Under similar heat treatment condition, the conductivity increases by threefold of magnitude and reaches a maximum value of $1.3 \times 10^{-3}$ S cm$^{-1}$ at room temperature. This enhancement in conductivity is caused by increase in crystallinity of the samples and grain contacts.

## 4.3.4 Glass-ceramics for glaze

Glass-ceramics can be prepared with a wide range of compositions yielding a variety of microstructures and properties for specific technological applications, for example, glazes for ceramic tiles. Ceramic tiles are the most familiar building material used for covering floors and walls, popularly known as "floor" and "wall" tiles, respectively. Glazed tiles are commonly manufactured from glass frits applying on the surface of green tiles and subsequent firing at a certain temperature depending on the chemical composition of the frits. The glaze, on firing, crystallizes and improves the mechanical

**Fig. 4.40** SEM photomicrograph of the fracture surface of the $Li_{1+x}Al_xTi_{2-x}(PO_4)_3$ glass-ceramics obtained by single stage heat treatment protocol of the precursor glass at (A) 700°C, (B) 800°C, (C) 900°C, and (D) 1000°C. The insets demonstrate the transparency or opacity of the respective samples.
Reproduced with permission from J.L. Narváez-Semanate, A.C.M. Rodrigues, Microstructure and ionic conductivity of $Li_{1+x}Al_xTi_{2-x}(PO_4)_3$ NASICON glass-ceramics, Solid State Ion. 181 (2010) 1197–1204, Copyright © Elsevier.

and chemical properties of the glazed tiles. A typical recipe (wt.%) of a glaze consists of 80–90 frit, 0–10 feldspar, 5–10 kaolin, 0.10–0.25 carboxymethyl cellulose (CMC), and 0.20–0.35 sodium tripolyphosphate (STTP).

In recent times, Casasola et al. [212] have scholarly reviewed the most up-to-date research carried out on glass-ceramic glazes used for ceramic tiles with silicates and other oxides as the main crystalline phases. Salem et al. [213] have described the techniques of characterization and establishing the technological process parameters of glass-ceramic glazes of ceramic tiles by the hot-stage microscopy (HSM). Table 4.8 summarizes the crystal phases, glass compositions, melting, and firing temperatures of glass-ceramics established for use as glazes [214–233]. It is found that the major crystal phases of the widely used glass-ceramics as glaze are zircon ($ZrSiO_4$), spinel ($MgAl_2O_4$), gahnite($ZnAl_2O_4$), mullite ($3Al_2O_3{\cdot}2SiO_2$), diopside ($CaO{\cdot}MgO{\cdot}SiO_2$), cordierite ($2MgO{\cdot}2Al_2O_3{\cdot}5SiO_2$), anorthite ($CaO{\cdot}Al_2O_3{\cdot}2SiO_2$), spodumene ($Li_2O{\cdot}Al_2O_3{\cdot}4SiO_2$), celsian ($BaO{\cdot}Al_2O_3{\cdot}2SiO_2$), leucite ($KAlSi_2O_6$), pyroxene ($(Ca,Mg,K,Na)_{0.84}(Mg,Al,Ti)(Si,Al)_2O_6$), and biotite mica ($(K(Mg,Fe^{2+})_3(Al,Fe^{3+})_3O_{10}(OH,F)_2)$). Glass melting and glaze firing temperature and duration vary in the ranges of 1450–1600°C for 1–2 h and 800–1300°C for 0.08–24 h, respectively. Some interesting findings are described as follows.

**Table 4.8** Crystal phases, glass compositions, melting, and firing temperatures of glass-ceramics for glazes

| Major crystal phase | Second major crystal phase | Glass system/composition (wt.%) | Glass melting temperature/time (°C $h^{-1}$) | Glaze firing temperature/time (°C $h^{-1}$) | Ref. |
|---|---|---|---|---|---|
| Zircon ($ZrSiO_4$) | Diopside (Ca(Mg, Al) (Si, Al)$_2O_6$) | $(100-x)$ZF–$x$GF + $y$$Fe_2O_3$ ($x$ = 0–50; $y$ = 1.8–8.9) [ZF = zircon frit, GF = granite frit] | 1450/1 | 1120/0.25 | [214] |
| Zircon ($ZrSiO_4$) | Calcium aluminum silicate ($Ca_3Al_6Si_2O_{16}$) | $Na_2O$–$K_2O$–MgO–CaO–SrO–ZnO–$Al_2O_3$–$B_2O_3$–$ZrO_2$–$SiO_2$ | 1450/1 | 1140/1.0 | [215] |
| Spinel ($MgAl_2O_4$) | Gahnite ($ZnAl_2O_4$), mullite ($3Al_2O_3{\cdot}2SiO_2$) | (49–53)$SiO_2$–(25–29)$Al_2O_3$–(8–9)MgO–3ZnO–4$TiO_2$–(6–9)$B_2O_3$–3$K_2O$–1$Na_2O$ | 1600/2 | 1100–1200/2 | [216] |
| Gahnite ($ZnAl_2O_4$) | β-Quartz solid solution (βqss) | 55$SiO_2$–(7–22)$Al_2O_3$–(10–18)ZnO–(0.6–11)CaO–(0.5–7.5)MgO–9.5$ZrO_2$–5.6$B_2O_3$–2.6$K_2O$–1$Na_2O$ | 1450/1 | 1180/0.17 | [217] |
| Mullite ($3Al_2O_3{\cdot}2SiO_2$) | – | 3CaO–9MgO–26$Al_2O_3$–53$SiO_2$–9$B_2O_3$ | 1650/2 | 1100–1190/0.08 | [218,219] |
| Diopside (CaO·MgO·$SiO_2$) | Wollastonite (CaO·$SiO_2$) | (45–77)$SiO_2$–(10–25)$Al_2O_3$–(5–30)CaO–(0.2–4)MgO–(2.2–5.4)$K_2O$–(1.8–4.6)$Na_2O$ | – | 1215–1260/1–24 | [220] |
| Diopside (CaO·MgO·$SiO_2$) | Zircon ($ZrSiO_4$) | 55.35$SiO_2$–(18.57–33.57)(MgO + CaO)–9.54$ZrO_2$–5.58$B_2O_3$–3.54($K_2O$ + $Na_2O$) | 1450/1 | 1180/0.17 | [221] |
| Diopside (CaO·MgO·$SiO_2$) | Hematite ($Fe_2O_3$) | $(100-x)$DF–$x$GF + $y$$Fe_2O_3$ ($x$ = 0–90; $y$ = 1.8–16.0) [ZF = diopside frit, GF = granite frit] | 1450/1 | 1120/0.25 | [222] |
| Diopside (CaO·MgO·$SiO_2$) | Zircon ($ZrSiO_4$) | $Na_2O$–$K_2O$–MgO–CaO–SrO–$Al_2O_3$–$B_2O_3$–$ZrO_2$–$SiO_2$–$Fe_2O_3$ | 1500/1 | 1197/1 | [223] |

| | | | | | |
|---|---|---|---|---|---|
| Cordierite ($2MgO{\cdot}2Al_2O_3{\cdot}5SiO_2$) | Mullite ($3Al_2O_3{\cdot}2SiO_2$) | $52SiO_2$–$36Al_2O_3$–10MgO–2CaO | 1650/2 | 950–1190/2 | [224] |
| Cordierite ($2MgO{\cdot}2Al_2O_3{\cdot}5SiO_2$) | Anorthite ($CaO{\cdot}Al_2O_3{\cdot}2SiO_2$) | $55SiO_2$–$25Al_2O_3$–15MgO–5CaO +(1–3)($TiO_2$, $B_2O_3$, $K_2O$, $Na_2O$) | 1600/2 | 775/(0–2) +1100/(0–2) | [225] |
| α-Cordierite ($2MgO{\cdot}2Al_2O_3{\cdot}5SiO_2$) | – | $55SiO_2$–$21.5Al_2O_3$–$(16.5-x)$ MgO–$x$CaO–$3.8TiO_2$–$2.9B_2O_3$ ($x=2.9$, 4.6, 6.5) | 1600/2 | 1100, 1160, 1190/0.08 | [226] |
| α-Cordierite ($2MgO{\cdot}2Al_2O_3{\cdot}5SiO_2$) | – | 4.66CaO–12.01MgO–$21.57Al_2O_3$–$55SiO_2$–$3.86TiO_2$–$2.90B_2O_3$ | 1600/2 | 1190/0.08 | [227] |
| α-Cordierite ($2MgO{\cdot}2Al_2O_3{\cdot}5SiO_2$) | Enstatite ($MgSiO_3$), forsterite ($Mg_2SiO_4$) | (50–62)$SiO_2$–(16–30)$Al_2O_3$–(14–20)MgO–(8–9)$TiO_2$–5CaO | 1450/1 | 1000–1300/(1–6) | [228] |
| Anorthite ($CaO{\cdot}Al_2O_3{\cdot}2SiO_2$) | Akermanite ($2CaO{\cdot}MgO{\cdot}2SiO_2$) | $51.51SiO_2$–$18.85Al_2O_3$–10.25CaO–2.91MgO–$4.73K_2O$–$1.46B_2O_3$–$0.83Na_2O$–$0.54TiO_2$–$1.93Fe_2O_3$ | – | 1150–1200/1 | [229] |
| Spodumene ($Li_2O{\cdot}Al_2O_3{\cdot}4SiO_2$), cordierite, diopside, anorthite | – | $Li_2O$–CaO–MgO–$Al_2O_3$–$SiO_2$ | – | 1000–1100/0.5 | [230] |
| Celsian ($BaO{\cdot}Al_2O_3{\cdot}2SiO_2$) | – | BaO–$Al_2O_3$–$SiO_2$ | – | 1000–1100/0.5 | [230] |
| Leucite ($KAlSi_2O_6$) | – | $Li_2O$–$K_2O$–$Al_2O_3$–$SiO_2$ | – | 1000–1100/0.5 | [230] |
| Pyroxene ((Ca,Mg,K,Na)$_{0.84}$ (Mg,Al,Ti)(Si,Al)$_2O_6$) | – | (55.5–60)$SiO_2$–(10.2–14)$Al_2O_3$–(9.3–10)MgO–(14.8–16)CaO–$3.7TiO_2$–$2.8B_2O_3$–$2.8K_2O$–$0.9Na_2O$ | 1600 | 800–1190/(0.08–24) | [231,232] |
| Biotite mica ((K(Mg,$Fe^{2+}$)$_3$(Al,$Fe^{3+}$)$_3$$O_{10}$(OH,F)$_2$) | Hematite ($Fe_2O_3$) | $58SiO_2$–$10.44Al_2O_3$–$8.67Fe_2O_3$–3.35CaO–1.89MgO–3.3ZnO–$2.28Na_2O$–$5.64K_2O$–6.43F | 1450/1 | 1120/0.25 | [233] |

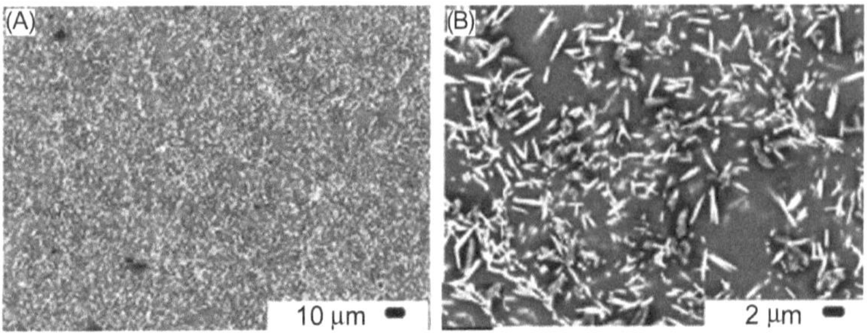


**Fig. 4.41** SEM photomicrograph of industrially fired zircon ($ZrSiO_4$)-based glaze in (A) low and (B) high magnifications.
Reproduced with permission from K. Pekkan, B. Karasu, Evaluation of borax solid wastes in production of frits suitable for fast single-fired wall tile opaque glass–ceramic glazes, Bull. Mater. Sci. 33 (2010) 135–144, Copyright © Springer.

Pekkan and Karasu [215] have studied the use of borax solid wastes in production of frits suitable for fast single-fired wall tile opaque glass-ceramic glazes in the $Na_2O$–$K_2O$–MgO–CaO–SrO–ZnO–$Al_2O_3$–$B_2O_3$–$ZrO_2$–$SiO_2$ glass system. Details are given in Table 4.8. Zircon ($ZrSiO_4$) is the main crystal phase that is the opacifier of glossy, opaque, and white colored tile glazes. Fig. 4.41 shows the SEM photomicrograph of an industrially fired zircon ($ZrSiO_4$)-based glaze in (A) low and (B) high magnifications. The zircon crystals are seen to have a tendency to become aligned through the surface of the glaze, and their elongated sizes guide to drop the extent of opacity. The amount of light reflected, that is, the opacity, is dependent on both the number and refractive index of the particles present in the glassy matrix.

Torres et al. [216] have studied the effect of some additives on the development of spinel-based glass-ceramic glazes for floor tiles. Details are given in Table 4.8. Here, spinel ($MgAl_2O_4$) along with mullite ($3Al_2O_3{\cdot}2SiO_2$) is grown in the glass

**Fig. 4.42** SEM photomicrograph of (A) glass-ceramic glaze consisting of spinel and mullite (glass E) obtained by heat treating at 1100°C/2 h (bar = 25 μm) in low magnification and (B) in high magnification (a) spinel (bar = 500 nm) and (b) mullite (bar = 1000 nm).
Reproduced with permission from F.J. Torres, E.R. de Sola, J. Alarcón, Effect of some additives on the development of spinel-based glass-ceramic glazes for floor-tiles, J. Non-Cryst. Solids 351 (2005) 2453–2461, Copyright © Elsevier.

composition (wt.%) of (49–53)$SiO_2$–(25–29)$Al_2O_3$–(8–9)MgO–3ZnO–4$TiO_2$–(6–9)$B_2O_3$–3$K_2O$–1$Na_2O$. Fig. 4.42 shows the SEM photomicrograph of (A) glass-ceramic glaze consisting of spinel and mullite (glass E) obtained by heat treating at 1100°C/2 h (bar = 25 μm) in low magnification and (B) in high magnification (a) spinel (bar = 500 nm) and (b) mullite (bar = 1000 nm). It is seen that the addition of small amount of $B_2O_3$ and alkaline oxides allows to enhance phase separation and the spinel nucleates and growth throughout the whole-heated glass. It is noted that a small amount of mullite appears in areas located at the center of the glass particles where some small glass pockets remain with no crystallization.

Torres et al. [218] have reported the mechanism of crystallization of fast-fired mullite-based glass-ceramic glazes for floor tiles in the glass composition (wt.%) of 3CaO–9MgO–26$Al_2O_3$–53$SiO_2$–9$B_2O_3$. Details are given in Table 4.8. Microscopic microstructural study made known that the crystallization preceded via intermediate the phase separation and crystallization that started around 900°C, which increased with rise in temperature yielding ~20 wt.% of mullite in between 1100°C and 1200°C. The glass-ceramic glaze heated at 1160°C/0.08 h composed of interlocked, well-shaped, acicular mullite crystals longer than 4 μm, submerged in the residual glassy phase. Fig. 4.43 depicts the FESEM photomicrograph of glass-ceramic glaze consisting of (A) interlocked mullite crystals (aspect ratio ≈10 and bar = 5 μm) and (B) acicular mullite (aspect ratio >20 and bar = 5 μm).

Rincón et al. [230] have reported the crystallization microstructure of various glass-ceramic glazes suitable for use in fast-firing ceramic tiles for pavements and wall tiles. Fig. 4.44 shows the SEM photomicrographs of glass-ceramic glaze consisting of (A) spodumene, (B) cordierite, (C) nepheline, and (D) leucite. The crystallization microstructure of the surface of these glass-ceramic glazes is dissimilar

**Fig. 4.43** FESEM photomicrograph of glass-ceramic glaze consisting of (A) interlocked mullite crystals (aspect ratio ≈ 10 and bar = 5 μm) and (B) acicular mullite (aspect ratio >20 and bar = 5 μm).
Reproduced with permission from F.J. Torres, E.R. de Sola, J. Alarcón, Mechanism of crystallization of fast fired mullite-based glass–ceramic glazes for floor-tiles, J. Non-Cryst. Solids 352 (2006) 2159–2165, Copyright © Elsevier.

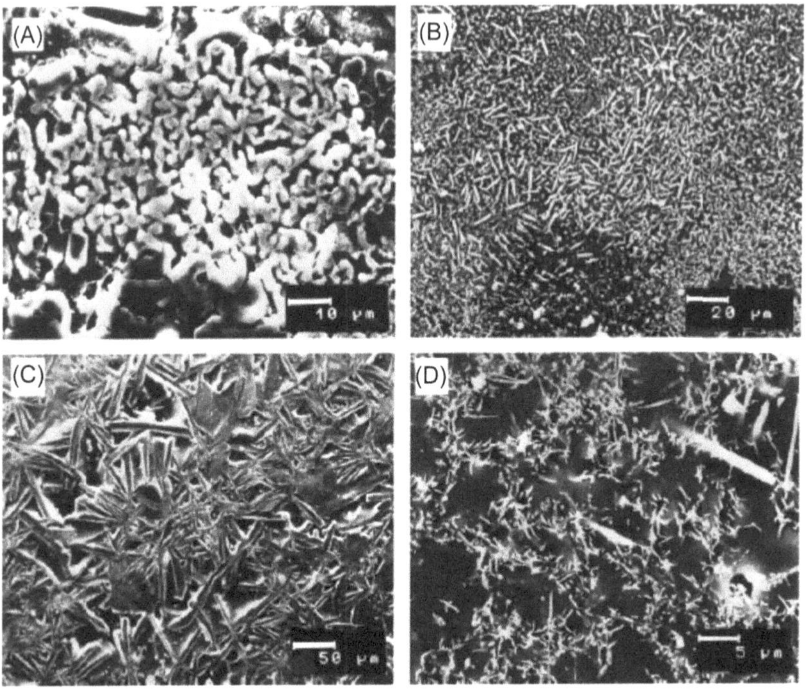


**Fig. 4.44** SEM photomicrograph of glass-ceramic glaze consisting of (A) spodumene, (B) cordierite, (C) nepheline, and (D) leucite.
Reproduced with permission from J.M. Rincón, M. Romero, J. Marco, V. Caballer, Some aspects of crystallization microstructure on new glass-ceramic glazes, Mater. Res. Bull. 33 (1998) 1159–1164, Copyright © Elsevier.

depending on the crystalline phase and precursor glass composition (see Table 4.8). For example, glazes after fast firing at 1000°C/0.5 h creating crystal phases of spodumene, cordierite, nepheline, and leucite display microstructure of rounded, grained, needle, and whisker crystals, respectively. It is also established that the new improved glass-ceramic glazes can be obtained by controlling the nucleation and crystallization processes at the interfaces.

Torres and Alarcón [231] have studied the mechanism of crystallization of pyroxene ($(Ca,Mg,K,Na)_{0.84}(Mg,Al,Ti)(Si,Al)_2O_6$)-based glass-ceramic glazes in the glass composition (wt.%) range of $(55.5–60)SiO_2–(10.2–14)Al_2O_3–(9.3–10)MgO–(14.8–16)CaO–3.7TiO_2–2.8B_2O_3–2.8K_2O–0.9Na_2O$ firing in the temperature range 800–1190°C for 0.08–24 h. Details are provided in Table 4.8. Fig. 4.45 displays the SEM photomicrographs of glass-ceramic glaze consisting of pyroxene obtained by heat treating (precursor glass F) at (A) 800°C/30 min (bar = 500 nm), (B) 800°C/24 h (bar = 2.5 μm), and (C) 1190°C/5 min (bar = 25 μm). The sample heat-treated at 800°C/30 min demonstrates the texture of phase separation in glass that consists of small droplets of glass immersed in a glassy matrix. The sample heat-treated at 800°C/24 h displays the crystallization of pyroxene and the increasing trend of its degree of crystallization with rise in heat treatment time. The sample heat-treated

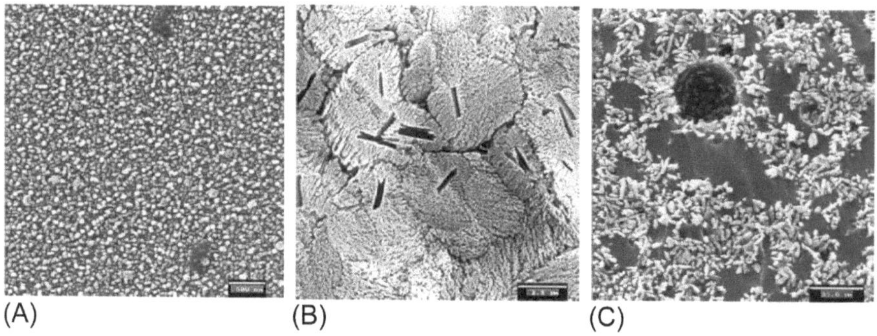

Fig. 4.45 SEM photomicrograph of glass-ceramic glaze consisting of pyroxene obtained by heat treating (precursor glass F) at (A) 800°C/30 min (bar = 500 nm), (B) 800°C/24 h (bar = 2.5 μm), and (C) 1190°C/5 min (bar = 25 μm).
Reproduced with permission from F.J. Torres, J. Alarcón, Mechanism of crystallization of pyroxene-based glass-ceramic glazes, J. Non-Cryst. Solids 347 (2004) 45–51, Copyright © Elsevier.

at 1190°C/5 min shows the final microstructure of the glass-ceramic glaze with small and well-formed prismatic crystals entrenched in the glassy matrix. It is observed that this glass-ceramic glaze also improves the hardness of ceramic tiles under industrial fast-firing schedule.

### 4.3.5 Ultralow expansion glass-ceramics

Glass-ceramics of the $Li_2O$–$Al_2O_3$–$SiO_2$ (LAS) system exhibit near-zero thermal expansion coefficient [234]. For this reason, they are used in various modern technologies such as mirror blanks for astronomical telescopes, stages and mirrors of lithography equipment, reference standards for precision measurement technology, ring laser gyroscope bodies, cookware, and cook top panel [234,7,235–238]. The main crystal phases in these glass-ceramics are the beta-quartz solid solution (β-quartz ss) or virgilite and β-spodumene. The β-quartz ss is based on the hexagonal β-quartz structure, while the β-spodumene ss derives from the tetragonal form of silica known as keatite. These aluminosilicate crystal phases consist of 3D networks of $SiO_4$ and $AlO_4$ tetrahedra and termed as "stuffed derivatives" because in the silica network some sites of $Si^{4+}$ are replaced by $Al^{3+}$ ion accompanied by a filling of interstitial vacancies by monovalent $Li^+$ ion to achieve charge balancing [239].

The commercial products of these glass-ceramics are Zerodur (Schott, Germany), Ceran (Schott, Germany), Vision (Corning, the United States), and Neoceram (NEG, Japan). The composition of the commercial glass-ceramics is complex and varies in the ranges (wt.%) of (55.4–68.8)$SiO_2$–(19.2–25.4)$Al_2O_3$–(2.7–4.2)$Li_2O$–(0.5–1.8)MgO–(1.0–1.6)ZnO–(0.8–2.2)BaO–(1.2–7.2)$P_2O_5$–(2.0–2.7)$TiO_2$–(1.5–2.3)$ZrO_2$ [7]. In these commercial glass-ceramics, the ultimate stoichiometric $Li^+$-stuffed derivative of β-quartz is the β-eucryptite ($LiAlSiO_4$). In this derivative, one-half of the $Si^+$ ions are replaced by $Al^{3+}$ and $Li^+$ ions and preferentially occupy the tetrahedral holes on the $SiO_4$ tetrahedral plane of alternating $SiO_4$ and $AlO_4$ tetrahedra. Divalent $Mg^{2+}$

and $Zn^{2+}$ ions can incompletely or totally replace $Li^+$ ions in some of these β-quartz structures. Here, the $Mg^{2+}$ and $Zn^{2+}$ ions occupy the octahedral sites between the abovesaid two tetrahedral sites like $Li^+$ ions. The general formula of these solid solutions is $Li_{2-2(x+y)}Mg_xZn_yO \cdot Al_2O_3 \cdot zSiO_2$, where $x+y \leq 1$ and $z>2$. Additionally, in the presence of $P_2O_5$, the β-quartz solid solutions are formed where isostructural $AlPO_4$ tetrahedra partially replaces the $SiO_4$ tetrahedra [235].

Karmakar et al. [240] have reported the dilatometric method for studying crystallization kinetics and mechanism of low expansion lithium aluminosilicate glass-ceramics of composition (wt.%) $56.9SiO_2$–$24.0Al_2O_3$–$7.2P_2O_5$–$3.7Li_2O$–$2.3TiO_2$–$1.8ZrO_2$–1.7ZnO–1.0MgO–$0.6K_2O$–$0.6As_2O_3$–$0.2Na_2O$. The activation energy ($E_a$) of crystallization was found to be around 500 kJ $mol^{-1}$, which is close to that of the diffusion of silicon and aluminum ions and metal-oxygen bond strengths, suggesting a three-dimensional (3D) diffusion-controlled reaction mechanism. Some selected properties of ultralow expansion transparent glass-ceramics of the above composition (close to that of ZERODUR) are listed in Table 4.9. Variation of $\Delta l/l_0$ of zero expansion glass-ceramic (ZERODUR) caused by alteration of ceramization temperature and time as a function of temperature is illustrated in Fig. 4.46A [7]. It is quite clear from this variation that controlling the ceramization parameters, the specific requirement of coefficient of thermal expansion (CTE) can be achieved. The transmission electron microscopy (TEM) image of zero expansion glass-ceramic (ZERODUR) is shown in Fig. 4.46B [235]. It is found from this photomicrograph that β-quartz ss of <100 nm and brick-like $ZrTiO_4$ nucleants are precipitated in the resultant glass-ceramics along with tiny crystals of $TiO_2$ along the grain boundaries. Photograph illustrating various shapes of zero expansion glass-ceramics (ZERODUR, Schott, Germany) is shown in Fig. 4.47A [236], whereas the photograph of transparent cookware (VISION, Corning, the United States) of β-quartz ss glass-ceramics is depicted in Fig. 4.47B.

The ultralow expansion material is not limited to the LAS glass-ceramics only, but this property can be achieved in other glass-ceramics and glass as well. MacDowell

**Table 4.9 Some selected properties of ultralow expansion transparent glass-ceramics of composition close to ZERODUR**

| Property | Corresponding value | Property | Corresponding value |
|---|---|---|---|
| Density (g $cm^{-3}$) | 2.535 | Transmission (%) at $\lambda=633$ nm, thickness = 5 mm | 88.2 |
| Coefficient of thermal expansion (CTE) ($\alpha \times 10^{-6}$ $K^{-1}$) | | Refractive index | |
| | | $n_e$ (at $\lambda_d=587.6$ nm) | 1.5399 |
| | | $n_e$ (at $\lambda_e=546.1$ nm) | 1.5421 |
| (i) 0–50°C | 0.00 | | |
| (ii) 0–100°C | −0.12 | Mean dispersion ($n_{F'}-n_{C'}$) | 0.00955 |
| | | Abbe number ($\nu_e$) | 56.78 |

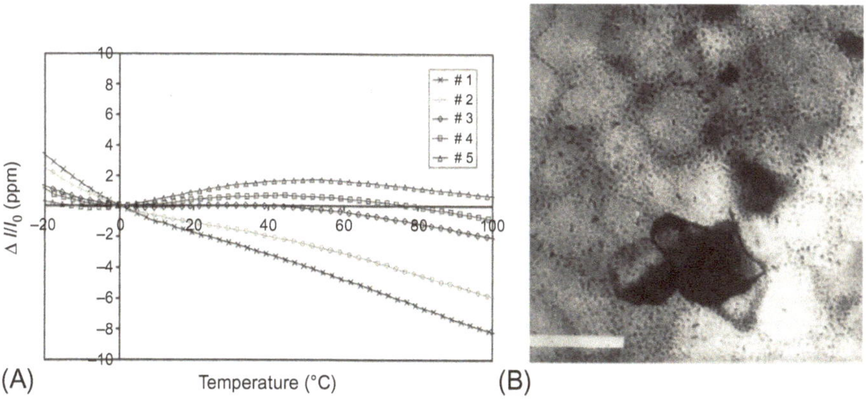


**Fig. 4.46** (A) Variation of $\Delta l/l_0$ of zero expansion glass-ceramic (ZERODUR) caused by alteration of ceramization temperature and time as a function of temperature. (B) Transmission electron microscopy (TEM) image of zero expansion glass-ceramic (ZERODUR).
(A) Reproduced with permission from W. Pannhorst, Recent developments for commercial applications of low expansion glass ceramics, Glass Technol. 45 (2004) 51-53, Copyright © Society of Glass Technology, UK. (B) Reproduced with permission from G.H. Beall, L.R. Pinckney, Nanophase glass-ceramics, J. Am. Ceram. Soc. 82 (1999) 5–16, Copyright © The American Ceramic Society.

(A) (B)

**Fig. 4.47** (A) Various shapes of zero expansion glass-ceramics (ZERODUR, Schott, Germany). (B) Transparent cookware (VISION, Corning, the United States) of β-quartz ss glass-ceramics.
(A) Courtesy: ZERODUR®, Zero Expansion Glass Ceramic, SCHOTT AG, Mainz, Germany. http://www.schott.com/advanced_optics/english/download/schott_zerodur_katalog_july_2011_en.pdf. (B) Reproduced with permission from G.H. Beall, L.R. Pinckney, Nanophase glass-ceramics, J. Am. Ceram. Soc. 82 (1999) 5–16, Copyright © The American Ceramic Society.

[241] has reported the low thermal expansivity of aluminoborate glass-ceramics in the glass systems of $RO–Al_2O_3–B_2O_3$ (where R = Ca, Sr, and Ba). These glasses on crystallization around 1000°C resulted in thermal expansion coefficients as low as $0.6 \times 10^{-6}$ °C$^{-1}$, $1.5 \times 10^{-6}$°C$^{-1}$, and $2.3 \times 10^{-6}$ °C$^{-1}$ for $SrAl_2B_2O_7$, $CaAl_2B_2O_7$, and $BaAl_2B_2O_7$ crystal phase containing glass-ceramics, respectively. The author has opined that this glass-ceramics can be used as electronic substrates (cofired multilayer packages for silicon chip-based circuitry), sealing frits for low expansion materials, coating for metals, and thermal-shock-resistant ceramics.

The most well-established glass system is $SiO_2–TiO_2$ [242–247]. Schultz and Smyth [243,244] and Duran et al. [242] have described the flame hydrolysis process for manufacturing the $SiO_2–TiO_2$ glasses. These glasses with 6–8 wt.% of $TiO_2$ in $SiO_2$ produce the near-zero expansion glasses. Beyond this limit, they exhibit negative coefficient of thermal expansion. Variation of coefficient of thermal expansion as a function of temperature in $SiO_2–TiO_2$ glasses with different compositions is illustrated in Fig. 4.48. The necessary experimental data were obtained from [247] for plotting. Evans [245] has established that at lower concentration of $TiO_2$ in $SiO_2–TiO_2$ glasses, $TiO_2$ acts as a network former with fourfold coordination ($TiO_4$) like silicon in silica glass ($SiO_4$) and forms the true solid solution (ss) resulting in the Si—O—Ti bonds up to 10.8 wt.% $TiO_2$. Schultz and Smyth [243] have suggested that the transverse vibrations of the oxygen atoms of the Si—O—Ti bonds bring about negative thermal expansion up to this concentration limit of $TiO_2$. With higher concentrations of this limit, $TiO_2$ precipitates out in the form of two immiscible liquids and/or 6-coordinated titanium ($TiO_6$) [243–245]. The photograph of ULE Corning Code 7972 ultralow expansion $SiO_2–TiO_2$ glass is available in [246]. This glass has several unique characteristics. For example, the density, refractive index ($n_D$ at 589 nm), Abbe number ($\nu_d$), and mean coefficient of thermal expansion (5–35°C) of ULE Corning Code 7972 ultralow expansion $SiO_2–TiO_2$ glass are 2.21 g cm$^{-3}$,

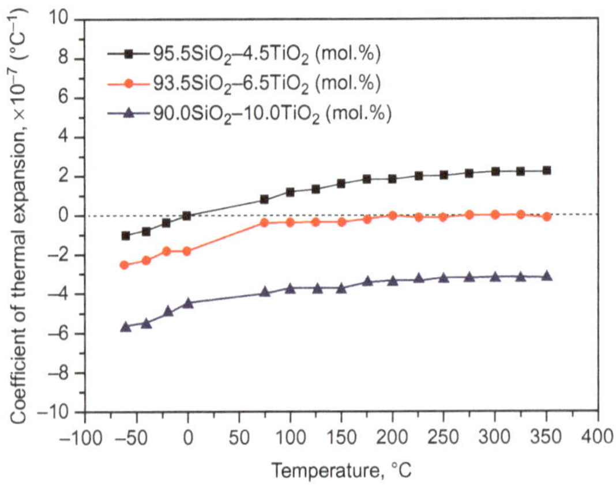


**Fig. 4.48** Variation of coefficient of thermal expansion as a function of temperature in $SiO_2–TiO_2$ glasses with different compositions.

1.4828, 53.1, and $0 \pm 0.03 \times 10^{-6}\,°C^{-1}$, respectively [246]. As a result of these unique properties, it is used in several specific applications such as lightweight mirror blanks for astronomical telescopes, space satellite, extreme ultraviolet (EUV), and reference blocks to solid.

## 4.3.6 Mica glass-ceramics

Glass-ceramics containing fluorophlogopite ($KMg_3AlSi_3O_{10}F_2$) mica and its derivatives have significant technical importance due to their several favorable properties such as machinability, high mechanical strength, heat resistance exceeding 800°C, and electric insulation properties. In 1970, Beall [248] developed fluorophlogopite mica glass-ceramics using the glass system of $SiO_2$–$B_2O_3$–$Al_2O_3$–MgO–$K_2O$–F at Corning, the United States, and received the US patent in 1972. Then after, considering the importance of this material for use in various technological applications, many investigations have been carried out. Salient features of some of these studies are listed in Table 4.10 [248–261].The commercial product of this glass-ceramics is the MACOR (Corning, the United States). It consists of 55% fluorophlogopite mica and 45% borosilicate glass. The photograph of various items fabricated from machinable mica glass-ceramic MACOR [262]. Some selected properties of MACOR brand fluorophlogopite machinable glass-ceramics are listed in Table 4.11 [262]. Because of these advantageous properties, they are used in a wide range of applications, for example, ultrahigh vacuum environments, constant vacuum applications, aerospace industry, spaceborne gamma radiation detector, boundary retainers on space shuttle, welding nozzles, precision electric insulators, and medical equipment. Fig. 4.49A shows the photograph of some rare-earth containing machinable glass-ceramics.

The machinability is explained as follows. The glass-ceramics contain the randomly oriented planes of weakness between the basal cleavage plane and the mica crystal-glass interfaces. This permits effortless propagation of fractures and mechanical displacement along the planes of weakness. Fracture propagation across the basal plane of mica is not easy. This causes fractures to follow the cleavage planes and the crystal boundaries, particularly when the mica crystals are interlocked. This results in detachment of mica crystals and enclosed glassy regions. Fig. 4.49B represents the FESEM photomicrograph showing the interlocking microstructure of fluorophlogopite mica glass-ceramics. Schematic illustration of layered structure of fluorophlogopite mica crystal with its hexagonal layer of $SiO_4$ and $AlO_4$ tetrahedra is shown in Fig. 4.50. The structure of fluorophlogopite consists of a two dimensional (2D) network of corner-shared $SiO_4$ and $AlO_4$ tetrahedra, which forms a hexagonal ring. The $Mg^{2+}$ ions are in octahedral coordination and located in the middle of two tetrahedral layers. The third layer is linked to each other by the relatively large $K^+$ ions. As a result of weak bonding of $K^+$ ions, the crack propagation takes place preferentially along the (001) cleavage plane.

Glass compositions, processing parameters, visual remarks, and crystal phases of various established mica glass-ceramics are summarized in Table 4.10. It is seen that the chemical composition of a precursor glass largely influence the crystallization

**Table 4.10 Glass compositions, processing parameters, visual remarks, and crystal phases of various established mica glass-ceramics**

| Glass composition (wt.%) | Temperature/time of glass melting and annealing (°C $h^{-1}$) | Crystallization temperature/ time (°C $h^{-1}$) | Visual remarks | Major crystal phase | Ref. |
|---|---|---|---|---|---|
| (35–48)$SiO_2$–(5–15)$B_2O_3$–(14–22)$Al_2O_3$–(15–25) MgO–(3–8)$K_2O$–(5–10)F | 1300–1450/6; 500–600/NS | 750–800/4 +950–1100/(4–6) | White, opaque | $K_{0.5}Mg_2Al_{0.83}BSi_3O_{10}F_2$ | [248] |
| (35–45)$SiO_2$–(6–14)$B_2O_3$–(14–19)$Al_2O_3$–(19.5–24) MgO–(4–6)$K_2O$–(6–10)F | 1450/5; 650/NS | 800–850/4 +1000–1075/4 | White, opaque | $K_{0.5}Mg_2Al_{0.83}BSi_3O_{10}F_2$ | [249] |
| (30–60)$SiO_2$–(5–26)$Al_2O_3$–(10–35)MgO–(8–30)SrO–(3–15)F | 1450–1500/(4–6); 600–650/NS | 725–800/4 +1050–1150/4 | Opaque | $Sr_xMg_yAlSi_3O_{10}F_2$ ($x=0.5$–1.0; $y=1.0$–2.5) | [250] |
| (35–55)$SiO_2$–(6–18) $Al_2O_3$–(12–27)MgO–(5–25)SrO–(0–20)BaO–(1–7)$K_2O$–(3–12)F | – | 1000–1150/(1–2) | Opaque | $(K,Sr_{0.5})Mg_3AlSi_3O_{10}F_2$ | [251] |
| 39$SiO_2$–16$Al_2O_3$–12MgO–10$K_2O$–11$B_2O_3$–12$MgF_2$+X (X=5$TiO_2$, 2.5$TiO_2$+2.5$ZrO_2$, 5$ZrO_2$) | 1550/2; 620/2 | 1050/2 | Opaque | $KMg_3AlSi_3O_{10}F_2$, $KMg_{2.75}Si_{3.5}Al_{0.5}O_{10}F_2$ | [252] |
| (mol.%) 41$SiO_2$–10$Al_2O_3$–7$K_2O$–10$B_2O_3$–12$MgF_2$–$(20-x)$MgO–$x$ZnO ($x=0$, 5, 10, 15, 20) | 1550/2; 620/2 | 1050/4 | Opaque | $KMg_3AlSi_3O_{10}F_2$, $Zn_2SiO4$ | [253] |
| 39$SiO_2$–16$Al_2O_3$–12MgO–10$K_2O$–11$B_2O_3$–12$MgF_2$+0.2(Cu, Ag, Au) | 1550/2; 620/2 | 1050/4 | Opaque | $KMg_3AlSi_3O_{10}F_2$, $KMg_{2.75}Si_{3.5}Al_{0.5}O_{10}F_2$ | [254] |
| 40.9$SiO_2$–23.8$Al_2O_3$–12.5MgO–7.3$K_2O$–9.7$MgF_2$+5($ZrO_2$, $La_2O_3$, $CeO_2$, $Yb_2O_3$, $V_2O_5$) | 1500–1600/2; 600/NS | 1100/2 | Opaque | $KMg_3AlSi_3O_{10}F_2$ | [255] |

| | | | | | |
|---|---|---|---|---|---|
| (mol.%) (33.8–49.5)$SiO_2$–(5.8–6.2)$B_2O_3$–(6.5–6.9)$Al_2O_3$–(13.4–31.4)MgO–(5.4–5.8)CaO–1.1$K_2O$–(16.0–17.1)$F^-$ | 1600/2; 600/NS | 1000–1100/2 | Opaque | $KMg_3AlSi_3O_{10}F_2$, $Ca_{0.5}Mg_3AlSi_3O_{10}F_2$, $MgAl_2O_4$, $Mg_2SiO_4$ | [256] |
| $Ca_{0.6}Mg_3Al_{1.2}Si_{2.8}O_{10}F_2$ or $K_{0.01}Ca_{0.595}Mg_3Al_{1.2}Si_{2.8}O_{10}F_2$ | 1450/2; $T_g$ +20/NS | 750–1000/2 | Transparent | $Ca_{0.5}Mg_3AlSi_3O_{10}F_2$ | [257] |
| (mol.%) (7.64–9.82)$Li_2O$–(45.85–58.93)$SiO_2$–(4.46–6.64)$Al_2O_3$–(17.86–26.58)MgO–(8.93–13.29)$MgF_2$ | 1450/2; $T_g$ +20/NS | 650–900/(2–48) | Transparent | $LiMg_3AlSi_3O_{10}F_2$, $LiAlSiO_4$, $Mg_5(SiO_4)_2F_2$, $LiAlSi_2O_6$ | [258] |
| 94.9 ($Li_{(1+x)}Mg_3AlSi_{3(1+x)}$ $O_{10+6.5x}F_2$) +5.1$MgF_2$ ($x$=0.5, 1.0) | 1450/2; $T_g$ +20/NS | 600–800/(1–6) | Transparent | $LiMg_3AlSi_3O_{10}F_2$, $LiAlSiO_4$, $Mg_5(SiO_4)_2F_2$ | [259] |
| 94.9$Li_{1.5}Mg_3AlSi_{4.5}O_{13.25}F_2$+5.1$MgF_2$+$x$ mol.% $Eu_2O_3$ ($x$=0, 0.05, 0.5, 5, 10, 20) | 1450/2; $T_g$ +20/NS | 600–800/1 | Transparent | $LiMg_3AlSi_3O_{10}F_2$, $LiAlSiO_4$, $Mg_5(SiO_4)_2F_2$ | [260] |
| 94.9$Li_{1.5}Mg_3AlSi_{4.5}O_{13.25}F_2$+5.1$MgF_2$+$x$ mol.% $CeO_2$ ($x$=0, 0.05, 0.5, 5, 10, 15) | 1450/2; $T_g$ +20/NS | 600–800/1 | Transparent | $LiMg_3AlSi_3O_{10}F_2$, $LiAlSiO_4$, $Mg_5(SiO_4)_2F_2$ | [261] |

*NS*, not specified.

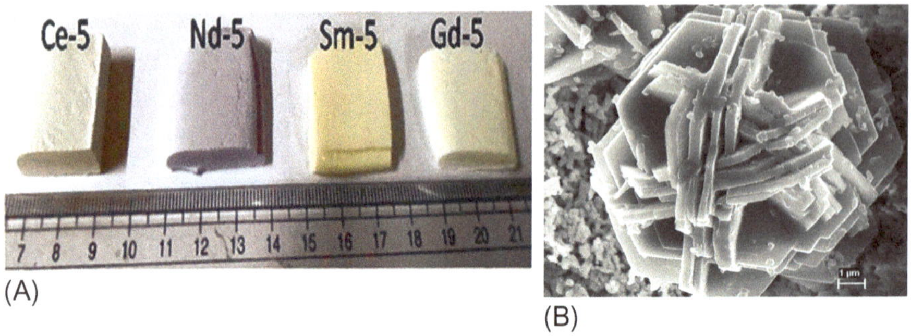


**Fig. 4.49** (A) Photograph of various rare-earth containing machinable mica glass-ceramics. (B) FESEM photomicrograph showing the interlocking microstructure of fluorophlogopite mica glass-ceramics.

**Table 4.11 Some selected properties of MACOR fluorophlogopite machinable glass-ceramics [262]**

| Property | Corresponding value | Property | Corresponding value |
|---|---|---|---|
| Density (g $cm^{-3}$) | 2.52 | Dielectric constant at 25°C and 1 kHz | 6.01 |
| Young's modulus at 25°C (GPa) | 66.9 | Thermal conductivity at 25°C (W $m^{-1}$°C) | 1.46 |
| Knoop hardness (100 g) (kg $mm^{-2}$) | 250 | Continuous operation temperature (°C) | 800 |
| Modulus of rupture at 25°C (MPa) | 94 | Coefficient of thermal expansion (CTE) ($\alpha \times 10^{-6}$ $K^{-1}$) | |
| | | (i) 25–300°C | 9.0 |
| | | (ii) 25–600°C | 11.2 |
| | | (iii) 25–600°C | 12.3 |

kinetics and microstructures of the glass-ceramics. Several recent studies make known that the different nucleating agents and additives have strong effects on the microstructure of the glass-ceramics. Evidences in support of these facts are illustrated in Figs. 4.51 and 4.52. FESEM photomicrographs of fluorophlogopite mica glass-ceramics heat-treated at 1050°C/2 h are shown in Fig. 4.51, which display the effects of (A) base glass without any nucleating agents or additives, the interlocking microstructure; (B) $TiO_2$ as nucleating agent, house-of-card like microstructure; (C) combination of $TiO_2$ and $ZrO_2$ as nucleating agent, house-of-platelet like microstructure; (D) $ZrO_2$ as nucleating agent interlocking card-like microstructure [252]; (E) $Al^{3+}$ ions totally replaced by $Cr^{6+}$ ions, regular octahedron and spherical particulate like microstructure [263]; and (F) MgO totally replaced by ZnO,

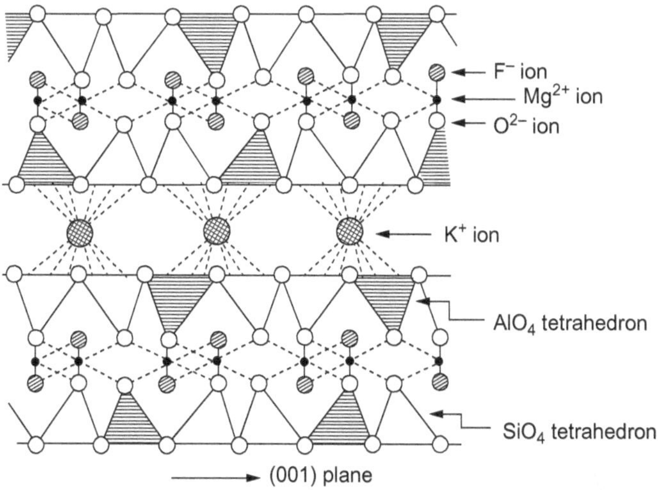


**Fig. 4.50** Schematic illustration of layered structure of fluorophlogopite mica crystal with its hexagonal layer of $SiO_4$ and $AlO_4$ tetrahedra.

backbone like microstructure showing the dendritic growth [253]. FESEM photomicrograph showing the influence of in situ-generated nanometals and different oxides on the microstructures of fluorophlogopite glass-ceramics heat-treated at 1050°C/2 h as depicted in Fig. 4.52 demonstrates the effects of (A) in situ-generated nanocopper (Cu), house-of-platelet like microstructure; (B) in situ-generated nanosilver (Ag), house-of-card like microstructure; and (C) in situ-generated nangold (Au), regular octahedron and elongated plate like microstructure [254]; (D) $CeO_2$, granular microstructure; (E) $Gd_2O_3$, spherical particulate like microstructure; and (F) PbO, cauliflower head microstructure [263].

The fluorophlogopite mica glass-ceramics are generally white and opaque [248–256]. Nevertheless, in recent times, Taruta et al. [257–261] have reported several transparent glass-ceramics containing calcium ($Ca_{0.5}Mg_3AlSi_3O_{10}F_2$) [257] and lithium ($LiMg_3AlSi_3O_{10}F_2$) [258–261] fluorophlogopite micas crystallized relatively at a lower temperature (700–800°C). Fig. 4.53 shows the (A) photograph of drilling test and (B) transmission electron microscopic (TEM) image of a transparent lithium fluorophlogopite glass-ceramics heat-treated at 700°C/4 h [259]. The TEM image demonstrates a tight interlocking microstructure lithium fluorophlogopite mica ($LiMg_3AlSi_3O_{10}F_2$) containing spherical particles of β-eucryptite (β-$LiAlSiO_4$).

Fig. 4.54A shows the emission spectra (excited at 254 nm) of a $Eu_2O_3$-doped transparent lithium fluorophlogopite glass and glass-ceramics heated at different temperature, (a) precursor glass, (b) 600°C/1 h, (c) 650°C/1 h, (d) 700°C/1 h, and (e) 750°C/1 h [260]. The emission peaks around 590 and 610 nm are originated from the $^5D_0 \rightarrow {}^7F_1$ and $^5D_0 \rightarrow {}^7F_2$ transitions, respectively, of $Eu^{3+}$ ions. Additionally, the weak, broad and discontinuous peak around 400 nm is originated from the $4f^65d \rightarrow 4f^7$ transition of $Eu^{2+}$ ion. This fact indicates that a part of $Eu^{3+}$ ions has been reduced to

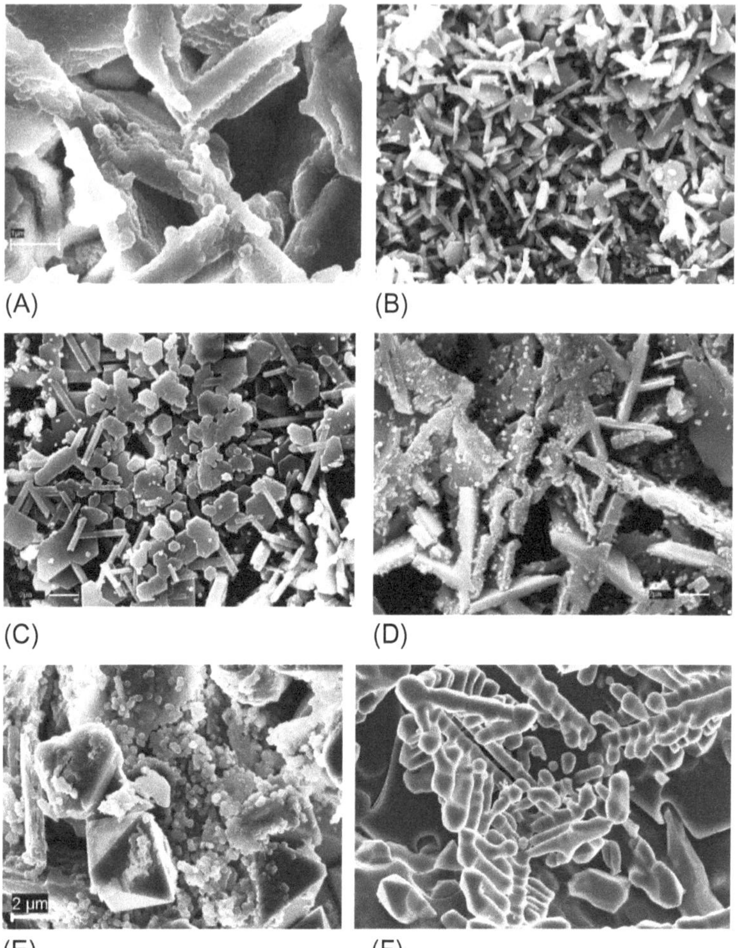


**Fig. 4.51** FESEM photomicrograph of fluorophlogopite mica glass-ceramics heat-treated at 1050°C/2 h. (A) Base glass, (B) 5 wt.% $TiO_2$ as nucleating agent, (C) 2.5 wt.% $TiO_2$+2.5 wt.% $ZrO_2$ as nucleating agent, and (D) 5 wt.% $ZrO_2$ as nucleating agent. (E) $Al^{3+}$ ions are totally replaced by $Cr^{6+}$ ions and (F) MgO is totally replaced by ZnO.
(A–D) Reproduced with permission from M. Garai, N. Sasmal, A.R. Molla, S.P. Singh, A. Tarafder, B. Karmakar, Effects of nucleating agents on crystallization and microstructure of fluorophlogopite mica-containing glass–ceramics, J. Mater. Sci. 49 (2014) 1612–1623, Copyright © Springer. (E) Courtesy: M. Garai, B. Karmakar, Effects of $CeO_2$, $Nd_2O_3$, $Sm_2O_3$, $Gd_2O_3$, $Cr_2O_3$, and PbO on crystallization and microstructure of K–Mg–B–Al–Si–F based glasses, Presented in the 102nd Indian Science Congress (ISC-2015), 3–7 January, University of Mysore, Mysore, 2016. (F) Reproduced with permission from M. Garai, N. Sasmal, A.R. Molla, B. Karmakar, Structural effects of $Zn^{+2}/Mg^{+2}$ ratios on crystallization characteristics and microstructure of fluorophlogopite mica-containing glass-ceramics, Solid State Sci. 44 (2015) 10–21, Copyright © Elsevier.

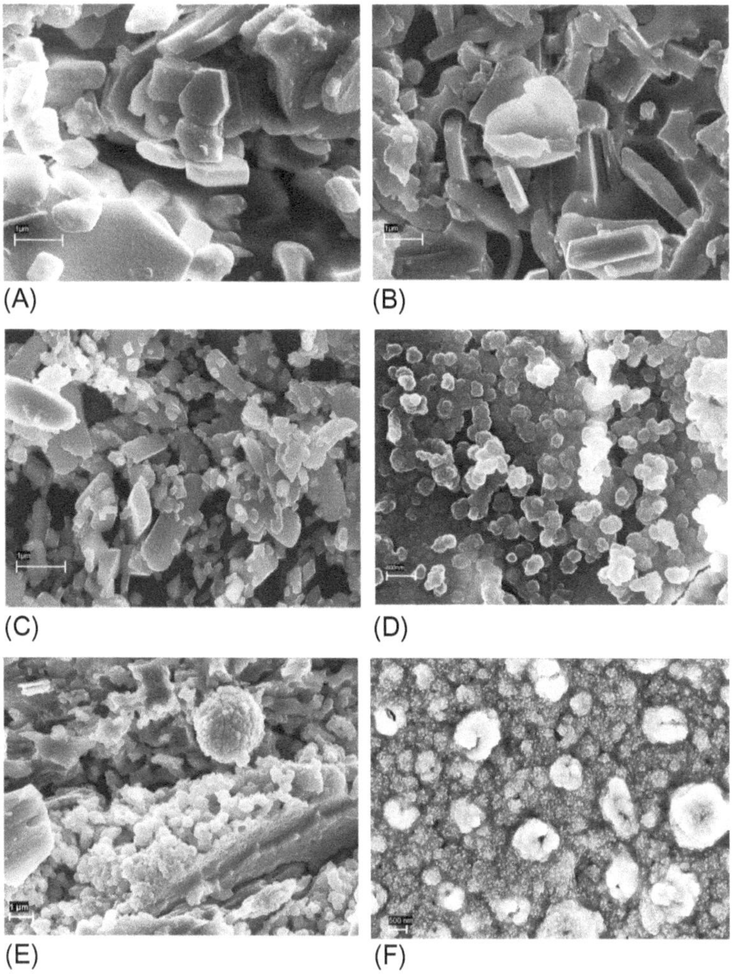


**Fig. 4.52** FESEM photomicrograph showing the influence of in situ-generated nanometals and different oxides on the microstructures of fluorophlogopite glass-ceramics heat-treated at 1050°C/2 h. (A) Copper (Cu), (B) silver (Ag), and (C) gold (Au). (D) $CeO_2$, (E) $Gd_2O_3$, and (F) PbO.
(A–C) Reproduced with permission from M. Garai, N. Sasmal, A.R. Molla, A. Tarafder, B. Karmakar, Effects of in-situ generated coinage nanometals on crystallization and microstructure of fluorophlogopite mica containing glass–ceramics, J. Mater. Sci. Technol. 31 (2015) 110–119, Copyright © Elsevier. (D–F) Courtesy: M. Garai, B. Karmakar, Effects of $CeO_2$, $Nd_2O_3$, $Sm_2O_3$, $Gd_2O_3$, $Cr_2O_3$, and PbO on crystallization and microstructure of K–Mg–B–Al–Si–F based glasses, Presented in the 102nd Indian Science Congress (ISC-2015), 3–7 January, University of Mysore, Mysore, 2016.

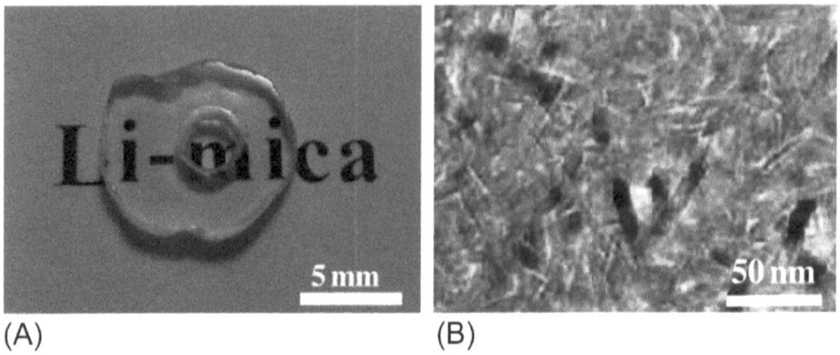


**Fig. 4.53** (A) Photograph of drilling test and (B) transmission electron microscopic (TEM) image of a transparent lithium fluorophlogopite glass-ceramics heat-treated at 700°C/4 h. Reproduced with permission from S. Taruta, M. Suzuki, T. Yamakami, T. Yamaguchi, K. Kitajima, Preparation and ionic conductivity of transparent glass-ceramics containing a large quantity of lithium-mica, J. Non-Cryst. Solids 354 (2008) 848–855, Copyright © Elsevier.

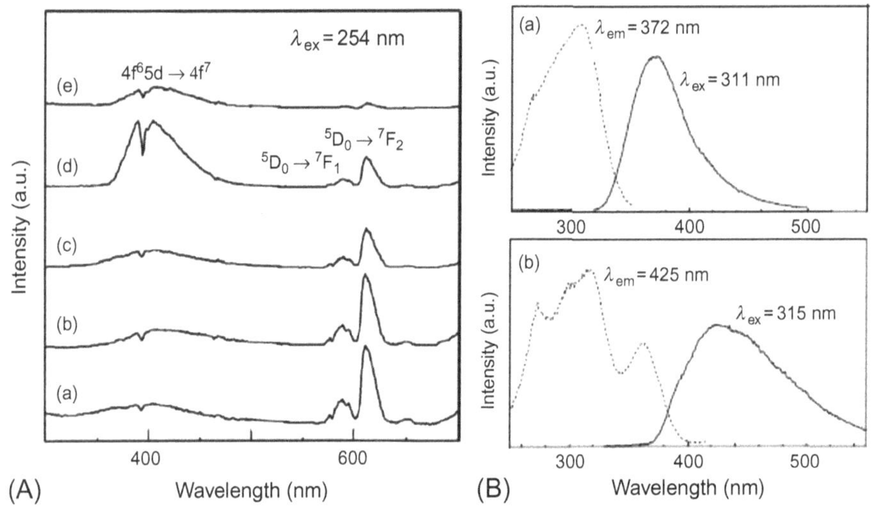


**Fig. 4.54** (A) Emission spectra (excited at 254 nm) of a $Eu_2O_3$-doped transparent lithium fluorophlogopite glass and glass-ceramics heated at different temperature. (a) Precursor glass, (b) 600°C/1 h, (c) 650°C/1 h, (d) 700°C/1 h, and (e) 750°C/1 h. (B) Emission *(solid line)* and excitation *(dotted line)* spectra of transparent lithium fluorophlogopite glass-ceramics heat-treated at 700°C/1 h doped with (a) 0.05 mol.% (Ce05) and (b) 10 mol.% (Ce10) $CeO_2$. (A) Reproduced with permission from S. Taruta, M. Matsuki, H. Nishikiori, T. Yamakami, T. Yamaguchi, K. Kitajima, Preparation and luminescent properties of Eu-doped transparent mica glass–ceramics, Ceram. Int. 36 (2010) 1303–1309, Copyright © Elsevier. (B) Reproduced with permission from S. Taruta, Y. Iwasaki, H. Nishikiori, T. Yamakami, T. Yamaguchi, K. Kitajima, K. Okada, Microstructures and luminescent properties of Ce-doped transparent mica glass-ceramics, Mater. Sci. Eng. B 177 (2012) 504–509, Copyright © Elsevier.

$Eu^{2+}$ ions with rise in heat treatment temperature from 600°C to 750°C. On the other hand, the emission bands around 590 and 610 nm become weaker with increase in heat treatment temperature. Consequently, the initial red emission gradually transforms into violet emission.

Fig. 4.54B shows the emission (solid line) and excitation (dotted line) spectra of transparent lithium fluorophlogopite glass-ceramics heat-treated at 700°C/1 h doped with (a) 0.05 mol.% (Ce05) and (b) 10 mol.% (Ce10) $CeO_2$ [261]. The emission bands around 310 and 420 nm originated from 5d to 4f transition of the $Ce^{3+}$ ions. Hence, the sample Ce10 emitted the blue light with increased concentration of $CeO_2$ under excitation in the near UV light at 315 nm.

## 4.4 Conclusions and future outlook

Here, functional glass-ceramics that are obtained by thermal treatment are described. Initially, with a brief introduction of glass-ceramics, its synthesis procedure is delineated with the relevant theories. Here, six major types of glass-ceramics, for example, photonic, ferroelectric, ultralow expansion, fluorophlogopite mica, and glass-ceramics for rechargeable battery and glaze, are considered. Glass-ceramics for down-conversion, quantum cutting, upconversion, WLEDs, and broadband emission are the topics in the domain of photonic glass-ceramics. In every case, associated glass compositions, processing parameters, properties, and applications are described with up-to-date progress achieved in these areas along with sufficient experimental data and literature support.

From the foregoing discussion, it is perfect that the development of high-performance functional materials through the process of glass-ceramics is very unique, and in some cases, glass-ceramics are the low cost alternative of the classical crystal growth process. At present, there are limited data and little understanding on functioning of these materials. Dependence of properties on nanoscale structure such as crystallite size, texture, nature of interface, and residual stresses are necessary to explore in a comprehensive manner. As far as device application is concerned, majority of these studies are in the preliminary stage. Therefore, more attention must be paid to establish much more factors toward device fabrication and demonstration. We believe that in the future in some cases, there would not be an alternative of glass-ceramics because of their superior performance over the conventional materials, for example, glass-ceramics for rechargeable battery and glass-ceramics for WLEDs.

## Exercises

**4.1** Define glass-ceramics. Why glass-ceramics is regarded as a unique type of material? Explain with example.

**4.2** What are photonic glass-ceramics? Why are they so-called? What is their significance in different advanced technologies?

**4.3** What is meant by "downconversion?" Explain this phenomenon with two examples where the glass-ceramics contain two and three rare-earth ions.

**4.4** What is meant by "quantum cutting?" Explain different mechanisms of this phenomenon with schematic energy-level diagrams. Explain this phenomenon with example of glass-ceramics containing rare-earth ions.

**4.5** What is meant by "upconversion?" Explain different mechanisms of this phenomenon with schematic energy-level diagrams. Explain this phenomenon with two examples where the glass-ceramics contain two and three rare-earth ions.

**4.6** Explain the mechanisms of white light emission with a suitable example of glass-ceramics and chromaticity diagram. Describe the design of a WLED with a schematic diagram. What are the advantages of the use of glass-ceramics in WLEDs in comparison with other similar materials?

**4.7** Explain the importance of broadband emission in advanced technologies. Justify your answer with a suitable example of glass-ceramics.

**4.8** Which type of crystals exhibits the ferroelectric properties? What is their structural uniqueness? What are the important uses of ferroelectric glass-ceramics?

**4.9** Explain the (i) crystal structures of $LiFePO_4$ and $LiGe_2(PO_4)_3$ and (ii) reasons of exhibiting high electric conductivity. What are the advantages of the use of glass-ceramics in rechargeable lithium-ion batteries?

**4.10** Why glass-ceramics are superior for use in floor and wall tiles? Describe different crystal phases which are created in glass-ceramics for use as glaze.

**4.11** What are the structural peculiarities of fluorophlogopite mica crystal which make these glass-ceramics as machinable glass-ceramics? Which type of mica crystal makes transparent glass-ceramics? Why?

# References

[1] A. Yasumori, Preparation of functional glasses and glass ceramics using stable immiscibility, J. Ceram. Soc. Jpn. 121 (2013) 471–479.

[2] E.D. Zanotto, Glass crystallization research—a 36-year retrospective. Part I, fundamental studies, Int. J. Appl. Glass Sci. 4 (2013) 105–116.

[3] E.D. Zanotto, Glass crystallization research—a 36-year retrospective. Part II, methods of study and glass-ceramics, Int. J. Appl. Glass Sci. 4 (2013) 117–124.

[4] E.D. Zanotto, A bright future for glass-ceramics, Am. Ceram. Soc. Bull. 89 (2010) 19–27.

[5] L.R. Pinckney, G.H. Beall, Microstructural evolution in some silicate glass–ceramics: a review, J. Am. Ceram. Soc. 91 (2008) 773–779.

[6] C. Rüssel, News from glass crystallization, J. Chem. Technol. Metall. 50 (2015) 357–366.

[7] W. Pannhorst, Recent developments for commercial applications of low expansion glass ceramics, Glass Technol. 45 (2004) 51–53.

[8] A. Sakamoto, S. Yamamoto, Glass–ceramics: engineering principles and applications, Int. J. Appl. Glass Sci. 1 (2010) 237–247.

[9] H. Jain, Transparent ferroelectric glass-ceramics, Ferroelectrics 306 (2004) 111–127.

[10] T. Komatsu, T. Honma, Optical active nano-glass-ceramics, Int. J. Appl. Glass Sci. 4 (2013) 125–135.

[11] Y. Teng, K. Sharafudeen, S. Zhou, J. Qiu, Glass-ceramics for photonic devices, J. Ceram. Soc. Jpn. 120 (2012) 458–466.

[12] D. Chen, W. Xiang, X. Liang, J. Zhong, H. Yu, M. Ding, H. Lu, Z. Ji, Advances in transparent glass–ceramic phosphors for white light-emitting diodes—a review, J. Eur. Ceram. Soc. 35 (2015) 859–869.

[13] J. Zhou, Y. Teng, S. Zhou, J. Qiu, Quantum cutting in luminescent glasses and glass ceramics, Int. J. Appl. Glass Sci. 3 (2012) 299–308.
[14] P.W. McMillan, Glass-Ceramics, second ed., Academic Press, London, 1979.
[15] W. Höland, G. Beall, Glass-Ceramic Technology, The American Ceramic Society, Westerville, OH, 2002.
[16] G. Lakshminarayana, R. Yang, M. Mao, J. Qiu, I.V. Kityk, Photoluminescence of $Sm^{3+}$, $Dy^{3+}$, and $Tm^{3+}$-doped transparent glass ceramics containing $CaF_2$ nanocrystals, J. Non-Cryst. Solids 355 (2009) 2668–2673.
[17] F. Yang, G. Chen, Z. You, C. Sun, J. Cao, Y. Ji, Y. Wang, Y. Wang, Z. Zhu, J. Li, C. Tu, The transformation of $Eu^{3+} \rightarrow Eu^{2+}$ in $SiO_2$–$Al_2O_3$–NaF–$YF_3$: Eu glasses and glass ceramics, Mater. Lett. 65 (2011) 1337–1339.
[18] Q. Luo, X. Qiao, X. Fan, S. Liu, H. Yang, X. Zhang, Reduction and luminescence of europium ions in glass ceramics containing $SrF_2$ nanocrystals, J. Non-Cryst. Solids 354 (2008) 4691–4694.
[19] K. Biswas, A.D. Sontakke, R. Sen, K. Annapurna, Luminescence properties of dual valence eu doped nano-crystalline $BaF_2$ embedded glass-ceramics and observation of $Eu^{2+} \rightarrow Eu^{3+}$ energy transfer, J. Fluoresc. 22 (2012) 745–752.
[20] X.Y. Liu, Y. Wei, H. Guo, Luminescence of Eu-doped transparent $LuPO_4$ glass-ceramics, J. Am. Ceram. Soc. 96 (2013) 369–371.
[21] G. Gao, S. Reibstein, M. Peng, L. Wondraczek, Tunable dual-mode photoluminescence from nanocrystalline Eu-doped $Li_2ZnSiO_4$ glass ceramic phosphors, J. Mater. Chem. 21 (2011) 3156–3161.
[22] Z. Cui, R. Ye, Y. Hua, D. Deng, S. Zhao, C. Li, S. Xu, Spectral analysis of a novel $Eu^{2+}$ activated $SiO_2$–BaO glass ceramics phosphor under UV-LED excitation, J. Non-Cryst. Solids 357 (2011) 2298–2301.
[23] D. Chen, Y. Wang, Y. Yu, P. Huang, Structure and optical spectroscopy of Eu-doped glass ceramics containing $GdF_3$ nanocrystals, J. Phys. Chem. C 112 (2008) 18943–18947.
[24] Y. Xu, X. Zhang, S. Dai, B. Fan, H. Ma, J. Adam, J. Ren, G. Chen, Efficient near-infrared down-conversion in $Pr^{3+}$-$Yb^{3+}$ codoped glasses and glass ceramics containing $LaF_3$ nanocrystals, J. Phys. Chem. C 115 (2011) 13056–13062.
[25] J.J. Velázquez, V.D. Rodríguez, A.C. Yanes, J. del-Castillo, J. Méndez-Ramos, Increase in the $Tb^{3+}$ green emission in $SiO_2$–$LaF_3$ nano-glass-ceramics by codoping with $Dy^{3+}$ ions, J. Appl. Phys. 108 (2010). 113530 (6 pages).
[26] G. Bai, L. Tao, K. Li, L. Hu, Y.H. Tsang, Enhanced ~2 μm and upconversion emission from Ho–Yb codoped oxyfluoride glass ceramics, J. Non-Cryst. Solids 361 (2013) 13–16.
[27] G. Dantelle, M. Mortier, D. Vivien, G. Patriarche, Effect of $CeF_3$ addition on the nucleation and up-conversion luminescence in transparent oxyfluoride glass-ceramics, Chem. Mater. 17 (2005) 2216–2222.
[28] D.L. Dexter, Possibility of luminescent quantum yields greater than unity, Phys. Rev. 108 (1957) 630–633.
[29] R.T. Wegh, H. Donker, K.D. Oskam, A. Meijerink, Visible quantum cutting in $LiGdF_4$: $Eu^{3+}$ through downconversion, Science 283 (1999) 663–666.
[30] S. Ye, B. Zhu, Y.X. Zhuang, X.F. Liu, J. Luo, L. Wang, J.R. Qiu, Spectral modification and quantum cutting in $RE^{3+}$-$Yb^{3+}$ (R = Pr, Tm, and Tb) codoped transparent glass-ceramics containing $CaF_2$ nanocrystals, IOP Conf. Ser. Mater. Sci. Eng. 1 (2009). 012008 (7 pages).
[31] L. Zhang, J. Xu, Y. Hu, G. Chen, Z. Wang, Near-infrared quantum cutting in $Pr^{3+}$-$Yb^{3+}$ codoped oxyfluoride glass ceramics containing $CaF_2$ nanocrystals, J. Wuhan Univ. Technol.-Mater. Sci. Ed. 28 (2013) 455–459.

[32] D. Chen, Y. Wang, Y. Yu, P. Huang, F. Weng, Near-infrared quantum cutting in transparent nanostructured glass ceramics, Opt. Lett. 33 (2008) 1884–1886.
[33] Z. Pan, G. Sekar, R. Akrobetu, R. Mu, S.H. Morgan, Visible to near-infrared down-conversion luminescence in $Tb^{3+}$ and $Yb^{3+}$ codoped lithium–lanthanum–aluminosilicate oxyfluoride glass and glass-ceramics, J. Non-Cryst. Solids 358 (2012) 1814–1817.
[34] D. Chen, Y. Yu, Y. Wang, P. Huang, F. Weng, Cooperative energy transfer up-conversion and quantum cutting down-conversion in $Yb^{3+}$:$TbF_3$ nanocrystals embedded glass ceramics, J. Phys. Chem. C 113 (2009) 6406–6410.
[35] F. Xin, S. Zhao, G. Jia, L. Huang, D. Deng, H. Wang, S. Xu, Up- and down-conversion luminescence in the oxyfluoride glass ceramics containing $Ba_{2+1.5x}Yb_{1-x}F_7$:$Tb^{3+}$ nanocrystals, Mater. Chem. Phys. 137 (2012) 177–181.
[36] H. Lin, D. Chen, Y. Yu, A. Yang, Y. Wang, Near-infrared quantum cutting in $Ho^{3+}$/$Yb^{3+}$ codoped nanostructured glass ceramic, Opt. Lett. 36 (2011) 876–878.
[37] X. Zhou, Y. Wang, X. Zhao, L. Li, Z. Wang, Q. Li, Near-infrared quantum cutting via downconversion energy transfers in $Ho^{3+}$/$Yb^{3+}$ codoped tellurite glass ceramics, J. Am. Ceram. Soc. 97 (2014) 179–184.
[38] S. Ye, B. Zhu, J. Luo, J. Chen, G. Lakshminarayana, J. Qiu, Enhanced cooperative quantum cutting in $Tm^{3+}$-$Yb^{3+}$ codoped glass ceramics containing $LaF_3$ nanocrystals, Opt. Express 16 (2008) 8989–8994.
[39] D.C. Yu, J.P. Zhang, Q.J. Chen, W.J. Zhang, Z.M. Yang, Q.Y. Zhang, Three-photon near-infrared quantum cutting in $Tm^{3+}$-doped transparent oxyfluoride glass ceramics, Appl. Phys. Lett. 101 (2012). 171108 (4 pages).
[40] F. Auzel, Upconversion and anti-Stokes processes with f and d ions in solids, Chem. Rev. 104 (2004) 139–173.
[41] H. Zheng, X.-J. Wang, M.J. Dejneka, W.M. Yen, R.S. Meltzer, Up-converted emission in $Pr^{3+}$-doped fluoride nanocrystals-based oxyfluoride glass ceramics, J. Lumin 108 (2004) 395–399.
[42] V. Lavín, I. Iparraguirre, J. Azkargorta, A. Mendioroz, J. González-Platas, R. Balda, J. Fernández, Stimulated and upconverted emissions of $Nd^{3+}$ in a transparent oxyfluoride glass-ceramic, Opt. Mater. 25 (2004) 201–208.
[43] D. Chen, Y. Wang, Y. Yu, F. Liu, P. Huang, Infrared to ultraviolet upconversion luminescence in $Nd^{3+}$ doped nano-glass-ceramic, J. Rare Earths 26 (2008) 428–431.
[44] A.S. Gouveia-Neto, E.B. da Costa, L.A. Bueno, S.J.L. Ribeiro, Intense red upconversion emission in infrared excited holmium-doped $PbGeO_3$–$PbF_2$–$CdF_2$ transparent glass ceramic, J. Lumin. 110 (2004) 79–84.
[45] Z. Hu, E. Ma, Y. Wang, D. Chen, Fluorescence property investigations on $Er^{3+}$-doped oxyfluoride glass ceramics containing $LaF_3$ nanocrystals, Mater. Chem. Phys. 100 (2006) 308–312.
[46] J. Wang, X. Qiao, X. Fan, M. Wang, Up-conversion and near-infrared emission of $Er^{3+}$ doped transparent glass ceramics containing $LaF_3$ nanocrystals, Physica B 353 (2004) 242–247.
[47] V.D. Rodríguez, J. Del Castillo, A.C. Yanes, J. Méndez-Ramos, M. Torres, J. Peraz, Luminescence of $Er^{3+}$-doped nanostructured $SiO_2$–$LaF_3$ glass-ceramics prepared by the sol–gel method, Opt. Mater. 29 (2007) 1557–1561.
[48] S. Chen, S. Zhao, F. Zheng, C. Zhang, L. Huang, S. Xu, Enhanced up-conversion luminescence of $Er^{3+}$:LaOF oxyfluoride borosilicate glass ceramics, Ceram. Int. 39 (2013) 2909–2913.
[49] X. Qiao, X. Fan, M. Wang, Luminescence behavior of $Er^{3+}$ in glass ceramics containing $BaF_2$ nanocrystals, Scripta Mater. 55 (2006) 211–214.

[50] D. Chen, Y. Wang, Y. Yu, E. Ma, L. Zhou, Microstructure and luminescence of transparent glass ceramic containing $Er^{3+}$:$BaF_2$ nano-crystals, J. Solid State Chem. 179 (2006) 532–537.
[51] D. Chen, Y. Wang, Y. Yu, E. Ma, F. Bao, Z. Hu, Y. Cheng, Influences of $Er^{3+}$ content on structure and upconversion emission of oxyfluoride glass ceramics containing $CaF_2$ nanocrystals, Mater. Chem. Phys. 95 (2006) 264–269.
[52] Z. Hu, Y. Wang, E. Ma, F. Bao, Y. Yu, D. Chen, Crystallization and spectroscopic properties investigations of $Er^{3+}$ doped transparent glass ceramics containing $CaF_2$, Mater. Res. Bull. 41 (2006) 217–224.
[53] X. Qiao, X. Fan, J. Wang, M. Wang, Luminescence behavior of $Er^{3+}$ ions in glass–ceramics containing $CaF_2$ nanocrystals, J. Non-Cryst. Solids 351 (2005) 357–363.
[54] Y. Yu, Y. Wang, D. Chen, F. Liu, Efficient upconversion luminescence of $Er^{3+}$:$SrF_2$–$SiO_2$–$Al_2O_3$ sol–gel glass ceramics, Ceram. Int. 34 (2008) 2143–2146.
[55] W.A. Pisarski, T. Goryczka, J. Pisarska, W. Ryba-Romanowski, Er-doped lead borate glasses and transparent glass ceramics for near-infrared luminescence and up-conversion applications, J. Phys. Chem. B 111 (2007) 2427–2430.
[56] F. Liu, E. Ma, D. Chen, Y. Yu, Y. Wang, Tunable red-green upconversion luminescence in novel transparent glass ceramics containing Er: $NaYF_4$ nanocrystals, J. Phys. Chem. B 110 (2006) 20843–20846.
[57] F. Xin, S. Zhao, L. Huang, D. Deng, G. Jia, H. Wang, S. Xu, Up-conversion luminescence of $Er^{3+}$-doped glass ceramics containing β-$NaGdF_4$ nanocrystals for silicon solar cells, Mater. Lett. 78 (2012) 75–77.
[58] F. Liu, Y. Wang, D. Chen, Y. Yu, E. Ma, L. Zhou, P. Huang, Upconversion emission of a novel glass ceramic containing $Er^{3+}$: $BaYF_5$ nano-crystals, Mater. Lett. 61 (2007) 5022–5025.
[59] X. Fan, J. Wang, X. Qiao, M. Wang, J.-L. Adam, X. Zhang, Preparation process and upconversion luminescence of $Er^{3+}$-doped glass ceramics containing $Ba_2LaF_7$ nanocrystals, J. Phys. Chem. B110 (2006) 5950–5954.
[60] C. Yu, J. Zhang, L. Wen, Z. Jiang, New transparent $Er^{3+}$-doped oxyfluoride tellurite glass ceramic with improved near infrared and up-conversion fluorescence properties, Mater. Lett. 61 (2007) 3644–3646.
[61] J. Méndez-Ramos, V. Lavín, I.R. Martín, U.R. Rodríguez-Mendoza, J.A. González-Almeida, V.D. Rodríguez, A.D. Lozano-Gorrín, P. Núñez, Optical properties of $Er^{3+}$ ions in transparent glass ceramics, J. Alloys Compd. 323–324 (2001) 753–758.
[62] A.S. Gouveia-Neto, L.A. Bueno, A.C.M. Afonso, J.F. Nascimento, E.B. Costa, Y. Messaddeq, S.J.L. Ribeiro, Upconversion luminescence in $Ho^{3+}$/$Yb^{3+}$- and $Tb^{3+}$/$Yb^{3+}$ codoped fluorogermanate glass and glass ceramic, J. Non-Cryst. Solids 354 (2008) 509–514.
[63] Y. Wei, J. Li, J.W. Yang, X.N. Chi, H. Guo, Enhanced green upconversion in $Tb^{3+}$–$Yb^{3+}$ codoped oxyfluoride glass ceramics containing $LaF_3$ nanocrystals, J. Lumin. 137 (2013) 70–72.
[64] D. Chen, Y. Yu, Y. Wang, P. Huang, F. Weng, Cooperative energy transfer up-conversion and quantum cutting down-conversion in $Yb^{3+}$:$TbF_3$ nanocrystals embedded glass ceramics, J. Phys. Chem. C113 (2009) 6406–6410.
[65] Y. Wei, X. Liu, X. Chi, R. Wei, H. Guo, Intense upconversion in novel transparent $NaLuF_4$:$Tb^{3+}$, $Yb^{3+}$ glass–Ceramics, J. Alloys Compd. 578 (2013) 385–388.
[66] Z. Shan, D. Chen, Y. Yu, P. Huang, F. Weng, H. Lin, Y. Wang, Upconversion luminescence of $Ho^{3+}$ sensitized by $Yb^{3+}$ in transparent glass ceramic embedding $BaYF_5$ nanocrystals, Mater. Res. Bull. 45 (2010) 1017–1020.

[67] J. Qiu, A. Makishima, Frequency up-conversion luminescence in $Yb^{3+}$–$Ho^{3+}$ co-doped $Pb_xCd_{1-x}F_2$ nano-crystals precipitated transparent oxyfluoride glass-ceramics, Sci. Technol. Adv. Mater. 5 (2004) 313–317.
[68] Y. Wei, H.M. Yang, X.M. Li, L.J. Wang, H. Guo, Elaboration, structure, and intense upconversion in transparent $KYb_2F_7$:$Ho^{3+}$ glass-ceramics, J. Am. Ceram. Soc. 97 (2014) 2012–2015.
[69] Y. Wei, X.N. Chi, X.Y. Liu, R.F. Wei, H. Guo, Novel upconversion behavior in $Ho^{3+}$-doped transparent oxyfluoride glass-ceramics containing $NaYbF_4$ nanocrystals, J. Am. Ceram. Soc. 96 (2013) 2073–2076.
[70] A. Biswas, G.S. Maciel, C.S. Friend, P.N. Prasad, Upconversion properties of a transparent $Er^{3+}$–$Yb^{3+}$ co-doped $LaF_3$–$SiO_2$ glass-ceramics prepared by sol–gel method, J. Non-Cryst. Solids 316 (2003) 393–397.
[71] X. Qiao, X. Fan, M. Wang, X. Zhang, Spectroscopic properties of $Er^{3+}$–$Yb^{3+}$ co-doped glass ceramics containing $BaF_2$ nanocrystals, J. Non-Cryst. Solids 354 (2008) 3273–3277.
[72] Y. Kishi, S. Tanabe, S. Tochino, G. Pezzotti, Fabrication and efficient infrared-to-visible upconversion in transparent glass ceramics of Er–Yb Co-doped $CaF_2$ nano-crystals, J. Am. Ceram. Soc. 88 (2005) 3423–3426.
[73] Z. Hu, Y. Wang, E. Ma, D. Chen, F. Bao, Microstructures and upconversion luminescence of $Er^{3+}$ doped and $Er^{3+}$/$Yb^{3+}$ co-doped oxyfluoride glass ceramics, Mater. Chem. Phys. 101 (2007) 234–237.
[74] F. Weng, D. Chen, Y. Wang, Y. Yu, P. Huang, H. Lin, Energy transfer and up-conversion luminescence in $Er^{3+}$/$Yb^{3+}$ co-doped transparent glass ceramic containing $YF_3$ nano-crystals, Ceram. Int. 35 (2009) 2619–2623.
[75] S. Zhao, S. Xu, G. Jia, D. Deng, L. Huang, H. Wang, $Er^{3+}$/$Yb^{3+}$ codoped oxyfluoride borosilicate glass ceramic containing $NaYF_4$ nanocrystals for amorphous silicon solar cells, Mater. Lett. 65 (2011) 2407–2409.
[76] Y. Ledemi, M.E. Amraoui, J.L. Ferrari, P.-L. Fortin, S.J.L. Ribeiro, Y. Messaddeq, Infrared to visible up-conversion emission in $Er^{3+}$/$Yb^{3+}$ codoped fluoro-phosphate glass–ceramics, J. Am. Ceram. Soc. 96 (2013) 825–832.
[77] K. Lou, Z. Song, J. Shang, Z. Yang, J. Qiu, Nano-crystals precipitation and upconversion luminescence properties in $Er^{3+}$/$Yb^{3+}$ co-doped fluoride phosphate glass-ceramics, J. Non-Cryst. Solids 357 (2011) 2251–2254.
[78] A.C. Yanes, A. Santana-Alonso, J. Méndez-Ramos, J. del-Castillo, V.D. Rodríguez, Novel sol–gel nano-glass–ceramics comprising $Ln^{3+}$-doped $YF_3$ nanocrystals: Structure and high efficient UV up-conversion, Adv. Funct. Mater. 21 (2011) 3136–3142.
[79] X. Qiao, X. Fan, Z. Xue, X. Xu, Short-wavelength upconversion luminescence of $Yb^{3+}$/$Tm^{3+}$ co-doped glass ceramic containing $SrF_2$ nanocrystals, J. Non-Cryst. Solids 357 (2011) 83–87.
[80] F.C. Guinhos, P.C. Nóbrega, P.A. Santa-Cruz, Compositional dependence of up-conversion process in Tm –Yb codoped oxyfluoride glasses and glass-ceramics, J. Alloys Compd. 323–324 (2001) 358–361.
[81] J. Zhang, D. He, Z. Duan, L. Zhang, S. Dai, L. Hu, Mechanisms and concentrations dependence of up-conversion luminescence in $Tm^{3+}$/$Yb^{3+}$ codoped oxyfluoride glass-ceramics, Phys. Lett. A 337 (2005) 480–486.
[82] Z. Duan, J. Zhang, W. Xiang, H. Sun, L. Hu, Multicolor upconversion of $Er^{3+}$/$Tm^{3+}$/$Yb^{3+}$ doped oxyfluoride glass ceramics, Mater. Lett. 61 (2007) 2200–2203.
[83] A.L. Patterson, The Scherrer formula for X-ray particle size determination, Phys. Rev. 56 (1939) 978–982.

[84] I. Akasaki, H. Amano, Y. Koide, K. Hiramatsu, N. Sawaki, Effects of ain buffer layer on crystallographic structure and on electrical and optical properties of GaN and Ga1–xAlxN(0<x 0.4) films grown on sapphire substrate by MOVPE, J. Cryst. Growth 98 (1989) 209–219.

[85] H. Amano, M. Kito, K. Hiramatsu, I. Akasaki, P-type conduction in Mg-doped GaN treated with low-energy electron beam irradiation (LEEBI), Jpn. J. Appl. Phys. 28 (1989) L2112–L2114.

[86] S. Nakamura, M. Senoh, N. Iwasa, S. Nagahama, High-brightness InGaN blue, green and yellow lighting-emitting diodes with quantum well structures, Jpn. J. Appl. Phys. 34 (1995) L797–L799.

[87] S. Nakamura, M. Senoh, N. Iwasa, S. Nagahama, High-power InGaN single-quantum well-structure blue and violet light-emitting diodes, Appl. Phys. Lett. 67 (1995) 114359–114362.

[88] S. Nakamura, G. Fasol, The Blue Laser Diode: GaN Based Light Emitters and Lasers, Springer, Berlin, 1997.

[89] F. Wang, Y. Lin, H. Shi, W. Wang, Z. Deng, J. Chen, X. Yuan, Y. Cao, Introduction on the fabrication technique of phosphor in glass by tape-casting and investigation on the chromaticity property, Opt. Express 22 (2014) A1355–A1362.

[90] D. Chen, Y. Chen, Transparent $Ce^{3+}$: $Y_3Al_5O_{12}$ glass ceramic for organic-resin-free white-light-emitting diodes, Ceram. Int. 40 (2014) 15325–15329.

[91] L. Wang, L. Mei, G. He, G. Liu, J. Li, L. Xu, Crystallization and fluorescence properties of Ce:YAG glass-ceramics with low $SiO_2$ content, J. Lumin. 136 (2013) 378–382.

[92] D. Chen, Y. Yu, P. Huang, H. Lin, Z. Shan, Y. Wang, Color-tunable luminescence of $Eu^{3+}$ in $LaF_3$ embedded nanocomposite for light emitting diode, Acta Mater. 58 (2010) 3035–3041.

[93] H. Guo, X.Y. Liu, F. Li, R.F. Wei, Y.L. Wei, C.G. Ma, Enhanced white luminescence in mixed-valence Eu-doped $BaAl_2Si_2O_8$ glass ceramics for W-LEDs, J. Electrochem. Soc. 159 (2012) J223–J226.

[94] H. Guo, F. Li, J. Li, H. Zhang, Luminescent properties of Eu-doped transparent glass–ceramics containing $YPO_4$ nanocrystals, J. Am. Ceram. Soc. 94 (2011) 1651–1653.

[95] T. Nakanishi, S. Tanabe, Novel $Eu^{2+}$-activated glass ceramics precipitated with green and red phosphors for high-power white LED, IEEE J. Sel. Topics Quantum Electron. 15 (2009) 1171–1176.

[96] T. Nakanishi, S. Tanabe, Preparation and luminescent properties of $Eu^{2+}$-activated glass ceramic phosphor precipitated with $\beta$-$Ca_2SiO_4$ and $Ca_3Si_2O_7$, Phys. Status Solidi A 206 (2009) 919–922.

[97] S. Nishiura, S. Tanabe, Preparation and luminescence properties of glass ceramics precipitated with $M_2MgSi_2O_7$: $Eu^{2+}$ (M = Sr, Ca) phosphor for white light source, IEEE J. Sel. Top. Quantum Electron. 15 (2009) 1177–1180.

[98] Z. Shan, D. Chen, Y. Yu, P. Huang, H. Lin, Y. Wang, Luminescence in rare earth-doped transparent glass ceramics containing $GdF_3$ nanocrystals for lighting applications, J. Mater. Sci. 45 (2010) 2775–2779.

[99] P. Babu, K.H. Jang, E.S. Kim, L. Shi, H.J. Seo, F. Rivera-López, U. R. Rodríguez-Mendoza, V. Lavín, R. Vijaya, C.K. Jayasankar, L.R. Moorthy, Spectral investigations on $Dy^{3+}$-doped transparent oxyfluoride glasses and nanocrystalline glass ceramics, J. Appl. Phys. 105 (2009). 013516 (6 pages).

[100] R. Bagga, V.G. Achanta, A. Goel, J.M.F. Ferreira, N.P. Singh, D.P. Singh, M. Falconierig, G. Sharma, $Dy^{3+}$-doped nano-glass ceramics comprising $NaAlSiO_4$

and $NaY_9Si_6O_{26}$ nanocrystals for white light generation, Mater. Sci. Eng. B 178 (2013) 218–224.
[101] Q. Luo, X. Qiao, X. Fan, H. Yang, X. Zhang, S. Cui, L. Wang, G. Wang, Luminescence behavior of $Ce^{3+}$ and $Dy^{3+}$ codoped oxyfluoride glasses and glass ceramics containing $LaF_3$ nanocrystals, J. Appl. Phys. 105 (2009). 043506 (5 pages).
[102] R. Ye, Z. Cui, Y. Hua, D. Deng, S. Zhao, C. Li, S. Xu, $Eu^{2+}/Dy^{3+}$ co-doped white light emission glass ceramics under UV light excitation, J. Non-Cryst. Solids 357 (2011) 2282–2285.
[103] X.F. Wang, X.H. Yan, Y.Y. Bu, J. Zhen, Y. Xuan, Fabrication, photoluminescence, and potential application in white light emitting diode of $Dy^{3+}$–$Tm^{3+}$ doped transparent glass ceramics containing $GdSr_2F_7$ nanocrystals, Appl. Phys. A 112 (2013) 317–322.
[104] H. Guo, F. Li, R.F. Wei, H. Zhang, C.G. Ma, Elaboration and luminescent properties of Eu/Tb Co-doped $GdPO_4$-based glass ceramics for white LEDs, J. Am. Ceram. Soc. 95 (2012) 1178–1181.
[105] H. Fu, X. Qiao, S. Cui, Q. Luo, J. Qian, X. Fan, X. Zhang, Tunable white light emission from glass-ceramics containing $Eu^{2+}$, $Tb^{3+}$, $Eu^{3+}$co-doped $SrLaF_5$ nanocrystals, Mater. Lett. 71 (2012) 15–17.
[106] F. Song, C. Ming, L. An, Q. Wang, Y. Yu, H. Yu, T. Sun, J. Tian, $Tm^{3+}/Tb^{3+}/Mn^{2+}$ tri-doped phosphate glass ceramic for enhanced white-light-emitting material, Mater. Lett. 65 (2011) 3140–3142.
[107] X.F. Wang, X.H. Yan, Y. Xuan, J. Zheng, W.Y. He, Green–white–yellow tunable luminescence from $Dy^{3+}$–$Tb^{3+}$–$Eu^{3+}$ doped transparent glass ceramics containing $GdSr_2F_7$ nanocrystals, Appl. Phys. A 113 (2013) 41–46.
[108] Y. Yu, F. Song, C. Ming, J. Zhang, F. Wang, Photoluminescence properties of $Tm^{3+}$/ $Tb^{3+}/Eu^{3+}$ tri-doped phosphate glass and glass ceramics for white-light-emitting diodes, Appl. Opt. 52 (2013) 5606–5610.
[109] Y. Xu, S. Cui, H. Fu, J. Qian, Q. Luo, X. Qiao, X. Fan, X. Zhang, White light generation of glass-ceramics containing $Ba_2LaF_7$: $Eu^{2+}$, $Tb^{3+}$ and $Sm^{3+}$ nanocrystals, Chin. Opt. Lett. 10 (2012). 031602 (4 pages).
[110] C. Li, S. Xu, R. Ye, S. Zhao, D. Deng, S. Zhuang, $Er^{3+}/Tm^{3+}/Yb^{3+}$ tridoped oxyfuoride glass ceramics with efficient upconversion white-light emission, Chin. Opt. Lett. 8 (2010) 66–69.
[111] D. Chen, Y. Wang, Y. Yu, P. Huang, F. Weng, Novel rare earth ions-doped oxyfluoride nano-composite with efficient upconversion white-light emission, J. Solid State Chem. 181 (2008) 2763–2767.
[112] J. del-Castillo, A.C. Yanes, J. Méndez-Ramos, V.K. Tikhomirov, V.V. Moshchalkov, V.D. Rodríguez, Sol–gel preparation and white up-conversion luminescence in rare-earth doped $PbF_2$ nanocrystals dissolved in silica glass, J. Sol-Gel Sci. Technol. 53 (2010) 509–514.
[113] A.C. Yanes, J.J. Velázquez, J. del-Castillo, J. Méndez-Ramos, V.D. Rodríguez, Colour tuneability and white light generation in $Yb^{3+}$–$Ho^{3+}$–$Tm^{3+}$ co-doped $SiO_2$–$LaF_3$ nano-glass-ceramics prepared by sol–gel method, J. Sol-Gel Sci. Technol. 51 (2009) 4–9.
[114] A. Santana-Alonso, J. Méndez-Ramos, A.C. Yanesa, J. del-Castillo, V.D. Rodríguez, White light up-conversion in transparent sol–gel derived glass-ceramics containing $Yb^{3+}$–$Er^{3+}$–$Tm^{3+}$ triply-doped $YF_3$ nanocrystals, Mater. Chem. Phys. 124 (2010) 699–703.
[115] H. Lin, D. Chen, Y. Yu, Z. Shan, P. Huang, Y. Wang, J. Yuan, $Nd^{3+}$-sensitized upconversion white light emission of $Tm^{3+}/Ho^{3+}$ bridged by $Yb^{3+}$ in $\beta$-$YF_3$ nanocrystals embedded transparent glass ceramics, J. Appl. Phys. 107 (2010) 103511 (4 pages).

[116] B. Wu, N. Jiang, S. Zhou, D. Chen, C. Zhu, J. Qiu, Transparent $Ni^{2+}$-doped silicate glass ceramics for broadband near-infrared emission, Opt. Mater. 30 (2008) 1900–1904.
[117] T. Suzuki, G.S. Murugan, Y. Ohishi, Optical properties of transparent $Li_2O$–$Ga_2O_3$–$SiO_2$ glass-ceramics embedding Ni-doped nanocrystals, Appl. Phys. Lett. 86 (2005). 131903 (3 pages).
[118] B. Wu, S. Zhou, J. Ren, D. Chen, X. Jiang, C. Zhu, J. Qiu, Broadband infrared luminescence from transparent glass–ceramics containing $Ni^{2+}$-doped $\beta$-$Ga_2O_3$ nanocrystals, Appl. Phys. B 87 (2007) 697–699.
[119] T. Suzuki, K. Horibuchi, Y. Ohishi, Structural and optical properties of ZnO–$Al_2O_3$–$SiO_2$ system glass–ceramics containing $Ni^{2+}$-doped nanocrystals, J. Non-Cryst. Solids 351 (2005) 2304–2309.
[120] B. Wu, J. Qiu, M. Peng, J. Ren, X. Jiang, C. Zhu, Transparent $Ni^{2+}$-doped ZnO–$Al_2O_3$–$SiO_2$ system glass-ceramics with broadband infrared luminescence, Mater. Res. Bull. 42 (2007) 762–768.
[121] S. Zhou, H. Dong, G. Feng, B.T. Wu, H. Zeng, J. Qiu, Broadband optical amplification in silicate glass ceramic containing $\beta$-$Ga_2O_3$:$Ni^{2+}$ nanocrystals, Opt. Express 15 (2007) 5477–5481.
[122] G. Gao, S. Reibstein, E. Spiecker, M. Peng, L. Wondraczek, Broadband NIR photoluminescence from $Ni^{2+}$-doped nanocrystalline Ba–Al titanate glass ceramics, J. Mater. Chem. 22 (2012) 2582–2588.
[123] D. Deng, H. Ma, S. Xu, Q. Wang, L. Huang, S. Zhao, H. Wang, C. Li, Broadband infrared luminescence of $Ni^{2+}$-doped silicate glass–ceramics containing lithium aluminate spinel nanocrystals, J. Non-Cryst. Solids 357 (2011) 1426–1429.
[124] H. Ma, P. Liu, D. Deng, S. Xu, Spectroscopy and crystal-field analysis of $Cr^{4+}$-doped transparent silicate glass-ceramics, J. Non-Cryst. Solids 357 (2011) 2294–2297.
[125] J.-S. Wang, F.-H. Shen, The development of $SiO_2$ resistant Cr-doped glass ceramics for high $Cr^{4+}$ emission, J. Non-Cryst. Solids 358 (2012) 246–251.
[126] N. Dai, H. Luan, Z. Liu, Y. Sheng, J. Peng, Z. Jiang, H. Li, L. Yang, J. Li, Broadband NIR luminescence of Bi-doped $Li_2O$-$Al_2O_3$-$SiO_2$ glass-ceramics, J. Non-Cryst. Solids 358 (2012) 2970–2973.
[127] M. Peng, D. Chen, J. Qiu, X. Jiang, C. Zhu, Bismuth-doped zinc aluminosilicate glasses and glass-ceramics with ultra-broad band infrared luminescence, Opt. Mater. 29 (2007) 556–561.
[128] K. Zhang, S. Zhou, Y. Zhuang, R. Yang, J. Qiu, Bandwidth broadening of near-infrared emission through nanocrystallization in Bi/Ni co-doped glass, Opt. Express 20 (2012) 8675–8680.
[129] B. Wu, S. Zhou, J. Ruan, Y. Qiao, D. Chen, C. Zhu, J. Qiu, Enhanced near-infrared emission from $Ni^{2+}$ in $Cr^{3+}$/$Ni^{2+}$ codoped transparent glass ceramics, Appl. Phys. Lett. 92 (2008). 151102 (3 pages).
[130] S. Zhou, N. Jiang, K. Miura, S. Tanabe, M. Shimizu, M. Sakakura, Y. Shimotsuma, M. Nishi, J. Qiu, K. Hirao, Simultaneous tailoring of phase evolution and dopant distribution in the glassy phase for controllable luminescence, J. Am. Chem. Soc. 132 (2010) 17945–17952.
[131] B. Wu, J. Ruan, J. Ren, D. Chen, C. Zhu, S. Zhou, J. Qiu, Enhanced broadband near-infrared luminescence in transparent silicate glass ceramics containing $Yb^{3+}$ ions and $Ni^{2+}$-doped $LiGa_5O_8$ nanocrystals, Appl. Phys. Lett. 92 (2008). 041110 (3 pages).
[132] D. Chen, Y. Wang, F. Bao, Y. Yu, Broadband near-infrared emission from $Tm^{3+}$/$Er^{3+}$ co-doped nanostructured glass ceramics, J. Appl. Phys. 101 (2007). 113511 (5 pages).

[133] V.K. Tikhomirov, K. Driesen, C. Görller-Walrand, M. Mortier, Mid-infrared emission in $Yb^{3+}$–$Er^{3+}$–$Tm^{3+}$ co-doped oxyfluoride glass-ceramics, Mater. Sci. Eng. B 146 (2008) 66–68.
[134] J.A. Duffy, M. Ingram, An interpretation of glass chemistry in terms of the optical basicity concept, J. Non-Cryst. Solids 21 (1976) 373–410.
[135] J.A. Duffy, M.D. Ingram, Comments on the application of optical basicity to glass, J. Non-Cryst. Solids 144 (1992) 76–80.
[136] J.A. Duffy, A review of optical basicity and its application to oxidic systems, Geochim. Cosmochim. Acta 57 (1993) 3961–3970.
[137] Y. Xi, Ferroelectric and ferromagnetic microcrystalline glass ceramics, Ferroelectrics 261 (2001) 3–13.
[138] P. Prapitpongwanich, R. Harizanova, K. Pengpat, C. Rüssel, Nanocrystallization of ferroelectric lithium niobate in $LiNbO_3$-$SiO_2$ glasses, Mater. Lett. 63 (2009) 1027–1029.
[139] Y. Hu, C.-L. Huang, Crystallization kinetics of the $LiNbO_3$-$SiO_2$-$Al_2O_3$ glass, J. Non-Cryst. Solids 278 (2000) 170–177.
[140] M. Todorović, L. Radonjić, Lithium-niobate ferroelectric material obtained by glass crystallization, Ceram. Int. 23 (1997) 55–60.
[141] M.P.F. Graça, M.A. Valente, M.G. Ferreira Da Silva, The electric behavior of a lithium-niobate-phosphate glass and glass-ceramics, J. Mater. Sci. 41 (2006) 1137–1144.
[142] N.S. Prasad, K.B.R. Varma, Evolution of ferroelectric $LiNbO_3$ phase in a reactive glass matrix ($LiBO_2$–$Nb_2O_5$), J. Non-Cryst. Solids 351 (2005) 1455–1465.
[143] R.S. Chaliha, K. Annapurna, A. Tarafder, V.S. Tiwari, P.K. Gupta, B. Karmakar, Structure, dielectric and optical properties of transparent $Nd^{3+}$: $KNbO_3$ nanocrystalline glass–ceramics, Opt. Mater. 32 (2010) 1202–1209.
[144] K. Shioya, T. Komatsu, H.G. Kim, R. Sato, K. Matusita, Optical properties of transparent glass-ceramics in $K_2O$-$Nb_2O_5$-$TeO_2$ glasses, J. Non-Cryst. Solids 189 (1995) 16–24.
[145] G.S. Maciel, N. Rakov, C.B. de Araújo, A.A. Lipovskii, D.K. Tagantsev, Optical limiting behavior of a glass–ceramic containing sodium niobate crystallites, Appl. Phys. Lett. 79 (2001) 584–586.
[146] U. Gangadharini, A.R. Molla, A. Tarafder, B. Karmakar, Synthesis and characterization of $Eu^{3+}$-doped transparent glass–ceramics containing nanocrystalline $Sr^{II}Nb^{IV}O_3$, J. Am. Ceram. Soc. 96 (2013) 2155–2162.
[147] P. Yongsiri, S. Eitssayeam, G. Rujijanagul, S. Sirisoonthorn, T. Tunkasiri, K. Pengpat, Fabrication of transparent lead-free KNN glassceramics by incorporation method, Nanoscale Res. Lett. 7 (2012). 136 (9 pages).
[148] O. Kalawa, R. Tipakontitikul, A. Niyompan, Phase formation and dielectric properties of ferroelectric glass-ceramics in $Na_2O$-$BaO$-$Nb_2O_5$-$SiO_2$ system doped with $Nd^{3+}$, J. Electroceram. 30 (2013) 87–92.
[149] Y. Zhou, Q. Zhang, J. Luo, Q. Tang, J. Du, Structural optimization and improved discharged energy density for niobate glass-ceramics by $La_2O_3$ addition, J. Am. Ceram. Soc. 96 (2013) 372–375.
[150] J.-J. Shyu, J.-R. Wang, Crystallization and dielectric properties of $SrO$–$BaO$–$Nb_2O_5$–$SiO_2$ tungsten-bronze glass-ceramics, J. Am. Ceram. Soc. 83 (2000) 3135–3140.
[151] C.-T. Cheng, M. Lanagan, B. Jones, J.-T. Lin, M.-J. Pan, Crystallization kinetics and phase development of $PbO$–$BaO$–$SrO$–$Nb_2O_5$–$B_2O_3$–$SiO_2$-based glass–ceramics, J. Am. Ceram. Soc. 88 (2005) 3037–3042.
[152] T. Liu, G. Chen, J. Song, C. Yuan, Crystallization kinetics and dielectric characterization of $CeO_2$-added $BaO$–$SrO$–$Nb_2O_5$–$B_2O_3$–$SiO_2$ glass-ceramics, Ceram. Int. 39 (2013) 5553–5559.

[153] N.S. Prasad, K.B.R. Varma, Nanocrystallization of $SrBi_2Nb_2O_9$ from glasses in the system. $Li_2B_4O_7$-SrO-$Bi_2O_3$-$Nb_2O_5$, Mater. Sci. Eng. B90 (2002) 246–253.
[154] N.S. Prasad, K.B.R. Varma, Y. Takahashi, Y. Benino, T. Fujiwara, T. Komatsu, Evolution and characterization of fluorite-like nano-$SrBi_2Nb_2O_9$ phase in the SrO–$Bi_2O_3$–$Nb_2O_5$–$Li_2B_4O_7$ glass system, J. Solid State Chem. 173 (2003) 209–215.
[155] N.S. Prasad, G.N. Subbanna, K.B.R. Varma, Glass nanocomposite of strontium bismuth niobate and lithium borate: structural and dielectric investigations, Mater. Lett. 47 (2001) 11–19.
[156] B.H. Venkataraman, T. Fujiwara, K.B.R. Varma, T. Komatsu, Influence of $Sm_2O_3$ doping on formation and structure of $SrBi_2Nb_2O_9$ nanocrystals in lithium borate glasses, Mater. Chem. Phys. 117 (2009) 244–249.
[157] K. Gerth, C. Rüssel, Crystallization of $Bi_3TiNbO_9$ from glasses in the system $Bi_2O_3$/ $TiO_2$/$Nb_2O_5$/ $B_2O_3$/$SiO_2$, J. Non-Cryst. Solids 243 (1999) 52–60.
[158] J. Chen, Y. Zhang, C. Deng, X. Dai, L. Li, Effect of the Ba/Ti ratio on the microstructures and dielectric properties of barium titanate-based glass–ceramics, J. Am. Ceram. Soc. 92 (2009) 1350–1353.
[159] A.K. Sahu, D. Kumar, O. Parkash, O.P. Thakur, C. Prakash, Effect of $K_2O$/BaO ratio on crystallization, microstructure and dielectric properties of strontium titanate borosilicate glass ceramics, Ceram. Int. 30 (2004) 477–483.
[160] S. Golezardi, V.K. Marghussian, A. Beitollahi, S.M. Mirkazemi, Crystallization behavior, microstructure and dielectric properties of lead titanate glass ceramics in the presence of $Bi_2O_3$ as a nucleating agent, J. Eur. Ceram. Soc. 30 (2010) 1453–1460.
[161] P. Sooksaen, I.M. Reaney, D.C. Sinclair, Crystallization and dielectric properties of borate-based ferroelectric $PbTiO_3$ glass-ceramics, J. Electroceram. 19 (2007) 221–228.
[162] J.J. Ruiz-Valdés, A.V. Gorokhovsky, J.I. Escalante-García, G. Mendoza-Suárez, Glass–ceramic materials with regulated dielectric properties based on the system BaO–PbO–$TiO_2$–$B_2O_3$–$Al_2O_3$, J. Eur. Ceram. Soc. 24 (2004) 1505–1508.
[163] J. Chen, Y. Zhang, C. Deng, X. Dai, Improvement in the microstructures and dielectric properties of barium strontium titanate glass–ceramics by $AlF_3$/$MnO_2$ addition, J. Am. Ceram. Soc. 92 (2009) 1863–1866.
[164] E.P. Gorzkowski, M.-J. Pan, B.A. Bender, C.C.M. Wu, Effect of additives on the crystallization kinetics of barium strontium titanate glass–ceramics, J. Am. Ceram. Soc. 91 (2008) 1065–1069.
[165] J. Chen, Y. Zhang, C. Deng, X. Dai, Crystallization kinetics and dielectric properties of barium strontium titanate based glass–ceramics, Mater. Chem. Phys. 121 (2010) 109–113.
[166] A.K. Sahu, D. Kumar, O. Parkash, Crystallisation of lead strontium titanate perovskite phase in [$(Pb_{1-x}Sr_x)O.TiO_2$]–[$2SiO_2.B_2O_3$]–[$K_2O$] glass ceramics, Br. Ceram. Trans. 102 (2003) 139–147.
[167] K. Gerth, C. Rüssel, Crystallization of $Bi_4Ti_3O_{12}$ from glasses in the system $Bi_2O_3$/$TiO_2$/ $B_2O_3$, J. Non-Cryst. Solids 221 (1997) 10–17.
[168] A.R. Molla, A. Tarafder, B. Karmakar, Synthesis and properties of glasses in the $K_2O$–$SiO_2$–$Bi_2O_3$–$TiO_2$ system and bismuth titanate ($Bi_4Ti_3O_{12}$) nano glass–ceramics thereof, J. Mater. Sci. 46 (2011) 2967–2976.
[169] A.R. Molla, A. Tarafder, S. Mukherjee, S.K. Mohanty, B. Karmakar, Processing and properties of $Eu^{3+}$-doped barium bismuth titanate ($BaBi_4Ti_4O_{15}$) glass–ceramic nanocomposites, J. Am. Ceram. Soc. 96 (2013) 2387–2395.
[170] V.N. Sigaev, S.Y. Stefanovich, P.D. Sarkisov, E.V. Lopatina, Stillwellite glass-ceramics with ferroelectric properties, Mater. Sci. Eng. B 32 (1995) 17–23.
[171] K. Pengpata, D. Holland, Ferroelectric glass-ceramics from the PbO–$GeO_2$–$Nb_2O_5$ system, J. Euro. Ceram. Soc. 24 (2004) 2951–2958.

[172] K. Pengpat, D. Holland, Glass-ceramics containing ferroelectric bismuth germanate ($Bi_2GeO_5$), J. Euro. Ceram. Soc. 23 (2003) 1599–1607.

[173] P. Pernice, A. Aronne, V.N. Sigaev, P.D. Sarkisov, V.I. Molev, S.Y. Stefanovich, Crystallization behavior of potassium niobium silicate glasses, J. Am. Ceram. Soc. 82 (1999) 3447–3452.

[174] S.Y. Stefanovich, V.N. Sigaev, S.V. Lotarev, E.V. Lopatina, A.V. Mosunov, A. G. Segalla, D.P. Chertin, Functional glass ceramic based on potassium niobate, Glass and Ceramics 70 (2013) 135–140 (Translated, Russian Original, Nos. 3–4, March–April, 2013).

[175] V.N. Sigaev, P.D. Sarkisov, S.Y. Stefanovich, P. Pernice, A. Aronne, Glass ceramic textures based on new ferroelectric complex oxides, Ferroelectrics 233 (1999) 165–185.

[176] A. Tarafder, K. Annapurna, R.S. Chaliha, V.S. Tiwari, P.K. Gupta, B. Karmakar, Processing and properties of $Eu^{3+}$:$LiTaO_3$ transparent glass–ceramic nanocomposites, J. Am. Ceram. Soc. 92 (2009) 1934–1939.

[177] A. Tarafder, K. Annapurna, R.S. Chaliha, S. Satapathy, V.S. Tiwari, P.K. Gupta, B. Karmaka, Second harmonic generation in ferroelectric $LiTaO_3$ and $KNbO_3$ containing bulk nano glass-ceramics, J. Nonlinear Opt. Phys. Mater. 20 (2011) 49–61.

[178] A. Tarafder, K. Annapurna, R.S. Chaliha, V.S. Tiwari, P.K. Gupta, B. Karmakar, Nanostructuring and fluorescence properties of $Eu^{3+}$:$LiTaO_3$ in $Li_2O$–$Ta_2O_5$–$SiO_2$–$Al_2O_3$ glass-ceramics, J. Mater. Sci. 44 (2009) 4495–4498.

[179] G.S. Murugan, K.B.R. Varma, Y. Takahashi, T. Komatsu, Nonlinear-optic and ferroelectric behavior of lithium borate–strontium bismuth tantalate glass–ceramic composite, Appl. Phys. Lett. 78 (2001). 4019 (3 pages).

[180] A. Tarafder, A.R. Molla, B. Karmakar, Enhanced photoluminescence and structure of $Dy^{3+}$-doped $SrBi_2Ta_2O_9$-containing transparent glass-ceramics, Opt. Mater. 35 (2013) 1549–1556.

[181] G.S. Murugan, K.B.R. Varma, Dielectric, linear and non-linear optical properties of lithium borate-bismuth tungstate glasses and glass-ceramics, J. Non-Cryst. Solids 279 (2001) 1–13.

[182] G.S. Murugan, K.B.R. Varma, Structural, dielectric and optical properties of lithium borate–bismuth tungstate glass-ceramics, Mater. Res. Bull. 34 (1999) 2201–2213.

[183] G.S. Murugan, G.N. Subbanna, K.B.R. Varma, Nanocrystallization of ferroelectric bismuth tungstate in lithium borate glass matrix, J. Mater. Sci. Lett. 18 (1999) 1687–1690.

[184] G.S. Murugan, K.B.R. Varma, Characterization of lithium borate–bismuth tungstate glasses and glass-ceramics by impedance spectroscopy, Solid State Ion. 139 (2001) 105–112.

[185] M.V. Shankar, K.B.R. Varma, Crystallization of ferroelectric bismuth vanadate in $Bi_2O_3$-$V_2O_5$-$SrB_4O_7$ glasses, J. Non-Cryst. Solids 226 (1998) 145–154.

[186] K. Hirose, T. Honma, Y. Benino, T. Komatsu, Glass–ceramics with $LiFePO_4$ crystals and oriented crystal line patterning in glass by YAG laser irradiation, Solid State Ion. 178 (2007) 801–807.

[187] K. Hirosea, T. Honmaa, Y. Doib, Y. Hinatsub, T. Komatsu, Mössbauer analysis of Fe ion state in lithium iron phosphate glasses and their glass-ceramics with olivine-type $LiFePO_4$ crystals, Solid State Commun. 146 (2008) 273–277.

[188] T. Nagakane, H. Yamauchi, K. Yuki, M. Ohji, A. Sakamoto, T. Komatsu, T. Honma, M. Zou, G. Park, T. Sakai, Glass-ceramic $LiFePO_4$ for lithium-ion rechargeable battery, Solid State Ionics 206 (2012) 78–83.

[189] T. Honma, K. Nagamine, T. Komatsu, Fabrication of olivine-type $LiMn_xFe_{1-x}PO_4$ crystals via the glass–ceramic route and their lithium ion battery performance, Ceram. Int. 36 (2010) 1137–1141.
[190] K. Nagamine, T. Honma, T. Komatsu, Selective synthesis of lithium ion-conductive β-$LiVOPO_4$ crystals via glass–ceramic processing, J. Am. Ceram. Soc. 91 (2008) 3920–3925.
[191] K. Nagamine, K. Hirose, T. Honma, T. Komatsu, Lithium ion conductive glass–ceramics with $Li_3Fe_2(PO_4)_3$ and YAG laser-induced local crystallization in lithium iron phosphate glasses, Solid State Ion. 179 (2008) 508–515.
[192] J. Fu, Fast $Li^+$ ion conducting glass-ceramics in the system $Li_2O$–$Al_2O_3$–$GeO_2$–$P_2O_5$, Solid State Ion. 104 (1997) 191–194.
[193] A.M. Cruz, E.B. Ferreira, A.C.M. Rodrigues, Controlled crystallization and ionic conductivity of a nanostructured $LiAlGePO_4$ glass–ceramic, J. Non-Cryst. Solids 355 (2009) 2295–2301.
[194] H. Chung, B. Kang, Increase in grain boundary ionic conductivity of $Li_{1.5}Al_{0.5}Ge_{1.5}(PO_4)_3$ by adding excess lithium, Solid State Ion. 263 (2014) 125–130.
[195] H.S. Jadhav, M.-S. Cho, R.S. Kalubarme, J.-S. Lee, K.-N. Jung, K.-H. Shin, C.-J. Park, Influence of $B_2O_3$ addition on the ionic conductivity of $Li_{1.5}Al_{0.5}Ge_{1.5}(PO_4)_3$ glass ceramics, J. Power Sources 241 (2013) 502–508.
[196] J. Yang, Z. Huang, B. Huang, Z. Zhou, X. Xu, Influence of phosphorus sources on lithium ion conducting performance in the system of $Li_2O$–$Al_2O_3$–$GeO_2$–$P_2O_5$ glass–ceramics, Solid State Ion. 270 (2015) 61–65.
[197] C.J. Leo, B.V.R. Chowdari, G.V. SubbaRao, J.L. Souquet, Lithium conducting glass ceramic with nasicon structure, Mater. Res. Bull. 37 (2002) 1419–1430.
[198] X. Xu, Z. Wen, X. Wu, X. Yang, Z. Gu, Lithium ion-conducting glass–ceramics of $Li_{1.5}Al_{0.5}Ge_{1.5}(PO_4)_3$ –$xLi_2O$ ($x=0$–$0.20$) with good electrical and electrochemical properties, J. Am. Ceram. Soc. 90 (2007) 2802–2806.
[199] B. Kumar, D. Thomas, J. Kumar, Space-charge-mediated superionic transport in lithium ion conducting glass–ceramics, J. Electrochem. Soc. 156 (2009) A506–A513.
[200] X.-X. Xu, Z.-Y. Wen, Z.-H. Gu, Z.-X. Lin, High lithium conductivity in $Li_{1.3}Cr_{0.3}Ge_{1.7}(PO_4)_3$ glass-ceramics, Mater. Lett. 58 (2004) 3428–3431.
[201] X. Xu, Z. Wen, Z. Gu, X. Xu, Z. Lin, Lithium ion conductive glass ceramics in the system $Li_{1.4}Al_{0.4}(Ge_{1-x}Ti_x)_{1.6}(PO_4)_3$ ($x=0$–$1.0$), Solid State Ion. 171 (2004) 207–213.
[202] B.V.R. Chowdari, G.V. Subba Rao, G.Y.H. Lee, XPS and ionic conductivity studies on $Li_2O$–$Al_2O_3$($TiO_2$ or $GeO_2$)–$P_2O_5$ glass–ceramics, Solid State Ion. 136–137 (2000) 1067–1075.
[203] J. Fu, Superionic conductivity of glass–ceramics in the system $Li_2O$–$Al_2O_3$-$TiO_2$–$P_2O_5$, Solid State Ion. 96 (1997) 195–200.
[204] X. Xu, Z. Wen, X. Yang, J. Zhang, Z. Gu, High lithium ion conductivity glass-ceramics in $Li_2O$–$Al_2O_3$–$TiO_2$–$P_2O_5$ from nanoscaled glassy powders by mechanical milling, Solid State Ion. 177 (2006) 2611–2615.
[205] J.L. Narváez-Semanate, A.C.M. Rodrigues, Microstructure and ionic conductivity of $Li_{1+x}Al_xTi_{2-x}(PO_4)_3$ NASICON glass-ceramics, Solid State Ion. 181 (2010) 1197–1204.
[206] J.S. Thokchom, B. Kumar, Ionically conducting composite membranes from the $Li_2O$–$Al_2O_3$–$TiO_2$–$P_2O_5$ glass–ceramic, J. Am. Ceram. Soc. 90 (2007) 462–466.
[207] J. Fu, Fast Li+ion conduction in $Li_2O$–$Al_2O_3$–$TiO_2$–$SiO_2$–$P_2O_5$ glass–ceramic, J. Am. Ceram. Soc. 80 (1997) 1901–1903.

[208] Y. Inda, T. Katoh, M. Baba, Development of all-solid lithium-ion battery using Li-ion conducting glass-ceramics, J. Power Sources 174 (2007) 741–744.

[209] https://crystallography365.wordpress.com/2014/04/29/lifepo4-the-unexpected-battery-success-story/ Accessed 10 February 2016.

[210] A.K. Padhi, K.S. Nanjundaswamy, J.B. Goodenough, Phospho-olivines as positive-electrode materials for rechargeable lithium batteries, J. Electrochem. Soc. 144 (1997) 1188–1194.

[211] S. Breuer, D. Prutsch, Q. Ma, V. Epp, F. Preishuber-Pflügl, F. Tietz, M. Wilkening, Separating bulk from grain boundary Li ion conductivity in the sol–gel prepared solid electrolyte $Li_{1.5}Al_{0.5}Ti_{1.5}(PO_4)_3$, J. Mater. Chem. A3 (2015) 21343–21350.

[212] R. Casasola, J.M. Rincón, M. Romero, Glass-ceramic glazes for ceramic tiles—a review, J. Mater. Sci. 47 (2012) 553–582.

[213] Sh. Salem, S.H. Jazayeri, F. Bondioli, A. Allahverdi, M. Shirvani, Characterizing thermal behavior of ceramic glaze containing nano-sized cobalt-aluminate pigment by hot stage microscopy, Thermochim. Acta 521 (2011) 191–196.

[214] M. Romero, J.M. Rincón, A. Acosta, Crystallisation of a zirconium-based glaze for ceramic tile coatings, J. Eur. Ceram. Soc. 23 (2003) 1629–1635.

[215] K. Pekkan, B. Karasu, Evaluation of borax solid wastes in production of frits suitable for fast single-fired wall tile opaque glass–ceramic glazes, Bull. Mater. Sci. 33 (2010) 135–144.

[216] F.J. Torres, E.R. de Sola, J. Alarcón, Effect of some additives on the development of spinel-based glass-ceramic glazes for floor-tiles, J. Non-Cryst. Solids 351 (2005) 2453–2461.

[217] B.E. Yekta, P. Alizadeh, L. Rezazadeh, Synthesis of glass-ceramic glazes in the ZnO–$Al_2O_3$–$SiO_2$–$ZrO_2$ system, J. Eur. Ceram. Soc. 27 (2007) 2311–2315.

[218] F.J. Torres, E.R. de Sola, J. Alarcón, Mechanism of crystallization of fast fired mullite-based glass–ceramic glazes for floor-tiles, J. Non-Cryst. Solids 352 (2006) 2159–2165.

[219] F.J. Torres, E.R. de Sola, J. Alarcón, Effect of boron oxide on the microstructure of mullite-based glass-ceramic glazes for floor-tiles in the CaO–MgO–$Al_2O_3$–$SiO_2$ system, J. Eur. Ceram. Soc. 26 (2006) 2285–2292.

[220] L. Fröberg, T. Kronberg, L. Hupa, M. Hupa, Influence of firing parameters on phase composition of raw glazes, J. Eur. Ceram. Soc. 27 (2007) 1671–1675.

[221] B.E. Yekta, P. Alizadeh, L. Rezazadeh, Floor tile glass-ceramic glaze for improvement of glaze surface properties, J. Eur. Ceram. Soc. 26 (2006) 3809–3812.

[222] M. Romero, J.M. Rincón, A. Acosta, Effect of iron oxide content on the crystallisation of a diopside glass–ceramic glaze, J. Eur. Ceram. Soc. 22 (2002) 883–890.

[223] G. Kaya, B. Karasu, A. Cakir, Characterization of diopside-based glass-ceramic porcelain tile glazes containing borax solid wastes, J. Ceram. Process. Res. 12 (2011) 135–139.

[224] S. Ghosh, K.S. Pal, N. Dandapat, J. Ghosh, S. Datta, Glass-ceramic glazes for future generation floor tiles, J. Eur. Ceram. Soc. 33 (2013) 935–942.

[225] F.J. Torres, J. Alarcón, Effect of additives on the crystallization of cordierite-based glass-ceramics as glazes for floor tiles, J. Eur. Ceram. Soc. 23 (2003) 817–826.

[226] F.J. Torres, J. Alarcón, Effect of MgO/CaO ratio on the microstructure of cordierite-based glass-ceramic glazes for floor tiles, Ceram. Int. 31 (2005) 683–690.

[227] F.J. Torres, J. Alarcón, Microstructural evolution in fast-heated cordierite-based glass-ceramic glazes for ceramic tile, J. Am. Ceram. Soc. 87 (2004) 1227–1232.

[228] A.M. Ferrari, L. Barbieri, C. Leonelli, T. Manfredini, C. Siligardi, A.B. Corradi, Feasibility of using cordierite glass-ceramics as tile glazes, J. Am. Ceram. Soc. 80 (1997) 1757–1766.

[229] H. Lu, M. He, Y. Liu, J. Guo, L. Zhang, D. Chen, H. Wang, H. Xu, R. Zhang, A preparation and performance study of glass-ceramic glazes derived from blast furnace slag and fly ash, J. Ceram. Process. Res. 12 (2011) 588–591.
[230] J.M. Rincón, M. Romero, J. Marco, V. Caballer, Some aspects of crystallization microstructure on new glass-ceramic glazes, Mater. Res. Bull. 33 (1998) 1159–1164.
[231] F.J. Torres, J. Alarcón, Mechanism of crystallization of pyroxene-based glass-ceramic glazes, J. Non-Cryst. Solids 347 (2004) 45–51.
[232] F.J. Torres, J. Alarcón, Pyroxene-based glass-ceramics as glazes for floor tiles, J. Eur. Ceram. Soc. 25 (2005) 349–355.
[233] M. Romero, J.M. Rincón, A. Acosta, Development of mica glass-ceramic glazes, J. Am. Ceram. Soc. 87 (2004) 819–823.
[234] H. Bach (Ed.), Low Thermal Expansion Glass Ceramics, Springer, Berlin, 1995.
[235] G.H. Beall, L.R. Pinckney, Nanophase glass-ceramics, J. Am. Ceram. Soc. 82 (1999) 5–16.
[236] ZERODUR®, Zero Expansion Glass Ceramic, SCHOTT AG, Mainz, Germany. http://www.schott.com/advanced_optics/english/download/schott_zerodur_katalog_july_2011_en.pdf.
[237] F.C. Serbena, V.O. Soares, O. Peitl, H. Pinto, R. Muccillo, E.D. Zanotto, Internal residual stresses in sintered and commercial low expansion $Li_2O$–$Al_2O_3$–$SiO_2$ glass–ceramics, J. Am. Ceram. Soc. 94 (2011) 1206–1214.
[238] L. Arnault, M. Gerland, A. Rivière, Microstructural study of two LAS-type glass-ceramics and their parent glass, J. Mater. Sci. 35 (2000) 2331–2345.
[239] M.J. Buerger, The stuffed derivatives of the silica structures, Am. Mineral. 39 (1954) 600–614.
[240] B. Karmakar, P. Kundu, S. Jana, R.N. Dwivedi, Crystallization kinetics and mechanism of low-expansion lithium aluminosilicate glass-ceramics by dilatometry, J. Am. Ceram. Soc. 85 (2002) 2572–2574.
[241] J.F. MacDowell, Aluminoborate glass-ceramics with low thermal expansivity, J. Am. Ceram. Soc. 73 (1990) 2287–2292.
[242] C. Duran, K.E. Hrdina, M.A. Mueller, Low expansivity, high transmission titania doped silica glass, US patent 8,541,325, 2013.
[243] P.C. Schultz, H.T. Smyth, Ultra-low-expansion glasses and their structure in the $SiO_2$-$TiO_2$ system, in: P.W. Douglas, B. Ellis (Eds.), Amorphous Materials, John Wiley and Sons Ltd., New York, NY, 1972, pp. 453–461.
[244] P.C. Schultz, Binary titania-silica glasses containing 10 to 20 wt% $TiO_2$, J. Am. Ceram. Soc. 59 (1976) 214–219.
[245] D.L. Evans, Solid solution of $TiO_2$ in $SiO_2$, J. Am. Ceram. Soc. 53 (1970) 418–419.
[246] ULE® Corning Code 7972 Ultra Low Expansion Glass, Corning Advanced Optics, New York, 2008. *www.corning.com/ule*.
[247] G.A. Pavlova, A.N. Amatuni, Physicochemical properties of glasses with a low thermal expansion coefficient in the system $SiO_2$-$TiO_2$, Inorg. Mater. 11 (1975) 1443–1445 (Engl. Transl. of Neorg. Mater. 11 (1975) 1686–1689).
[248] G.H. Beall, Mica glass-ceramics, US patent 3,689,293, 1972.
[249] G.H. Beall, C.-K. Chyung, H.J. Watkins, Mica glass-ceramics, US patent 3,801,295, 1974.
[250] G.H. Beall, Strontium fluormica glass-ceramics, US patent 3,756,838, 1973.
[251] G.H. Beall, High strength machinable glass-ceramics, US patent 20090318277A1, 2009.
[252] M. Garai, N. Sasmal, A.R. Molla, S.P. Singh, A. Tarafder, B. Karmakar, Effects of nucleating agents on crystallization and microstructure of fluorophlogopite mica-containing glass–ceramics, J. Mater. Sci. 49 (2014) 1612–1623.

[253] M. Garai, N. Sasmal, A.R. Molla, B. Karmakar, Structural effects of $Zn^{+2}/Mg^{+2}$ ratios on crystallization characteristics and microstructure of fluorophlogopite mica-containing glass-ceramics, Solid State Sci. 44 (2015) 10–21.
[254] M. Garai, N. Sasmal, A.R. Molla, A. Tarafder, B. Karmakar, Effects of in-situ generated coinage nanometals on crystallization and microstructure of fluorophlogopite mica containing glass–ceramics, J. Mater. Sci. Technol. 31 (2015) 110–119.
[255] P. Wang, L. Yu, H. Xiao, Y. Cheng, S. Lian, Influence of nucleation agents on crystallization and machinability of mica glass–ceramics, Ceram. Int. 35 (2009) 2633–2638.
[256] L. Yu, H. Xiao, Y. Cheng, Influence of magnesia on the structure and properties of $MgO-Al_2O_3-SiO_2-F^-$ glass-ceramics, Ceram. Int. 34 (2008) 63–68.
[257] S. Taruta, M. Sakata, T. Yamaguchi, K. Kitajima, Crystallization process and some properties of novel transparent machinable calcium-mica glass-ceramics, Ceram. Int. 34 (2008) 75–79.
[258] S. Taruta, T. Ichinose, T. Yamaguchi, K. Kitajima, Preparation of transparent lithium-mica glass-ceramics, J. Non-Cryst. Solids 352 (2006) 5556–5563.
[259] S. Taruta, M. Suzuki, T. Yamakami, T. Yamaguchi, K. Kitajima, Preparation and ionic conductivity of transparent glass-ceramics containing a large quantity of lithium-mica, J. Non-Cryst. Solids 354 (2008) 848–855.
[260] S. Taruta, M. Matsuki, H. Nishikiori, T. Yamakami, T. Yamaguchi, K. Kitajima, Preparation and luminescent properties of Eu-doped transparent mica glass–ceramics, Ceram. Int. 36 (2010) 1303–1309.
[261] S. Taruta, Y. Iwasaki, H. Nishikiori, T. Yamakami, T. Yamaguchi, K. Kitajima, K. Okada, Microstructures and luminescent properties of Ce-doped transparent mica glass-ceramics, Mater. Sci. Eng. B 177 (2012) 504–509.
[262] MACOR®, Lighting & Materials, Corning Incorporated, New York, NY. www.corning.com.
[263] M. Garai, B. Karmakar, Effects of $CeO_2$, $Nd_2O_3$, $Sm_2O_3$, $Gd_2O_3$, $Cr_2O_3$, and PbO on crystallization and microstructure of K-Mg-B-Al-Si-F based glasses, in: Presented in the 102nd Indian Science Congress (ISC-2015), 3–7 January, 2016, University of Mysore, Mysore, 2016.

# Functionalization of glasses by laser irradiation microfabrication

# 5

## Chapter Outline



## 5.1 Introduction

In glass, the atomic arrangement, lacking in long-range periodic order, has higher configuration entropy and higher free energy than that of a crystalline material of the same composition. For that reason, thermodynamic glass is a metastable (i.e., non-equilibrium) material stuck between the molten liquid state and the crystalline state. It remains untransformed to its most stable state because of hindrance of the atomic rearrangement during the process of glass formation [1]. Motivated by this fact, in 1995, Hirao [2] proposed the "induced structure" as the idea of basic research. Because of the metastable state of glass, it can be transformed into other states by intense external electromagnetic field. Therefore, by monitoring the external electromagnetic field, it is possible to induce crystallization, alteration of microstructure, patterning (1D, 2D, and 3D), refractive index change, change of valence state of active ions, color center defects, etc. in glasses. Thus, novel functional properties of the glasses are expected as a result of these modifications with promising applications. The different types of external electromagnetic fields, which can create microscopic modifications to the glass structure, may be any one of the laser, electron beam, ultraviolet (UV) light, and X-ray and gamma ($\gamma$-) ray. However, we shall restrict our discussion in this chapter to the use of different types of lasers only with respect to microfabrication of glasses.

**Functional Glasses and Glass-Ceramics. http://dx.doi.org/10.1016/B978-0-12-805056-9.00005-2**

Laser irradiation microfabrication induces nonlinear processes (e.g., multiphoton, tunneling, and avalanche ionizations), complicated dynamics (melting, redistribution of ions, evaporation, and thermal plasma generation), refractive index change, space-selective control of valence state of active ions (rare-earth and transition metal ions), precipitation, and control of metal and functional crystals in glasses during irradiation. Therefore, different microoptical components can be fabricated using laser irradiations. Some of these optical components are waveguides, gratings, microlens, fiber attenuators, 3D optical memory, couplers, periodic nanosized voids, photonic crystals, amplifiers, and lasers [3–6].The laser-irradiated structures have applications in optical data storage, telecommunications, biosensing, bioimaging, etc. The laser irradiation microfabrication technique can be used for writing simple lines for waveguides and can also be extended for more complex device fabrication. The technique is simple and can be achieved by simply scanning the glass along single or multiple axes according to the device design.

An attempt was made using the Web of Science (Version 5.17) as described in Chapter 1 to provide a global publication trend and quantitative growth in the area of laser irradiation of glass. Here, year-wise data were extracted for the last 25 years (1990–2014) by searching, using the topic keyword "glass" followed by refining the document type as "article" published in the journals and meetings. It was further refined using the term "laser irradiation." The term "laser irradiation" is found either in the title or in the abstract of the article. This plot is shown in Fig. 5.1. It gives us a

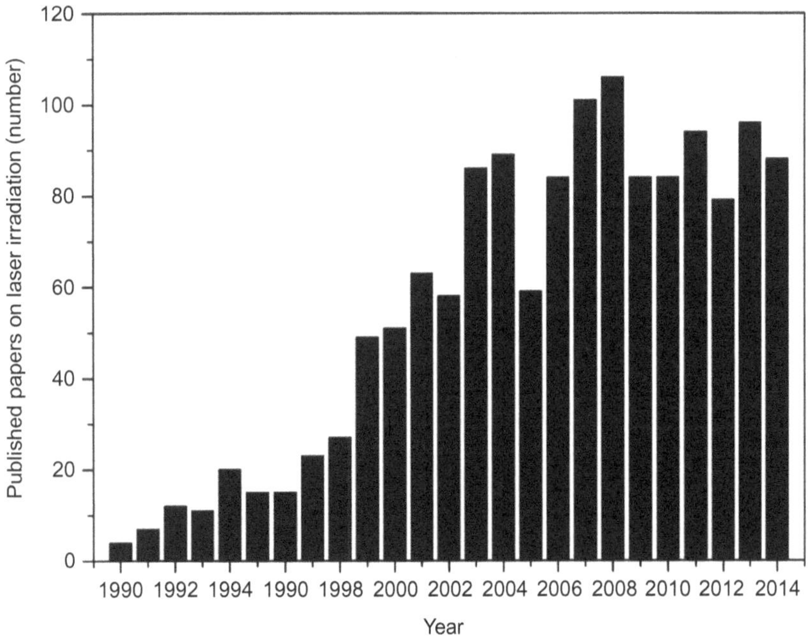


**Fig. 5.1** Year-wise published papers on glass processed by laser irradiation during the last 25 years (1990–2014).

clear publication trend and quantitative growth of published papers on laser irradiation of glass during the last 25 years (1990–2014). It is seen that before 1997, the publication growth rate is poor but after that the publication growth rate is gigantic.

Glass compositions, irradiation lasers, crystal phases, redox states, and laser patterning features of different glasses functionalized by laser irradiation microfabrication are listed in Table 5.1 [7–40]. These will be briefly described in subsequent sections. Before that, we shall briefly describe some important aspects of functionalization of glasses by laser irradiation microfabrication technique.

## 5.2 Technique of laser irradiation microfabrication of glass

The glasses of predecided composition are prepared using the conventional melt-quenching technique followed by annealing near the glass transition temperature ($T_g$) to release the residual thermal stresses. The annealed glasses are cut down to the required sizes of plate-shaped and polished to mirror finish with $CeO_2$ powders. The polished glasses are used for microfabrication by laser irradiation. Various types of lasers are used for this purpose as per convenience such as continuous-wave CW Yb:$YVO_4$ fiber laser ($\lambda = 1080$ nm), CW Nd:YAG laser ($\lambda = 1064$ nm), femtosecond (fs) pulsed Ti sapphire laser ($\lambda = 800$ or 1030 nm), CW Ar laser ($\lambda = 488$ nm), CW $CO_2$ laser ($\lambda = 10.6$ μm), and CW helium-neon laser ($\lambda = 632.8$ nm) and diode laser. Schematic representation of the setup used for laser-induced patterning in glass is shown in Fig. 5.2A [24]. An example of typical equipment used for this purpose is also shown in Fig. 5.2B [19]. The polished surface of the glass is irradiated by laser using objective lens for magnification. The glasses is placed on the platform and mechanically moved during laser irradiations to construct patterns. Laser-irradiated parts can be observed with polarization optical microscope (POM) or confocal scanning laser microscope (CSLM). Other properties can be evaluated by suitable instruments.

Incorporation of new functionality in glasses by laser irradiation originates from creation of modifications inside the glass by the lasers. When lasers are focused inside a glass, a large amount of light energy is absorbed by electrons, and the energy is transferred to the lattice; these phenomena result in rise in temperature up to several thousand degrees Celsius (°C) around the photoexcited region [41]. Thus, heat accumulation by laser irradiation generates crystallization, ion migration, color center (electron, $e^-$ and hole, $h^+$), etc. around the laser focal volume inside the glass. These are enlightened with the following two examples of CW Yb:$YVO_4$ fiber and femtosecond (fs) pulsed lasers.

Fig. 5.3 shows (left) the POM photomicrograph of $Li_2Si_2O_5$ crystal line pattern created by irradiations of CW Yb:$YVO_4$ fiber laser (power = 2.1 W and scanning speed = 9 μm $s^{-1}$) in 1CuO–33.3$Li_2O$–66.7$SiO_2$ glass and (right) the schematic model of temperature gradient along the cross section of the crystal line as shown in the left [28]. A homogeneous line of 6 μm width is patterned. Honma et al. [42] have proposed a rise in temperature of 700–800°C in the laser-irradiated region from the laser irradiation scanning speed, homogeneous growth of $Li_2Si_2O_5$ crystal, crystal orientation,

**Table 5.1 Glass compositions, irradiation lasers, crystal phases, redox states, and laser patterning features of different glasses functionalized by laser irradiation microfabrication**

| Glass composition (mol.%) | Irradiation laser[a] | Crystal phase/redox state | Laser patterning feature | References |
|---|---|---|---|---|
| ***Crystal formation by laser irradiation*** | | | | |
| $47.5BaO–5Al_2O_3–47.5B_2O_3$ | FS 800 | $\beta$-$BaB_2O_4$ | Dot | [7] |
| $33.3BaO–16.7TiO_2–50SiO_2$ | FS 800 | $Ba_2TiSi_2O_8$ | Dot | [7] |
| $Nd_{0.2}La_{0.8}BGeO_5$ | FS 800 | $Nd_{0.2}La_{0.8}BGeO_5$ | Line, dot | [8] |
| $LaBGeO_5$ | FS 800 | $LaBGeO_5$ | Line | [9] |
| $34Li_2O–33Nb_2O_5–33SiO_2$ | FS 1030 | $LiNbO_3$ | – | [10] |
| (wt.%) $63BaO–26B_2O_3–11TiO_2$ | CW $CO_2$ | $\beta$-$BaB_2O_4$ | Surface | [11] |
| $42BaO–42B_2O_3–16SiO_2$ | CW $CO_2$ | $\beta$-$BaB_2O_4$ | Dot | [12] |
| ***Doped crystal formation by laser irradiation*** | | | | |
| $40BaO–20TiO_2–40SiO_2 + 1.5\ Nd_2O_3$ | CW Ar 488 | $Nd^{3+}:Ba_2TiSi_2O_8$ | Dot | [13] |
| $40BaO–20TiO_2–40SiO_2 + 1.5Dy_2O_3$ | CW Ar 488 | $Dy^{3+}:Ba_2TiSi_2O_8$ | – | [14] |
| $5Nd_2O_3–11.25SrO–11.25BaO–22.5Nb_2O_5–50B_2O_3$ | CW Ar 488 | $Nd^{3+}:Sr_{1-x}Ba_xNb_2O_6$ | Dot | [15] |
| $5Er_2O_3–11.25SrO–11.25BaO–22.5Nb_2O_5–50B_2O_3$ | FS 800 | $Er^{3+}:Sr_{1-x}Ba_xNb_2O_6$ | Dot | [16] |
| $2.5Ho_2O_3–11.25SrO–11.25BaO–22.5Nb_2O_5–52.5B_2O_3$ | CW Ar 488 | $Ho^{3+}:Sr_{1-x}Ba_xNb_2O_6$ | Dot | [17] |
| $50SiO_2–20Al_2O_3–30CaF_2–5ErF_3$ | FS 800 | $Er^{3+}:CaF_2$ | Dot | [7] |
| $44.9SiO_2–20GeO_2–16.5Al_2O_3–1.6AlF_3–12.7LaF_3–4.3Na_2CO_3 + 0.06ErF_3$ | CW Ar 244 | $Er^{3+}:LaF_3$ | Dot | [18] |

| | | | | |
|---|---|---|---|---|
| ***Crystallization by REAH processing laser irradiation*** | | | | |
| $8Sm_2O_3$–$42BaO$–$50B_2O_3$ | CW Yb:$YVO_4$ | β-$BaB_2O_4$ | Line | [19] |
| $10Sm_2O_3$–$40BaO$–$50B_2O_3$ | Q-Nd:YAG | β-$BaB_2O_4$ | Line | [20] |
| $10Dy_2O_3$–$40BaO$–$50B_2O_3$ | | | | |
| $3Sm_2O_3$–$18.25Gd_2O_3$–$63.75MoO_3$–$15B_2O_3$ | CW Yb:$YVO_4$ | β′-$Gd_2(MoO_4)_3$ | Line | [19] |
| $10Sm_2O_3$–$35Bi_2O_3$–$55B_2O_3$ | Q-Nd:YAG | $Sm_xBi_{1-x}BO_3$ | Line | [20,21] |
| $10.625Sm_2O_3$–$10.625Gd_2O_3$–$63.75MoO_3$–$15B_2O_3$ | CW Nd:YAG | β′-$Gd_{2-x}Sm_x(MoO_4)_3$ | Line | [22] |
| $10.625Sm_2O_3$–$10.625Dy_2O_3$–$63.75MoO_3$–$15B_2O_3$ | CW Nd:YAG | β′-$Dy_{2-x}Sm_x(MoO_4)_3$ | Line | [22] |
| $35Li_2O$–$30Nb_2O_5$–$20SiO_2$–$15B_2O_3+5Sm_2O_3$ | CW Yb:$YVO_4$ | $Sm^{3+}$:$LiNbO_3$ | Line | [23,24] |
| $11.1Sm_2O_3$–$44.4BaO$–$44.4B_2O_3$ | CW Nd:YAG | $Sm^{3+}$:β-$BaB_2O_4$ | Dot | [25] |
| $5Sm_2O_3$–$5Gd_2O_3$–$35Bi_2O_3$–$55B_2O_3$ | CW Yb:$YVO_4$ | $Sm_{0.11}Gd_{0.11}Bi_{0.78}BO_3$ | Surface | [26] |
| ***Crystallization by TMAH processing laser irradiation*** | | | | |
| $26Li_2O$–$43FeO$–$5Nb_2O_5$–$26P_2O_5$ | CW Nd:YAG | $LiFePO_4$ | Line | [27] |
| $1CuO$–$33.3Li_2O$–$66.7SiO_2$ | CW Yb:$YVO_4$ | $Li_2Si_2O_5$ | Line | [28] |
| $3NiO$–$50BaF_2$–$25Al_2O_3$–$25B_2O_3$ | CW Yb:$YVO_4$ | $BaAlBO_3F_2$ | Line | [28] |
| $40Li_2O$–$32Nb_2O_5$–$10B_2O_3$–$20SiO_2+0.5CuO$ | CW Yb:$YVO_4$ | $LiNbO_3$ | Line | [29] |
| $1CuO$–$40Li_2O$–$32Nb_2O_5$–$28SiO_2$ | CW Nd:YAG, CW Yb:$YVO_4$, DL | $LiNbO_3$ | Line | [30] |
| $33.3BaO$–$16.7TiO_2$–$50GeO_2+(0.3$–$1.0)(NiO, Fe_2O_3, V_2O_5)$ | CW Nd:YAG | $Ba_2TiGe_2O_8$ | Line | [31] |
| $33.3BaO$–$16.7TiO_2$–$50SiO_2+(0.3$–$1.0)(NiO, Fe_2O_3, V_2O_5)$ | CW Nd:YAG | $Ba_2TiSi_2O_8$ | Line | [31] |
| $40Li_2O$–$32Nb_2O_5$–$28SiO_2+0.5CuO+0.3Er_2O_3$ | CW Yb:$YVO_4$ | $Er^{3+}$:$LiNbO_3$ | Line | [23,24] |
| $0.5ErF_3$–$3NiO$–$20CaF_2$–$16NaF$–$4CaO$–$20Al_2O_3$–$10B_2O_3$–$30SiO_2$ | CW Yb:$YVO_4$ | $Er^{3+}$:$CaF_2$ | Line | [32] |

*Continued*

**Table 5.1 Continued**

| Glass composition (mol.%) | Irradiation laser[a] | Crystal phase/redox state | Laser patterning feature | References |
|---|---|---|---|---|
| ***Reduction-oxidation by laser irradiation*** | | | | |
| $10Na_2O$–$5Al_2O_3$–$85B_2O_3+0.05Sm_2O_3$ | FS 800 | $Sm^{3+} \rightarrow Sm^{2+}$ (photoreduction) | – | [33] |
| $53ZrF_4$–$20BaF_2$–$3.9LaF_3$–$3AlF_3$–$20NaF$–$0.1EuF_3$ | FS 800 | $Eu^{3+} \rightarrow Eu^{2+}$ (photoreduction) | – | [34] |
| $xWO_3$–$(100-x)[3B_2O_3$–$PbO]$, where $0 \leq x \leq 40$ | CW He-Ne | $W^{6+} \rightarrow W^{5+}$ ($W^{4+}$) (photoreduction) | – | [35] |
| $52Bi_2O_3$–$23B_2O_3$–$4SiO_2$–$17PbO$–$4Sb_2O_3$ | FS 800 | $Bi^{3+} \rightarrow Bi^{+}$ (photoreduction) | – | [36] |
| $70SiO_2$–$20Na_2O$–$8CaO$–$2PbO+0.01Au_2O_3$ | FS 800 | $Au^{3+} \rightarrow Au^{0}$ (photoreduction) | – | [37] |
| $70SiO_2$–$10CaO$–$20Na_2O$–$0.1Ag_2O$ | FS 800 | $Ag^{+} \rightarrow Ag^{0}$ (photoreduction) | – | [38] |
| $70SiO_2$–$10CaO$–$20Na_2O+0.1Au_2O_3$ | FS 800+550°C/0.5 h | $Au^{3+} \rightarrow Au^{0}$ (reduction) | | [39] |
| $70SiO_2$–$10CaO$–$20Na_2O+0.05Fe_2O_3+0.1MnO$ | FS 800 | $Mn^{2+} \rightarrow Mn^{3+}$ (photooxidation) | – | [40] |

[a] *FS800*, Femtosecond pulsed Ti sapphire laser ($\lambda = 800$ nm); *FS 1030*, femtosecond pulsed laser (1030 nm); *CW Yb:YVO$_4$*, continuous-wave Yb:$YVO_4$fiber laser ($\lambda = 1080$ nm); *CW Nd:YAG*, continuous-wave Nd:YAG laser ($\lambda = 1064$ nm); *CW Ar 488*, continuous-wave Ar laser ($\lambda = 488$ nm); *Q-Nd:YAG*, Q-switched pulse Nd:YAG laser ($\lambda = 1064$ nm); *DL*, diode laser ($\lambda = 795$ nm); *CW CO$_2$*, continuous-wave $CO_2$ laser ($\lambda = 10.6$ μm); *CW He-Ne*, continuous-wave helium-neon laser ($\lambda = 632.8$ nm); *CW Ar244*, continuous-wave Ar laser ($\lambda = 244$ nm).

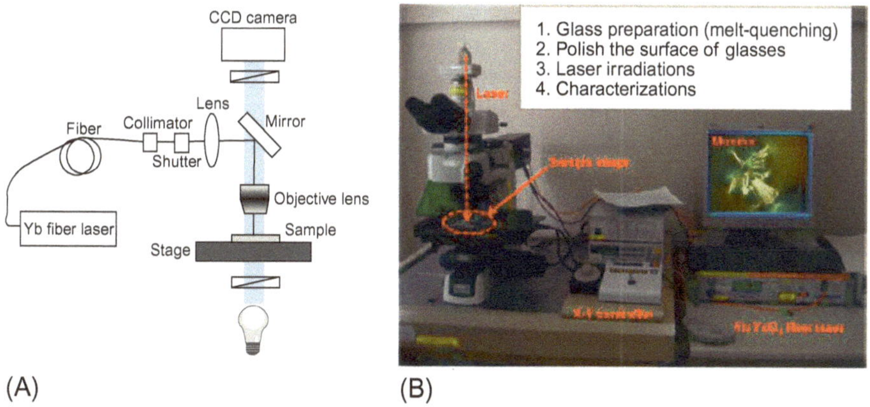


**Fig. 5.2** (A) Schematic representation of setup used for laser-induced patterning in glass. (B) Equipment used for laser-induced patterning in glass.
(A) Reproduced with permission from T. Honma, T. Komatsu, D. Zhao, H. Jain, Writing of rare-earth ion doped lithium niobate line patterns in glass by laser scanning, IOP Conf. Series: Mater. Sci. Engg. 1 (2009) 012006 (6 pages), Copyright © IOP. (B) Courtesy: T. Komatsu, F. Suzuki, K. Ogawa, T. Honma, Recent progress in laser patterning of crystals in glasses, Trans. GIGAKU 1 (2012) 01017 (6 pages).

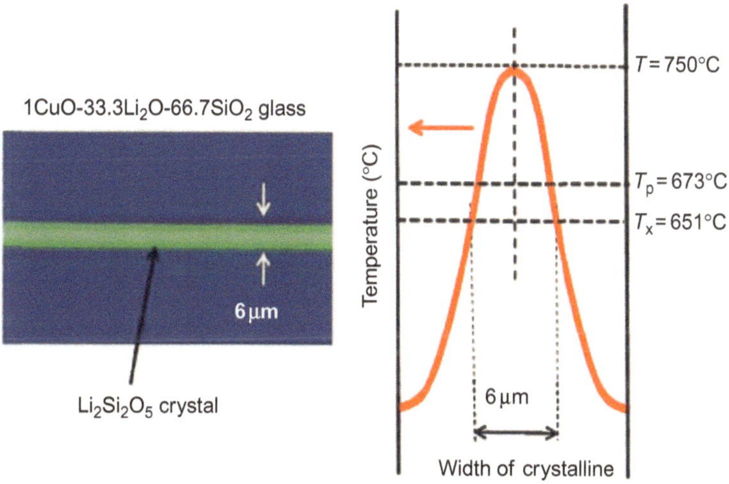


**Fig. 5.3** (Left) Polarization optical microscope (POM) photomicrograph of $Li_2Si_2O_5$ crystal line pattern created by irradiations of CW Yb:$YVO_4$fiber laser in 1CuO–33.3$Li_2O$–66.7$SiO_2$ glass and (right) the schematic model of temperature gradient along the cross section of the crystal line as shown in the left.
Reproduced with permission from T. Komatsu, T. Honma, Laser patterning and characterization of optical active crystals in glasses, J. Asian Ceram. Soc. 1 (2013) 9–16, Copyright © Elsevier.

and thermal properties of the precursor glass (glass transition temperature, $T_g = 454°C$; onset of crystallization, $T_x = 651°C$; and crystallization peak temperature, $T_p = 673°C$). These authors considered the generation of a steep temperature gradient of about 17 K $\mu m^{-1}$ along the laser-irradiated spot based on some previous reports carried out using a temperature gradient furnace. Therefore, the key point of patterning of crystals in glass by laser irradiation is the generation of steep temperature gradient. The great advantage of crystallization of glass by laser irradiation is that local crystallization can be carried out within a specified region of the glass that is not possible in conventional glass-ceramics processing by heating in a furnace because in this case the entire glass is crystallized. Comparing the morphologies created by CW Nd:YAG, CW Yb:$YVO_4$ fiber, and femtosecond (fs) pulsed lasers in glasses, it is observed that the former two lasers can generate both the dot and line patterning whereas the later only the dot. The temperature distribution inside the glass during irradiation by femtosecond laser pulses is described as follows:

$$R_b(T_S) = R_0 + \left(\frac{A}{T_{out} - T_S} - B\right)^{1/2} \tag{5.1}$$

$$T(r, T_S) = \frac{A}{(r - R_0)^2 + B} + T_S \tag{5.2}$$

Sakakura et al. [41] evaluated the temperature distribution inside a soda-lime silicate glass by analyzing the relationship between the radius of modification ($R_b$) and glass temperature ($T_S$) those occurred during irradiation by femtosecond laser pulses. The radius of modification ($R_b$) and glass temperature ($T_S$) are expressed by Eqs. (5.1) and (5.2), respectively.

At the boundary of the modified region where $r = R_b$ (as shown in Fig. 5.4A), the temperature is identical to the characteristic temperature ($T_{out}$), that is, $T$($R_b$ and $T_S$) = $T_{out}$. Fitting the relation between $R_b$ and $T_S$ by Eq. (5.1), the evaluated $T_{out}$ was found to be around 560°C for the soda-lime silicate glass used under this investigation. Fig. 5.4B (a) shows the relation between radius of modified region ($R_b$) and temperature of glass ($T_S$) at various pulse energies. The solid lines are fitting curves as given by Eq. (5.1). Fig. 5.4B (b) depicts the temperature distribution curves calculated from the relation as given by Eq. (5.2).

## 5.3 Processing, functional properties, and applications of laser-irradiated glasses

In this section, we shall describe the processing and functional properties of some selected examples of refractive index modification, crystal formation, rare-earth (RE) or transition metal (TM) ion-doped crystal formation, and happening of reduction-oxidation (redox) inside glasses by laser irradiation. In addition, laser irradiation of glasses involving rare-earth atom heat (REAH) and transition metal atom heat (TMAH) processing are also discussed.

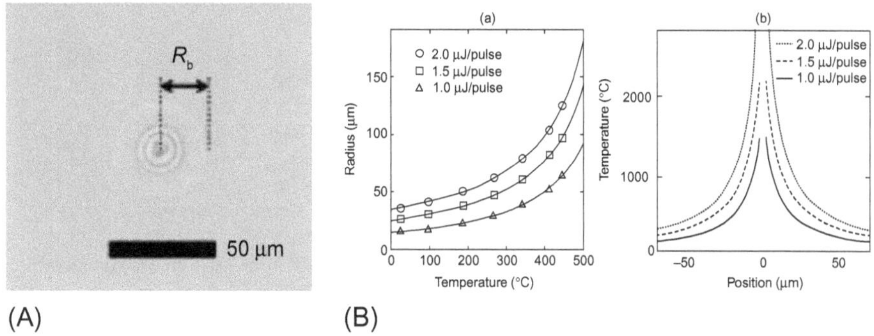


**Fig. 5.4** (A) Optical microscope photograph showing the modifications of glass surface created by 250 kHz femtosecond laser irradiation. (B) (a) Relation between radius of modified region ($R_b$) and temperature of glass ($T_S$) at various pulse energies. The solid lines are fitting curves by as given in Eq. (5.1) and (b) temperature distribution curves calculated from the relation as given by Eq. (5.2).
Reproduced with permission from M. Sakakura, M. Shimizu, Y. Shimotsuma, K. Miura, K. Hirao, Temperature distribution and modification mechanism inside glass with heat accumulation during 250 kHz irradiation of femtosecond laser pulses, Appl. Phys. Lett. 93 (2008) 231112 (3 pages), Copyright © AIP.

## 5.3.1 Refractive index modification by laser irradiation

Bhardwaj et al. [43] have reported a comprehensive study of refractive index modification by femtosecond (fs) laser irradiation of various multicomponent glasses such as borate, borosilicate, aluminosilicate, and phosphate. They have observed positive, negative, and nonuniform changes in refractive index of glass by laser irradiation. Measured refractive index profile (average of 20 scan lines) of (A) alkaline earth aluminosilicate and (B) alkali aluminosilicate glasses of femtosecond laser irradiations is shown in Fig. 5.5. Laser parameters are 50 fs and 100 kHz. It is seen that alkaline earth aluminosilicate glass shows a positive refractive index change of $\Delta n \approx +0.003 \pm 0.0002$, whereas the alkali aluminosilicate glass displays nonuniform refractive index changes at the edge islands of positive $\Delta n \approx +0.0004$ and in the center negative of $\Delta n \approx -0.0012 \pm 0.0002$. The 2D microreflectivity scans of the modified area in alkali aluminosilicate glass are shown in Fig. 5.5C where the dark and bright areas correspond to negative and positive refractive index changes, respectively. Kanehira et al. [44] have demonstrated the ion-exchange process in glass after femtosecond laser irradiation, which is found to be responsible for modification of refractive index. Krol [4] has observed that the fs laser-irradiated area of fused silica has a higher refractive index than the original glass, while in phosphate glass, the index of the exposed area is lower. He correlated this fact with the cooling rate of glass melt because in fused silica the refractive index increases with cooling rate, whereas it decreases in phosphate glass. Raman microscopy results indicate that the irradiated area of fused silica has a higher concentration of three- and four-membered rings, which results in increase in density and refractive index. In the case of phosphate glass, the irradiated

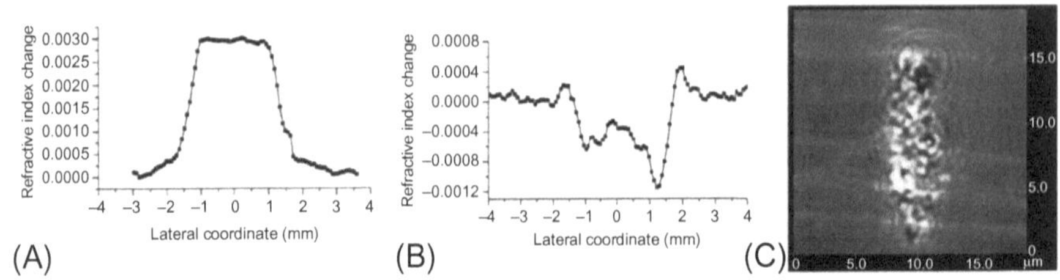


**Fig. 5.5** Measured refractive index profile (average of 20 scan lines) of (A) alkaline earth aluminosilicate and (B) alkali aluminosilicate glasses. (C) 2D microreflectivity scans of the modified area in alkali aluminosilicate glass. Laser parameters are 50 fs and 100 kHz. Reproduced with permission from V.R. Bhardwaj, E. Simova, P.B. Corkum, D.M. Rayner, C. Hnatovsky, R.S. Taylor, B. Schreder, M. Kluge, J. Zimmer, Femtosecond laser-induced refractive index modification in multicomponent glasses, J. Appl. Phys. 97 (2005) 083102 (9 pages), Copyright © AIP.

area has an expansion of structure as result of stress creation, therefore the lower refractive index. Applications of negative refractive index-modified glasses are precursors of microfluidic channels and photonic crystals, writing depressed-index core photonic devices (e.g., doughnut-shaped waveguides), etc.

### *5.3.2 Crystal formation by laser irradiation*

Transparent glasses embedded with functional nanocrystals are very important as a result of their applications in different optical devices, for example, nonlinear optical devices, optical storage, 3D displays, and optical amplifiers. Laser irradiation-induced crystallization of glass is a dominant tool to create the space-selective precipitation of many nanocrystals inside glasses and has potential applications in the fabrication of microoptical devices. Various ferroelectric crystals such as β-$BaB_2O_4$, $Ba_2TiSi_2O_8$, $LaBGeO_5$, and $LiNbO_3$ have been precipitated inside transparent glasses by irradiation with different lasers. Some of these investigations are listed in Table 5.1 [7–12]. Teng et al. [7] have described the space-selective crystallization of β-$BaB_2O_4$ and $Ba_2TiSi_2O_8$ in 47.5BaO–$5Al_2O_3$–$47.5B_2O_3$ (mol.%) and 33.3BaO–$16.7TiO_2$–$50SiO_2$ (mol.%) glasses, respectively, by femtosecond laser ($\lambda = 800$ nm) irradiation. Gupta et al. [8] have demonstrated the creation of tailored micron-size structures (line and dot) of $Nd_{0.2}La_{0.8}BGeO_5$ crystals in glass by femtosecond laser irradiation. Stone et al. [9] have described the femtosecond laser irradiation fabrication technique of complex, highly oriented three dimensional ferroelectric single crystal architecture of $LaBGeO_5$ crystal lines and bends in lanthanum borogermanate glass. Fan et al. [10] have demonstrated the space-selective nucleation and crystal growth of $LiNbO_3$ in silica-based glass by femtosecond laser ($\lambda = 1030$ nm) irradiation at a high-repetition rate. Adjusting the laser irradiation parameters, oriented crystals aligned with or perpendicular to the laser scanning direction, is invented.

In contrast, crystallization of β-$BaB_2O_4$ phase by $CO_2$ laser ($\lambda = 10.6$ μm) irradiation is demonstrated in 63BaO–$26B_2O_3$–$11TiO_2$ (wt.%) and 42BaO–$42B_2O_3$–$16SiO_2$

(mol.%) glasses by Maciente et al. [11] and Avansi et al. [12], respectively. Fig. 5.6A shows the optical micrograph of the pattern of surface crystallization of β-$BaB_2O_4$ phase in BaO–$B_2O_3$–$TiO_2$ glass obtained by $CO_2$ laser irradiation using a nominal laser power of 1.7 W [11]. It is observed that the use of a lower power density and a shorter exposure time produces this pattern with crystal size varies in the range of 3–7 μm. The morphology of the crystallites was found to be similar to that obtained after heat treatment at 620°C/1 h. However, a completely surface-crystallized layer is obtained only after 30 s.

Fig. 5.6B shows the photograph of six different $CO_2$ laser-irradiated spots in BaO–$B_2O_3$–$SiO_2$ glass exhibiting second harmonic generation (SHG) of green light (532 nm) by the crystallized β-$BaB_2O_4$ phase on excitation with a Nd:YAG-pulsed laser operating at 1064 nm [12]. It seen that the intensity of green light due to SHG is almost uniform in the six different irradiated regions.

## 5.3.3 Doped crystal formation by laser irradiation

Rare-earth doped nanocrystals are the promising materials for use in various photonic devices, for instance, laser, optical fiber amplifier, planer optical waveguide, frequency converter, 3D laser display, and 3D optical data storage. Similarly, rare-earth doped ferroelectric crystals have find applications in the area of photorefractive memories and

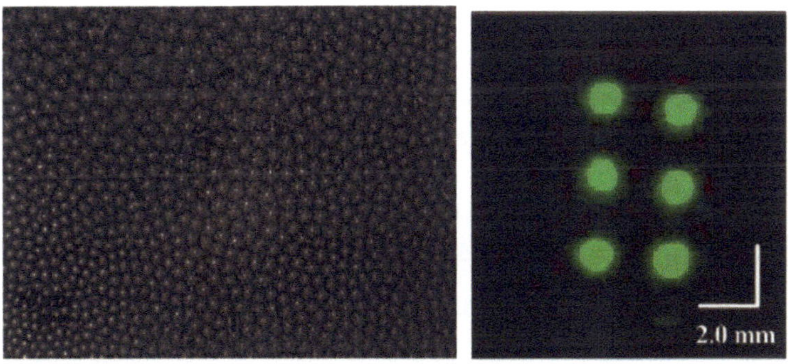


(A) (B)

**Fig. 5.6** (A) Optical micrograph showing the pattern of surface crystallization of β-$BaB_2O_4$ phase in BaO–$B_2O_3$–$TiO_2$ glass obtained by $CO_2$ laser irradiation. (B) Photograph of six $CO_2$ laser-irradiated spots in BaO–$B_2O_3$–$SiO_2$ glass exhibiting second harmonic generation of green light (532 nm) by the crystallized β-$BaB_2O_4$ phase on excitation with a Nd:YAG-pulsed laser operating at 1064 nm.

(A) Reproduced with permission from A.F. Maciente, V.R. Mastelaro, A.L. Martinez, A.C. Hernandes, C.A.C. Carneiro, Surface crystallization of β-$BaB_2O_4$ phase using a $CO_2$ laser source, J. Non-Cryst. Solids 306 (2002) 309–312, Copyright © Elsevier. (B) Reproduced with permission from W. Avansi, V.R. Mastelaro, M.R.B. Andreeta, Surface modification and crystallization of the BaO–$B_2O_3$–$SiO_2$ glassy system using $CO_2$ laser irradiation, J. Non-Cryst. Solids 354 (2008) 279–283, Copyright © Elsevier.

self-frequency converter solid-state laser materials. Consequently, rare-earth doped ferroelectric and nonferroelectric crystals along with their patterns have been created inside different transparent glasses by continuous-wave or pulsed laser irradiations. Some of these investigations are summarized in Table 5.1 [13–18]. Martin et al. [13,14] have studied the formation of $Nd^{3+}$- and $Dy^{3+}$-doped $Ba_2TiSi_2O_8$ nanocrystals in $40BaO–20TiO_2–40SiO_2$ (mol.%) glasses by continuous-wave CW Ar laser ($\lambda = 488$ nm). Their photoluminescence property measurements confirm the incorporation of the rare-earth ions into the $Ba_2TiSi_2O_8$ nanocrystals, which results in enhancement of luminescence intensity. Therefore, these materials can be used in optical amplifications. Haro-González et al. [15–17] have investigated the formation of $Nd^{3+}$-, $Er^{3+}$-, and $Ho^{3+}$-doped $Sr_{1-x}Ba_xNb_2O_6$ nanocrystals in $SrO–BaO–Nb_2O_5–B_2O_3$ glasses by continuous-wave CW Ar laser ($\lambda = 488$ nm) and femtosecond pulsed laser ($\lambda = 800$ nm) irradiations. They have also studied the photoluminescence of these materials using a confocal microluminescence setup. This setup is shown in Fig. 5.7A [14]. Fig. 5.7B shows the emission spectra of a laser-irradiated area of the $5Nd_2O_3–11.25SrO–11.25BaO–22.5Nb_2O_5–50B_2O_3$ glass obtained by using CW Ar 488 laser at low (400 mW, *dashed line*) and high (1500 mW, *solid line*) power excitation. Inset shows the variation of integrated area of the emission bands peaking at 810 (*square symbol*) and 880 (*circle symbol*) nm as a function of CW Ar 488 laser pump power [15]. It is seen that when the irradiation intensity (also the excitation power) increases, the emission band at 810 nm starts to growth faster than that of the 880 nm band. This observation gives us a clear indication of altering structure under laser irradiation. Teng et al. and Mekhlouf et al. have reported the formation and photoluminescence properties of $Er^{3+}$:$CaF_2$ crystals in $50SiO_2–20Al_2O_3–30CaF_2–5ErF_3$ (mol.%) glass [7] and $Er^{3+}$:$LaF_3$ crystals in $44.9SiO_2–20GeO_2–16.5Al_2O_3–1.6AlF_3–12.7LaF_3–4.3Na_2CO_3+0.06ErF_3$ (mol.%) glass [18]. The photoluminescence property measurements support the partitioning of rare-earth ions into the $LaF_3$ and $CaF_2$ nanocrystals.

### 5.3.4 Crystallization by REAH processing laser irradiation

It is understood from the foregoing discussion that the laser-induced crystallization in glass is attractive for patterning active crystals in spatially selected part of the glass keeping other parts unaltered, which is not possible by conventional glass-ceramic preparation technique through heat treatment. Various types of lasers, such as CW Nd:YAG laser ($\lambda = 1064$ nm), CW Yb:$YVO_4$ fiber laser ($\lambda = 1080$ nm), and femtosecond pulsed laser ($\lambda = 800$ or 1030 nm) or diode laser ($\lambda = 795$ nm) are used for this purpose. The reactions occurring during laser irradiation are governed by the type of laser and composition of glass. There are several advantages of Nd:YAG and Yb:$YVO_4$ lasers over other lasers, for example, argon or femtosecond pulsed laser. Consequently, Komatsu and his group [19–26] developed a novel technique called "rare-earth atom heat (REAH) processing" and applied to various glass systems. In this technique, a combination of CW Nd:YAG laser ($\lambda = 1064$ nm) or CW Yb:$YVO_4$fiber laser ($\lambda = 1080$ nm) and hugely rare-earth ion-doped (more than 5 mol.% $Sm_2O_3$ or $Dy_2O_3$) glasses are used to increase the temperature adequately in the

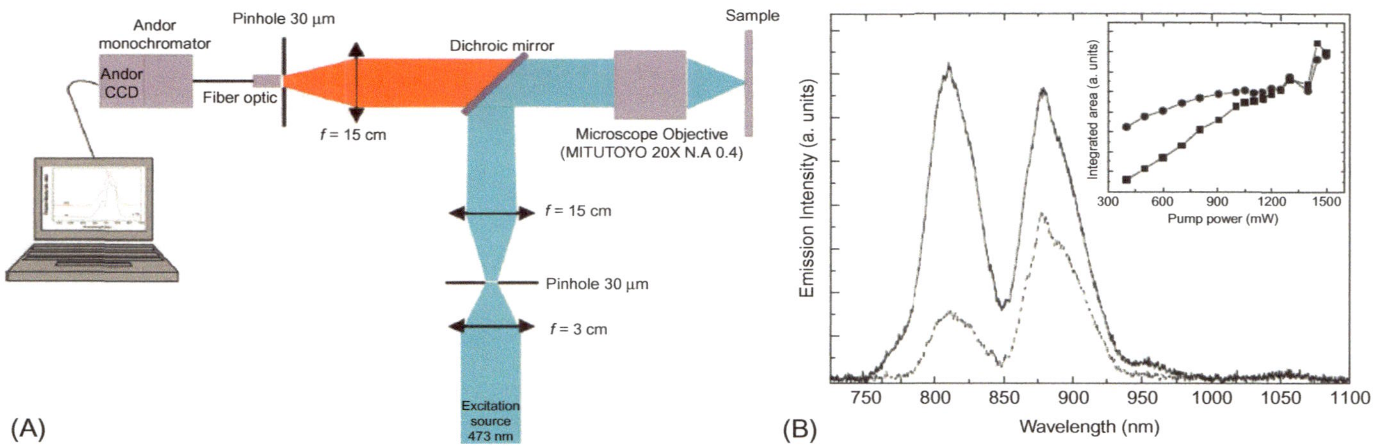


**Fig. 5.7** (A) Confocal microluminescence setup used to characterize photoluminescence property of the laser-irradiated glass samples. (B) Emission spectra of a laser-irradiated area of the $5Nd_2O_3$–$11.25SrO$–$11.25BaO$–$22.5Nb_2O_5$–$50B_2O_3$ glass obtained by using CW Ar 488 laser at low (400 mW, *dashed line*) and high (1500 mW, *solid line*) power excitation. Inset shows the variation of integrated area of the emission bands peaking at 810 *(square symbol)* and 880 *(circle symbol)* nm as a function of CW Ar 488 laser pump power.
(A) Reproduced with permission from L.L. Martín, P. Haro-González, I.R. Martín, D. Puerto, J. Solís, J.M. Cáceres, N.E. Capuj, Local devitrification of $Dy^{3+}$ doped $Ba_2TiSi_2O_8$ glass by laser irradiation, Opt. Mater. 33 (2010) 186–190, Copyright © Elsevier. (B) Reproduced with permission from P. Haro-González, L.L. Martín, S. González-Pérez, I.R. Martín, Formation of $Nd^{3+}$ doped strontium barium niobatenanocrystals by two different methods, Opt. Mater. 32 (2010) 1389–1392, Copyright © Elsevier.

laser-exposed area through the process of absorption of laser energy by the rare-earth ions due to f-f transitions.

For example, the absorption spectrum of $Sm^{3+}$ ions in $11.1Sm_2O_3$–$44.4BaO$–$44.4B_2O_3$ glass is shown in Fig. 5.8A [25]. The CW Yb:$YVO_4$ fiber laser ($\lambda = 1080$ nm) excitation and nonradiative relaxation process of the f-f transition in the $Sm^{3+}$ ion-containing glasses is also shown in Fig. 5.8B [19]. It is seen that the irradiated laser energy is absorbed by the $Sm^{3+}$ ions as a result of f-f transitions, and the absorbed energy is transferred to the thermal energies through nonradiative relaxation process (electron-phonon coupling). Line or dot pattern of various ferroelectric crystals has been created in different type of glass compositions [19–26].

Fig. 5.9A illustrates the polarization optical micrograph showing the morphology of line pattern obtained in $10Sm_2O_3$–$35Bi_2O_3$–$55B_2O_3$ glass by Q-Nd:YAG laser irradiation varying power in the range of 0.6–0.9 W at the scanning speed of 10 μm $s^{-1}$, while the Fig. 5.9B shows the (a) polarization optical micrograph and (b) SHG microscope observation for the curved lines (here the bird) on the surface of glass written by Q-Nd:YAG laser irradiation [20]. The laser-irradiated regions produce the crystal phase of $Sm_xBi_{1-x}BO_3$ as revealed in X-ray analysis. The $Sm_xBi_{1-x}BO_3$ is a nonlinear optical crystal, and consequently, it exhibits strong SHG. The values of glass transition temperature ($T_g$) and onset of crystallization ($T_x$) are 474°C and 574°C, respectively, for $10Sm_2O_3$–$35Bi_2O_3$–$55B_2O_3$ glass. Therefore, it is reasonable to believe that the laser-irradiated regions were heated up to the temperature greater than 580°C.

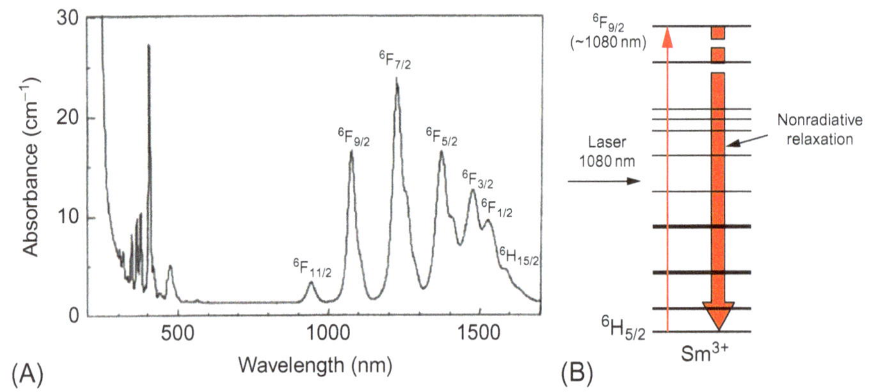


**Fig. 5.8** (A) Absorption spectrum of $Sm^{3+}$ ions in $11.1Sm_2O_3$–$44.4BaO$–$44.4B_2O_3$ glass. (B) CW Yb:$YVO_4$ fiber laser ($\lambda = 1080$ nm) excitation and nonradiative relaxation process of the f-f transition in the $Sm^{3+}$ ion-containing glasses.

(A) Reproduced with permission from H. Tanaka, T. Honma, Y. Benino, T. Fujiwara, T. Komatsu, YAG laser-induced β-$BaB_2O_4$ crystalline dot formation in $Sm_2O_3$–$BaO$–$B_2O_3$ glasses, J. Phys. Chem. Solids 64 (2003) 1179–1184, Copyright © Elsevier. (B) Courtesy: T. Komatsu, F. Suzuki, K. Ogawa, T. Honma, Recent progress in laser patterning of crystals in glasses, Trans. GIGAKU 1 (2012) 01017 (6 pages).

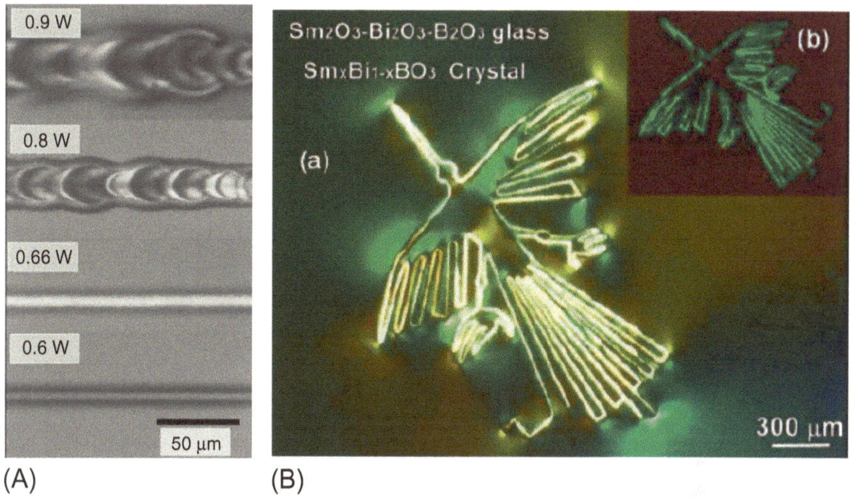


**Fig. 5.9** (A) Polarization optical micrograph showing morphology of line pattern obtained in $10Sm_2O_3$–$35Bi_2O_3$–$55B_2O_3$ glass by Q-Nd:YAG laser irradiation varying power in the range of 0.6–0.9 W at the scanning speed of 10 μm $s^{-1}$. (B) (a) Polarization optical micrograph and (b) second harmonic generation microscope observation for the curved lines (here the *bird*) on the surface of glass written by Q-Nd:YAG laser irradiation.
Reproduced with permission from T. Komatsu, R. Ihara, T. Honma, Y. Benino, R. Sato, H.G. Kim, T. Fujiwara, Patterning of non-linear optical crystals in glass by laser-induced crystallization, J. Am. Ceram. Soc. 90 (2007) 699–705, Copyright © The American Ceramic Society.

## 5.3.5 Crystallization by TMAH processing laser irradiation

In view of the success of "rare-earth atom heat (REAH) processing," it is believed that doping glasses by other type of ions would be suitable for laser-induced crystallization in glass. Accordingly, Komatsu and his group [27–32] have proposed the technique of "transition metal atom heat (TMAH) processing" for laser-induced crystallization in glass. They have found that the combination of transition metal (TM) ion (e.g., $Ni^{2+}$, $V^{4+}$, $Fe^{2+}$, and $Cu^{2+}$) doped (about 1 mol.%) glasses and CW Nd:YAG laser ($\lambda = 1064$ nm) or CW Yb:$YVO_4$ fiber laser ($\lambda = 1080$ nm) induces effective crystallization. A fraction of the incident laser light energy used for irradiation is absorbed by transition metal ions due to d-d transitions. Through the process of nonradiative relaxation between the excited state and the ground state, the absorbed laser energy by the transition metal ions is converted into the thermal energy (as a result of lattice vibration) inducing heating of surrounding of transition metal ions. Hence, the crystallization of glass will occur if the temperature of the irradiated region reaches the temperature of crystallization.

Fig. 5.10A shows the absorption spectrum of 1.0 mol.% NiO-doped 33.3BaO–$16.7TiO_2$–$50GeO_2$ glass [20]. The broad absorption band around 750–1400 nm is attributed to the ${}^3A_2 \rightarrow {}^3T_2$ transition of octahedral $Ni^{2+}$ ion. This glass has an

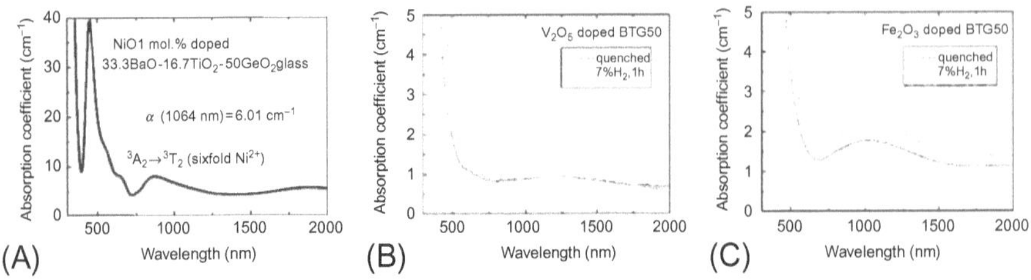


**Fig. 5.10** (A) Absorption spectrum of 1.0 mol.% NiO-doped 33.3BaO–16.7$TiO_2$–50$GeO_2$ glass. Absorption spectra of as-quenched and heat-treated (670°C/1 h in 7%$H_2$–93% Ar-mixed gas atmosphere) 33.3BaO–16.7$TiO_2$–50$GeO_2$ glass samples doped with (B) 0.7 mol.% $V_2O_5$ and (c) 0.3 mol.% $Fe_2O_3$.
(A) Reproduced with permission from T. Komatsu, R. Ihara, T. Honma, Y. Benino, R. Sato, H.G. Kim, T. Fujiwara, Patterning of non-linear optical crystals in glass by laser-induced crystallization, J. Am. Ceram. Soc. 90 (2007) 699–705, Copyright © The American Ceramic Society. (B) Reproduced with permission from T. Honma, Y. Benino, T. Fujiwara, T. Komatsua, Transition metal atom heat processing for writing of crystal lines in glass, Appl. Phys. Lett. 88 (2006) 231105 (3 pages), Copyright © AIP.

absorption coefficient of $\alpha = 6.01\ cm^{-1}$ at 1064 nm, which is comparable with that of 10 mol.% $Sm_2O_3$ content glasses in the case of REAH processing. Therefore, it is expected that the glass would be heated up and induce crystallization due to high absorption of 1064 nm laser light in the irradiated region of the glass. Fig. 5.10 also shows the absorption spectra of as-quenched and heat-treated (670°C/1 h in 7% $H_2$-93% Ar-mixed gas atmosphere) 33.3BaO–16.7$TiO_2$–50$GeO_2$ glass samples doped with (B) 0.7 mol.% $V_2O_5$ and (C) 0.3 mol.% $Fe_2O_3$ [31]. Here, the valence states were altered by heat treatment in the mixed gas atmosphere because of the multivalence states of $Fe^{2+}/Fe^{3+}$ and $V^{4+}/V^{5+}$. In both cases, the broad absorption bands are observed in the range of 750–1400 nm, which is suitable for exploiting in TMAH processing.

Fig. 5.11 shows the photographs obtained by (A) POM and (B) CSLM (CSLM). Fig. 5.11C shows the microphotoluminescence spectra of the 0.5$ErF_3$–3NiO–20$CaF_2$–16NaF–4CaO–20$Al_2O_3$–10$B_2O_3$–30$SiO_2$ glass and the $Er^{3+}$:$CaF_2$-crystallized line obtained by CW Yb:$YVO_4$ fiber laser ($\lambda = 1080$ nm) with 1.0 W power at the scanning speed of 2 μm $s^{-1}$ [32]. It is understood from the Fig. 5.11A and B that the patterned line has a smooth and homogeneous surface with the width and height of 4 and 3 μm, respectively. The microphotoluminescence spectrum of the patterned line part displays the strong emissions of the $Er^{3+}$ ion in crystalline host-like feature (Stark splitting) conforming to the f-f transitions of ${}^2H_{11/2} \to {}^4I_{15/2}$ around 525 nm, ${}^4S_{3/2} \to {}^4I_{15/2}$ in the range of 540–570 nm, and ${}^4F_{9/2} \to {}^4I_{15/2}$ at around 660 nm, whereas the nonlaser-irradiated part (i.e., the glassy part) shows broad emission band only around 525 and 550 nm. Therefore, the $Er^{3+}$ ions are incorporated into $CaF_2$ crystals during laser irradiation, which was facilitated by the use of laser light absorption through $Ni^{2+}$ ion in glass. This is analogous to the emission spectra

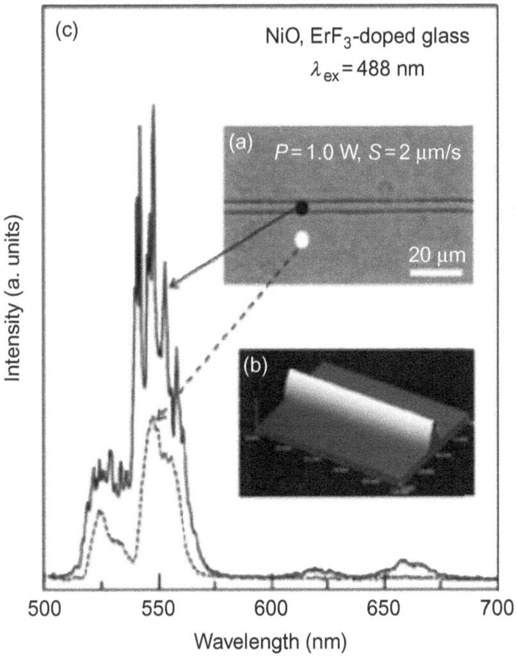


**Fig. 5.11** Photographs obtained by (a) polarization optical microscope and (b) confocal scanning laser microscope. (c) Microphotoluminescence spectra of the $0.5ErF_3$–3NiO–$20CaF_2$–16NaF–4CaO–$20Al_2O_3$–$10B_2O_3$–$30SiO_2$ glass and the $Er^{3+}$:$CaF_2$ crystallized line obtained by CW Yb:$YVO_4$fiber laser ($\lambda = 1080$ nm) with 1.0 W power at the scanning speed of 2 μm $s^{-1}$. Reproduced with permission from T. Komatsu, T. Honma, Optical active nano-glass-ceramics, Internat.J. Appl. Glass Sci. 4 (2013) 125–135, Copyright © The American Ceramic Society.

observed in $Pr^{3+}$-doped $LaF_3$ nanocrystals (also in other crystals) created in glass by crystallization by means of heat treatment [45].

### 5.3.6 Reduction-oxidation by laser irradiation

So far, we have found refractive index modification and crystallization of glasses by laser irradiation. Laser irradiation can also perform reduction-oxidation (redox) of some selective components of the glass during irradiation. This is due to the creation of electrons ($e^-$) and holes ($h^+$) in the glass through multiphoton ionization, Joule heating, and collisional (avalanche) ionization processes [46–48]. Holes are generally trapped by the nonbridging oxygen (NBO) ions, and the electrons are trapped by the easily (requires little amount of energy) reducible or oxidizable species such as $Sm^{3+}$, $Eu^{3+}$, $Ag^+$, $Au^{3+}$, and $Cu^{2+}$ of the glass. Some examples of this type of phenomenon are summarized in Table 5.1 [33–40].

Qiu et al. [33] have reported the permanent photo reduction of $Sm^{3+}$ to $Sm^{2+}$ inside a transparent and colorless sodium aluminoborate glass, $10Na_2O$–$5Al_2O_3$–$85B_2O_3+0.05Sm_2O_3$ (mol.%), by an infrared ($\lambda = 800$ nm) femtosecond (fs) pulsed laser. Fig. 5.12A shows the emission spectra of these glasses excited at 514.5 nm by $Ar^+$ laser (a) before and (b) after femtosecond (fs) laser irradiation. In the unirradiated glass, the emission bands peaking at 560, 600, 645, and 705 nm are caused by the f-f transitions of $Sm^{3+}$. Whereas the emission bands peaking at 683, 700, 724, and 760 nm, in the irradiated glass sample, are caused by the f-f transitions of

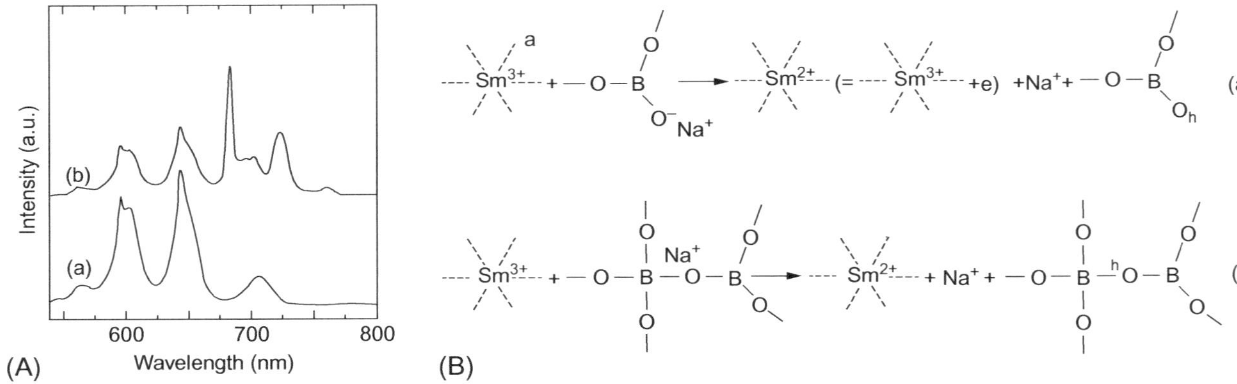


**Fig. 5.12** (A) Emission spectra of $10Na_2O$–$5Al_2O_3$–$85B_2O_3+0.05Sm_2O_3$ glasses excited at 514.5 nm by $Ar^+$ laser, (a) before and (b) after femtosecond (fs) laser irradiation. (B) Mechanism of photoreduction of $Sm^{3+}$ to $Sm^{2+}$ ion in the above glass by fs laser irradiation. Reproduced with permission from J. Qiu, K. Miura, T. Suzuki, T. Mitsuyu, K. Hirao, Permanent photoreduction of $Sm^{3+}$ to $Sm^{2+}$ inside a sodium aluminoborate glass by an infrared femtosecond pulsed laser, Appl. Phys. Lett. 74 (1999) 10–12, Copyright © AIP.

$Sm^{2+}$. Therefore, the $Sm^{3+}$ ions have been reduced to $Sm^{2+}$ ions during fs laser irradiation. The authors have also described the mechanism of photoreduction of $Sm^{3+}$ to $Sm^{2+}$ ion in the above glass by fs laser irradiation [33]. This is illustrated in Fig. 5.12B. In accordance with Stuart et al. [46], electrons ($e^-$) and holes ($h^+$) were produced in the glass during laser irradiation as a result of multiphoton ionization, Joule heating, and collisional (avalanche) ionization processes. The holes ($h^+$) were trapped by the non-bridging oxygen (NBO = B–$O^-$) ions and also by the tetrahedrally coordinated boron atoms ($BO_4$); at the same time, the electrons ($e^-$) were trapped by the $Sm^{3+}$ ions leading to the formation of $Sm^{2+}$. The reactions are articulated as (a) and (b) in Fig. 5.12B.

Peng et al. [36] have reported the generation of emission centers for broadband NIR luminescence in bismuthate glass by femtosecond laser irradiation. Fig. 5.13 illustrates the emission spectra of the $52Bi_2O_3$–$23B_2O_3$–$4SiO_2$–17PbO–$4Sb_2O_3$ (mol.%) glass excited at 785 nm by a laser diode: (A) before femtosecond (fs) laser irradiation and after fs laser irradiation, (B) 1 μJ and (C) 2.5 μJ. It is observed that on increasing the laser irradiation energy up to 2.5 μJ, the intensity of the broadband emission in the range of 1000–1500 nm is enhanced by sixfold keeping the band position unaltered. This enhancement is caused by laser-induced photoreduction of $Bi^{3+}$ ions to $Bi^+$ ions that are expressed by the following reactions:

$$Bi^{3+} + e^- \rightarrow Bi^{2+} \tag{5.3}$$

$$Bi^{2+} + e^- \rightarrow Bi^+ \tag{5.4}$$

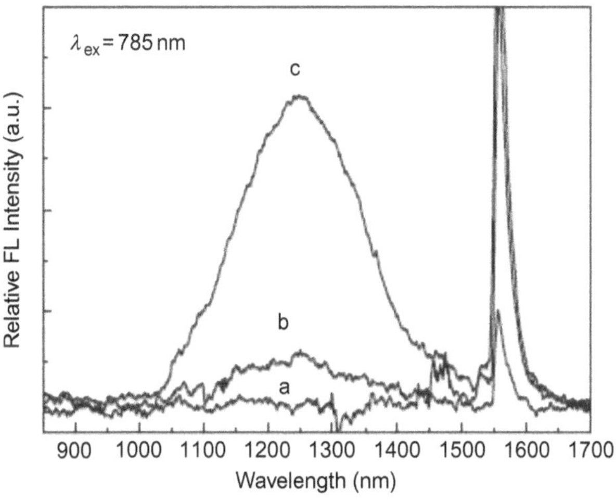


**Fig. 5.13** Emission spectra of $52Bi_2O_3$–$23B_2O_3$–$4SiO_2$–17PbO–$4Sb_2O_3$ glass excited at 785 nm by a laser diode, (A) before femtosecond (fs) laser irradiation and after fs laser irradiation, (B) 1 μJ and (C) 2.5 μJ.

Reproduced with permission from M. Peng, Q. Zhao, J. Qiu, L. Wondraczek, Generation of emission centers for broadband NIR luminescence in bismuthate glass by femtosecond laser irradiation, J. Am. Ceram. Soc. 92 (2009) 542–544, Copyright © The American Ceramic Society

At higher laser intensity, electrons concurrently absorb multiple photons to facilitate them to pass from the valence band to the conduction band. Just once, an electron is upheld to the conduction band, it nucleates a process of avalanche ionization by impact ionization, and the excess energy is transferred to another electron, resulting in two electrons in the conduction band. Increased laser pulse energy amplifies the number of these electrons, and by this means, the probability of permanent reduction gets better. This fact results in increasing the concentration of $Bi^+$ ion and subsequently increases the emission intensity in the range of 1000–1500 nm [36].

Zeng et al. [37] have described the precipitation of fs laser-induced gold nanoparticles inside silicate glasses of composition (mol.%) $70SiO_2$–$20Na_2O$–8CaO–2PbO +$0.01Au_2O_3$. Fig. 5.14A depicts the absorption spectra of these glasses: (a) after fs laser irradiation (*dashed line*, sample 1) and (b) after fs laser irradiation and heat-treated at 550°C/1 h (*solid line*, sample 2). During laser irradiation, reduction of gold ions ($Au^{3+}$) to gold atoms ($Au^0$) occurs by capturing the electrons (Eq. 5.5). Subsequent annealing at 550°C/1 h promotes the diffusion and clustering of gold atoms ($Au^0$) into gold nanoparticle (Eq. 5.6) and then grows. These are represented by the following reactions:

$$Au^{3+} + 3e^- \rightarrow Au^0 \tag{5.5}$$

$$nAu^0 \rightarrow \left(Au^0\right)_n \tag{5.6}$$

Qiu et al. [39] have reported the optical properties of a structurally modified silicate glass doped with $Au^{3+}$ ions by fs laser irradiation of the composition $70SiO_2$–10CaO–$20Na_2O$+$0.1Au_2O_3$ (mol.%). Fig. 5.14B shows the photograph of butterfly images

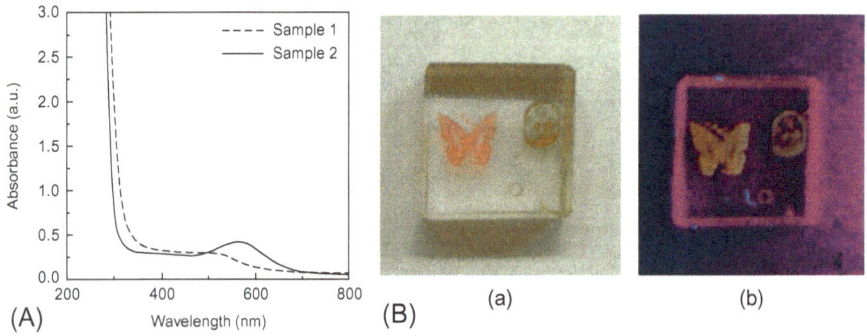


**Fig. 5.14** (A) Absorption spectra of $70SiO_2$–$20Na_2O$–8CaO–2PbO+$0.01Au_2O_3$ glasses, (a) after fs laser irradiation (*dashed line*, sample 1) and (b) after fs laser irradiation and heat-treated at 550°C/1 h (*solid line*, sample 2). (B) Photograph of butterfly images written by fs laser irradiation inside a $70SiO_2$–10CaO–$20Na_2O$+$0.1Au_2O_3$ glass when excited by 365 nm UV light, (a) before heat treatment at 550°C/0.5 h and (b) after heat treatment at 550°C/0.5 h. (A) Reproduced with permission from H. Zeng, G. Chen, J. Qiu, X. Jiang, C. Zhu,F. Gan, Effect of PbO on precipitation of laser-induced gold nanoparticles inside silicate glasses, J. Non-Cryst. Solids 354 (2008) 1155–1158, Copyright © Elsevier. (B) Reproduced with permission from J. Qiu, X. Jiang, C. Zhu, H. Inouye, J. Si, K. Hirao, Optical properties of structurally modified glasses doped with gold ions, Opt. Lett. 29 (2004) 370–372, Copyright © OSA.

written by fs laser irradiation inside the above glass when excited by 365 nm UV light: (a) before heat treatment at 55°C/0.5 h and (b) after heat treatment at 550°C/0.5 h. The laser-irradiated part before and after heat treatment at 550°C/0.5 h exhibits yellowish-white and orange-yellow emissions, respectively, under excitation at 365 nm UV light. For the gold nanoparticle (Au NP) created glass sample by heat treatment, the emission became orange yellow as a consequence of the surface plasmon resonance (SPR) absorption band of Au NPs around 530 nm.

## 5.4 Summary

Here, functionalization of various glasses by different types of laser irradiation microfabrication has been described. Different types of microoptical components such as waveguides, gratings, microlens, fiber attenuators, 3D optical memory, couplers, periodic nanosized voids, photonic crystals, amplifiers, and lasers can be fabricated by this technique. They have applications in optical data storage, telecommunications, biosensing, bioimaging, etc. Initially, the fundamental basis of this fabrication has been delineated. It is followed by the descriptions of processing, functional properties, and applications of laser-irradiated various kinds of glasses with specific examples, important findings, and interpretation of results. All these descriptions have been made with reference to refractive index modification, reduction-oxidation phenomena, and doped and undoped crystal formation. In addition to this, crystallization and patterning by rare-earth atom heat (REAH) and transition metal atom heat (TMAH) processing are also accounted.

## 5.5 Future outlook

In this chapter, we have discussed how laser irradiation of glass can be used to add new functionality to glasses. It has the ability to modify the refractive index; induce crystallization, valence state of active ions (rare-earth and transition metal ions); and form nanoparticles, nanograting, and 3D writing with great spatial selectivity and control. It can pattern crystals in glass with desired morphologies, orientation, and epitaxial growth. Therefore, this technique is useful in the fabrication of different microoptical devices, optical memory, 3D colored industrial art objects, etc. It has great applications in photonics and bio- and medical technology. However, the laser-induced functionality creation in glass is largely dependent on laser irradiation conditions, type of lasers (CW or pulse), wavelength of lasers, glass composition, etc. Much more research is necessary to methodically recognize the mechanisms associated with the laser conversion processes and device fabrication. The laser-induced crystallization indicates that the science and technology of crystallization in glass are now in a new phase, and its exploration just started; hence, there are plenty of new challenges. We believe that the laser irradiation microfabrication will open up many new possibilities in functionalization of glasses and glass-ceramics.

# Exercises

**5.1** What are the effects of laser irradiation on glass?

**5.2** Describe briefly the technique of laser irradiation microfabrication of glass with a suitable schematic diagram.

**5.3** Describe briefly temperature distribution and modification mechanism inside a model glass with heat accumulation during irradiation of continuous-wave Yb:$YVO_4$ fiber laser ($\lambda = 1080$ nm) or continuous-wave Nd:YAG laser ($\lambda = 1064$ nm).

**5.4** Describe briefly temperature distribution and modification mechanism inside a model glass with heat accumulation during irradiation of femtosecond laser pulses with relevant mathematical expressions.

**5.5** Explain the mechanism of modification of positive and negative refractive index of glass by laser irradiation with suitable example of glass.

**5.6** What is the basis of rare-earth atom heat (REAH) processing laser irradiation of glass? Describe with two suitable examples.

**5.7** What is the basis of transition metal atom heat (TMAH) processing laser irradiation of glass? Describe with two suitable examples.

**5.8** What is the basis of exhibiting reduction-oxidation (redox) phenomena by laser irradiation of glass?

**5.9** A glass having composition of $10Na_2O–5Al_2O_3–85B_2O_3 + 0.05Sm_2O_3$ (mol.%) is irradiated with a femtosecond (fs) pulsed laser of wavelength 800 nm. Explain the emission properties of these glasses after excitation at 514.5 nm by an $Ar^+$ laser (a) before and (b) after fs laser irradiation. Also explain the pertaining reaction mechanism.

**5.10** A glass having composition of $70SiO_2–10CaO–20Na_2O + 0.1Au_2O_3$ (mol.%) is irradiated with a femtosecond (fs) pulsed laser of wavelength 800 nm. Explain the emission properties of these glasses after excitation at 365 nm UV light (a) before heat treatment at 550°C/0.5 h and (b) after heat treatment at 550°C/0.5 h. Also explain the pertaining reaction mechanism.

# References

[1] W.H. Zachariasen, The atomic arrangement in glass, J. Am. Chem. Soc. 54 (1932) 3841–3851.

[2] K. Hirao, Ceram. Jpn. 30 (1995) 689 (in Japanese).

[3] J. Qiu, Femtosecond laser-induced microstructures in glasses and applications in micro-optics, Chem. Rec. 4 (2004) 50–58.

[4] D.M. Krol, Femtosecond laser modification of glass, J. Non-Cryst. Solids 354 (2008) 416–424.

[5] J. Qiu, K. Miura, K. Hirao, Femtosecond laser-induced microfeatures in glasses and their applications, J. Non-Cryst. Solids 354 (2008) 1100–1111.

[6] F. Luo, B. Qian, G. Lin, J. Xu, Y. Liao, J. Song, H. Sun, B. Zhu, J. Qiu, Q. Zhao, Z. Xu, Redistribution of elements in glass induced by a high-repetition-rate femtosecond laser, Opt. Express 18 (2010) 6262–6269.

[7] Y. Teng, J. Zhou, K. Sharafudeen, S. Zhou, K. Miura, J. Qiu, Space-selective crystallization of glass induced by femtosecond laser irradiation, J. Non-Cryst. Solids 383 (2014) 91–96.

[8] P. Gupta, H. Jain, D.B. Williams, J. Toulouse, I. Veltchev, Creation of tailored features by laser heating of $Nd_{0.2}La_{0.8}BGeO_5$ glass, Opt. Mater. 29 (2006) 355–359.
[9] A. Stone, M. Sakakura, Y. Shimotsuma, G. Stone, P. Gupta, K. Miura, K. Hirao, V. Dierolf, H. Jain, Directionally controlled 3D ferroelectric single crystal growth in $LaBGeO_5$ glass by femtosecond laser irradiation, Opt. Express 17 (2009) 23284–23289.
[10] C. Fan, B. Poumellec, M. Lancry, X. He, H. Zeng, A. Erraji-Chahid, Q. Liu, G. Chen, Three-dimensional photoprecipitation of oriented $LiNbO_3$-like crystals in silica-based glass with femtosecond laser irradiation, Opt. Lett. 37 (2012) 2955–2957.
[11] A.F. Maciente, V.R. Mastelaro, A.L. Martinez, A.C. Hernandes, C.A.C. Carneiro, Surface crystallization of $\beta$-$BaB_2O_4$ phase using a $CO_2$ laser source, J. Non-Cryst. Solids 306 (2002) 309–312.
[12] W. Avansi, V.R. Mastelaro, M.R.B. Andreeta, Surface modification and crystallization of the $BaO$–$B_2O_3$–$SiO_2$ glassy system using $CO_2$ laser irradiation, J. Non-Cryst. Solids 354 (2008) 279–283.
[13] L.L. Martin, S. Ríos, I.R. Martín, P. Haro-González, J.M. Cáceres, A. Hernández-Creus, Nanocrystal formation using laser irradiation on $Nd^{3+}$ doped barium titanium silicate glasses, J. Alloy. Compd. 553 (2013) 35–39.
[14] L.L. Martín, P. Haro-González, I.R. Martín, D. Puerto, J. Solís, J.M. Cáceres, N.E. Capuj, Local devitrification of $Dy^{3+}$ doped $Ba_2TiSi_2O_8$ glass by laser irradiation, Opt. Mater. 33 (2010) 186–190.
[15] P. Haro-González, L.L. Martín, S. González-Pérez, I.R. Martín, Formation of $Nd^{3+}$ doped strontium barium niobatenanocrystals by two different methods, Opt. Mater. 32 (2010) 1389–1392.
[16] P. Haro-González, I.R. Martín, S. González-Pérez, L.L. Martin, F. Lahoz, D. Puerto, J. Solís, Structural changes induced on strontium barium niobate glass by femtosecond laser irradiation, Appl. Phys. A 98 (2010) 879–884.
[17] P. Haro-González, I.R. Martín, A. Hernández-Creus, Nanocrystals distribution inside the writing lines in a glass matrix using argon laser irradiation, Opt. Express 18 (2010) 582–590.
[18] S.E. Mekhlouf, A. Boukenter, M. Ferrari, F. Goutaland, N. Ollier, Y. Ouerdane, UV assisted local crystallization in $Er^{3+}$ doped oxy-fluoride glass, J. Non-Cryst. Solids 353 (2007) 506–509.
[19] T. Komatsu, F. Suzuki, K. Ogawa, T. Honma, Recent progress in laser patterning of crystals in glasses, Trans. GIGAKU 1 (2012) 01017 (6 pages).
[20] T. Komatsu, R. Ihara, T. Honma, Y. Benino, R. Sato, H.G. Kim, T. Fujiwara, Patterning of non-linear optical crystals in glass by laser-induced crystallization, J. Am. Ceram. Soc. 90 (2007) 699–705.
[21] R. Ihara, T. Honma, Y. Benino, T. Fujiwara, R. Sato, T. Komatsu, Writing of two-dimensional crystal curved lines at the surface of $Sm_2O_3$–$Bi_2O_3$–$B_2O_3$ glass by samarium atom heat processing, Solid State Commun. 136 (2005) 273–277.
[22] R. Nakajima, M. Abe, Y. Benino, T. Fujiwara, H.G. Kim, T. Komatsu, Laser-induced crystallization of $\beta'$-$RE_2(MoO_4)_3$ ferroelectrics (RE: Sm, Gd, Dy) in glasses and their surface morphologies, J. Non-Cryst. Solids 353 (2007) 85–93.
[23] T. Honma, Laser-induced crystal growth of nonlinear optical crystal on glass surface, J. Ceram. Soc. Jpn. 118 (2010) 71–76.
[24] T. Honma, T. Komatsu, D. Zhao, H. Jain, Writing of rare-earth ion doped lithium niobate line patterns in glass by laser scanning, IOP Conf. Ser. Mater. Sci. Eng. 1 (2009) 012006 (6 pages).

[25] H. Tanaka, T. Honma, Y. Benino, T. Fujiwara, T. Komatsu, YAG laser-induced β-$BaB_2O_4$ crystalline dot formation in $Sm_2O_3$–BaO–$B_2O_3$ glasses, J. Phys. Chem. Solids 64 (2003) 1179–1184.
[26] K. Shinozaki, K. Hashimoto, T. Honma, T. Komatsu, TEM analysis for crystal structure of metastable $BiBO_3$(II) phase formed in glass by laser-induced crystallization, J. Eur. Ceram. Soc. 35 (2015) 2541–2546.
[27] K. Hirose, T. Honma, Y. Benino, T. Komatsu, Glass–ceramics with $LiFePO_4$ crystals and crystal line patterning in glass by YAG laser irradiation, Solid State Ionics 178 (2007) 801–807.
[28] T. Komatsu, T. Honma, Laser patterning and characterization of optical active crystals in glasses, J. Asian Ceram. Soc. 1 (2013) 9–16.
[29] T. Honma, T. Komatsu, Patterning of two-dimensional planar lithium niobate architectures on glass surface by laser scanning, Opt. Express 18 (2010) 8019–8024.
[30] T. Komatsu, K. Koshiba, T. Honma, Preferential growth orientation of laser-patterned $LiNbO_3$ crystals in lithium niobium silicate glass, J. Solid State Chem. 184 (2011) 411–418.
[31] T. Honma, Y. Benino, T. Fujiwara, T. Komatsua, Transition metal atom heat processing for writing of crystal lines in glass, Appl. Phys. Lett. 88 (2006) 231105 (3 pages).
[32] T. Komatsu, T. Honma, Optical active nano-glass-ceramics, Int. J. Appl. Glass Sci. 4 (2013) 125–135.
[33] J. Qiu, K. Miura, T. Suzuki, T. Mitsuyu, K. Hirao, Permanent photoreduction of $Sm^{3+}$ to $Sm^{2+}$ inside a sodium aluminoborate glass by an infrared femtosecond pulsed laser, Appl. Phys. Lett. 74 (1999) 10–12.
[34] J. Qiu, K. Kojima, K. Miura, T. Mitsuyu, K. Hirao, Infrared femtosecond laser pulse–induced permanent reduction of $Eu^{3+}$ to $Eu^{2+}$ in a fluorozirconate glass, Opt. Lett. 24 (1999) 786–788.
[35] M. Rada, S. Rada, E. Culea, Structural properties of the tungsten–lead–borate glasses before and after laser irradiation, J. Non-Cryst. Solids 357 (2011) 2024–2028.
[36] M. Peng, Q. Zhao, J. Qiu, L. Wondraczek, Generation of emission centers for broadband NIR luminescence in bismuthate glass by femtosecond laser irradiation, J. Am. Ceram. Soc. 92 (2009) 542–544.
[37] H. Zeng, G. Chen, J. Qiu, X. Jiang, C. Zhu, F. Gan, Effect of PbO on precipitation of laser-induced gold nanoparticles inside silicate glasses, J. Non-Cryst. Solids 354 (2008) 1155–1158.
[38] Y. Shimotsuma, K. Hirao, J. Qiu, K. Miura, Nanofabrication in transparent materials with a femtosecond pulse laser, J. Non-Cryst. Solids 352 (2006) 646–656.
[39] J. Qiu, X. Jiang, C. Zhu, H. Inouye, J. Si, K. Hirao, Optical properties of structurally modified glasses doped with gold ions, Opt. Lett. 29 (2004) 370–372.
[40] J. Qiu, C. Zhu, T. Nakaya, J. Si, K. Kojima, F. Ogura, K. Hirao, Space-selective valence state manipulation of transition metal ions inside glasses by a femtosecond laser, Appl. Phys. Lett. 79 (2001) 3567–3569.
[41] M. Sakakura, M. Shimizu, Y. Shimotsuma, K. Miura, K. Hirao, Temperature distribution and modification mechanism inside glass with heat accumulation during 250 kHz irradiation of femtosecond laser pulses, Appl. Phys. Lett. 93 (2008) 231112 (3 pages).
[42] T. Honma, P.T. Nguyen, T. Komatsu, Crystal growth behavior in CuO-doped lithium disilicate glasses by continuous-wave fiber laser irradiation, J. Ceram. Soc. Jpn. 116 (2008) 1314–1318.

[43] V.R. Bhardwaj, E. Simova, P.B. Corkum, D.M. Rayner, C. Hnatovsky, R.S. Taylor, B. Schreder, M. Kluge, J. Zimmer, Femtosecond laser-induced refractive index modification in multicomponent glasses, J. Appl. Phys. 97 (2005) 083102 (9 pages).

[44] S. Kanehira, K. Miura, K. Hirao, Ion exchange in glass using femtosecond laser irradiation, Appl. Phys. Lett. 93 (2008) 023112 (3 pages).

[45] X. Wang, S.H. Huang, R. Reeves, W. Wells, M.J. Dejneka, R.S. Meltzer, W.M. Yen, Studies of the spectroscopic properties of $Pr^{3+}$ doped $LaF_3$ nanocrystals/glass, J. Lumin. 94–95 (2001) 229–233.

[46] B.C. Stuart, M.D. Feit, A.M. Rubenchik, B.W. Shore, M.D. Perry, Laser-induced damage in dielectrics with nanosecond to subpicosecond pulses, Phys. Rev. Lett. 74 (1995) 2248–2251.

[47] B. Rethfeld, Free-electron generation in laser-irradiated dielectrics, Contrib. Plasma Phys. 47 (2007) 360–367.

[48] J.M.P. Almeida, L. De Boni, W. Avansi, C. Ribeiro, E. Longo, A.C. Hernandes, C. R. Mendonca, Generation of copper nanoparticles induced by fs-laser irradiation in borosilicate glass, Opt. Express 20 (2012) 15106–15113.

# Functionalization of glasses by incorporation of semiconductor

6

## Chapter Outline



## 6.1 Introduction

According to the band theory, semiconductors are those materials which have the bandgap energy, $E_g$ (0.1–9.0 eV) in between that of insulators (>9 eV) and metals (no energy gap), where the thermal energy (~25 meV) at 300 K is sufficient to promote the electrons from the valence band to the conduction band. These semiconductors are of two types, elemental and compound. The examples of elemental semiconductors are silicon (Si), germanium (Ge), etc. Compound semiconductors may be binary, ternary, quaternary, and so on. Binary compound semiconductors are made up of elements of different groups of the periodic table. As a result, they are named accordingly such as I–VII (e.g., CuCl, CuBr, CuI), II–IV (e.g., ZnS, ZnSe, ZnTe, CdS, CdSe, CdTe), III–V (e.g., AlP, GaAs, InSb), IV–VI (e.g., PbS, PbSe, PbTe), etc. Here we shall consider only the II–VI and IV–VI semiconductors.

Semiconductor nanocrystals (NCs) or quantum dots (QDs) exhibit intermediate behavior between a bulk crystal and an isolated molecule [1,2] due to the quantum confinement effect. Glass may be utilized as a nanochamber where semiconductor NCs or QDs can be grown. The properties of the semiconductor NCs or QDs are determined not only by the confinements of the host material (here glass) but also by the properties of the system including internal and external surface properties the NCs or QDs. Glasses doped with semiconductor NCs demonstrate active properties which allows modulation of light, optical nonlinearity, optical gain, and lasing. Consequently, semiconductor NCs incorporated glasses are used in nonlinear optics (NLO), optical switches, optoelectronic devices, telecommunications, infrared (IR)

**Functional Glasses and Glass-Ceramics. http://dx.doi.org/10.1016/B978-0-12-805056-9.00006-4**

optics, lasers, etc. [1–3]. In this chapter, various aspects of functionalization of glasses by incorporation of semiconductor are described.

An attempt was made using the Web of Science (Version 5.17) as described in Chapter 1 to provide a global publication trend and quantitative growth in the area of semiconductor incorporated glasses. Here, year-wise data were extracted for the last 25 years (1990–2014) by searching, using the search keyword "*glass*" followed by refining the document type as "*article*" published in the journals and meetings. The search was further refined using the term "*semiconductor*." This plot is shown in Fig. 6.1. It gives us a clear publication trend and quantitative growth of published papers on semiconductor incorporated glasses during the last 25 years (1990–2014). It is observed that there was a slow growth of the published papers during the period 1990–2014 with many ups and downs.

## 6.2 Processing and properties of II–VI semiconductor incorporated glasses

The most important II–VI compound semiconductors pertaining to incorporation in glasses are ZnO, ZnS, ZnSe, ZnTe, CdS, CdSe, and CdTe. Some of their properties are provided in Table 6.1 [4]. Their bandgap energy and wavelength vary in the range of 1.471–3.68 (3.91) eV and 337 (317)–841 nm, respectively. Therefore, the optical properties of the glasses can be tuned by incorporation of these semiconductors. Further, nanometer-sized crystals have distinguishing optical properties due to the

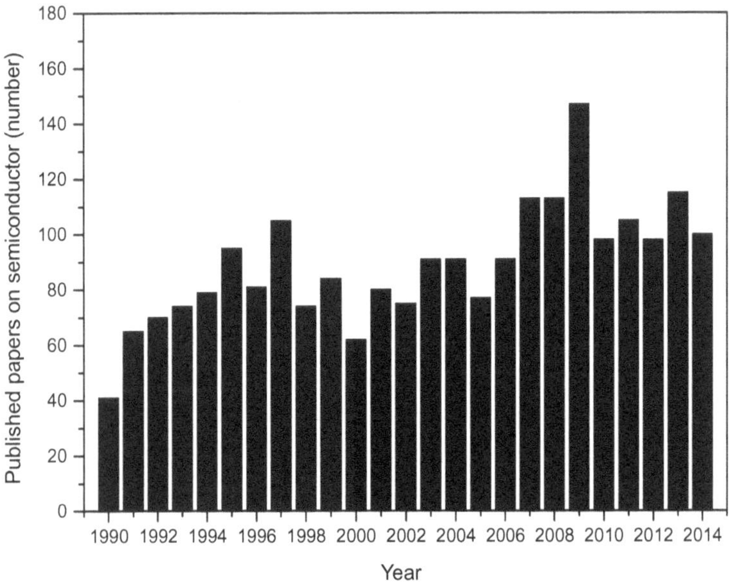


**Fig. 6.1** Year-wise published papers on semiconductor containing glasses during the last 25 years (1990–2014).

**Table 6.1** **Some properties of the wide-bandgap II–VI compound semiconductors [4]**

| Material property | ZnO | ZnS | ZnSe | ZnTe | CdS | CdSe | CdTe |
|---|---|---|---|---|---|---|---|
| Melting point (°C) | 1975 | 1765 (WZ) | 1524 | 1240 | 1750 (WZ) | 1350 | 1041 |
| Density (g cm$^{-3}$) | 5.61 | 4.11 | 5.26 | 5.65 | 4.87 | 5.66 | 5.86 |
| Refractive index | 2.029 | 2.368/2.378 | 2.5 | 2.72 | 2.529 | 2.5 | 2.72 |
| Phonon energy (cm$^{-1}$) | 576 | 350 | 252 | 210 | 303 | 211 | 170 |
| Bandgap energy, $E_g$ (eV) | 3.4 | 3.68 (ZB)/3.91 (WZ) | 2.71 | 2.394 | 2.50 | 1.751 | 1.474 |
| Bandgap wavelength (nm) | 365 | 337/317 | 458 | 518 | 496 | 708 | 841 |
| Bandgap type | Direct | Direct | Direct | Direct | Direct | Direct | Direct |

*WZ*, wurtzite structure; *ZB*, zinc blende structure.

confinement of charged particles in all three space dimensions. Consequently, the bandgap can be adjusted to optimize the optical properties of the nanostructures by controlling their size.

Ömür et al. [5] described the growth of $CdS_xSe_{1-x}$ nanoparticles or QDs in a borosilicate glass by a twostep heat treatment process from a melt-quenched color filter glass RG695 (Schott). These authors first melted the as-received filter glass RG695 (borosilicate glass doped with $CdS_{0.08}Se_{0.92}$) at around 1000°C for 15 min to dissolve the particles and quenched the melt rapidly to room temperature and then annealed the glass at 450°C for 5 h (primary heat treatment) to release stress in glass. Nucleation and growth of nanoparticles were carried out at different temperatures around the glass transition temperature (600–675°C) for different durations (secondary heat treatment). Fig. 6.2 shows the absorption spectra of $CdS_{0.08}Se_{0.92}$ doped glasses that were primary heat-treated at 450°C/5 h and secondary heat-treated at (1) 600°C/0.5 h, (2) 625°C/1 h, (3) 625°C/2 h, (4) 625°C/4 h, (5) 625°C/8 h, (6) 650°C/2 h, (7) 650°C/4 h, (8) 650°C/8 h, (9) 675°C/2 h, and (10) 675°C/2 h. The colors in the samples are

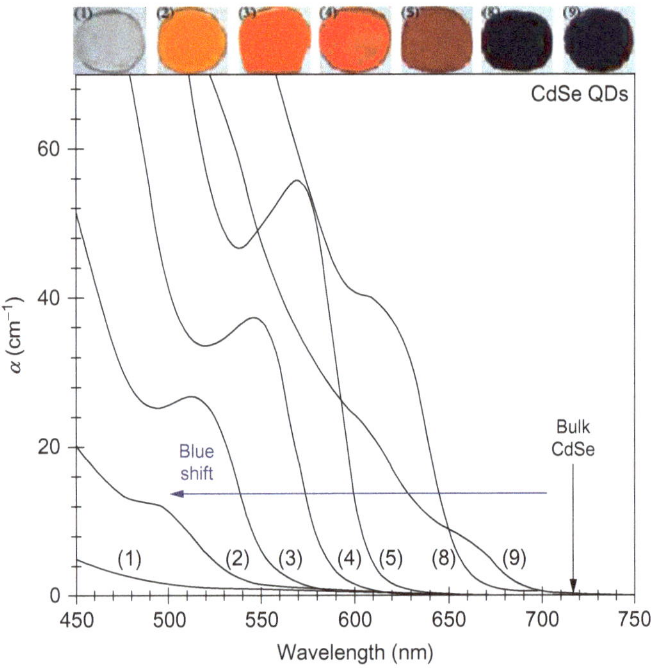


**Fig. 6.2** Absorption spectra of $CdS_{0.08}Se_{0.92}$ doped glasses primary heat-treated at 450°C/5 h and secondary heat-treated at (1) 600°C/0.5 h, (2) 625°C/1 h, (3) 625°C/2 h, (4) 625°C/4 h, (5) 625°C/8 h, (6) 650°C/2 h, (7) 650°C/4 h, (8) 650°C/8 h, (9) 675°C/2 h, and (10) 675°C/2 h. The colors in the samples are indicated at the top of the figure.
Reproduced with permission from B.C. Ömür, A. Aşkoğlu, Ç. Allahverdi, M.H. Yükselici, $CdS_xSe_{1-x}$ quantum dots studied through optical absorption, steady-state photoluminescence and resonant Raman spectroscopies, J. Mater. Sci. 45 (2010) 112–116, Copyright © Springer.

presented at the top of the figure. With the progress of color development, that is growth of QDs, an absorption peak of the QDs appears by reason of the first exciton peak as a result of transition from a hole state to an electron state. A shift of around 200 nm in the exciton peak position was observed with the progress of heat treatment at specified temperature and time. It is seen that the longer the heat treatment time or the higher the temperature, the larger is the growth of the nanocrystals. After thermal treatment for a long time at a given temperature, the optical absorption band edge approaches that of bulk crystal of the same composition.

This phenomenon can be understood more clearly from the studies carried out by Kolobkova [6] with specific features of growth of CdSe nanocrystals in fluorophosphate glasses as shown in Fig. 6.3. Similarly, the optical absorption edge of the $CdS_xSe_{1-x}$ mixed semiconductor nanocrystals embedded glasses can be tuned within the spectral range 350–700 nm by controlling the compositional parameter $x$, temperature, and time [7,8]. This orderliness makes it possible to select the composition and processing parameters to obtain the sharp cut-off commercial glass filters. Fig. 6.4 shows shifting of the absorption edges towards near infrared (NIR) region of sharp cut-off commercial filters (Hoya Corporation) with decrease of $x$ in the mixed semiconductor microcrystals of $CdS_xSe_{1-x}$ in glasses [9]. Ciolek and Krashkevich [10], and Lee [11] described the processing of this category of glass filters in detail. A preferred chemical composition (wt.%) range is as follows: 60–73$SiO_2$, 4–7$B_2O_3$, 3.5–5.0 $K_20$, 12–14$Na_2O$, 5–9CaO, 0.5–10.0ZnO, 0.1–1.5CdS, 0.1–3.0ZnS, 0–1.0CdSe, and 0–0.5CdTe. Glass melting and semiconductor crystallization temperatures were monitored for the ranges 1450–1550°C and 600–650°C, respectively.

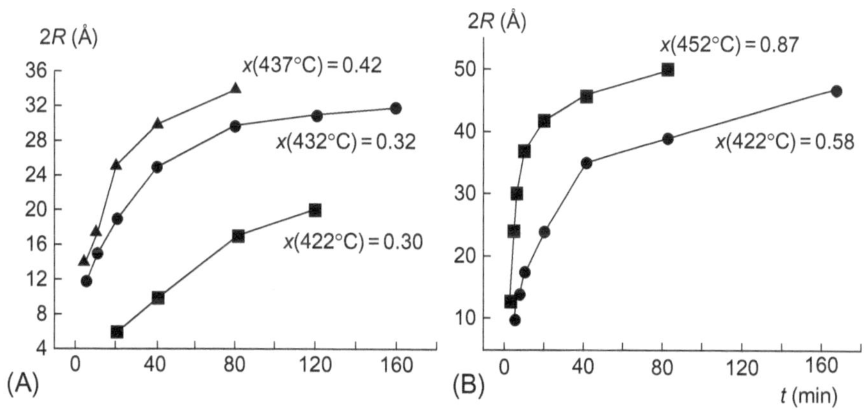


**Fig. 6.3** Growth in CdSe nanocrystal size in a fluorophosphate glass of composition $0.25Na_2O–0.5P_2O_5–0.1Ga_2O_3–0.05AlF_3–0.1ZnO$ as a function of heat treatment time at (A) 422°C, 432°C, and 437°C, (B) 422°C and 452°C.
Reproduced with permission from E.V. Kolobkova, Specific features of growth of CdSe nanocrystals in fluorophosphate glasses, Glass Phys. Chem. 32 (2006) 404–411, Copyright © Pleiades Publishing, Inc.

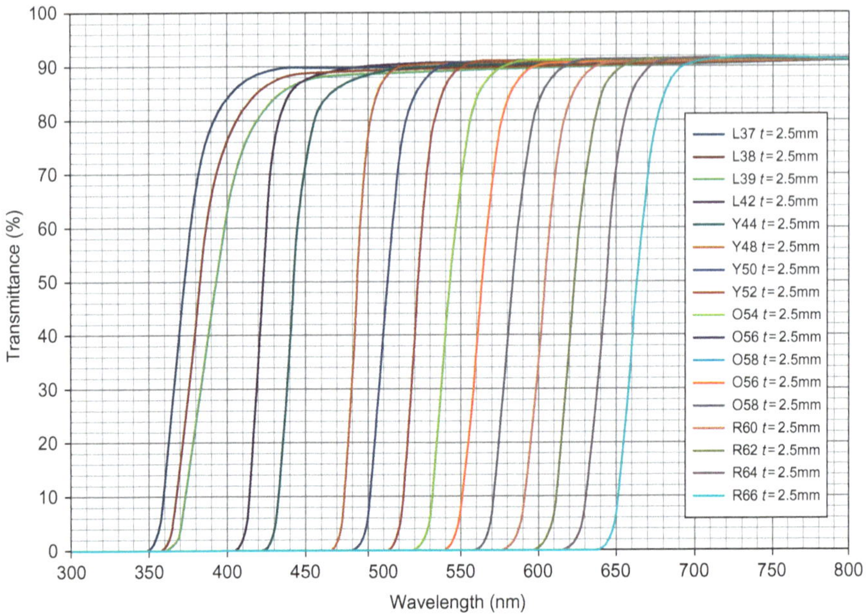


**Fig. 6.4** Shifting of absorption edges towards near infrared (NIR) region of sharp cut-off filters with decrease of $x$ in the mixed semiconductor microcrystals of $CdS_xSe_{1-x}$ in glasses. Courtesy: Hoya Corporation USA, Santa Clara, USA. Web: www.hoyaoptics.com, © Hoya Corporation USA.

Ushakova et al. [12] studied the formation and optical properties of CdSSe nanocrystals in a silicate glass matrix. Fig. 6.5A shows the Raman scattering spectra of CdSSe nanocrystals in glass obtained by heat treatment of precursor glass at (a) 470°C/45 h, (b) 500°C/20 h, and (c) 550°C/8 h. The crystal grains have the composition $CdS_{0.58}Se_{0.42}$. It is important to note that an increase in the heat treatment temperature does not change the stoichiometric composition of the semiconductor crystal grains. Consequently, there is no shifting of the Raman scattering peak position. TEM image of CdSSe nanocrystals in the glass obtained by heat treatment of precursor glass at 550°C for 20 h is shown in Fig. 6.5B. It is seen that the crystal grain size is predominantly retained in the range of 2–5 nm along with several larger grains. Fig. 6.6A illustrates the photoluminescence (PL) spectra (excitation at $\lambda_{exc} = 415.6$ nm by He–Cd laser) of CdSSe nanocrystals in the glass obtained by heat treatment of precursor glass at (a) 470°C/45 h, (b) 500°C/20 h, and (c) 550°C/8 h. It is seen that the spectrum of all the samples consists of two broad bands with maxima at energies of 1.84–1.62 and 1.38 eV, that is, 674–765 and 899 nm, respectively. The long wavelength band is observed in the spectra of all the samples corresponding to the luminescent centers of the defect structures of the silicate glass matrix whereas that of the short wavelength is attributed to CdSSe crystal grains. The short

(A)

(B)

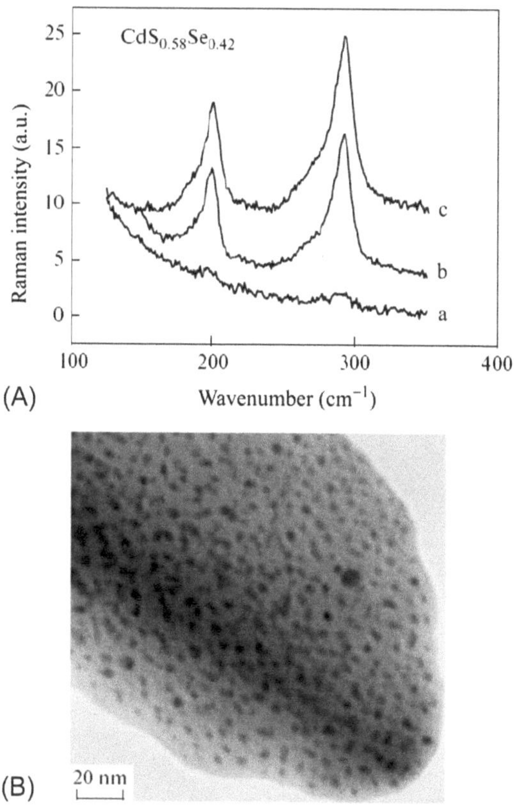


**Fig. 6.5** (A) Raman scattering spectra of CdSSe nanocrystals in glass obtained by heat treatment of precursor glass at (a) 470°C/45 h, (b) 500°C/20 h, and (c) 550°C/8 h. (B) TEM image of CdSSe nanocrystals in glass obtained by heat treatment of precursor glass at 550°C for 20 h.
Reproduced with permission from V.V. Ushakova, A.S. Aronin, V.A. Karavanskii, A.A. Gippius, Formation and optical properties of CdSSe nanocrystals in silicate glass matrix, Phys. Solid State 51 (2009) 2161–2165, Copyright © Pleiades Publishing, Ltd.

wavelength band exceeding the value of the bandgap of the bulk material shifts toward lower energies (1.84–1.73–1.62 eV) as the size of the nanocrystals increases. This is due to quantum confinement effect and interband transitions in semiconductor nanocrystals.

For any clarification on these phenomena, schematic representations of the main photo-physical processes of nano-sized semiconductors excited by light with energy equal to or higher than the bandgap energy, $E_g$ are depicted in Fig. 6.6B [13].

Recently, Karmakar and coworkers [14,15] have described the synthesis and PL properties of CdSe and ZnSe nanocrystals ingrained borosilicate glass of the $B_2O_3$–$SiO_2$–$Al_2O_3$–$Na_2O$–$K_2O$ system. Fig. 6.7A shows the PL spectra (excitation at 446 nm by diode laser) of undoped (BG) and CdSe nanocrystal ingrained borosilicate glass nanocomposites with increased concentration of CdSe (NC-1 to NC-5). Inset shows variation of PL intensity with concentration of CdSe. The increase in CdSe concentration lowers the PL intensity as a result of formation of larger NPs. However, the trap state or defect state emission transition peaks around 665, 834, 867, and 968 nm do not shift with concentration of CdSe. Fig. 6.7B depicts (a) the

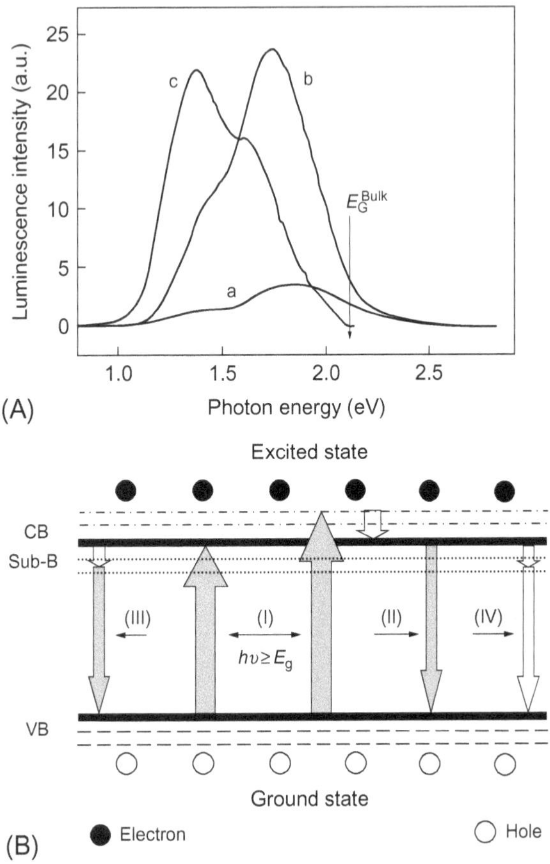


**Fig. 6.6** (A) Photoluminescence spectra (excitation at $\lambda_{exc} = 415.6$ nm by He–Cd laser) of CdSSe nanocrystals in a silicate glass obtained by heat treatment of precursor glass at (a) 470°C/45 h, (b) 500°C/20 h, and (c) 550°C/8 h. (B) Schematic representation of main photo-physical processes of nano-sized semiconductors excited by light with energy equal to or higher than the bandgap energy, $E_g$ (*VB*, valence band; CB, conduction band; *I*, photo-excited process; *II*, band-to-band PL process; *III*, excitonic PL process; *IV*, nonradiative transition process).
(A) Reproduced with permission from V.V. Ushakova, A.S. Aronin, V.A. Karavanskii, A.A. Gippius, Formation and optical properties of CdSSe nanocrystals in silicate glass matrix, Phys. Solid State 51 (2009) 2161–2165, Copyright © Pleiades Publishing, Ltd. (B) Reproduced with permission from J. Liqiang, Q. Yichun, W. Baiqi, L. Shudan, J. Baojiang, Y. Libin, F. Wei, F. Honggang, S. Jiazhong, Review of photoluminescence performance of nano-sized semiconductor materials and its relationships with photocatalytic activity, Sol. Energy Mater. Sol. Cells, 90 (2006) 1773–1786, Copyright © Elsevier.

CIE chromaticity diagram of the emitted light from sample NC-3 on excitation at 446 nm. The point "Q" represents the red luminescence color coordinate (0.649, 0.290) of the emitted light. In support of red light emission, the luminescence light emission photographs of the samples BG and NC-3 under laser excitation are shown in Fig. 6.7B (b) and (c), respectively.

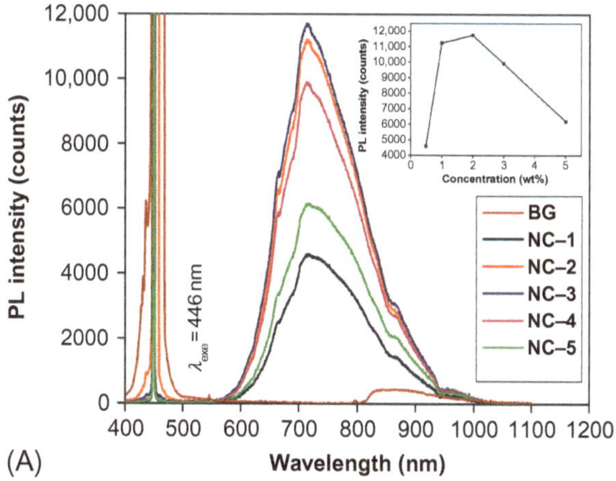


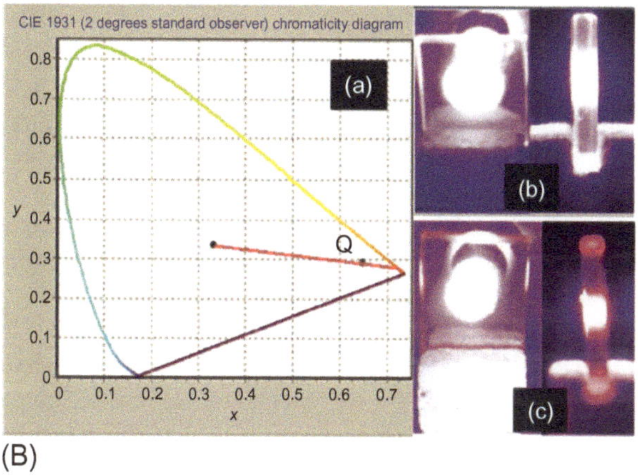


**Fig. 6.7** (A) Photoluminescence (PL) spectra (excitation at 446 nm by diode laser) of undoped (BG) and CdSe nanocrystal ingrained borosilicate glass nanocomposites with increased concentration of CdSe (NC-1 to NC-5). Inset shows variation of PL intensity with concentration of CdSe. (B) (a) CIE chromaticity diagram of the emitted light from sample NC-3 on excitation at 446 nm. The point "Q" represents the color coordinate (0.649, 0.290) of the emitted light. Luminescence photographs of samples (b) BG and (c) NC-3 under laser excitation are also shown. Reproduced with permission from C. Dey, M. Goswami, B. Karmakar, CdSe nanocrystals ingrained dielectric nanocomposites: synthesis and photoluminescence properties, Mater. Res. Express 2 (2015) 015014 (12 pages), Copyright © IOP.

## 6.3 Processing and properties of rare-earth doped II–VI semiconductor incorporated glasses

II–VI semiconductors are known to exhibit low phonon energies (see Table 6.1) which reduce the extent of electron-phonon interaction and improve the emission performance. Therefore, the II–VI semiconductors doped glasses when codoped with rare-earth (RE) ions exhibit enhanced PL emission properties. This enhancement is caused by the partition of the RE into the low phonon energy lattice environment of semiconductors which reduces the multiphonon relaxation rate. In addition, this enhancement is a result of phonon assisted energy transfer from electron-hole recombination of the semiconductor nanocrystals to the RE ion.

In recent times, Karmakar and coworkers [16] have described synthesis and properties of ZnTe and $Eu^{3+}$ ion codoped glass nanocomposites, a $SiO_2$–$K_2O$–CaO–BaO–$B_2O_3$ borosilicate glass system. Here, glass nanocomposites were prepared by the melt quenching technique followed by controlled heat treatment. Fig. 6.8A displays the PL spectra (excitation at 394 nm) of ZnTe and $Eu^{3+}$ ion codoped borosilicate glass nanocomposites heat-treated at 610°C for 0–100 h. It is seen that the intensity of both the emission bands at 592 nm due to magnetic dipole (MD) transition of $^5D_0 \rightarrow {}^7F_1$ and electric dipole (ED) transition of $^5D_0 \rightarrow {}^7F_2$ of $Eu^{3+}$ ion at 612 nm increases with increasing heat treatment time. It is interesting to note that this increment is around sixfold of the band at 612 nm on heat treatment at 610°C/0.5 h. It is attributed to the separation of $Eu^{3+}$ ions into the low phonon energy ZnTe crystallites (210 $cm^{-1}$) and as the size of the nanocrystals is smaller than the size of the exciton, the quantum confinement effect is visible. Further increase in heat treatment duration led to decrease in luminescence intensity due to the growth of larger size ZnTe crystals as well as concentration quenching as a result of increase in concentration of $Eu^{3+}$ in the ZnTe crystallites. Fig. 6.8B shows the TEM image and SAED pattern of ZnTe and $Eu^{3+}$ ion codoped borosilicate glass nanocomposites heat-treated at 610°C for 0.5 h (sample ZnTe–Eu–0.5 h). TEM micrograph demonstrates the formation of 5–10 nm sized crystal at 610°C/0.5 h and the SAED pattern confirms the formation of ZnTe semiconductor in the glass matrix.

An example of such enhancement is demonstrated by Jose et al. [17] with reference to $Eu^{3+}$ ion in CdSe codoped silica glass prepared by the sol-gel technique. Fig. 6.9A displays the emission spectra of $Eu^{3+}$ ion (excited at 393 nm) in CdSe codoped silica glass (heat-treated at 500°C) prepared by the sol-gel technique. It is seen that the emission intensity of the electric dipole (ED) transition $^5D_0 \rightarrow {}^7F_2$ of $Eu^{3+}$ ion is enhanced significantly in the presence of CdSe nanocrystals. Similar enhancement is also observed by Jyothy et al. [18] in $Tb^{3+}$/CdS nanoparticles doped silica xerogels. Fig. 6.9B depicts the emission spectra of $Tb^{3+}$ ion (3 wt.%) in CdS (A: 1, B: 3, C: 5, and D: 7 wt.%) codoped silica xerogels prepared by the sol-gel technique. It is seen that the emission intensity increases up to 5 wt.% CdS doping then decreases with further doping.

Another example of enhancement of fluorescence intensities of $Eu^{3+}$ ions in $SnO_2$ nanocrystals-precipitated sol-gel derived silica glass higher than 150 times than that of glass without $SnO_2$ nanocrystals was demonstrated by Nogami et al. [19]. Fig. 6.10A shows the fluorescence spectra (excited at 337 nm) of (a) Eu–1$SnO_2$, (b) Eu–3$SnO_2$, and (c) Eu–5$SnO_2$ in sol-gel derived silica glass heat-treated at 700°C. The spectra of (a) and (b) were enlarged by 150 and 5 times respectively to vertical axis for comparison. The inset shows the relation between the integrated intensities versus $SnO_2$ content. This enhancement is caused by the energy transfer due to the recombination of electrons and holes (as shown in Fig. 6.6B) excited in $SnO_2$ semiconductor nanocrystals [20]. This energy transfer mechanism is schematically represented in

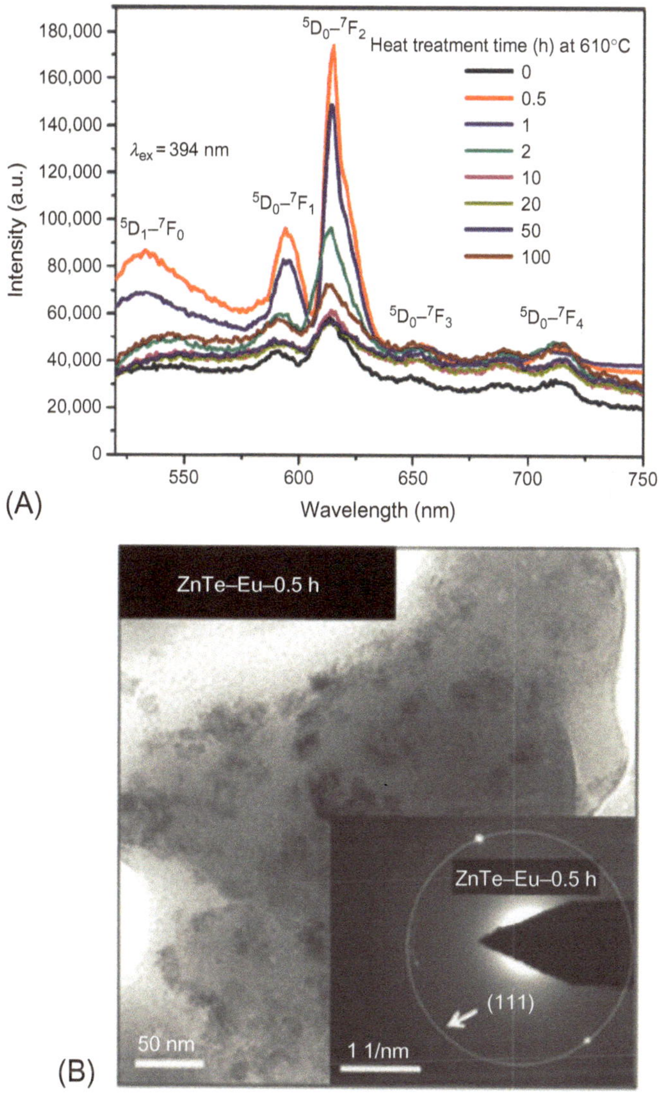


**Fig. 6.8** (A) Photoluminescence (PL) spectra (excitation at 394 nm) of ZnTe and $Eu^{3+}$ ion codoped borosilicate glass nanocomposites heat-treated at 610°C for 0–100 h. (B) TEM image and SAED pattern of ZnTe and $Eu^{3+}$ ion codoped borosilicate glass nanocomposites heat-treated at 610°C for 0.5 h (sample ZnTe–Eu–0.5 h).

Reproduced with permission from A.R. Molla, A. Tarafder, C. Dey, B. Karmakar, Synthesis and properties of ZnTe and $Eu^{3+}$ ion co-doped glass nanocomposites, J. Appl. Phys. 116 (2014) 163510 (13 pages), Copyright © AIP.

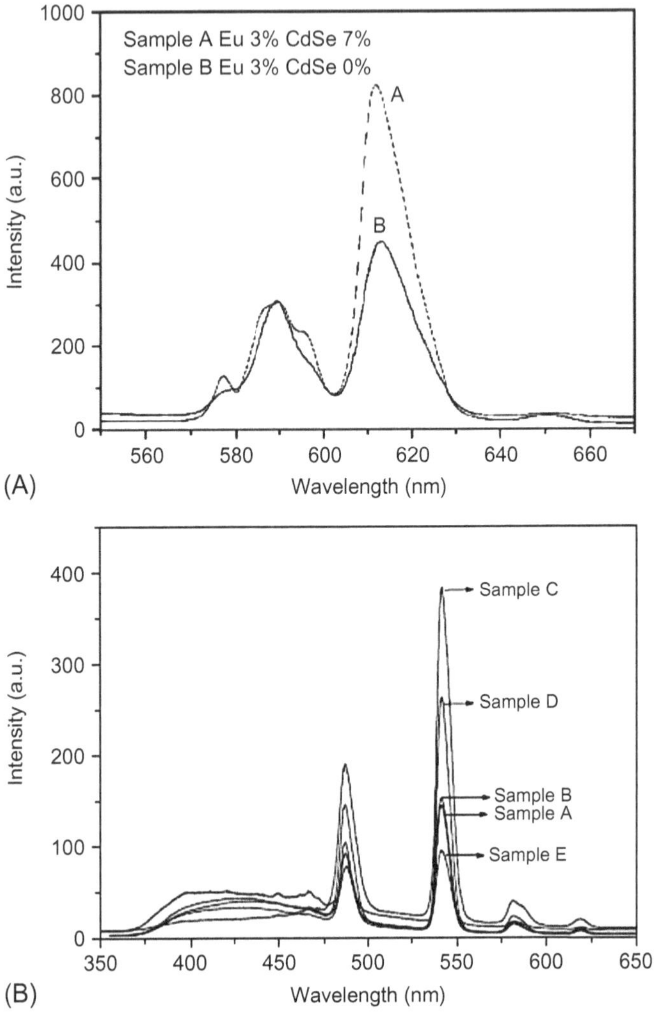


**Fig. 6.9** (A) Emission spectra of $Eu^{3+}$ ion (excited at 393 nm) in CdSe codoped silica glass (heat-treated at 500°C) prepared by the sol-gel technique. (B) Emission spectra of $Tb^{3+}$ ion (3 wt.%) in CdS (A: 1, B: 3, C: 5, and D: 7 wt.%) codoped silica xerogels prepared by the sol-gel technique.

(A) Reproduced with permission from G. Jose, K.A. Amrutha, T.F. Toney, V. Thomas, C. Joseph, M.A. Ittyachen, N.V. Unnikrishnan, Structural and optical characterization of $Eu^{3+}$/CdSe nanocrystal containing silica glass, Mater. Chem. Phys. 96 (2006) 381–386, Copyright © Elsevier. (B) Reproduced with permission from P.V. Jyothy, K.A. Amrutha, J. Gijo, N.V. Unnikrishnan, Fluorescence enhancement in $Tb^{3+}$/CdS nanoparticles doped silica xerogels, J. Fluoresc. 19 (2009) 165–168, Copyright © Springer.

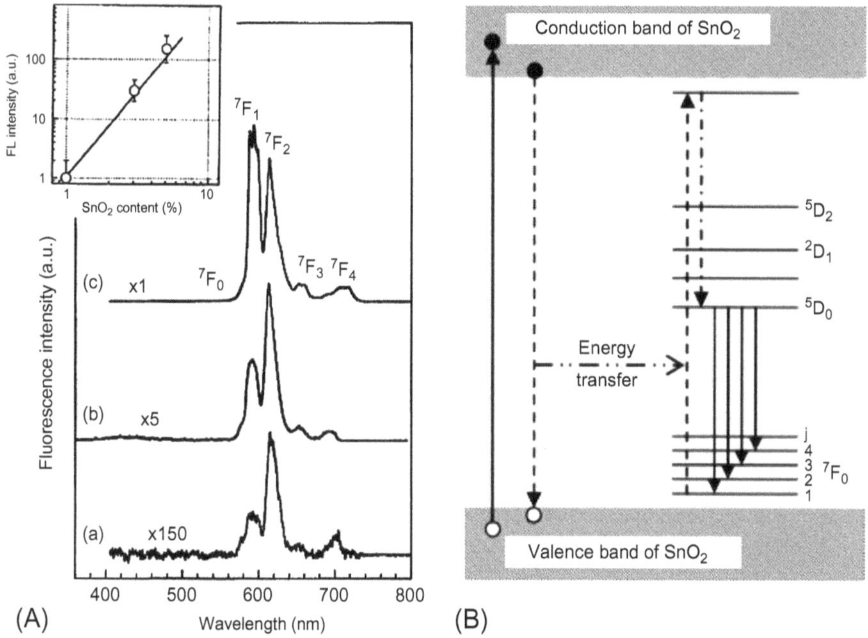


**Fig. 6.10** (A) Fluorescence spectra (excited at 337 nm) of (a) Eu–1$SnO_2$, (b) Eu–3$SnO_2$, and (c) Eu–5$SnO_2$ in sol-gel derived silica glass heat-treated at 700°C. The spectra of (a) and (b) are enlarged by 150 and 5 times respectively to vertical axis for comparison. The inset shows the relation between the integrated intensities versus $SnO_2$ content. (B) Energy transfer process from $SnO_2$ nanocrystals to $Eu^{3+}$ ions in nano-crystallized $SnO_2$–$SiO_2$ glass-ceramics.
(A) Reproduced with permission from M. Nogami, T. Enomoto, T. Hayakawa, Enhanced fluorescence of $Eu^{3+}$ induced by energy transfer from nanosized $SnO_2$ crystals in silica glass, J. Lumin. 97 (2002) 147–152, Copyright © Elsevier. (B) Reproduced with permission from V. Gueu, H. You, T. Hayakawa, M. Nogami, $Eu^{3+}$-fluorescence properties in nano-crystallized $SnO_2$-$SiO_2$ glass-ceramics, J. Sol-Gel Sci. Technol. 41 (2007) 231–236, Copyright © Springer.

Fig. 6.10B. Similar energy transfer processes are also demonstrated in $Er^{3+}$ ions doped nano-crystallized $SnO_2$–$SiO_2$ sol-gel derived glass-ceramics [21].

Of late, Karmakar and coworkers [22] have demonstrated the white light emission in $Ho^{3+}$-doped CdS nanocrystal ingrained glass nanocomposites. Fig. 6.11A illustrates the PL spectra (excitation at 405 nm by diode laser) of undoped (CS-1) and $Ho^{3+}$ion doped (CS-2) CdS nanocrystal ingrained borosilicate glass nanocomposites heat-treated at 560°C for 5–10 h exhibiting white light emission. The zoomed-in portion is displayed as inset showing different transitions of $Ho^{3+}$ ion. Fig. 6.11B displayed the CIE chromaticity diagram of samples CS-2-5 h and CS-2-10 h with excited sample photographs showing white light emission. Color coordinates of samples CS-2-5 h and CS-2-10 h are "*p*" (0.309, 0.326) and "*q*" (0.308, 0.338),

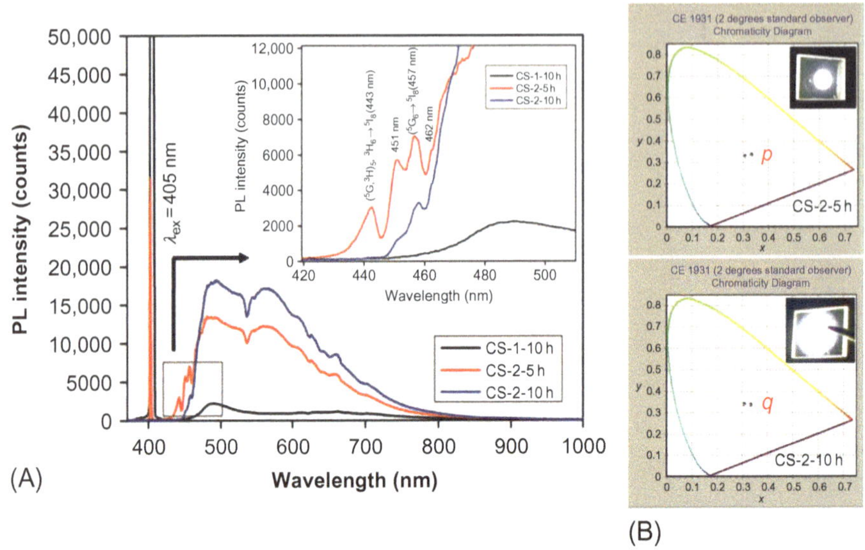


**Fig. 6.11** (A) Photoluminescence spectra (excitation at 405 nm by diode laser) of undoped (CS-1) and $Ho^{3+}$ion doped (CS-2) CdS nanocrystal ingrained borosilicate glass nanocomposites heat-treated at 560°C for 5–10 h exhibiting white light emission. The zoomed-in portion is displayed as inset showing different transitions of $Ho^{3+}$ ion. (B) CIE chromaticity diagram of samples CS-2-5 h and CS-2-10 h with excited sample photographs showing white light emission. Color coordinates of samples CS-2-5 h and CS-2-10 h are "*p*" (0.309, 0.326) and "*q*" (0.308, 0.338), respectively.
Reproduced with permission from C. Dey, M. Goswami, B. Karmakar, White light emitting $Ho^{3+}$-doped CdS nanocrystal ingrained glass nanocomposites, Appl. Phys. Lett. 106 (2015) 083106 (4 pages), Copyright © AIP.

respectively. The near visible 405 nm diode laser excited white light emission is produced by tuning the blue emission from the $Ho^{3+}$ ions, green band edge, and orange-red surface-state emissions of the nanocrystalline CdS in glasses, which are further controlled by the size of the CdS nanocrystals.

## 6.4 Processing and properties of IV–VI semiconductor incorporated glasses

PbS, PbSe, and PbTe are examples of IV–VI semiconductors. These semiconductor QDs in glasses demonstrate strong quantum confinement effect when the particle sizes are smaller than their bulk exciton Bohr radius. These QDs with discrete electron and hole states exhibit size-tunable optical transitions resulting in absorption and PL bands in the NIR wavelength region. Consequently, their functional applications are in biological labeling, light emitting diodes, solar cells, lasers, and telecommunication [23,24].

Savitski et al. [23] reported the PbS nanocrystallite size dependence NIR absorption properties in a phosphate glass of composition (mol.%) $56P_2O_5$–$25Na_2O$–5ZnO–$2AlF_3$–$12Ga_2O_3$. Fig. 6.12 illustrates the absorption spectra of phosphate glasses doped with PbS-QDs of diameter (#1) 4.7, (#2) 4.8, (#3) 5.1, (#4) 5.4, (#5) 5.6, and (#6) 5.8 nm. PbS having large bulk exciton Bohr radius ($a_B = 18$ nm) and narrow bandgap (0.41 eV) permits obtaining strong quantum confinement with relatively large crystallites, and to move the absorption peak resonance (first excitonic resonance) position in a wide spectral range from 1 to 3 μm by varying the size of QDs through the heat treatment of the synthesized glasses.

Xu and Heo [24] described the synthesis and properties of functional glasses of compositions (mol.%) $50SiO_2$–$35Na_2O$–10ZnO–$5Al_2O_3$ containing PbSe-QDs. Fig. 6.13 shows the (A) absorption and (B) PL spectra of this silicate glass containing

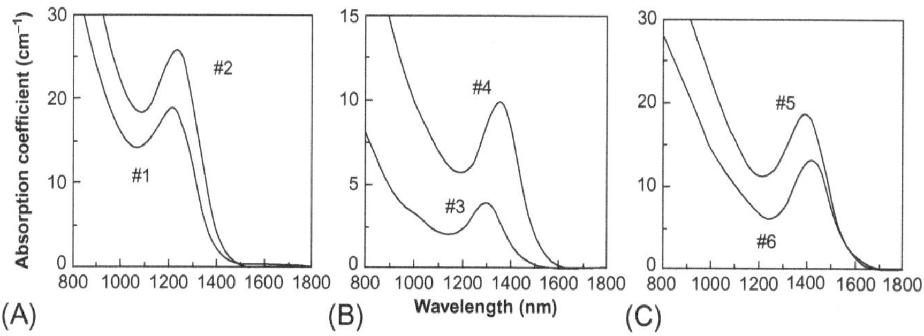


**Fig. 6.12** Absorption spectra of phosphate glasses doped with PbS-QDs of diameter: (A) (#1) 4.7 and (#2) 4.8 nm; (B) (#3) 5.1 and (#4) 5.4 nm; and (C) (#5) 5.6 and (#6) 5.8 nm. Reproduced with permission from V.G. Savitski, N.N. Posnov, P.V. Prokoshin, A.M. Malyarevich, K.V. Yumashev, M.I. Demchuk, A.A. Lipovskii, PbS-doped phosphate glasses saturable absorbers for 1.3-μm neodymium lasers, Appl. Phys. B 75, 841–846 (2002), Copyright © Springer.

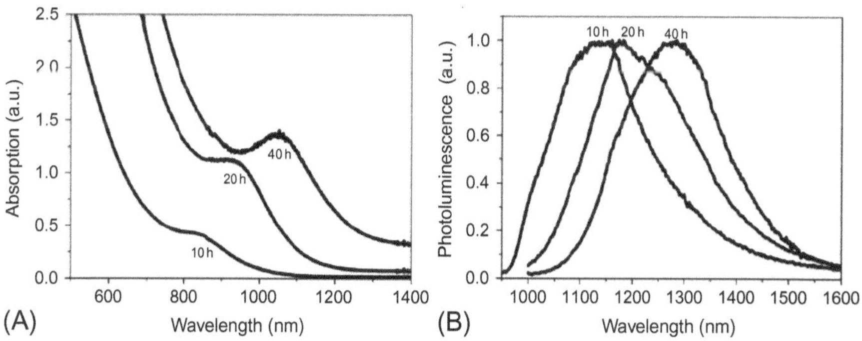


**Fig. 6.13** (A) Absorption and (B) photoluminescence spectra of silicate glass containing PbSe-QDs heat-treated at 450°C for 10–40 h.
Reproduced with permission from K. Xu, J. Heo, New functional glasses containing semiconductor quantum dots, Phys. Scr. T139 (2010) 014062 (4 pages), Copyright © IOP.

PbSe-QDs heat-treated at 450°C for 10–40 h. The broad absorption bands peaked at 846, 935, and 1048 nm for the glasses heat-treated at 450°C for 10, 20, and 40 h, respectively. Since the bandgap energy of PbSe is 0.27 eV (4593 nm), there is a large blue shift in these glasses; therefore, it clearly gives an idea about the strong quantum confinement effect. Photoluminescence spectra of the glasses also exhibit a similar trend with thermal treatment. The emission peaks shift from 1130 nm to 1200 and 1281 nm.

## 6.5 Functional applications of semiconductor incorporated glasses

It is understandable from the foregoing discussion that the optical properties of semiconductor nanoparticles are out of the ordinary trend when their sizes are comparable or lower than the exciton Bohr radius ($a_B$). These particles are called the quantum-dots and exhibit the quantum confinement effects. If the dot sizes are small enough, significant blue-energy shift of the absorption and emission band is found to be induced by the strong quantum confinement of excitons in the interior of the quantum-dots. This unique characteristic of the QDs is an asset for their technological applications in various devices of optoelectronic, nonlinear, photonic, biomedical, and photovoltaic systems [25]. Nanocrystalline compound semiconductors have applications in solar cells, light emitting diodes (LEDs), photo diodes, solid state lasers, and photo detectors [5]. Glass-embedded semiconductor nanocrystals are of interest in applications such as glass filters, active media for solid-state lasers, units for the second harmonic generation, memory elements in new generation computers, optoelectronics, and nonlinear optic devices. The ability to produce rare-earth-doped semiconductor incorporated glasses offering IR and visible luminescence, excitable by different means, opens up the possibility of fabricating next-generation light emitting diodes, solar cells, solid-state lasers, detectors, phosphors, photonic, and biophotonic devices,

## 6.6 Summary

Functionalization of glasses by incorporation of different types of semiconductors was described. Predominantly the II–VI (e.g., ZnS, ZnSe, ZnTe, CdS, CdSe, CdTe) and IV–VI (e.g., PbS, PbSe, PbTe) semiconductors incorporated silicate, borosilicate, phosphate, and fluorophosphate glasses were addressed. Their processing protocol was explained mainly with reference to the melt quenching technique followed by heat treatment. Sometimes sol-gel technique was also referenced. Optical absorption and PL properties were described in detail. In addition to this, Raman spectroscopy and transmission electron microscopy (TEM) were also discussed. Processing and properties of rare-earth doped II–VI semiconductor incorporated glasses were described elaborately including white light emission and application in LEDs. Various functional applications of semiconductor incorporated glasses and exercises were provided.

## Exercises

**6.1** Describe the band theory and bandgap. What is the significance of bandgap?

**6.2** Define semiconductor and describe different types of semiconductors with examples.

**6.3** Explain the terms with example: QD, quantum confinement, exciton Bohr radius, and blue-shifting.
**6.4** Describe one method of preparation of semiconductor incorporated glasses.
**6.5** Describe absorption properties of $CdS_xSe_{1-x}$ semiconductor incorporated glasses with reference to heat-treatment temperature and time.
**6.6** Describe the growth of CdSe nanocrystal size in a fluorophosphate glass as a function of heat-treatment temperature and time.
**6.7** Write a brief account on the processing and absorption spectra of sharp cut-off glass filters.
**6.8** Describe major photo-physical processes of nano-sized semiconductors excited by light with energy equal to or higher than the bandgap energy.
**6.9** What are the effects of rare-earth ions on semiconductor incorporated glasses? Describe them with various examples.
**6.10** How do you prepare and characterize white light emitting rare-earth doped semiconductor incorporated glass?
**6.11** Describe processing and absorption properties of PbS incorporated fluorophosphate glass.
**6.12** Describe processing with absorption and PL spectra of PbSe incorporated silicate glass.
**6.13** Illustrate various functional applications of semiconductor incorporated glasses with reference to their properties.

# References

[1] R. Reisfeld, Nanosized semiconductor particles in glasses prepared by the sol–gel method: their optical properties and potential uses, J. Alloys Compd. 341 (2002) 56–61.
[2] A.F.G. Monte, Semiconductor doped glasses, J. Ceram. Soc. Jpn. 116 (2008) 1033–1039.
[3] A.A. Lipovskii, E.V. Kolobkova, A. Olkhovets, V.D. Petrikov, F. Wise, Synthesis of monodisperse PbS quantum dots in phosphate glass, Physica E 5 (2000) 157–160.
[4] O. Madelung, Semiconductors—Basic Data, second ed., Springer-Verlag, Berlin, 1996.
[5] B.C. Ömür, A. Aşkoğlu, Ç. Allahverdi, M.H. Yükselici, $CdS_xSe_{1-x}$ quantum dots studied through optical absorption, steady-state photoluminescence and resonant Raman spectroscopies, J. Mater. Sci. 45 (2010) 112–116.
[6] E.V. Kolobkova, Specific features of growth of CdSe nanocrystals in fluorophosphate glasses, Glass Phys. Chem. 32 (2006) 404–411.
[7] A.A. Lipovskii, E.V. Kolobkova, V.D. Petrikov, Growth and optical properties of $CdS_xSe_{1-x}$ nanocrystals embedded in a novelphosphate glass, J. Mater. Sci. 34 (1999) 887–890.
[8] T. Miyoshi, T. Nakatsuka, N. Matsuo, Determination of composition of $CdS_xSe_{1-x}$ microcrystals in semiconductor-doped glasses using Raman scattering, Jpn. J. Appl. Phys. 34 (1995) 1835–1836.
[9] Hoya Corporation USA, Santa Clara, USA. Web: *www.hoyaoptics.com*.
[10] S. Ciolek, D. Krashkevich, UV attenuated colored glasses, US patent 5,059,561, 1991.
[11] H.-C. Lee, Moving disks made of semiconductor nanocrystallite embedded glass, US patent 5,308,804, 1994.
[12] V.V. Ushakova, A.S. Aronin, V.A. Karavanskii, A.A. Gippius, Formation and optical properties of CdSSe nanocrystals in silicate glass matrix, Phys. Solid State 51 (2009) 2161–2165.
[13] J. Liqiang, Q. Yichun, W. Baiqi, L. Shudan, J. Baojiang, Y. Libin, F. Wei, F. Honggang, S. Jiazhong, Review of photoluminescence performance of nano-sized semiconductor

materials and its relationships with photocatalytic activity, Sol. Energy Mater. Sol. Cells 90 (2006) 1773–1786.

[14] C. Dey, M. Goswami, B. Karmakar, CdSe nanocrystals ingrained dielectric nanocomposites: synthesis and photoluminescence properties, Mater. Res. Express 2 (2015) 015014 (12 pages).

[15] C. Dey, M. Goswami, B. Karmakar, Structural and optical properties of ZnSe quantum dots in glass nanocomposites, Mater. Chem. Phys. 163 (2015) 554–561.

[16] A.R. Molla, A. Tarafder, C. Dey, B. Karmakar, Synthesis and properties of ZnTe and $Eu^{3+}$ ion codoped glass nanocomposites, J. Appl. Phys. 116 (2014) 163510 (13 pages).

[17] G. Jose, K.A. Amrutha, T.F. Toney, V. Thomas, C. Joseph, M.A. Ittyachen, N. V. Unnikrishnan, Structural and optical characterization of $Eu^{3+}$/CdSe nanocrystal containing silica glass, Mater. Chem. Phys. 96 (2006) 381–386.

[18] P.V. Jyothy, K.A. Amrutha, J. Gijo, N.V. Unnikrishnan, Fluorescence enhancement in $Tb^{3+}$/CdS nanoparticles doped silica xerogels, J. Fluoresc. 19 (2009) 165–168.

[19] M. Nogami, T. Enomoto, T. Hayakawa, Enhanced fluorescence of $Eu^{3+}$ induced by energy transfer from nanosized $SnO_2$ crystals in silica glass, J. Lumin. 97 (2002) 147–152.

[20] V. Gueu, H. You, T. Hayakawa, M. Nogami, $Eu^{3+}$-fluorescence properties in nanocrystallized $SnO_2$-$SiO_2$ glass-ceramics, J. Sol-Gel Sci. Technol. 41 (2007) 231–236.

[21] J. del-Castillo, V.D. Rodríguez, A.C. Yanes, J. Méndez-Ramos, Energy transfer from the host to $Er^{3+}$ dopants in semiconductor $SnO_2$ nanocrystals segregated in sol-gel silica glasses, J. Nanopart. Res. 10 (2008) 499–506.

[22] C. Dey, M. Goswami, B. Karmakar, White light emitting $Ho^{3+}$-doped CdS nanocrystal ingrained glass nanocomposites, Appl. Phys. Lett. 106 (2015) 083106 (4 pages).

[23] V.G. Savitski, N.N. Posnov, P.V. Prokoshin, A.M. Malyarevich, K.V. Yumashev, M. I. Demchuk, A.A. Lipovskii, PbS-doped phosphate glasses saturable absorbers for 1.3-μm neodymium lasers, Appl. Phys. B 75 (2002) 841–846.

[24] K. Xu, J. Heo, New functional glasses containing semiconductor quantum dots, Phys. Scr. T139 (2010) 014062 (4 pages).

[25] M.H. Yükselicia, Ç. Allahverdia, H. Athalin, Zinc incorporation into CdTe quantum dots in glass, Mater. Chem. Phys. 119 (2010) 218–221.

# Glasses and glass-ceramics for biomedical applications

# 7

## Chapter Outline



## 7.1 Introduction

The human body and its various organs are supported and protected by 206 (in adulthood) bones of the skeleton. The human skeleton and some of its constituent bones are illustrated in Fig. 7.1. The inorganic chemical composition of the bone is primarily hydroxyapatite ($Ca_{10}(PO_4)_6(OH)_2$). It is well-known that from the age of 30 years onward, all (hard and soft) the tissues gradually deteriorate. Other reasons of loss of these tissues are damages caused by diseases or accidents. Unfortunately, when we were born, we do not have any "*spare parts*" for its maintenance and service. Thus, we require "*spare parts*" of teeth, hips, knees, ears, etc. to repair, replace, and restore their functions to achieve lengthened life span and improved quality of life. Multidisciplinary researchers (e.g., materials scientists, ceramists, chemists, engineers, dentists, and surgeons) are trying to find out the solutions of these demands for a long time. Some glasses and glass-ceramics of appropriate compositions after suitable processing are found to be functionalized as befitting spares for several hard and soft tissues. Consequently, several commercial glass and glass-ceramic products emerged

**Functional Glasses and Glass-Ceramics. http://dx.doi.org/10.1016/B978-0-12-805056-9.00007-6**

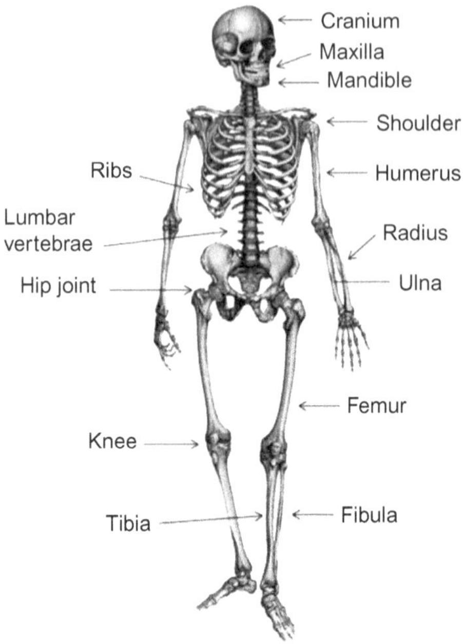


**Fig. 7.1** The human skeleton and some of its important bones.

as bone substitutes in the human skeleton, for example, teeth, middle ear, vertebrae (cervical, thoracic, and dorsal), maxilla, knee, shoulder, and fingers. In this chapter, we shall describe various aspects of these biomedical glasses and glass-ceramics.

An attempt was made using the Web of Science (version 5.17) as described in Chapter 1 to provide a global publication trend and quantitative growth in the area of bioactive glasses and glass-ceramics. Here, year-wise data were extracted for the last 25 years (1990–2014) by searching, using the topic keyword "*glass*" followed by refining the document type as "*article*" published in the journals and meetings. It was further refined using the term "*bioactive glass*." This plot is shown in Fig. 7.2. It gives us a clear publication trend and quantitative growth of published papers on bioactive glasses and glass-ceramics during the last 25 years (1990–2014). It is seen that there is a slow growth of the published papers during the period 1990–99 compared with the rapid growth during the period 2000–14. The rapid growth of the published papers during last 15 years clearly signifies the importance of this topic.

## 7.2 History of biomedical glasses and glass-ceramics

In the beginning, bioinert materials such as ceramics, metals, and organic polymers are used as the alternative artificial materials to repair or replace the damaged bones. These artificial materials implanted in place of damaged bones are encapsulated by fibrous tissues and do not form bond to the living bones. Consequently, tissue

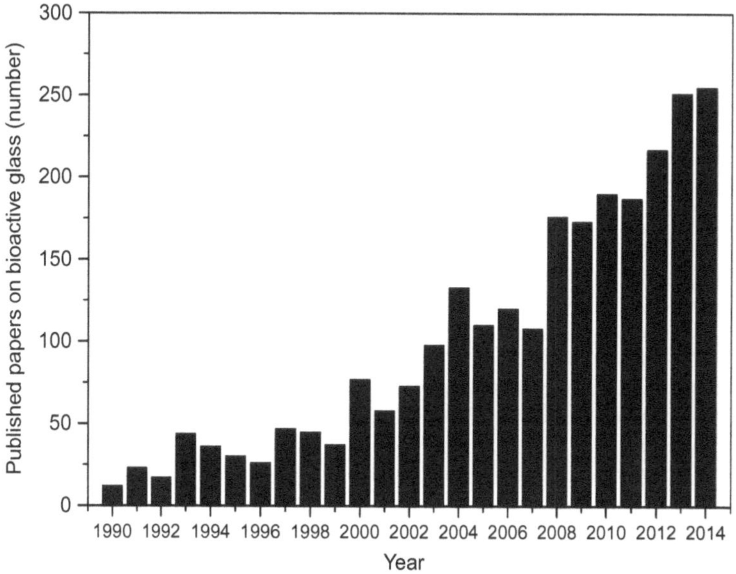


**Fig. 7.2** Year-wise published papers on bioactive glasses during the last 25 years (1990–2014).

breakdown and loosening over time is a common mode of failure of the bioinert implants. For this reason, bioactive materials received major attention to solve this problem since these materials directly create bonds with the living bones after implantation in the physiological environment. It is noteworthy that bioactivity is defined as the property of a material that permits it to form bond to the living tissue. Guided by this hypothesis, Hench and coworkers [1] discovered bone bonding to bioactive 45S5 Bioglass based on the $Na_2O$–$CaO$–$P_2O_5$–$SiO_2$ glass system in 1969, which was published in 1971. Some chronological developmental events of biomedical glasses and glass-ceramics are listed in Table 7.1. In 1971, Wilson and Kent [2] first published the report on glass ionomer cement (GIC) as an acid-base reaction product of basic fluoroaluminosilicate glass powder and aqueous solution of polycarboxylic acid, which was essentially developed in 1967. In 1972, Greenlee Jr. et al. [3] first reported bone bonding to bioactive glass-ceramic implant containing 45% $SiO_2$ with a Ca/P ratio of 5. In 1976, Griss et al. [4] reported bone bonding of Bioglass-coated fully dense alumina total hip implants in sheep. In 1981, Wilson et al. [5] discovered bonding of soft connective tissues of rats and rabbits to 45S5 Bioglass.

In 1982, Kokubo et al. [6] first reported the tight bone bonding in rabbits of bioactive apatite-wollastonite (A-W) glass-ceramic implant developed in the $MgO$–$CaO$–$SiO_2$–$P_2O_5$–$CaF_2$ system. Höland et al. [7] developed machinable bioactive apatite-fluorophlogopite glass-ceramics (Bioverit) in $SiO_2$–$Al_2O_3$–$MgO$–$Na_2O$–$K_2O$–$F$–$CaO$–$P_2O_5$ system in 1983. In 1995, Höland et al. [8] developed leucite ($KAlSi_2O_6$) glass-ceramics in the $SiO_2$–$Al_2O_3$–$K_2O$–$Na_2O$–$B_2O_3$–$CeO_2$–$CaO$–$BaO$–$TiO_2$ system for restorative dental applications.

**Table 7.1 Some chronological developmental events of biomedical glasses and glass-ceramics**

| Event | Year | Author | Ref. |
|---|---|---|---|
| Discovery of bone bonding to bioactive 45S5 Bioglass based on $Na_2O$–$CaO$–$P_2O_5$–$SiO_2$ system in 1969, which was published in 1971 | 1971 | Hench et al. | [1] |
| First report on glass ionomer cement (GIC) as an acid-base reaction product of basic fluoroaluminosilicate glass powder and aqueous solution of polycarboxylic acid, which was essentially developed in 1967 | 1971 | Wilson and Kent | [2] |
| First report of bone bonding to bioactive glass-ceramic implant containing 45% $SiO_2$ with a Ca/P ratio of 5 | 1972 | Greenlee Jr. et al. | [3] |
| Bone bonding of Bioglass-coated fully dense alumina total hip implants in sheep | 1976 | Griss et al. | [4] |
| Discovery of bonding of soft connective tissues of rats and rabbits to 45S5 Bioglass | 1981 | Wilson et al. | [5] |
| First report of tight bone bonding in rabbits of bioactive apatite-wollastonite (A-W) glass-ceramic implant developed in the $MgO$–$CaO$–$SiO_2$–$P_2O_5$–$CaF_2$ system | 1982 | Kokubo et al. | [6] |
| Developed machinable bioactive apatite-fluorophlogopite glass-ceramics (Bioverit) in $SiO_2$–$Al_2O_3$–$MgO$–$Na_2O$–$K_2O$–$F$–$CaO$–$P_2O_5$ system | 1983 | Höland et al. | [7] |
| Developed leucite ($KAlSi_2O_6$) glass-ceramics in the $SiO_2$–$Al_2O_3$–$K_2O$–$Na_2O$–$B_2O_3$–$CeO_2$–$CaO$–$BaO$–$TiO_2$ system for restorative dental applications | 1995 | Höland et al. | [8] |
| Developed antibacterial $Ag_2O$-doped bioactive glasses by sol-gel technique | 2000 | Bellantone et al. | [9] |
| Enhanced osteogenesis with ionic dissolution products of Bioglass 45S5 in gene-expression profiling of human osteoblasts | 2001 | Xynos et al. | [10] |
| Groups led by Hench and Boccaccini simultaneously developed bioactive glass derived glass-ceramic scaffolds for bone tissue engineering | 2006 | Jones et al. and Chen et al. | [11,12] |
| A group led by Inoue developed new TiZrCuPd quaternary bulk metallic glass for biomedical applications | 2007 | Zhu et al. | [13] |
| A group led by Xiao developed mesoporous bioactive glasses for drug delivery and bone tissue engineering | 2011 | Wu et al. | [14] |

In 2000, Bellantone et al. [9] developed antibacterial $Ag_2O$-doped bioactive glasses by sol-gel technique. Xynos et al. [10] discovered enhanced osteogenesis with ionic dissolution products of Bioglass 45S5 in gene-expression profiling of human osteoblasts in 2001. In 2006, groups led by Hench [11] and Boccaccini [12] simultaneously developed bioactive glass-derived glass-ceramic scaffolds for bone tissue engineering. A group led by Inoue developed new TiZrCuPd quaternary bulk metallic glass for biomedical applications in 2007 [13]. In 2011, a group led by Xiao developed mesoporous bioactive glasses for drug delivery and bone tissue engineering [14].

Of late, Hench [15] has classified the development of biomedical materials into four generations. The first generation biomaterials used for tissue replacement were the "*bioinert*" materials where there was no chemical bonding between the implants and the existing living bones. The second generation biomaterials were the "*bioactive*" glasses and glass-ceramics that form the chemical bonding between the living tissues and the implant surfaces. The third generation biomaterials are the "*bioactive resorbable glasses*" used for tissue regeneration. The fourth generation biomaterials are the "*bioactive glass particulates*" used for the prevention of oral disease and tissue damage.

In addition to the above, very recently, Hench [16,17] has divided the discovery and development of bioactive glasses during the period from 1969 to 2014 into four eras such as the following:

(i) *Era of discovery (1969–79)*: The major discovery in this era by Hench, Splinter, Allen, and Greenlee [1] was some glass compositions in the $Na_2O$–$CaO$–$P_2O_5$–$SiO_2$ system that forms strong chemical bond with the living tissues creating hydroxyl carbonate apatite (HCA) crystals.
(ii) *Era of clinical application (1980–95)*: The major discovery in this era is bonding of Bioglass to soft tissues by Wilson et al. [5], which opens up clinical applications of bioactive glasses.
(iii) *Era of tissue regeneration (1985–05)*: The major discovery in this era is the use of bioactive glass particulates for regeneration of bones [18,19].
(iv) *Era of innovation (2000–20)*: The major discovery in this era until now is the development of bioactive glass-derived glass-ceramic scaffolds for bone tissue engineering [11,12].

## 7.3 Different types of biomedical glasses and glass-ceramics

Various kinds of biomedical glasses and glass-ceramics have been developed. Nowadays, they become an integral and vital part of our current health care supply system. They are shown in Fig. 7.3. These glasses and glass-ceramics are briefly described in the subsequent sections.

### 7.3.1 Bioactive glasses

The "bioinert" materials generate a nonadherent fibrous capsule of variable thickness at the interface of the implant materials and the host tissue, which results in interfacial mobility. In order to suppress this interfacial mobility of the implants, in 1969, Hench,

Fig. 7.3 Various types of biomedical glasses and glass-ceramics.

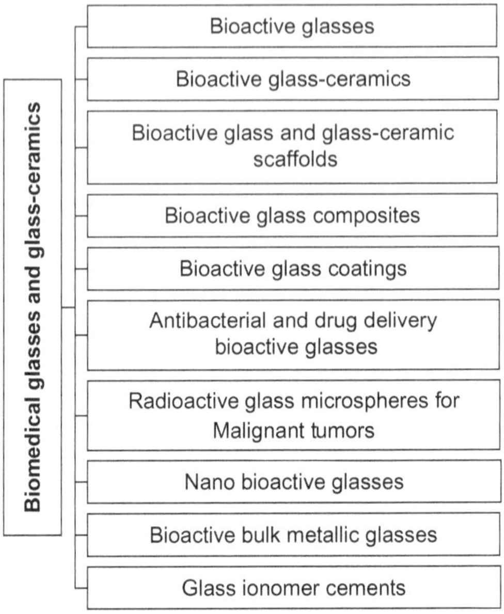


Splinter, Allen, and Greenlee [1] discovered several glass compositions of the $Na_2O$–CaO–$P_2O_5$–$SiO_2$ system, which create strong, adherent bonds to bone. This is the birth of "bioactive" glass that implicates the interaction with the physiological environment toward chemical bonding between the living tissue and the implant surface; thus, it surmounts interfacial mobility. Compositions of some commercial bioactive glasses are provided in Table 7.2 [20]. Compositional dependence of bonding of bone and soft tissue to bioactive glasses is illustrated in Fig. 7.4 [16]. From these compositions of bioactive glasses, it is flawless that the presence of both CaO and $P_2O_5$ is the primary condition for bonding with living bone made up of hydroxyapatite ($Ca_{10}(PO_4)_6(OH)_2$). The Bioglass is designated by 45S5 to signify the weight percent (wt.%) of glass former $SiO_2$ (S) and a fivefold ratio of Ca/P. In Fig. 7.4, region I is for class B (slow rate of bonding) bone-bonding glass compositions comprising of 44–60 wt.% of $SiO_2$, region II is for nonbonding (reactivity is too low), region III is nonbonding (reactivity is too high), region IV is for nonglass forming, and region

Table 7.2 **Composition of some commercial bioactive glasses [20]**

| Bioactive glass | Composition (wt.%) | | | |
|---|---|---|---|---|
| | $Na_2O$ | CaO | $P_2O_5$ | $SiO_2$ |
| 45S5 Bioglass (NovaBone) | 24.5 | 24.5 | 6.0 | 45.0 |
| S53P4 (AbminDent1) | 23.0 | 20.0 | 4.0 | 53.0 |

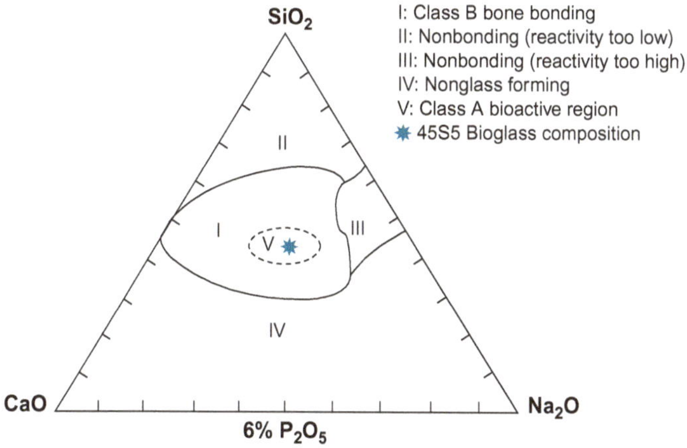


**Fig. 7.4** Compositional (wt.%) dependence of bonding of bone and soft tissue to bioactive glasses.
Courtesy: L.L. Hench, Opening paper 2015—Some comments on Bioglass: four eras of discovery and development, Biomed. Glasses 1 (2015) 1–11, Copyright © De Gruyter Open.

V is for class A bioactive glass compositions including 45S5 Bioglass where bone and soft tissue bonding is possible at 52 wt.% of $SiO_2$.

The ability of bioactivity of a material is examined by "in vitro" tests determining the hydroxyl carbonate apatite (HCA) layer formation on the surface of material soaked in the simulated body fluid (SBF, also called Kokubo solution) solution at 37°C [21,22]. Composition and concentration (mM) of ions in simulated body fluid (SBF) in aqueous solution are very close to that of the human blood plasma. This is provided in Table 7.3 [21]. The pH and buffer compositions of the SBF are 7.25 and $(CH_2OH)_3CNH_2$ 50 mM plus HCI 45 mM, respectively. The progress of HCA layer formation with time of immersion in SBF solution is monitored by Fourier transform infrared reflection spectroscopy (FTIRRS) study. The HCA layer formation on the surface of a bioactive glass continues through five stages as described below [23,24].

*Stage I*: Leaching through the rapid ion exchange of protons ($H^+$ or $H_3O^+$) of SBF in aqueous solution with easily displaceable network modifying ions (e.g., $Na^+$ and $K^+$).

**Table 7.3 Composition and concentration (mM) of ions in simulated body fluid (SBF) in aqueous solution and human blood plasma [21]**

| Ion (mM) | $Na^+$ | $K^+$ | $Mg^{2+}$ | $Ca^{2+}$ | $Cl^-$ | $HCO_3^-$ | $HPO_4^{2-}$ | $SO_4^{2-}$ |
|---|---|---|---|---|---|---|---|---|
| Simulated body fluid | 142.0 | 5.0 | 1.5 | 2.5 | 148.8 | 4.2 | 1.0 | 0.5 |
| Human blood plasma | 142.0 | 5.0 | 1.5 | 2.5 | 103.0 | 27.0 | 1.0 | 0.5 |

$$\equiv Si-O-Na^{+}+H^{+}+OH^{-}\rightarrow \equiv Si-OH+Na^{+}(solution)+OH^{-} \quad (7.1)$$

In this stage, the reaction rate and the formation of silanol groups ($\equiv$Si—OH) at the interface of bioactive glass/SBF solution are diffusion controlled. The ion exchange process increases hydroxyl ion ($OH^-$) concentration resulting in the rise in pH up to 10.5.

*Stage II*: The rise in pH facilities the dissolution of silica network ($\equiv$Si—O—Si$\equiv$) forming additional silanol ($\equiv$Si—OH) groups according to the following reaction and loss of soluble silica in the form of $Si(OH)_4$ that passes into the solution. Hence, this stage is controlled by the interfacial reactions.

$$\equiv Si-O-Si\equiv+H_2O\rightarrow 2[\equiv Si-OH] \quad (7.2)$$

*Stage III*: Polymerization of the $SiO_2$-rich layer through condensation of the neighboring silanol ($\equiv$Si—OH) groups as shown below, which produces an amorphous silica-rich layer on the surface.

$$\equiv Si-OH+HO-Si\equiv\rightarrow \equiv Si-O-Si\equiv+H_2O \quad (7.3)$$

*Stage IV*: Migration of $Ca^{2+}$ and $PO_4^{3-}$ ions to the surface through the $SiO_2$-rich layer forming a film rich in CaO and $P_2O_5$ on the top of the $SiO_2$-rich layer, followed by the thickening of the amorphous CaO–$P_2O_5$-rich film by incorporation of soluble $Ca^{2+}$ and $PO_4^{3-}$ ions from the solution.

*Stage V*: Crystallization of the amorphous CaO–$P_2O_5$ film takes place by incorporation of $OH^-$, $CO_3^{2-}$, $F^-$, or $Cl^-$ ions from the solution, to form mixed layers of the hydroxyl carbonate apatite (HCA), $Ca_{10}(PO_4,CO_3,OH)_6(OH)_2$; hydroxyl carbonate fluorapatite (HCFA), $Ca_{10}(PO_4,CO_3,OH)_6F_2$; or hydroxyl carbonate chlorapatite (HCCA), $Ca_{10}(PO_4,CO_3,OH)_6Cl_2$.

The reaction sequence as stated in stages I–V is monitored by the Fourier transform infrared reflection spectroscopy (FTIRRS). For example, the FTIR reflection spectra of a highly bioactive glass (wt.%) 47.5$SiO_2$–23.2$Na_2O$–23.2CaO–6.0$P_2O_5$ after immersion in SBF solution for 0–53 h are depicted in Fig. 7.5 [22]. Two noticeable features of these spectra are (i) the original peaks at 1050, 910, and 490 $cm^{-1}$ of the unreacted glass are decomposed into several narrow peaks and (ii) a new peak appears at 575 $cm^{-1}$. These are related with the various chemical changes as stated in stages I–V. The formation of the $SiO_2$-rich layer in stage III is confirmed by two Si—O—Si stretching peaks at 1250 and 1095 $cm^{-1}$ and by sharpening of the Si—O—Si bending peaks at 470 $cm^{-1}$. The new peak developed at 780 $cm^{-1}$ is attributed to Si—O—Si tetrahedral vibration between two neighboring $SiO_4$ tetrahedrons. As a final point, the band around 930 $cm^{-1}$ appeared due to Si—O stretching vibration, of the non-bridging oxygen (NBO) group, disappeared with repolymerization in stage III. Formation of amorphous CaO–$P_2O_5$-rich film can be observed by appearance of a broad peak at 560 $cm^{-1}$. As the reaction time increases, initially, this peak sharpens and finally divided into two peaks that appeared at 602 and 560 $cm^{-1}$. This is a characteristic of crystalline apatite. Since $CO_3^-$ is incorporated from the solution producing a broad peak at 870 $cm^{-1}$ due to C—O stretching vibration and the P—O stretching vibration peak at 1050 $cm^{-1}$ after 9 h, the crystallized phase is hydroxyl carbonate apatite (HCA), $Ca_{10}(PO_4,CO_3,OH)_6(OH)_2$ [23,24].

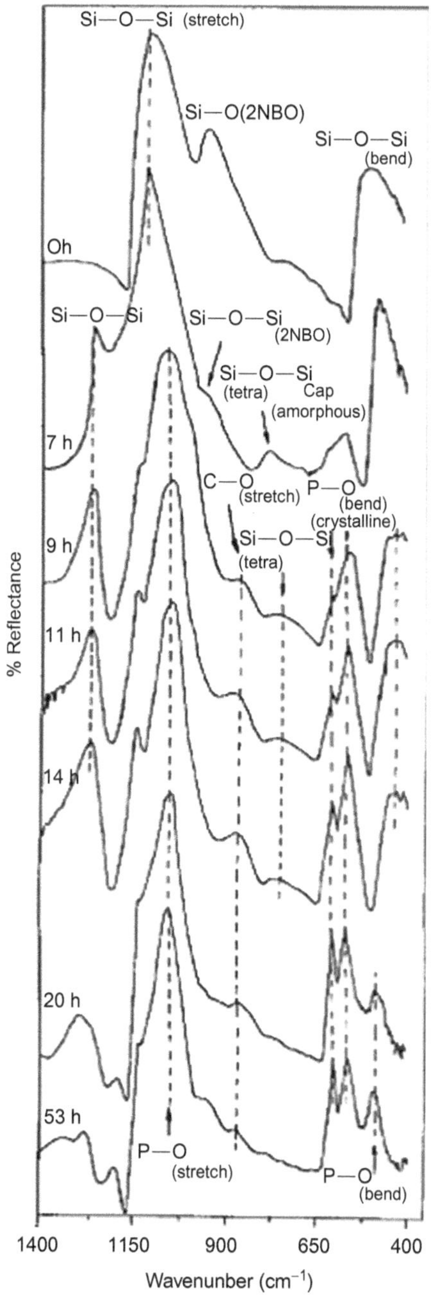


**Fig. 7.5** FTIR reflection spectra of a highly bioactive glass after immersion in SBF solution for 0–53 h.

Reproduced with permission from O. Peitl, E.D. Zanotto, L.L. Hench, Highly bioactive $P_2O_5$–$Na_2O$–CaO–$SiO_2$ glass-ceramics, J. Non-Cryst. Solids 292 (2001) 115–126, Copyright © Elsevier.

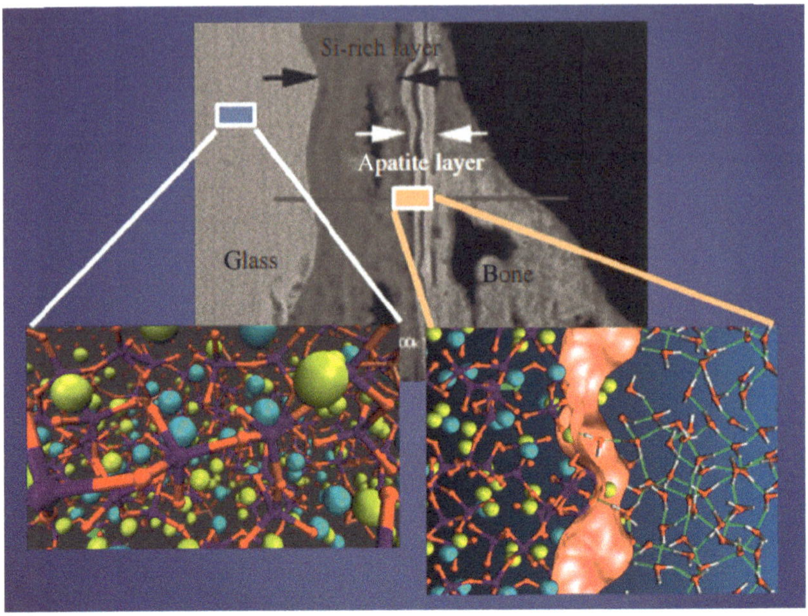


**Fig. 7.6** SEM photomicrograph (top) of bioglass-bone interface with atomistic models of bioactive glass (bottom left) and interface (flesh color) with water (bottom right). Reproduced with permission from A.N. Cormack, A. Tilocca, Structure and biological activity of glasses and ceramics, Philos. Trans. R. Soc. A (2012) 370, 1271–1280, Copyright © The Royal Society.

Fig. 7.6 shows the SEM photomicrograph (top) of bioglass-bone interface with atomistic models of bioactive glass (bottom left) and interface (flesh color) with water (bottom right) [25–28]. Recent molecular dynamics (MD) simulation studies have revealed the first atomistic views of the hydrated [27] and dry [28] Bioglass surface. Fig. 7.7A depicts the SEM photomicrograph of particulate $CaO–P_2O_5–TiO_2–Na_2O$ bioactive glass-ceramics after 7 days of soaking in SBF solution [29]. It is seen that numerous bone-like apatite particles and layers are formed covering almost the entire surface of the bioactive glass-ceramic particle. This rapid apatite particle and layer formation indicates that the bioactive glasses and glass-ceramics particulates can be used as nucleus for quick repairing of bone defects or damaged bones. Fig. 7.7B represents the photograph showing the bone tissue (stained in red) formation in an excavated bioactive glass particle [30].

Fine particles of bioactive glasses are also used for oral care for treatment of hypersensitivity in dentistry. Fig. 7.8 shows the SEM photomicrographs of human dentine (bar = 1 μm) (A) untreated and (B) 5 days after application of a toothpaste (Sensodyne with NovaMin) containing fine particulates of 45S5 Bioglass [31]. Comparing Fig. 7.8A with Fig. 7.8B, it is seen that there is a good signature of healing of dentine defects after 5 days of application of toothpaste containing fine particulates of 45S5 Bioglass, which is caused by stimulating mineralization (calcium phosphate deposition)

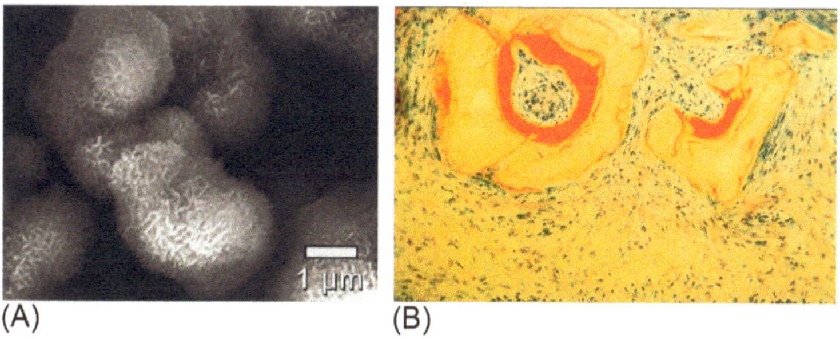


**Fig. 7.7** (A) SEM photomicrograph of particulate $CaO–P_2O_5–TiO_2–Na_2O$ bioactive glass-ceramics after 7 days of soaking in SBF solution, (B) Photograph showing the bone tissue (stained in red) formation in excavated bioactive glass particles.
(A) Reproduced with permission from T. Kasuga, Bioactive calcium pyrophosphate glasses and glass-ceramics, Acta Biomater. 1 (2005) 55–64, Copyright © Elsevier. (B) Reproduced with permission from P. Ducheyne, Q. Qiu, Bioactive ceramics: the effect of surface reactivity on bone formation and bone cell function, Biomaterials 20 (1999) 2287–2303, Copyright © Elsevier.

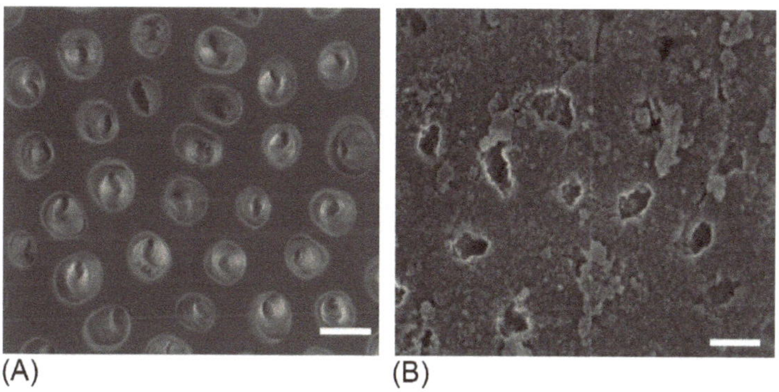


**Fig. 7.8** SEM photomicrographs of human dentine (bar = 1 μm) (A) untreated and (B) 5 days after application of a toothpaste (Sensodyne with NovaMin) containing fine particulates of 45S5 Bioglass.
Reproduced with permission from J.R. Jones, Review of bioactive glass: from Hench to hybrids, Acta Biomater. 9 (2013) 4457–4486, Copyright © Elsevier.

over the dentine tubules. It is also reported that the Bioglass 45S5 particle polishing results in around 44% reduction in tooth sensitivity and makes teeth brighter [31].

### *7.3.2 Bioactive glass-ceramics*

Glass-ceramics comprise nano- and/or micrograined polycrystalline phases embedded in residual glass phase, which are prepared by controlled crystallization (devitrification or ceramization) of a precursor glass. They are prepared by the well-controlled

Table 7.4 **Composition and crystal phase of some commercial bioactive glass-ceramics [24]**

| Component (wt.%) | Cerabone A-W | Bioverit | Ceravital |
|---|---|---|---|
| $Na_2O$ | 0 | 3–8 | 5–10 |
| $K_2O$ | 0 | 0 | 0.5–3.0 |
| MgO | 4.6 | 2–21 | 2.5–5.0 |
| CaO | 44.7 | 10–34 | 30–35 |
| $Al_2O_3$ | 0 | 8–15 | 0 |
| $SiO_2$ | 34.0 | 19–54 | 40–50 |
| $P_2O_5$ | 6.2 | 2–10 | 10–50 |
| $CaF_2$ | 0.5 | 3–23 | 0 |
| Phases | Apatite, β-wollastonite, and glass | Apatite, fluorophlogopite mica, and glass | Apatite and glass |

heat treatment program following the progressions of nucleation and crystal growth within the glass matrix. Bioactive glass-ceramics are those types of glass-ceramics that have the ability to form chemical bond with bone. The general characteristics of these type of glass-ceramics are that they contain CaO and $P_2O_5$ in their chemical composition, and the glass-ceramics contain apatite $Ca_{10}(PO_4)_6(F,Cl,OH)_2$ as one of the crystal phases.

Composition and crystal phases of some commercial bioactive glass-ceramics are provided in Table 7.4 [24].

The Cerabone A-W glass-ceramics contains mixed fluorapatite, $Ca_{10}(PO_4)_6(O,F_2)$, and β-wollastonite (β-$CaSiO_3$) crystal phases along with some residual glass. This bioactive glass-ceramics was developed by Kokubo et al. [6] for prosthetic applications in 1982 and has high mechanical strengths. The Young's modulus, Vickers hardness, bending strength, and compressive strength of the glass-ceramics are 103 GPa, $7.16 \pm 0.84$ GPa, $157 \pm 8$ MPa, and $1060 \pm 60$ MPa, respectively. This glass-ceramic possesses good compatibility with the bone and forms strong chemical bond with bones. The Bioverit bioactive glass-ceramics was developed in 1983 by Höland and coworkers [7] in the $SiO_2$–$Al_2O_3$–MgO–$Na_2O$–$K_2O$–F–CaO–$P_2O_5$ glass system. It contains mixed fluorapatite, $Ca_{10}(PO_4)_6(O,F_2)$, and fluorophlogopite mica, ($Na/KMg_3(AlSi_3)O_{10}F_2$), crystal phases along with some residual glass. As a result of the presence of interlocked mica in this glass-ceramics, it can be machined with standard tools and even renovated in the operation theater. The Ceravital bioactive glass-ceramics was developed in 1973 by Brömer and coworkers [32]. It contains apatite crystals and some residual glass. Perhaps, this is the earliest bioactive glass-ceramics used clinically. Its flexural and compressive strengths are 100–150 and 500 MPa, respectively.

The bioactive leucite-fluorapatite glass-ceramics was developed in the $SiO_2$–$Al_2O_3$–$Na_2O$–$K_2O$–CaO–$ZrO_2$–$TiO_2$–$P_2O_5$–$Li_2O$–$B_2O_3$–$CeO_2$–F glass system for

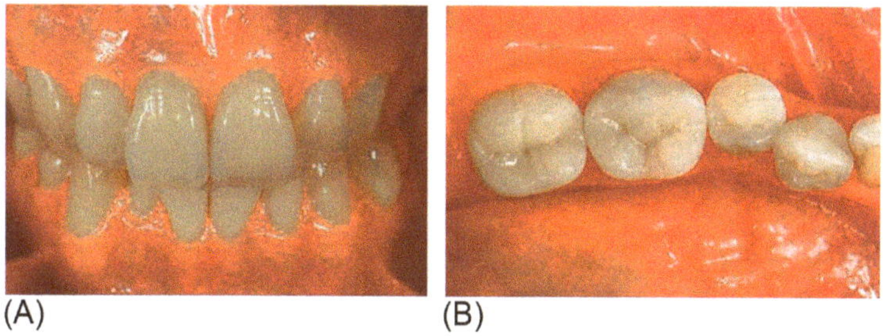

Fig. 7.9 (A, B) Photographs showing clinical application of bioactive leucite-fluorapatite glass-ceramics in dentistry.
Reproduced with permission from W. Höland, V. Rheinberger, E. Apel, C. van't Hoen, M. Höland, A. Dommann, M. Obrecht, C. Mauth, U. Graf-Hausner, Clinical applications of glass-ceramics in dentistry, J. Mater. Sci. Mater. Med. 17 (2006) 1037–1042, Copyright © Springer.

clinical applications in dentistry [33]. This is illustrated in Fig. 7.9. This clinical application of this glass-ceramic has been proved to be suitable for veneering material on metal frameworks for single unit and for large dental bridges involving more than three units.

## 7.3.3 Bioactive glass and glass-ceramic scaffolds

The bioactive glass and glass-ceramic scaffolds are used for tissue engineering that has emerged as a promising approach for repair and regeneration of tissues and organs lost or damaged as a result of trauma, injury, disease, or aging [34]. Thus, the scaffolds can surmount the problem of deficiency of living tissues and organs required for transplantation. The general purpose of tissue engineering based on scaffolds is to provide the temporary support structure for the tissue forming cells to synthesize new tissues of the desired shape and dimensions. Generally, scaffolds are highly porous and three dimensional (3D) templates possessing tailored pore size, porosity, and controlled interconnectivity. The properties of an ideal scaffold are as follows: (i) it is biocompatible (nontoxic), promoting cell adhesion and proliferation; (ii) it has matching mechanical properties of the tissue to be replaced; (iii) it has three dimensional porous structure to permit cell proliferation, vascularization, and diffusion of nutrients; (iv) it degrades at a matching rate of production of the new tissue; and (v) it is suitable for processing economically, anatomically, and clinically.

Details of bone formation by bioactive glasses and glass-ceramics are provided in Section 7.3.1. The advantages of the use of the bioactive glasses and glass-ceramics as scaffold materials are that their structure and chemistry can be tailored over a considerable range by changing either composition, thermal, or processing history. Recently, different eminent research groups very excellently reviewed the area of scaffolds based on bioactive glasses and glass-ceramics; interested readers are referred to those publications [34,31,35,36].

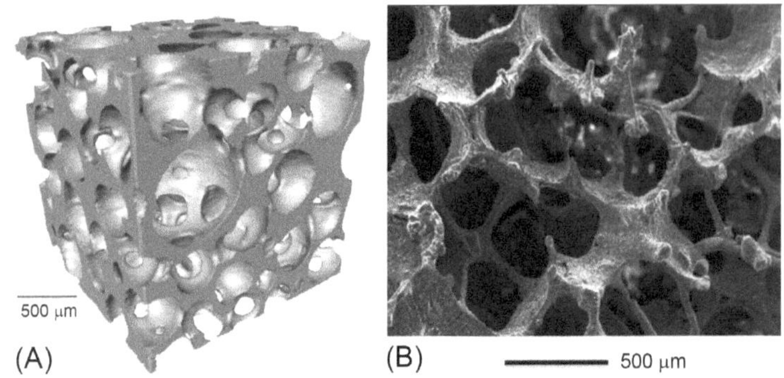


**Fig. 7.10** (A) X-ray microcomputer tomography (μCT) image of a bioactive glass foam scaffold. (B) SEM photomicrograph of a porous bioactive glass-ceramic scaffold made of 45S5 Bioglass.
(A) Reproduced with permission from J.R. Jones, New trends in bioactive scaffolds: the importance of nanostructure, J. Eur. Ceram. Soc. 29 (2009) 1275–1281, Copyright © Elsevier. (B) Reproduced with permission from J.R. Jones, Review of bioactive glass: From Hench to hybrids, Acta Biomater. 9 (2013) 4457–4486, Copyright © Elsevier.

Fig. 7.10A illustrates the X-ray microcomputer tomography (μCT) image of a bioactive glass foam scaffold [37]. It is seen that the pore structure is similar to that of the trabecular bone with many interconnects that may perhaps facilitate tissue ingrowths. Fig. 7.10B shows the SEM photomicrograph of a porous bioactive glass-ceramic scaffold made of 45S5 Bioglass [31]. The porous structure of the scaffolds can be optimized by monitoring the processing parameters. For example, by choosing an optimal polymer foam content and the amount of 13–93 bioactive glass ($54.6SiO_2$–22.1CaO–$6.0Na_2O$–$1.7P_2O_5$–$7.9K_2O$–7.7MgO, mol.%) particles (<10 μm) used in the slurry (35 vol.%), scaffolds of compressive strength of 11 MPa were obtained with 85% porosity and pore sizes ranging from 100 to 500 μm. Fig. 7.11 depicts the scanning

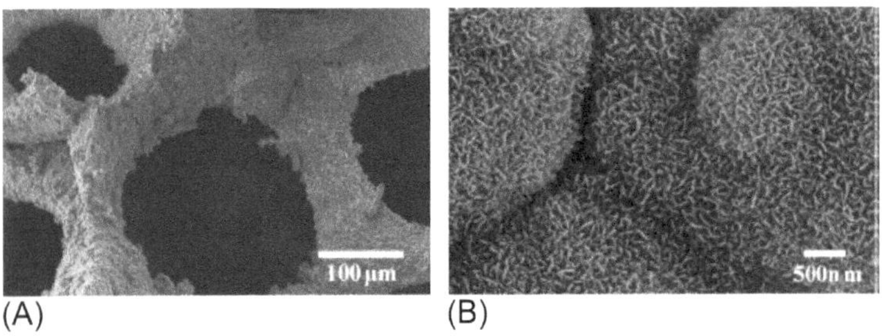


**Fig. 7.11** SEM photomicrographs of the surface of 13–93 bioactive glass foamed scaffold after immersion in SBF solution for 7 days: (A) lower magnification and (B) higher magnification showing fine needle-like hydroxyapatite crystals.
Reproduced with permission from Q. Fu, M.N. Rahaman, B.S. Bal, R.F. Brown, D.E. Day, Mechanical and in vitro performance of 13-93 bioactive glass scaffolds prepared by a polymer foam replication technique. Acta Biomater. 4 (2008) 1854–1864, Copyright © Elsevier.

electron microscopy (SEM) photomicrographs of the surface of 13–93 bioactive glass-foamed scaffold after immersion in SBF solution for 7 days: (A) lower magnification and (B) higher magnification showing fine needle-like hydroxyapatite crystals [38]. It is observed that a nanostructured hydroxyapatite (HA) layer is formed on the surface of the porous 13–93 bioactive glass scaffolds within 7 days, indicating the in vitro bioactivity of the scaffolds. Such HA nanocrystals are found in human bone and believed to be beneficial for increased cell adhesion, proliferation, and greater tissue growth into the scaffold. The cell culture results and SEM observations confirm an excellent attachment and subsequent proliferation of preosteoblastic cells, both on the surface and in the interior of the scaffolds.

### *7.3.4 Bioactive glass composites*

The major shortcoming of bioactive glasses and glass-ceramics for their wider use is derived from their inappropriate mechanical properties, for instance, higher brittleness and lower flexibility than natural bone. Various approaches have been made to match the mechanical properties of the glass materials with natural bones. Ohtsuki and coworkers [39] have attempted to prepare organic-inorganic hybrids using methacryloxy propyltrimethoxy silane (MPS) and 2-hydroxyethyl methacrylate (HEMA) with the addition of a calcium salt. MPS has alkoxysilane groups that provide silanol (Si—OH) groups after hydrolysis, whereas HEMA provides a hydrophilic polymer matrix in the hybrid. This type of MPS-HEMA hybrids has been found to possess the ability to form bone-like apatite layer in SBF solution. You et al. [40] have reported similar bioactivity (i.e., formation of apatite layer) of the tetraethoxysilane-polyvinyl alcohol (TEOS-PVA) hybrid composites in SBF solution. Juhasz et al. [41] have reported the improvement of mechanical properties of bioactive glass-ceramic A-W-polyethylene composites. Ducheyne and Hench [42] have reported the composite of Bioglass 45S5 and metal fibers of stainless steel, which has the improved mechanical strength, Young's modulus, and ductility comparable with that of human cortical bone. Müller-Mai et al. [43] have reported titanium particle-reinforced Ceravital bioactive composites.

### *7.3.5 Bioactive glass coatings*

High density alumina ($\alpha$-$Al_2O_3$), partially stabilized zirconia (PSZ, yttria- or calcia-stabilized) and different metallic alloys are widely used in load-bearing hip prostheses and dental implants owing to their good combination of excellent corrosion resistance, good biocompatibility, high wear resistance, and high strength. They do not form chemical bond with existing living tissues ("bioinert"); this causes these implants detachment over time through the formation of thick fibrous tissues around it. To improve their adherence, in the late 1960s, the concept of biological fixation of load-bearing implants using bioactive material coatings was proposed for invasive bone-prosthesis fixation.

Gomez-Vega et al. [44] have reported a simple enameling technique to develop bioactive glass ($67.7SiO_2$–$8.3Na_2O$–$2.2K_2O$–$10.1CaO$–$5.7MgO$–$6.0P_2O_5$, wt.%) coating on titanium alloy (Ti6Al4V) substrates to promote increased bioactivity.

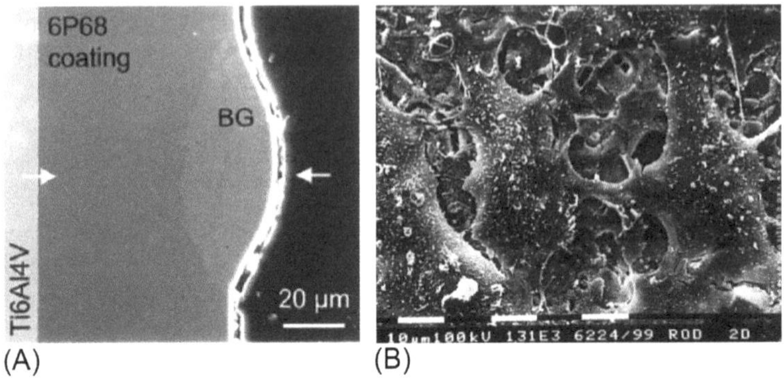


**Fig. 7.12** SEM photomicrographs of (A) cross section of the bioactive glass (BG, $67.7SiO_2$–$8.3Na_2O$–$2.2K_2O$–$10.1CaO$–$5.7MgO$–$6.0P_2O_5$, wt.%) coating on Ti implants showing the embedded particle and interface. (B) Bioactive glass ($SiO_2$–$CaO$–$P_2O_5$–$Na_2O$–$K_2O$–$MgO$–$CaF_2$–$Ta_2O_5$–$La_2O_3$ glass system) coated $Al_2O_3$ implants after 7 days of cell culture showing compact structure of osteoblasts with dorsal ruffles and filopodia.
(A) Reproduced with permission from J.M. Gomez-Vega, E. Saiz, A.P. Tomsia, G.W. Marshall, S.J. Marshall, Bioactive glass coatings with hydroxyapatite and Bioglass® particles on Ti-based implants. 1. Processing, Biomaterials 21 (2000) 105–111, Copyright © Elsevier. (B) Reproduced with permission from P. Torricelli, E. Verné, C.V. Brovarone, P. Appendino, F. Rustichelli, A. Krajewski, A. Ravaglioli, G. Pierini, M. Fini, G. Giavaresi, R. Giardino, Biological glass coating on ceramic materials: in vitro evaluation using primary osteoblast cultures from healthy and osteopenic rat bone, Biomaterials 22 (2001) 2535–2543, Copyright © Elsevier.

Fig. 7.12A shows the SEM photomicrograph of cross section of the bioactive glass (BG) coating on Ti implants showing the embedded particle and interface. It discloses some infiltration of the BG into the glass coating at the BG/glass interface. Torricelli et al. [45] have reported the in vitro evaluation of bioactive glass-coated ceramic material ($Al_2O_3$ and $ZrO_2$) substrates using primary osteoblast cultures from healthy and osteopenic rat bone. Fig. 7.12B represents the SEM photomicrographs of bioactive glass ($SiO_2$–$CaO$–$P_2O_5$–$Na_2O$–$K_2O$–$MgO$–$CaF_2$–$Ta_2O_5$–$La_2O_3$ glass system) coated $Al_2O_3$ implants after 7 days of cell culture showing compact structure of osteoblasts with dorsal ruffles and filopodia [45].

### 7.3.6 Antibacterial and drug delivery bioactive glasses

Treatment of infected bones and bacterial contamination in implanted orthopedic prostheses is a serious problem. This treatment requires prolonged systematic therapy involving controlled release of antibacterial ions and local delivery of drugs [9,46–49]. This type of treatment is carried out using antibacterial ion and/or drug-loaded (i) biodegradable bioactive glasses and (ii) bioactive scaffolds. Bioactive glasses (45S5 Bioglass and S53P4; see Table 7.2) are exclusive in their ability to degrade in the body fluid, liberating ions into the medium, simultaneously convert to hydroxyapatite, and bond strongly to bones and soft tissues. Bellantone et al.

[9,46] reported the broad-spectrum bactericidal bioactivity of $Ag_2O$-doped 45S5 Bioglass based on the studies carried out using *Escherichia coli*, *Pseudomonas aeruginosa*, and *Staphylococcus aureus* test microorganisms. A complete bactericidal effect was obtained within the first hour of incubation of $Ag_2O$-doped 45S5 Bioglass and attributed exclusively to the leaching of $Ag^+$ ions from the glass matrix. Other $Ag_2O$-doped silicate-based bioactive glasses are found to exhibit antibacterial activity.

Recently, $Ag^+$ ion-doped borate bioactive glass has shown the ability to inhibit bacterial growth in vitro and to eradicate bone infection in rabbit tibiae [47]. In addition, it can also be used for local delivery of antibiotics. Fig. 7.13 illustrates the mechanism of antibacterial activity and local delivery of antibiotics using bioactive glass particulates: (A) release of alkali and alkaline-earth metal ions (e.g., $Na^+$ and $Ca^{2+}$) in the local physiological environment, (B) release of antibacterial $Ag^+$ ion from $Ag^+$-doped bioactive borate glass, and (C) release of antibiotics from composite of an antibiotic-loaded biodegradable matrix and bioactive glass particles [47]. The supreme advantage of bioactive glass-based carrier systems is that they simultaneously eradicate infection and regenerate bone in loaded or nonloaded sites, thereby eliminating the need for subsequent bone reconstruction.

Jones et al. [48] reported the synthesis and antibacterial properties of $Ag_2O$-doped bioactive glass scaffolds using sol-gel foaming technique. It was observed that the rate of release of ions was more rapid for pore structures with a larger modal pore diameter

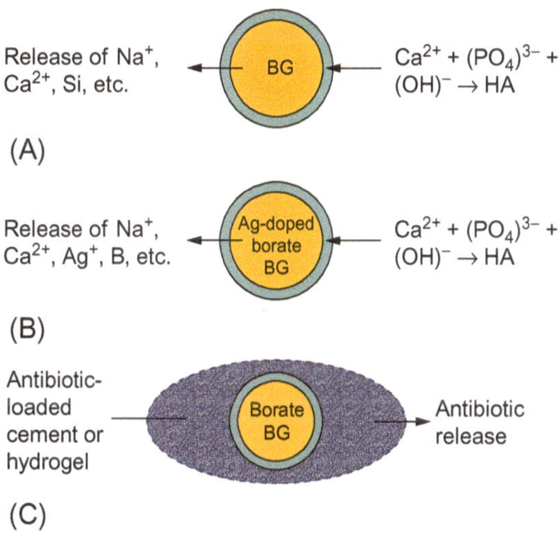


**Fig. 7.13** Mechanism of antibacterial activity and local delivery of antibiotics using bioactive glass particulates: (A) release of alkali and alkaline earth metal ions (e.g., $Na^+$ and $Ca^{2+}$) in the local physiological environment, (B) release of antibacterial $Ag^+$ ion from $Ag^+$-doped bioactive borate glass, and (C) release of antibiotics from composite of an antibiotic-loaded biodegradable matrix and bioactive glass particles.
Reproduced with permission from M.N. Rahaman, B.S. Bal, W. Huang, Review: Emerging developments in the use of bioactive glasses for treating infected prosthetic joints, Mater. Sci. Eng. C 41 (2014) 224–231, Copyright © Elsevier.

of the scaffolds. Thanyaphoo and Kaewsrichan [49] reported the synthesis and evaluation of novel drug-loaded glass-ceramic scaffolds for drug delivery systems in osteomyelitis. The scaffolds containing the bioactive glass-ceramics were prepared by chitosan cross-linking and loaded with vancomycin (VCM). Their exhibited properties were found to be very close to those of an ideal material for implantable drug delivery system and so seem suitable for use to cure bone diseases including osteomyelitis.

### 7.3.7 Radioactive glass microspheres for malignant tumors

Inoperable and deadly form of primary liver cancer, called hepatocellular carcinoma (HCC), is considered as the sixth most common cancer in the world (1 million new cases annually). Recently, treatment of HCC tumors with radioactive yttrium aluminosilicate (YAS) glass microspheres (diameter, 20–40 μm) is recognized as a safe and effective technique compared with other available procedures [20,50–52]. In this treatment, beta ($\beta^-$) emissions (energy, 2.27 MeV) of $^{90}Y$ radionuclide (half life, 64.1 h) are used for liver tumors. The energy of beta radiation (0.76–2.30 MeV) is found to be suitable for radiotherapy owing to its short range irradiation in tissue (0.3–1.1 cm maximum); hence, the radiation is localized within a specific treatment site circumventing the use of high dose (2000–35,000 rads (cGy)) external beam irradiation that is harmful. The glass microspheres deliver large doses (>15,000 rads (cGy)) of localized beta radiation to the targeted organs, which is 5–7 times more effective than the external beam irradiation. In this technique, only the tissue of the malignant tumors is sensitive to the ionizing beta radiation and undergoes cell division, whereas the healthy liver tissue resists and remains unaffected by the ionizing beta radiation. Fabrication of glass microspheres and treatment of malignant tumors in the liver are as follows.

The YAS glass ($40Y_2O_3$–$20Al_2O_3$–$40SiO_2$, wt.%) is initially prepared using high-purity oxides as raw materials by melting in a platinum crucible at 1650°C for 6–12 h followed by casting in air. The YAS glass thus obtained is crushed to a powder of <45 μm size, and each particle is spheroidized by melting in an oxygen-acetylene flame following the flame spheroidization technique to obtain 20–40 μm size glass microspheres. In the last step of fabrication of the glass microspheres for use in radiotherapy, they are made radioactive by neutron (${}_0n^1$) bombardment in a nuclear reactor to convert $^{89}Y$ isotope of the glass into $^{90}Y$ radionuclide according to the radioactive reaction as shown below:

$$ {}_{39}Y^{89} + {}_0n^1 \rightarrow {}_{39}Y^{90} + \gamma \tag{7.4} $$

This procedure minimizes the expense, potential hazards, and special precautions involved in handling radioactive materials and reduces the exposure of personnel to unwanted radiation.

For treatment, these radioactive YAS glass microspheres are injected into the hepatic artery of the patients using a catheter, so they are distributed to the tumors and permanently entrapped within the vascular bed. Fig. 7.14 illustrates the schematic diagram showing the injection of radioactive YAS glass microspheres into the hepatic

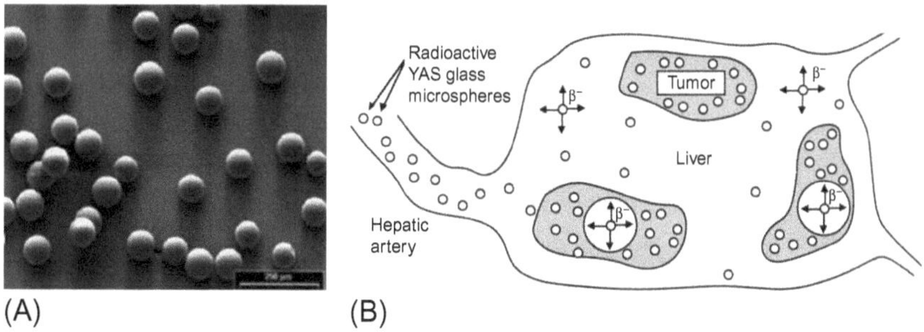


**Fig. 7.14** (A) Yttrium aluminosilicate (YAS) glass microspheres. (B) Schematic diagram showing the injection of radioactive YAS glass microspheres into the hepatic artery and the distribution of microspheres in the liver for treatment of malignant tumors in the liver.
(A) Reproduced with permission from M.B. Bortota, S. Prastalob, M. Pradoa, Production and characterization of glass microspheres for hepatic cancer treatment, Procedia Mater. Sci. 1 (2012) 351–358, Copyright © Elsevier. (B) Reproduced with permission from J.E. White, D.E. Day, Rare earth aluminosilicate glasses for in vivo radiation delivery, in: J.E. Shelby (Ed.), Rare Elements in Glasses, Key Engineering Materials, vols. 94–95, 1994, pp. 181–208, Copyright © Trans Tech Publications.

artery and the distribution of microspheres in the liver for the treatment of malignant tumors in the liver [50]. The glass microspheres deliver the beta ($\beta^-$) radiation to the malignant tumors according to the following radioactive reaction:

$$^{90}Y \rightarrow {}^{90}Zr + \beta^- (2.27\,MeV) \tag{7.5}$$

A large dose (>15,000 rads (cGy)) of the radiation ($\beta^-$) is injected into the liver of cancer patients without any detectable radiation leakage from the YAS glass microspheres. This is due to very high chemical durability of the YAS glass microspheres in the physiological environment with an extremely low total glass dissolution rate of $\leq 10^{-9}$ g/cm$^2$/min.

The YAS microsphere radiotherapy is not only useful for the treatment of the HCC (liver cancer) tumors but also suitable for the treatment of the kidney, ovarian, and prostate tumors.

### 7.3.8 *Nano bioactive glasses*

Silica nanospheres have diverse biomedical applications, for example, cell tracking and intercellular delivery of molecules, therapeutic agents, proteins, and DNA [31]. The sol-gel technique is one of the best for synthesis of silica nanospheres. Karmakar and coworkers [53,54] have demonstrated the synthesis of silica spheres of wide size ranges by acid hydrolysis of tetraethyl orthosilicate (TEOS) following the sol-gel process, whereas Stöber et al. [55] used the base (ammonium hydroxide) catalyzed hydrolysis of tetraethyl orthosilicate (TEOS) to obtain diverse sizes of silica spheres. Zhao et al. [56] reported the preparation of mesoporous silica spheres from silica/hyperbranched polyester and silica/glycidoxy-propyl trimethoxysilane using the sol-gel process. Similarly, tailored porosity can be obtained playing surfactant nature, concentration, and synthesis protocol. These small silica spheres are found to be suitable for use in cell labeling and drug delivery.

A broad range of bioactive glass nanoparticles were developed for various biomedical applications. They have several advantages compared with the bulk- or microsized-based materials. Their major advantages originated from their superior specific surface area that leads to higher dissolution rate and therefore faster apatite formation and ion or drug delivery [57]. Recently, Wu et al. [58] demonstrated a facile hydrothermal method for synthesis of mesoporous bioactive glass (MBG) nanospheres with high specific surface area. These nanospheres are found to possess excellent apatite-mineralization ability and high loading efficiency (90%) of anticancer drug, doxorubicin hydrochloride (DOX). It is observed that the release of DOX from MBG nanospheres significantly inhibited the sustainability of bone cancer cells. Their studies have established MBG nanospheres as a promising drug carrier for the therapy of bone cancer.

Another important area is the development of nanobioactive glass composite scaffold for use in bone tissue engineering applications. Recently, Gao et al. [59] reported the improved mechanical properties and enhanced bone tissue engineering performance of graphene/nano-58S bioactive glass composite scaffolds. The compressive strength and fracture toughness of the composite scaffolds were found to be improved by 105% and 38%, respectively. Similarly, simulated body fluid (SBF) solution and cell culture tests indicated enhanced bioactivity and biocompatibility of the composite scaffold.

### 7.3.9 Bioactive bulk metallic glasses

Titanium alloys are widely used in orthopedic prostheses and dental implants due to their high corrosion resistance and good biocompatibility. Their long-term clinical performance is negotiated due to their different set of mechanical properties compared with those of bones. This mismatching results in particle disease, stress shielding, and fatigue failures, whereas the bulk metallic glasses (BMGs) without long-range atomic order possess high yield strength, low Young's modulus, high corrosion, and fatigue resistance. As a result of combination of these properties, titanium BMGs have the potential for use as a superior biomaterial. In 2007, Inoue and coworkers [13] developed a new series of TiZrCuPd quaternary bulk metallic glasses for biomedical applications because of the absence of toxic elements such as Be and Ni. Fukushima et al. [60] reported the hydrothermal (H), electrochemical (E), and hydrothermal-electrochemical (H-E) processes in NaOH solution for synthesizing bioactive nanomesh structural layer on $Ti_{40}Zr_{10}Cu_{36}Pd_{14}$ bulk metallic glass. The spontaneous formation of hydroxyapatite in simulated body fluid (SBF) was observed in all of the samples prepared by these processes and the H-E process was found to be the most appropriate for actual biomedical applications.

Fig. 7.15 shows the SEM photomicrographs of hydrothermal-electrochemical treated $Ti_{40}Zr_{10}Cu_{34}Pd_{14}Sn_2$ bulk metallic glass: (A) nanomesh surface morphology before immersion in Hanks' solution and (B) formation of calcium phosphate particles

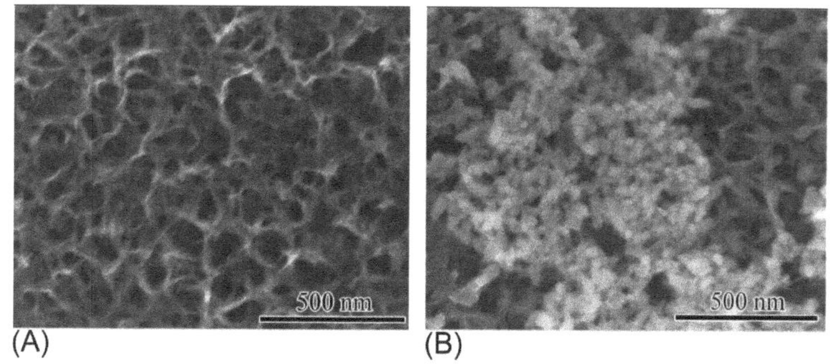


**Fig. 7.15** SEM photomicrographs of hydrothermal-electrochemical treated $Ti_{40}Zr_{10}Cu_{34}Pd_{14}Sn_2$ bulk metallic glass: (A) nanomesh surface morphology before immersion in Hanks' solution and (B) formation of calcium phosphate particles on nanomesh surface of the glass after immersion in Hanks' solution for 48 h.
Reproduced with permission from F. Qin, S. Zhu, Z. Dan, A. Kawashima, G. Xie, Stress corrosion cracking and bioactivity of Ti-based bulk metallic glass, J. Alloys Compd. 615 (2014) S123–S127, Copyright © Elsevier.

on nanomesh surface of the glass after immersion in Hanks' solution for 48 h [61]. The composition of Hanks' solution (g/L) used in this research was 8.00 NaCl, 0.40 KCl, 0.35 $NaHCO_3$, 0.19 $CaCl_2{\cdot}2H_2O$, 0.09 $Na_2HPO_4{\cdot}7H_2O$, 0.20 $MgSO_4{\cdot}7H_2O$, 0.06 $KH_2PO_4$, and 1.00 glucose. The formation of calcium phosphate layer on the nanomesh surfaces of the BMG reveals its surface bioactivity. Likewise, Pang et al. [62] and Oak et al. [63] demonstrated the bioactivity of Ti-based $Ti_{47}Cu_{38}Zr_{7.5}Fe_{2.5}Sn_2Si_1Ag_2$ (at.%) and $Ti_{42}Hf_{11}Cu_{11}Pd_{36}$ (at.%) bulk metallic glasses (BMGs), respectively.

### *7.3.10 Glass ionomer cements*

In 1971, Wilson and Kent [2] published the first report on glass ionomer cement (GIC) as an acid-base reaction product of basic fluoroaluminosilicate glass powder and aqueous solution of polycarboxylic acid, which was actually developed in 1967. Glass ionomer cements (GICs) are biomedical "smart" materials obtained from the combination of "glass powder" and "water-based solution of polyacrylic acid," which hardens through an acid-base reaction between them. The term "glass ionomer cement (GIC)" was coined by Wilson and Kent [2] and is widely used. However, in the chemical sense, it is not a true ionomer; thus, McLean et al. [64] proposed "glass polyalkenoate cement (GPC)" as a more accurate term for this material and is accepted by the International Organization for Standardization (ISO). Since it is extensively used in dental practice to replace the dentin; therefore, other given names are dentin substitute, manmade dentin, and artificial dentin. Other uses of these cements are for orthograde root canal sealing, treatment of vertically fractured teeth, maintenance of coronal seal, and repair of perforations and root resorption defects [65]. GICs are also used in medicine, mainly in otologic and reconstructive surgery and orthopedics.

The powder of GIC is a fluoride ion leachable acid soluble basic calcium fluoroaluminosilicate glass of a typical composition (wt%) $41.9SiO_2$–$28.6Al_2O_3$–$1.6AlF_3$–$15.6CaF_2$–$7.3NaF$–$3.8AlPO_4$. A typical composition (wt%) of the liquid is polyacrylic acid 40, tartaric acid 10, water 45, and itaconic, maleic, and tricarballylic acids 5. The GICs are formulated by mixing glass powder with the liquid. The setting time varies in the range of several seconds depending on powder-to-liquid ratio. Eugenol (4-allyl-2-methoxyphenol) possesses analgesic and antiinflammatory properties; thus, eugenol moieties are sometimes incorporated into the polyacrylic acid for the treatment of irritated or diseased tooth pulp [66]. The GICs could also be transformed into bioactive restorative materials. This is done by mixing with bioactive glass powder. Yli-Urpo et al. [67] demonstrated the enhanced tooth mineralization effects of GICs containing S53P4 bioactive glass [67]. Recently, Khoroushi and Keshani [68] have reviewed various features of bioactive glass ionomer for use as the "smart" materials in dentistry.

## 7.4 Functional applications

Functional applications of various types of biomedical glasses and glass-ceramics are mentioned in the preceding sections under their respective headings. In this section, their functional applications are summarized in Table 7.5. More information can also be obtained in the publications cited under references [20,23,33,31,65,69–71].

**Table 7.5 Functional applications of biomedical glasses and glass-ceramics**

| Product | Functional applications |
|---|---|
| Bioactive glasses | • Orthopedic prostheses • Dental implants • Maxillofacial reconstruction • Middle ear reconstruction • Cranial reconstruction • Junction of spinal vertebrae • Percutaneous access devices • Periodontal pocket obliteration • Otolaryngological • Alveolar ridge augmentations • Promote angiogenesis (formation of the blood vessel) • Soft tissue repair • Osteochondral tissue engineering |
| Particulate bioactive glasses | • Treatment of bone defects • Bone graft • Treatment of dentin hypersensitivity • Dentin remineralization • Antibacterial treatment • Root canal sterilization tools • Repair of enamel sensitivity • Filler around implants • Filling of bone after cyst or tumor removal • Treatment of infected prosthetic joints with antibiotics |

**Table 7.5 Continued**

| | |
|---|---|
| Bioactive glass-ceramics | • Dental implants • Maxillofacial reconstruction • Vertebral prosthesis devices • Iliac crest prostheses • Percutaneous access devices • Spinal surgery |
| Porous bioactive glasses and glass-ceramics | • Scaffolds for tissue engineering • Drug delivery • Bone graft • Repair and regeneration of large bone defects |
| Bioactive glass and glass-ceramic coated bioinert ceramic and metallic alloy implants | • Bone prostheses • Maxillofacial reconstruction • Periodontal pocket obliteration • Hip articulation prostheses • Attachment of artificial members to bone • Junction of spinal vertebrae • Middle ear reconstruction |
| Bioactive glass composites | • Bone and soft tissue joining • Bone defect repairing |
| Glass ionomer cements | • Root canal sealing • Root end filling • Root perforation repair • Root resorption cavity repair • Treatment of vertically fractured teeth • Surgical endodontics • Antibacterial anticariogenic dental cements |

## 7.5 Summary

In this chapter, glasses and glass-ceramics that are used in various biomedical applications are discussed. Initially, the publication trend on bioactive glasses and glass-ceramics during the last 25 years (1990–2014), and a brief chronological history of biomedical glasses and glass-ceramics are provided. These are followed by a brief account of bioactive glasses, bioactive glass-ceramics, bioactive glass and glass-ceramic scaffolds, bioactive glass composites, bioactive glass coatings, antibacterial and drug delivery bioactive glasses, radioactive glass microspheres for malignant tumors, nanobioactive glasses, bioactive bulk metallic glasses, and glass ionomer cements. Some of the important topics of these descriptions are composition of bioactive glasses and glass-ceramics, bone bonding mechanism with bioactive glasses through the formation of hydroxyl carbonate apatite layers with experimental evidences, mechanism of antibacterial activity and local delivery of drugs by bioactive glass particulates, distribution of radioactive glass microspheres in the liver for treatment of malignant tumors, etc. Finally, a summarized list of functional applications of various types of biomedical glasses and glass-ceramics along with future outlook is offered.

## 7.6 Future outlook

We have seen that bioactive glasses and glass-ceramics have a unique set of biomedical properties that are not found in any other type of materials; however, there is still potential for advances in this area. Future research should focus on the following:

- The limiting inconvenient effects of the brittleness through innovative scaffold design and processing, particularly for repair for load bearing bones without sacrificing the bioactivity
- Improvement of porous bioactive synthetic scaffolds with matching bone pore structure, mechanical strength, and toughness
- Improvement of scaffold function that is related to surface modification
- Development of porous scaffolds with bioactivity and local drug delivery
- Improvement of scaffolds of bioactive glasses and glass-ceramics in combination with biodegradable polymers and stem cells
- Enhanced bioactivity in terms of gene activation
- Improvement in the performance of biomedical coatings in terms of their adherence and mechanical stability
- Development of smart bioactive glasses and glass-ceramics capable of combining sensing with bioactivity
- Development of improved biomimetic composites

## Exercises

**7.1** Define bioactive and bioinert materials with examples.

**7.2** Who discovered the bioactive glass? What is the chemical composition of 45S5 Bioglass? What is the meaning of 45S5?

**7.3** Describe the bone-bonding mechanism with bioactive glasses through the formation of hydroxyl carbonate apatite layers in simulated body fluid (SBF).

**7.4** What are bioactive glass-ceramics? Describe three bioactive glass-ceramics with chemical compositions and crystal phases.

**7.5** What are bioactive glass and glass-ceramic scaffolds? Describe the important features of bioactive glass and glass-ceramic scaffolds.

**7.6** What is the importance of bioactive glass coatings? Describe important features of bioactive glass and glass-ceramic coatings on implants.

**7.7** How do you prepare antibacterial and drug delivery bioactive glasses? Describe the mechanism of antibacterial activity and local delivery of drugs by bioactive glass particulates.

**7.8** What are radioactive glass microspheres? Describe the distribution of radioactive glass microspheres in the liver for treatment of malignant tumors.

**7.9** Write brief notes on bioactive glass composites, nanobioactive glasses, bioactive bulk metallic glasses, and glass ionomer cements.

**7.10** Describe the functional applications of bioactive glasses, particulate bioactive glasses, bioactive glass-ceramics, porous bioactive glasses and glass-ceramics, bioactive glass and glass-ceramic coated bioinert ceramic and metallic alloy implants, bioactive glass composites, and glass ionomer cements.

# References


[1] L.L. Hench, R.J. Splinter, W.C. Allen, T.K. Greenlee, Bonding mechanisms at the interface of ceramic prosthetic materials, J. Biomed. Mater. Res. 5 (1971) 117–141.

[2] A.D. Wilson, B.E. Kent, The glass-ionomer cement: a new transluscent dental filling material, J. Appl. Chem. Biotechnol. 21 (1971) 313.

[3] T.K. Greenlee Jr., C.A. Beckham, A.R. Crebo, J.C. Malmorg, Glass-ceramic bone implants—a light microscopic study, J. Biomed. Mater. Res. 6 (1972) 235–244.

[4] P. Griss, D.C. Greenspan, G. Heimke, B. Krempien, R. Buchinger, L.L. Hench, G. Jentschura, Evaluation of a bioglass-coated $Al_2O_3$ total hip prosthesis in sheep, J. Biomed. Mater. Res. 10 (1976) 511–518.

[5] J. Wilson, G.H. Pigott, F.J. Schoen, L.L. Hench, Toxicology and biocompatibility of bioglasses, J. Biomed. Mater. Res. 15 (1981) 805–817.

[6] T. Kokubo, M. Shigematsu, Y. Nagashima, M. Tashiro, T. Nakamura, T. Yamamuro, S. Higashi, Apatite- and wollastonite-containing glass-ceramics for prosthetic applications, Bull. Inst. Chem. Res. Kyoto Univ. 60 (1982) 260–268.

[7] W. Höland, J. Naumann, W. Vogel, J. Gummel, Z. Wiss, Maschinellbearbeitbare bioactive glaskeramiken, Friedrich-Schiller Univ. Jena Math. Naturwiss. Reihe 32 (1983) 571–580.

[8] W. Höland, M. Frank, V. Rheinberger, Surface crystallization of leucite in glasses, J. Non-Cryst. Solids 180 (1995) 292–307.

[9] M. Bellantone, L.L. Hench, Bioactive behaviour of sol-gel derived antibacterial bioactive glass, Key Eng. Mater. 192–195 (2000) 617–660.

[10] I.D. Xynos, A.J. Edgar, L.D. Buttery, L.L. Hench, J.M. Polak, Gene-expression profiling of human osteoblasts following treatment with the ionic dissolution products of Bioglass® 45S5 dissolution, J. Biomed. Mater. Res. 55 (2001) 151–157.

[11] J.R. Jones, L.M. Ehrenfried, L.L. Hench, Optimising bioactive glass scaffolds for bone tissue engineering, Biomaterials 27 (2006) 964–973.

[12] Q.Z. Chen, I.D. Thompson, A.R. Boccaccini, 45S5 Bioglass-derived glass-ceramic scaffolds for bone tissue engineering, Biomaterials 27 (2006) 2414–2425.

[13] S.L. Zhu, X.M. Wang, F.X. Qin, M. Yoshimura, A. Inoue, New TiZrCuPd quaternary bulk glassy alloys with potential of biomedical applications, Mater. Trans. 48 (2007) 2445–2448.

[14] C. Wu, J. Chang, Y. Xiao, Mesoporous bioactive glasses as drug delivery and bone tissue regeneration platforms, Therap. Deliv. 2 (2011) 1189–1198.

[15] L.L. Hench, Chronology of bioactive glass development and clinical applications, New J. Glass Ceram. 3 (2013) 67–73.

[16] L.L. Hench, Opening paper 2015—Some comments on Bioglass: four eras of discovery and development, Biomed. Glasses 1 (2015) 1–11.

[17] L.L. Hench, J.R. Jones, Bioactive glasses: frontiers and challenges, Front. Bioeng. Biotechnol. 3 (2015) 194 (12 pages).

[18] L.L. Hench, I.D. Xynos, J.M. Polak, Bioactive glasses for in situ tissue regeneration, J. Biomater. Sci. Polym. Ed. 15 (2004) 543–562.

[19] L.L. Hench, Genetic design of bioactive glass, J. Eur. Ceram. Soc. 29 (2009) 1257–1265.

[20] L.L. Hench, D.E. Day, W. Höland, V.M. Rheinberger, Glass and medicine, Int. J. Appl. Glass Sci. 1 (2010) 104–117.

[21] T. Kokubo, H. Kushitani, S. Sakka, T. Kitsugi, T. Yamamuro, Solutions able to reproduce in vivo surface-structure changes in bioactive glass-ceramic A-W, J. Biomed. Mater. Res. 24 (1990) 721–734.
[22] O. Peitl, E.D. Zanotto, L.L. Hench, Highly bioactive $P_2O_5$-$Na_2O$-CaO-$SiO_2$ glass-ceramics, J. Non-Cryst. Solids 292 (2001) 115–126.
[23] L.L. Hench, Bioceramics: from concept to clinic, J. Am. Ceram. Soc. 74 (1991) 1487–1510.
[24] P.N. De Aza, A.H. De Aza, P. Pena, S. De Aza, Bioactive glasses and glass-ceramics, Bol. Soc. Esp. Ceram. V 46 (2007) 45–55.
[25] A.N. Cormack, A. Tilocca, Structure and biological activity of glasses and ceramics, Phil. Trans. R. Soc. A 370 (2012) 1271–1280.
[26] S. Fujibayashi, M. Neo, H.-M. Kim, T. Kokubo, T. Nakamura, A comparative study between in vivo bone ingrowth and in vitro apatite formation on $Na_2$-CaO-$SiO_2$ glasses, Biomaterials 24 (2003) 1349–1356.
[27] A. Tilocca, A.N. Cormack, Modeling the water–bioglass interface by *ab initio* molecular dynamics simulations, ACS Appl. Mater. Interfaces 1 (2009) 1324–1333.
[28] A. Tilocca, A.N. Cormack, Surface signatures of bioactivity: MD simulations of 45S and 65S silicate glasses, Langmuir 26 (2010) 545–551.
[29] T. Kasuga, Bioactive calcium pyrophosphate glasses and glass-ceramics, Acta Biomater. 1 (2005) 55–64.
[30] P. Ducheyne, Q. Qiu, Bioactive ceramics: the effect of surface reactivity on bone formation and bone cell function, Biomaterials 20 (1999) 2287–2303.
[31] J.R. Jones, Review of bioactive glass: from Hench to hybrids, Acta Biomater. 9 (2013) 4457–4486.
[32] H. Brömer, E. Pfeil, M. Kos, German patent no. 2,326,100, 1973.
[33] W. Höland, V. Rheinberger, E. Apel, C. van't Hoen, M. Höland, A. Dommann, M. Obrecht, C. Mauth, U. Graf-Hausner, Clinical applications of glass-ceramics in dentistry, J. Mater. Sci. Mater. Med. 17 (2006) 1037–1042.
[34] M.N. Rahaman, D.E. Day, B.S. Bal, Q. Fu, S.B. Jung, L.F. Bonewald, A.P. Tomsia, Bioactive glass in tissue engineering, Acta Biomater. 7 (2011) 2355–2373.
[35] L.-C. Gerhardt, A.R. Boccaccini, Bioactive glass and glass-ceramic scaffolds for bone tissue engineering, Materials 3 (2010) 3867–3910.
[36] Q. Fu, E. Saiz, M.N. Rahaman, A.P. Tomsia, Bioactive glass scaffolds for bone tissue engineering: state of the art and future perspectives, Mater. Sci. Eng. C 31 (2011) 1245–1256.
[37] J.R. Jones, New trends in bioactive scaffolds: the importance of nanostructure, J. Eur. Ceram. Soc. 29 (2009) 1275–1281.
[38] Q. Fu, M.N. Rahaman, B.S. Bal, R.F. Brown, D.E. Day, Mechanical and in vitro performance of 13-93 bioactive glass scaffolds prepared by a polymer foam replication technique, Acta Biomater. 4 (2008) 1854–1864.
[39] C. Ohtsuki, T. Miyazaki, M. Kamitakahara, M. Tanihara, Design of novel bioactive materials through organic modification of calcium silicate, J. Eur. Ceram. Soc. 27 (2007) 1527–1533.
[40] C. You, T. Miyazaki, E. Ishida, M. Ashizuka, C. Ohtsuki, M. Tanihara, Fabrication of poly(vinyl alcohol)–apatite hybrids through biomimetic process, J. Eur. Ceram. Soc. 27 (2007) 1585–1588.
[41] J.A. Juhasz, S.M. Best, R. Brooks, M. Kawashita, N. Miyata, T. Kokubo, T. Nakamura, W. Bonfield, Mchanical properties of glass-ceramic A-W-polyethylene composites: effect of filler content and particle size, Biomaterials 25 (2004) 949–955.

[42] P. Ducheyne, L.L. Hench, The processing of static mechanical properties of metal fibre-reinforced Bioglass, J. Mater. Sci. 17 (1982) 595–606.
[43] C. Müller-Mai, H.J. Schmitz, V. Strunz, G. Fuhrmann, T. Fritz, U.M. Gross, Tissues at the surface of the new composite material titanium/glass-ceramic for replacement of bone and teeth, J. Biomed. Mater. Res. 23 (1989) 1149–1161.
[44] J.M. Gomez-Vega, E. Saiz, A.P. Tomsia, G.W. Marshall, S.J. Marshall, Bioactive glass coatings with hydroxyapatite and Bioglass® particles on Ti-based implants. 1. Processing, Biomaterials 21 (2000) 105–111.
[45] P. Torricelli, E. Verné, C.V. Brovarone, P. Appendino, F. Rustichelli, A. Krajewski, A. Ravaglioli, G. Pierini, M. Fini, G. Giavaresi, R. Giardino, Biological glass coating on ceramic materials: in vitro evaluation using primary osteoblast cultures from healthy and osteopenic rat bone, Biomaterials 22 (2001) 2535–2543.
[46] M. Bellantone, H.D. Williams, L.L. Hench, Broad-spectrum bactericidal activity of $Ag_2O$-doped bioactive glass, Antimicrob. Agents Chemother. 46 (2002) 1940–1945.
[47] M.N. Rahaman, B.S. Bal, W. Huang, Review: Emerging developments in the use of bioactive glasses for treating infected prosthetic joints, Mater. Sci. Eng. C 41 (2014) 224–231.
[48] J.R. Jones, L.M. Ehrenfried, P. Saravanapavan, L.L. Hench, Controlling ion release from bioactive glass foam scaffolds with antibacterial properties, J. Mater. Sci. Mater. Med. 17 (2006) 989–996.
[49] S. Thanyaphoo, J. Kaewsrichan, Synthesis and evaluation of novel glass ceramics as drug delivery systems in osteomyelitis, J. Pharm. Sci. 101 (2012) 2870–2882.
[50] J.E. White, D.E. Day, Rare earth aluminosilicate glasses for in vivo radiation delivery, in: J.E. Shelby (Ed.), Rare Elements in Glasses, Key Engineering Materials, vols. 94–95, Trans Tech Publications Ltd., Switzerland, 1994, pp. 181–208.
[51] M.B. Bortota, S. Prastalob, M. Pradoa, Production and characterization of glass microspheres for hepatic cancer treatment, Procedia Mater. Sci. 1 (2012) 351–358.
[52] H. Poorbaygi, S.M.R. Aghamiri, S. Sheibani, A. Kamali-asl, E. Mohagheghpoor, Production of glass microspheres comprising $^{90}Y$ and $^{177}Lu$ for treating of hepatic tumors with SPECT imaging capabilities, Appl. Radiat. Isot. 69 (2011) 1407–1414.
[53] B. Karmakar, G. De, D. Kundu, D. Ganguli, Silica microspheres from the system tetraethyl orthosilicate-acetic acid-water, J. Non-Cryst. Solids 135 (1991) 29–36.
[54] B. Karmakar, G. De, D. Ganguli, Dense silica microspheres from organic and inorganic acid hydrolysis of TEOS, J. Non-Cryst. Solids 272 (2000) 119–126.
[55] W. Stöber, A. Fink, E. Bohn, Controlled growth of monodisperse silica spheres in the micron size range, J. Colloid Interface Sci. 26 (1968) 62–67.
[56] Y. Zhao, J. Zou, W. Shi, L. Tang, Preparation and characterization of mesoporous silica spheres with bimodal pore structure from silica/hyperbranched polyester nanocomposites, Microporous Mesoporous Mater. 92 (2006) 251–258.
[57] C. Vichery, J.-M. Nedelec, Bioactive glass nanoparticles: from synthesis to materials design for biomedical applications, Materials 9 (2016) 288 (17 pages).
[58] C. Wu, W. Fan, J. Chang, Functional mesoporous bioactive glass nanospheres: synthesis, high loading efficiency, controllable delivery of doxorubicin and inhibitory effect on bone cancer cells, J. Mater. Chem. B 1 (2013) 2710–2718.
[59] C.D. Gao, T.T. Liu, C.J. Shuai, S.P. Peng, Enhancement mechanisms of graphene in nano-58S bioactive glass scaffold: mechanical and biological performance, Sci. Rep. 4 (2014) 4712 (10 pages).
[60] Y. Fukushima, K. Katsumata, Z. Shengli, X. Guoqiang, M. Niinomi, K. Okada, N. Matsushita, Solution process for synthesizing bioactive nano-mesh layer on Ti-based bulk metallic glasses, Mater. Trans. 54 (2013) 1343–1346.

[61] F. Qin, S. Zhu, Z. Dan, A. Kawashima, G. Xie, Stress corrosion cracking and bioactivity of Ti-based bulk metallic glass, J. Alloys Compd. 615 (2014) S123–S127.
[62] S. Pang, Y. Liu, H. Li, L. Sun, Y. Li, T. Zhang, New Ti-based Ti–Cu–Zr–Fe–Sn–Si–Ag bulk metallic glass for biomedical applications, J. Alloys Compd. 625 (2015) 323–327.
[63] J.-J. Oak, A. Inoue, K.V. Rao, H.-H. Chun, Y.H. Park, Surface modified Ti based metallic glasses for bioactivation by electrochemical treatment technique, J. Alloys Compd. 615 (2014) S136–S141.
[64] J.W. McLean, J.W. Nicholson, A.D. Wilson, Proposed nomenclature for glass-ionomer dental cements and related materials, Quint. Int. 25 (1994) 587.
[65] M.A.A. De Bruyne, R.J.G. De Moor, The use of glass ionomer cements in both conventional and surgical endodontics, Int. Endod. J. 37 (2004) 91–104.
[66] L. Rojo, B. Vázquez, J.S. Román, S. Deb, Eugenol functionalized poly(acrylic acid) derivatives in the formation of glass-ionomer cements, Dent. Mater. 24 (2008) 1709–1716.
[67] H. Yli-Urpo, M. Närhi, T. Närhi, Compound changes and tooth mineralization effects of glass ionomer cements containing bioactive glass (S53P4), an in vivo study, Biomaterials 26 (2005) 5934–5941.
[68] M. Khoroushi, F. Keshani, A review of glass-ionomers: from conventional glass-ionomer to bioactive glass-ionomer, Dent. Res. J. 10 (2013) 411–420.
[69] W. Cao, L.L. Hench, Bioactive materials, Ceram. Int. 22 (1996) 493–507.
[70] M. Humagain, D.G. Nayak, A.S. Uppoor, A clinical evaluation of bioactive glass particulate in the treatment of mandibular class II furcation defects, Braz. J. Oral Sci. 6 (2007) 1450–1456.
[71] M. Sharma, P.E. Murray, D. Sharma, K. Parmar, S. Gupta, P. Goyal, Modern approaches to use bioactive materials and molecules in medical and dental treatments, Int. J. Curr. Microbiol. App. Sci. 2 (2013) 429–437.

# Glass and glass-ceramics for solid oxide fuel cell (SOFC) sealants

8

## Chapter Outline



## 8.1 Introduction

Solid oxide fuel cell (SOFC) is one the most important solutions of exhausting the energy sources of our planet [1]. SOFC is an energy conversion unit that converts gaseous fuels (e.g., hydrogen, carbon monoxide, and methane) into electric energy and heat by electrochemical combination of fuel with oxidant (air) [2]. It is operated based on the electrochemical principle and, therefore, is not limited by the heat engine principle of Carnot cycle. SOFC is considered as an effective process to produce electricity from chemical components, because of its high conversion efficiency and environmental friendliness. SOFC, therefore, materializes to be one the most promising technology to produce clean energy for a wide range (small units to large scale power plant) of applications. Fig. 8.1 shows the 5-kW SOFC power system [3]. The SOFC consists of a number of components such as cathode, anode, electrolyte (8 mol.% $Y_2O_3$–$ZrO_2$, CTE $= 10$–$11 \times 10^{-6}$ °C$^{-1}$), interconnect (Crofer-22 APU, CTE $= 11$–$12 \times 10^{-6}$ °C$^{-1}$), and sealing materials. Popularly, glasses and glass-ceramics are used as SOFC sealing materials to accomplish several important functions. In this chapter, we shall describe different aspects of glasses and glass-ceramics for use as SOFC sealing materials.

An attempt was made using the Web of Science (version 5.17) as described in Chapter 1 to provide a global publication trend and quantitative growth in the area

Functional Glasses and Glass-Ceramics. http://dx.doi.org/10.1016/B978-0-12-805056-9.00008-8

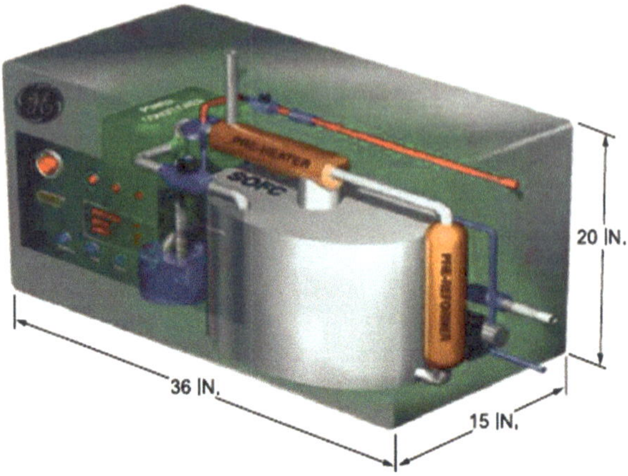


**Fig. 8.1** 5-kW SOFC power system.
Reproduced with permission from N. Q. Minh, Solid oxide fuel cell technology—features and applications, Solid State Ionics 174 (2004) 271–277, Copyright © Elsevier.

of SOFC. Here, year-wise data were extracted for the last 25 years (1990–2014) by searching, using the topic keyword "*glass*" followed by refining the document type as "*article*" published in the journals and meetings. It was further refined using the term "solid oxide fuel cell." This plot is shown in Fig. 8.2. It gives us a clear publication trend and quantitative growth of published papers on SOFC during the last 25 years (1990–2014). It is seen that growth of the published papers on SOFC sealant up to 2002 is negligible compared with the rapid growth during the period 2003–14. The rapid growth of the published papers during last 12 years clearly signifies the importance of this topic.

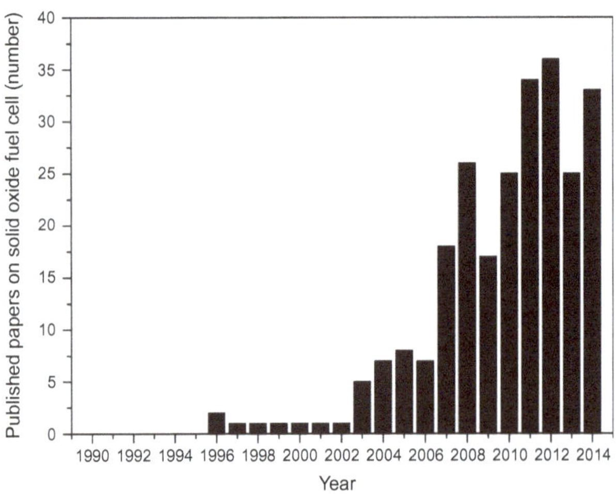


**Fig. 8.2** Year-wise published papers on solid oxide fuel cell (SOFC) during the last 25 years (1990–2014).

## 8.2 Requirements for use glasses and glass-ceramics as SOFC sealant

Fig. 8.3 illustrates the schematic of a SOFC showing locations of possible seal [4]. Glasses and glass-ceramics that are to be used as SOFC sealant should have the following mechanical, chemical, and electric properties and functions:

Mechanical
- Minimal CTE mismatching with the adjacent SOFC components
- Thermal cycle stability
- Acceptable bonding strength
- Near hermetic or hermetic
- Shock resistance

Chemical
- Long term chemical stability under oxidizing and wet environments
- Resistance to hydrogen embrittlement and corrosion

Electric
- Nonconductive (insulating)

Fabrication
- Low cost
- Sealing condition should be compatible with other SOFC components
- Highly reliable to form hermetic seal

Functions
- Seal prevents mixing of fuel and oxidant within stack
- Seal prevents leaking of fuel and oxidant from stack
- Seal isolates cells electrically in stack

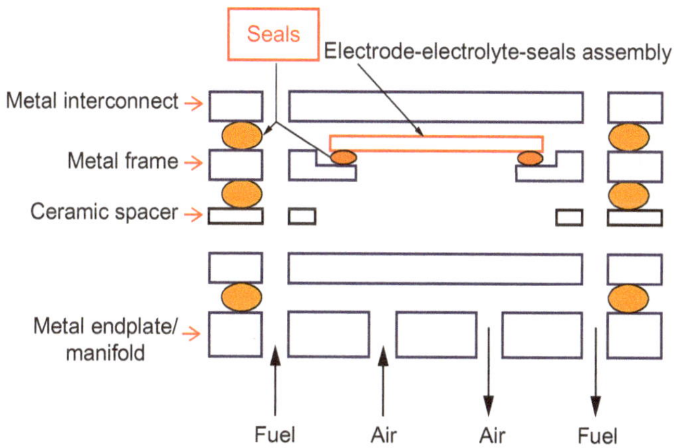


**Fig. 8.3** Schematic of a solid oxide fuel cell showing possible seal locations. Reproduced with permission from R.N. Singh, Sealing technology for solid oxide fuel cells (SOFC), Int. J. Appl. Ceram. Technol. 4 (2) (2007) 134–144, Copyright © The American Ceramic Society.

To know the viscosity of a glassy material at a particular temperature and its extent of change with temperature are very important for the processes that involve flow of glass, for example, sealing at elevated temperature and forming. For these reasons, the viscosity versus temperature curves of three commercial glasses in the temperature range 900–1600°C are provided here. Fig. 8.4 shows the comparison of characteristic viscosity points determined by hot-stage microscopy (experimental points) and Vogel-Fulcher-Tamman (VFT) curves from high-temperature viscometry (solid line) of different commercial glasses [5]. Fig. 8.5 demonstrates the photomicrographs of the characteristic points of viscosity as defined by Scholze, (A) initial shape of the sample; (B) first shrinkage ($\log \eta = 8.0 \pm 0.3$ P); (C) maximum contraction ($\log \eta = 8.2 \pm 0.5$ P); (D) softening point ($\log \eta = 6.1 \pm 0.2$ P); (E) half ball ($\log \eta = 4.6 \pm 0.1$ P); and (F) flow point ($\log \eta = 4.1 \pm 0.1$ P) [5]. For good cyclability, the sealing glass and its crystallized product should have minimal CTE mismatching with the adjacent SOFC components such as electrolyte (8 mol.% $Y_2O_3$–$ZrO_2$, CTE = $10–11 \times 10^{-6}$°$C^{-1}$) and interconnect (Crofer-22 APU, CTE = $11–12 \times 10^{-6}$°$C^{-1}$). Fig. 8.6 shows the variation of coefficient of thermal expansion (CTE) of sealing glass (G50) with time at 800°C in comparison with that of SOFC components, 430 stainless steel (430 SS) and yttria-stabilized zirconia (YSZ) [6].

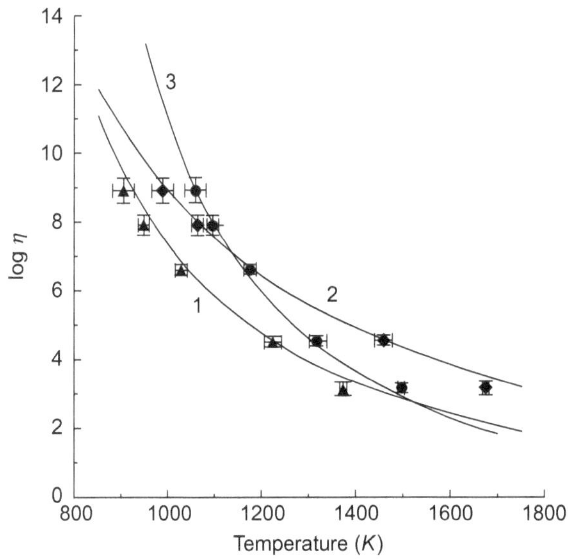


**Fig. 8.4** Comparison of characteristic viscosity points determined by hot-stage microscopy (experimental points) and Vogel-Fulcher-Tamman (VFT) curves from high-temperature viscometry *(solid line)* of different commercial glasses ((1) ▲, float glass; (2) ◆, pyrex glass; and (3) •, E glass).

Reproduced with permission from M.J. Pascual, L. Pascual, A. Durán, Determination of the viscosity-temperature curve for glasses on the basis of fixed viscosity points determined by hot stage microscopy, Phys. Chem. Glasses 42 (2001) 61–66, Copyright © The Society of Glass Technology, UK.

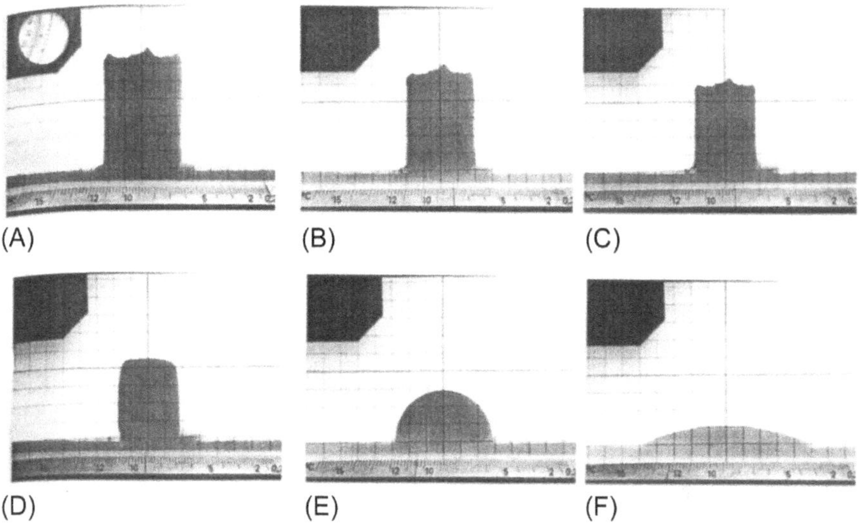


**Fig. 8.5** Photomicrographs of the characteristic points of viscosity as defined by Scholze, (A) initial shape of the sample; (B) first shrinkage ($\log \eta = 8.0 \pm 0.3$ P); (C) maximum contraction ($\log \eta = 8.2 \pm 0.5$ P); (D) softening point ($\log \eta = 6.1 \pm 0.2$ P); (E) half ball ($\log \eta = 4.6 \pm 0.1$ P); and (F) flow point ($\log \eta = 4.1 \pm 0.1$ P).
Reproduced with permission from M.J. Pascual, L. Pascual, A. Durán, Determination of the viscosity-temperature curve for glasses on the basis of fixed viscosity points determined by hot stage microscopy, Phys. Chem. Glasses 42 (2001) 61–66, Copyright © The Society of Glass Technology, UK.

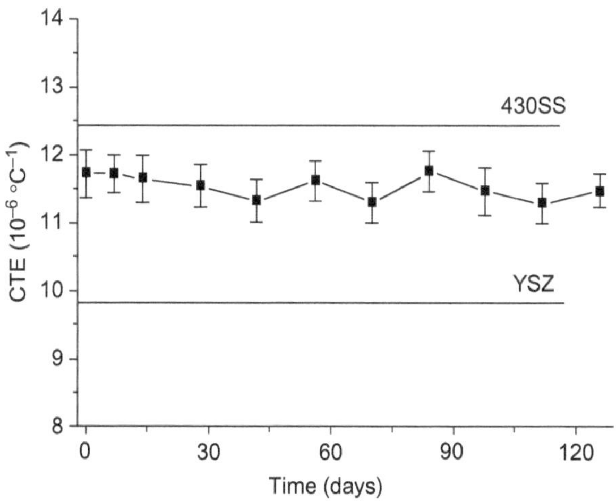


**Fig. 8.6** Variation of coefficient of thermal expansion (CTE) of sealing glass (G50) with time at 800°C in comparison with that of SOFC components, 430 stainless steel (430 SS) and yttria-stabilized zirconia (YSZ).
Reproduced with permission from S.T. Reis, M.J. Pascual, R.K. Brow, C.S. Ray, T. Zhang, Crystallization and processing of SOFC sealing glasses, J. Non-Cryst. Solids 356 (2010) 3009–3012, Copyright © Elsevier.

# 8.3 Classification of glasses and glass-ceramics based on type of SOFC sealant

There are several types of glasses developed for use as SOFC sealant based on glass former. $P_2O_5$-based glasses are unsuitable for SOFC use due to volatility and reaction with the anode. Similarly, $B_2O_3$-based glasses are also unsuitable for SOFC use due to volatility in wet fuel gas. $SiO_2$-based glasses are best possible candidates for SOFC use where the volatility has been avoided. Among the silicate glass family, alkaline earth aluminosilicate glasses were found to be the best choice, considering the properties such as high electric resistivity, high thermal expansion, high glass transition temperature, and good glass-ceramic formation.

Fig. 8.7 shows a classification of glass and glass-ceramics based on type of SOFC sealants. Compositions and properties of glasses of various types of SOFC sealant are provided in Table 8.1. These glass and glass-ceramic SOFC sealants are described as follows:

# 8.4 Processing and properties of different glasses and glass-ceramics-based SOFC sealant

## 8.4.1 Rigid bonded seals

### 8.4.1.1 Self-healing seals

Development of thermally cyclable sealants is at or near the top of the list of technical problems that hamper the progression of SOFCs. The sealants are to be thermally and chemically stable in oxidizing and reducing environment and should remain electrically insulating and mechanically strong for long term operation for a period of several thousand hours. The problem of seal cracking becomes unavoidable during thermal cycle when different SOFC components with dissimilar coefficient of thermal expansion (CTE) are sealed together. The sealants have to survive several thousand thermal

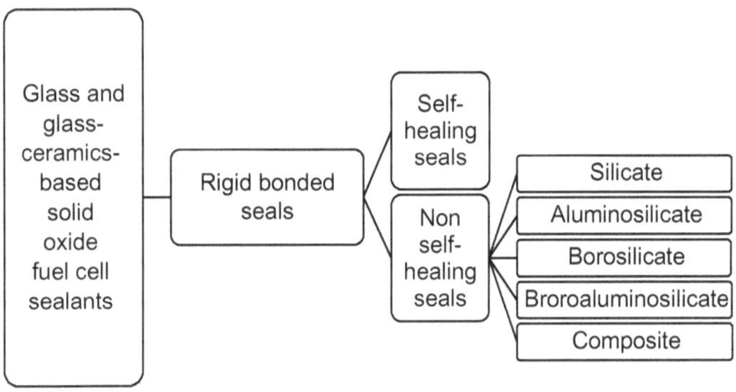


**Fig. 8.7** A classification of glass and glass-ceramics based on type of solid oxide fuel cell sealants.

**Table 8.1 Compositions and properties of glasses of various types of SOFC sealant**

| Sealant type | Composition (mol.%) | | | | | | | Property | | | | Ref. |
|---|---|---|---|---|---|---|---|---|---|---|---|---|
| | $SiO_2$ | $Al_2O_3$ | $B_2O_3$ | BaO | CaO | MgO | Others | CTE[a] ($\times 10^{-6}\,°C^{-1}$) | $T_g$[b] (°C) | $T_d$[c] (°C) | $T_s$[d] (°C) | |
| Self-healing seal | 22.1[e] | 5.4[e] | 7.3[e] | 56.4[e] | 8.8[e] | | 20 vol.% VB[f] | | 750 | | | [7] |
| | 40 | | 20 | 30 | | | 10 ($La_2O_3$) | 9.72–8.69 | 682 | | | [8] |
| | | | 67 | | 33 | | | | 655 | | | [9,10] |
| Silicate seal | 55.0 | | | 27.0 | | 18.0 | | 8.1 ± 0.5 | 715 ± 3 | 750 ± 5 | | [11] |
| Aluminosilicate seal | 38.7 | 12.9 | | | 38.7 | 9.7 | | 9.7 ± 0.1 | 732 ± 3 | | 817 ± 3 | [12] |
| Borosilicate seal | 40.0 | | 5.0 | 27.0 | | 18.0 | 8.0 (PbO) | 12.3 ± 0.5 | | 512 ± 5 | | [11] |
| | 50.0 | | 5.0 | 27.0 | | 8.0 | 8.0 (ZnO) | 8.9 ± 0.5 | | 656 ± 5 | | [11] |
| | 30 | | 30 | 20 | | | 20 (SrO) | 11.7 | 650 | 663 | | [13] |
| Boroaluminosilicate seal | 35.0 | 5.0 | 8.0 | 35.0 | 15.0 | | | 11.8 | 619 | 682 | | [14] |
| | 39.2 | 1.96 | 1.96 | | 25.48 | | 3.92 (ZnO), 2.00 ($TiO_2$), 25.48 (SrO) | 11.5 | 690 | | | [6] |
| | 34.3 | 4.9 | 5.9 | 29.3 | 5.9 | 5.9 | 9.8 (SrO), 2.0 ($Y_2O_3$), 2.0 ($La_2O_3$) | 11.2 | 672 | 739 | | [15] |
| | 36.0 | 9.0 | 6.0 | 37.0 | | | 3.0 ($Ta_2O_5$), 9.0 (SrO) | 12.4 | 673 | 736 | | [15] |
| | 22.1[e] | 5.4[e] | 7.3[e] | 56.4[e] | 8.8[e] | | | 11.2–11.4 | 534 | | 701 | [16] |
| Composite seal | 45.0 | 9.0 | 8.0 | | 18.0 | | 18.0 ($Na_2O$) | 9.3 | 580 | | 740 | [17] |
| | 40.0 | 9.0 | 8.0 | | 18.0 | | 23.0 ($Na_2O$) | 11.2 | 545 | | 680 | [17] |
| | 35.0 | 5.2 | 9.4 | 35.8 | 12.6 | 0.9 | 1.2 ($ZrO_2$) | 12.5 | 630 | | 685 | [18,19] |

[a] *CTE*, coefficient of thermal expansion (20–800°C).
[b] $T_g$, glass transition temperature (viscosity ≈ $10^{11.3}$ Pa s).
[c] $T_d$, dilatometric deformation (softening) temperature (viscosity = $10^8$–$10^9$ Pa s).
[d] $T_s$, Littleton softening point (viscosity = $10^{6.6}$ Pa s).
[e] Composition in wt.%.
[f] *VB*, vanadium boride ($VB_2$).

cycles during its service period. Different authors have developed diverse approaches for SOFC sealants to address this problem. Singh [4] has proposed and developed the concept of self-healing SOFC glass seals. He has also made this type of seals for potential application in SOFC to increase cyclability and life of the SOFC. Fig. 8.8 shows the self-healing behavior of a self-repairable glass, indicating different stages of self-restoring of cracks created by the microindentation technique (×400) magnification [4]. In this glass, disappearance of cracks as a result of flow of the glass phase after heating at high operating temperature. Self-healing glasses, therefore, heal (repair, cure) cracks generated during SOFC operation, enhancing life and reliability of the seals.

Several authors [7–10] have reported different composition for self-healing SOFC sealants. These are summarized in Table 8.1. Coillot et al. [7] have developed self-healing glassy composites prepared from the mixture of a glass of composition (wt.%) 56.4BaO–8.8CaO–5.4$Al_2O_3$–7.3$B_2O_3$–22.1$SiO_2$ and 20 vol.% vanadium boride ($VB_2$) particles. A complete and rapid self-healing was observed only under oxidizing conditions, which indicated the self-healing mechanism as a consequence of the vanadium boride ($VB_2$) active particle oxidation and subsequent pouring of fluid oxides into the cracks. Fig. 8.9 shows the "healing ratio" plotted as a function of heat treatment duration [7,20]. Kaur et al. [8] have established the self-healing behavior of barium-lanthanum-borosilicate glass of composition (mol.%) 30BaO–10$La_2O_3$–20$B_2O_3$–40$SiO_2$. This glass has the glass transition temperature ($T_g$) of 682°C and CTE of 9.72–8.69 × $10^{-6}$°$C^{-1}$. Zhang et al. [9,10] have demonstrated that the crystalline products such as $CaB_4O_7$ and $CaB_2O_4$ of the glass having composition (mol.%) 33CaO–67$B_2O_3$ exhibit the self-healing behavior. This glass has the glass transition temperature ($T_g$) of 655°C.

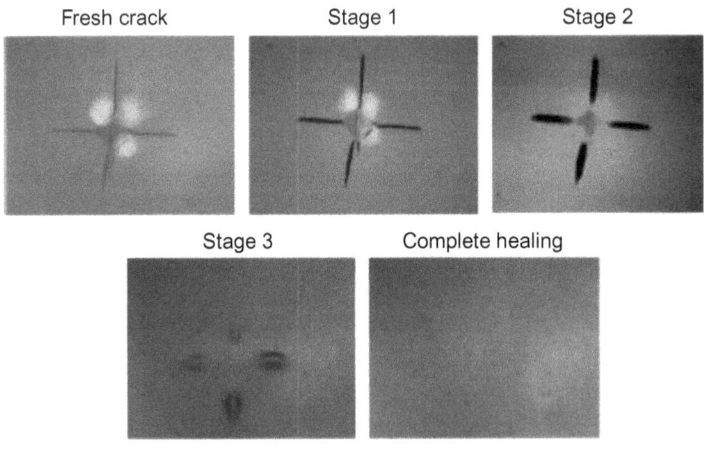


**Fig. 8.8** Self-healing behavior of a self-repairable glass indicating different stages of self-restoring of cracks created by the microindentation technique (×400) magnification. Reproduced with permission from R.N. Singh, Sealing technology for solid oxide fuel cells (SOFC), Int. J. Appl. Ceram. Technol. 4 (2) (2007) 134–144, Copyright © The American Ceramic Society.

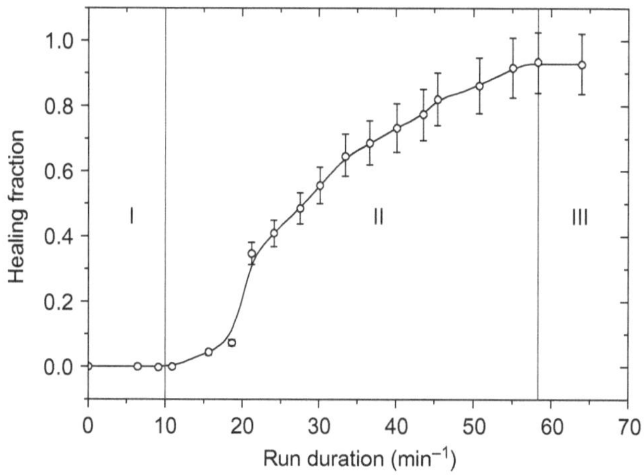


**Fig. 8.9** "Healing ratio" plotted as a function of heat treatment duration. Reproduced with permission from D. Coillot, R. Podor, F.O. Méar, L. Montagne, Characterization of self-healing glassy composites by high-temperature environmental scanning electron microscopy (HT-ESEM), J. Electron Microsc. 59 (2010) 359–366.

### *8.4.1.2 Nonself-healing seals*

#### 8.4.1.2.1 Silicate seals

Pascual et al. [11] have demonstrated that the glass-ceramics with composition (mol.%) $55SiO_2$–27BaO–18MgO presents suitable properties for application as a sealant material in SOFC. It exhibits good unions with the electrolyte 8YSZ and with the aluminum-based interconnect material FeCr alloy. This glass has the glass transition temperature ($T_g$) of $715 \pm 3$°C and CTE of $8.1 \pm 0.5 \times 10^{-6}$°$C^{-1}$. However, sealing delaminates when applied to chromium-based ferritic stainless steel Crofer 22 (CTE of $12.8 \times 10^{-6}$ °$C^{-1}$), one of the most used interconnect materials in SOFC, due to the precipitation of $BaCrO_4$ (CTE of $18–20 \times 10^{-6}$ °$C^{-1}$) at the interface. It is seen that there is a huge mismatching between the CTE of glass (CTE of $8.1 \pm 0.5 \times 10^{-6}$ °$C^{-1}$) and $BaCrO_4$ (CTE of $18–20 \times 10^{-6}$ °$C^{-1}$), the crack generated resulting in failure.

#### 8.4.1.2.2 Aluminosilicate seals

Reddy et al. [12] have studied the aluminosilicate glass and glass-ceramic sealant for SOFCs in the calcium-magnesium-aluminum-silicate (CMAS) system. The composition (mol.%) of the glass is 38.7CaO–9.7MgO–$12.9Al_2O_3$–$38.7SiO_2$. This glass has the glass transition temperature ($T_g$) of $732 \pm 3$°C and CTE of $9.7 \pm 0.1 \times 10^{-6}$ °$C^{-1}$.

#### 8.4.1.2.3 Borosilicate seals

Pascual et al. [11] have developed borosilicate glass and glass-ceramic SOFC seals replacing $SiO_2$ component of the silicate glass $55SiO_2$–27BaO–18MgO (mol.%) seals by $B_2O_3$ and PbO. Thus, we get the borosilicate glass and glass-ceramic SOFC sealants having composition (mol.%) of $40SiO_2$–27BaO–18MgO–$5B_2O_3$–10PbO with CTE of $12.3 \pm 0.5 \times 10^{-6}$ °C$^{-1}$ and $T_d$ 512±5°C. Similarly, replacing $SiO_2$ by $B_2O_3$ and MgO by ZnO, we get another borosilicate glass and glass-ceramic SOFC sealant of composition (mol.%) $50SiO_2$–27BaO–10MgO–$5B_2O_3$–8ZnO with CTE of $8.9 \pm 0.5 \times 10^{-6}$ °C$^{-1}$ and $T_d$ 656±5°C. Both the glasses offer the suitable CTE for use in SOFCs. The adding up of $B_2O_3$ gives better wettability of the glasses on the steel and a better unification, which inhibit the direct reaction between chromium oxide and the barium-rich crystalline phases to form $BaCrO_4$. It minimizes the CTE mismatching (as indicated under silicate seals), thereby cracking of the seals. Recently, Zhang and Zou [13] have reported the tuning of the thermal properties of borosilicate glass-ceramic seals for SOFCs. Its optimized composition, CTE, $T_g$, and $T_d$ are $30SiO_2$–20BaO–20SrO–$30B_2O_3$ (mol.%), $11.7 \times 10^{-6}$ °C$^{-1}$, 650°C, and 663°C, respectively.

#### 8.4.1.2.4 Boroaluminosilicate seals

The performance of boroaluminosilicate SOFC seals is found to be the best compared with the other types of silicate-based SOFC sealants. The composition and thermal properties of several boroaluminosilicate SOFC seals are provided in Table 8.1 [6,14–16]. It is seen that in these glasses or glass-ceramics, the glass modifiers are alkaline earth metal ions. This is due to the fact that the alkaline earth metal ions have different chemical properties such as field strength, ionic radius, and electronegativity, which strongly influence CTE, $T_g$, $T_s$, crystallization behavior, electric conductivity, and reactivity of the glasses [21]. Tulyaganov et al. [21] have reported the influence of BaO, SrO, ZnO, $La_2O_3$, $B_2O_3$, etc. on various properties of the resultant glasses. For example, $B_2O_3$ decreases viscosity and crystallization tendency of the silicate glasses (both are favorable in SOFC sealant application), whereas $Al_2O_3$ increases viscosity and improves chemical durability.

Luo et al. [16] have reported the application of BaO–CaO–$Al_2O_3$–$B_2O_3$–$SiO_2$ glass-ceramic seals in large size planar IT-SOFC including interfacial microstructure with SOFC components and variation of open circuit voltage (OCV). Fig. 8.10A shows the interfacial microstructure of BCAS glass seal with YSZ electrolyte and Ni-YSZ anode tested at 750°C/200 h, and Fig. 8.10B shows the variation of open circuit voltage (OCV) of a large size single cell ($10 \times 10$ cm$^2$; active area, $9 \times 9$ cm$^2$) as a function of time of thermal cycling [16].

#### 8.4.1.2.5 Composite seals

Smeacetto et al. [17] have developed two sodium-calcium-aluminosilicate glasses (SACN1, $45SiO_2$–$9Al_2O_3$–$10B_2O_3$–18CaO–$18Na_2O$, and SACN2, $40SiO_2$–$9Al_2O_3$–$10B_2O_3$–18CaO–$23Na_2O$ (mol.%)) for use as glassy seals and glass-ceramic composite seals for joining of YSZ to metallic interconnect in SOFCs. Their thermal properties are listed in Table 8.1.

The sealants have to survive several thousands of thermal cycles during its life service. To enhance thermal cyclability of the sealants, different approaches have been investigated. In the conventional compressive mica seals, it is identified that

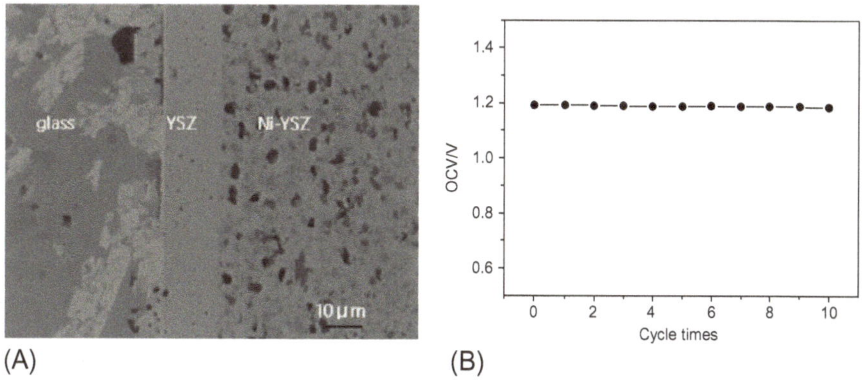


**Fig. 8.10** (A) Interfacial microstructure of BCAS glass seal with YSZ electrolyte and Ni-YSZ anode tested at 750°C/200 h. (B) Variation of open circuit voltage (OCV) of a large size single cell ($10 \times 10$ cm$^2$; active area, $9 \times 9$ cm$^2$) as function of time of thermal cycling.
Reproduced with permission from L. Luo, Y. Lin, Z.Z. Huang, Y. Wu, L. Sun, L. Cheng, J. Shi, Application of BaO–CaO–$Al_2O_3$–$B_2O_3$–$SiO_2$ glass–ceramic seals in large size planar IT-SOFC, Ceram. Int. 41 (2015) 9239–9243, Copyright © Elsevier.

the major leak paths are at the interfaces between the mica and mating materials. This observation encouraged to propose the concept of "hybrid" mica seal by adding glass interlayers at these interfaces. Chou et al. [18] have observed that the leak rates were reduced hundreds to thousands of times using this hybrid approach as compared with the conventional mica seals. Fig. 8.11 shows the schematic representation of a glass-mica composite seal [18]. The composition (mol.%) of the glass is 35.0$SiO_2$–5.2$Al_2O_3$–9.4$B_2O_3$–35.8BaO–12.6CaO–0.9MgO–1.2$ZrO_2$ with CTE of $12.5 \times 10^{-6}$ °C$^{-1}$, $T_g$ of 630°C, and $T_s$ of 685°C.

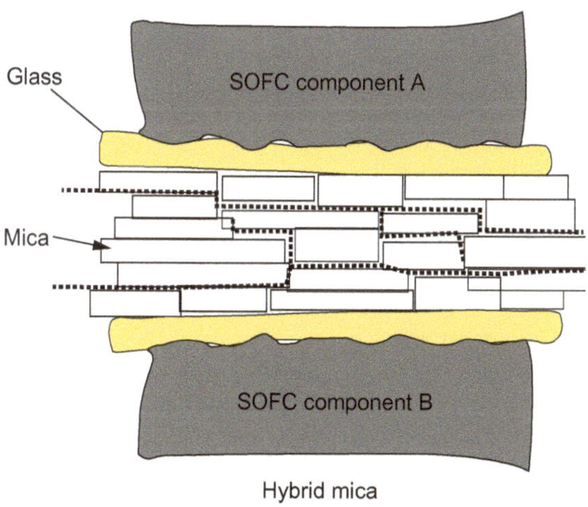


**Fig. 8.11** Schematic of glass-mica composite seal. Reproduced with permission from Y.-S. Chou, J.W. Stevenson, P. Singh, Thermal cycle stability of a novel glass–mica composite seal for solid oxide fuel cells: Effect of glass volume fraction and stresses, J. Power Sources 152 (2005) 168–174., Copyright © Elsevier.

## 8.5 Summary

In this chapter, different aspects of glasses and glass-ceramics for use as SOFC sealant were described. Year-wise published papers on SOFC during the last 25 years (1990–2014) were found out using the Web of Science software. Requirements of properties for the use of glasses and glass-ceramics as SOFC sealant are delineated. An attempt was made for classification of glasses and glass-ceramics based on the type of SOFC sealant. Processing and properties of different glasses and glass-ceramics based on different types of SOFC sealants, such as self-healing and nonself-healing (e.g., silicate seals, aluminosilicate seals, borosilicate seals, boroaluminosilicate seals, and composite seals), were described.

## Exercises

**8.1** The growth of year-wise published papers on SOFC during last 12 years is very rapid. Explain the significance.
**8.2** What are the required properties of glasses and glass-ceramics to be the good SOFC sealant?
**8.3** Explain the function of self-healing SOFC sealants.
**8.4** Describe composition and thermal properties of a boroaluminosilicate SOFC seal.
**8.5** Describe composition and thermal properties of a composite SOFC seal.

## References

[1] V. Kozhukharov, N. Brashkova, M. Ivanova, J. Carda, M. Machkova, Ceramic materials for SOFCs: current status, Bol. Soc. Esp. Cerám. Vidrio 41 (2002) 471–480.
[2] N. Laosiripojana, W. Wiyaratn, W. Kiatkittipong, A. Arpornwichanop, A. Soottitantawat, S. Assabumrungrat, Reviews on solid oxide fuel cell technology, Eng. J. 13 (2009) 65–83.
[3] N.Q. Minh, Solid oxide fuel cell technology—features and applications, Solid State Ionics 174 (2004) 271–277.
[4] R.N. Singh, Sealing technology for solid oxide fuel cells (SOFC), Int. J. Appl. Ceram. Technol. 4 (2) (2007) 134–144.
[5] M.J. Pascual, L. Pascual, A. Durán, Determination of the viscosity-temperature curve for glasses on the basis of fixed viscosity points determined by hot stage microscopy, Phys. Chem. Glasses 42 (2001) 61–66.
[6] S.T. Reis, M.J. Pascual, R.K. Brow, C.S. Ray, T. Zhang, Crystallization and processing of SOFC sealing glasses, J. Non-Cryst. Solids 356 (2010) 3009–3012.
[7] D. Coillot, R. Podor, F.O. Méar, L. Montagne, Characterisation of self-healing glassy composites by high-temperature environmental scanning electron microscopy (HT-ESEM), J. Electron Microsc. 59 (2010) 359–366.
[8] G. Kaur, O.P. Pandey, K. Singh, Self-healing behavior of barium–lanthanum–borosilicate glass and its reactivity with different electrolytes for SOFC applications, Int. J. Appl. Ceram. Technol. 11 (2014) 136–145.
[9] T. Zhang, D. Tang, H. Yang, Can crystalline phases be self-healing sealants for solid oxide fuel cells? J. Power Sources 196 (2011) 1321–1323.

[10] T. Zhang, Q. Zou, J. Zhang, D. Tang, H. Yang, Development of ceramic sealant for solid oxide fuel cell application: self-healing property, mechanical stability and thermal stability, J. Power Sources 204 (2012) 122–126.
[11] M.J. Pascual, A. Guillet, A. Durán, Optimization of glass–ceramic sealant compositions in the system MgO–BaO–$SiO_2$ for solid oxide fuel cells (SOFC), J. Power Sources 169 (2007) 40–46.
[12] A.A. Reddy, A. Goel, D.U. Tulyaganov, S. Kapoor, K. Pradeesh, M.J. Pascual, J.M. F. Ferreira, Study of calcium-magnesium-aluminum-silicate (CMAS) glass and glass-ceramic sealant for solid oxide fuel cells, J. Power Sources 231 (2013) 203–212.
[13] T. Zhang, Q. Zou, Tuning the thermal properties of borosilicate glass ceramic seals for solid oxide fuel cells, J. Eur. Ceram. Soc. 32 (2012) 4009–4013.
[14] N.P. Bansal, E.A. Gamble, Crystallization kinetics of a solid oxide fuel cell seal glass by differential thermal analysis, J. Power Sources 147 (2005) 107–115.
[15] A. Flügel, M.D. Dolan, A.K. Varshneya, Y. Zheng, N. Coleman, M. Hall, D. Earl, S. T. Misture, Development of an improved devitrifiable fuel cell sealing glass. I. Bulk properties, J. Electrochem. Soc. 154 (2007) B601–B608.
[16] L. Luo, Y. Lin, Z.Z. Huang, Y. Wu, L. Sun, L. Cheng, J. Shi, Application of BaO–CaO–$Al_2O_3$–$B_2O_3$–$SiO_2$ glass–ceramic seals in large size planar IT-SOFC, Ceram. Int. 41 (2015) 9239–9243.
[17] F. Smeacetto, M. Salvo, M. Ferraris, V. Casalegno, P. Asinari, Glass and composite seals for the joining of YSZ to metallic interconnect in solid oxide fuel cells, J. Eur. Ceram. Soc. 28 (2008) 611–616.
[18] Y.-S. Chou, J.W. Stevenson, P. Singh, Thermal cycle stability of a novel glass–mica composite seal for solid oxide fuel cells: Effect of glass volume fraction and stresses, J. Power Sources 152 (2005) 168–174.
[19] K.D. Meinhardt, L.R. Pederson, US patent 6,430,966, 2001.
[20] D. Coillot, R. Podor, F.O. Méar, L. Montagne, Characterization of self-healing glassy composites by high-temperature environmental scanning electron microscopy (HT-ESEM), J. Electron Microsc. 59 (2010) 359–366.
[21] D.U. Tulyaganov, A.A. Reddy, V.V. Kharton, J.M.F. Ferreira, Aluminosilicate-based sealants for SOFCs and other electrochemical applications—a brief review, J. Power Sources 242 (2013) 486–502.

# Functional glasses and glass-ceramics from solid waste materials

9

## Chapter Outline



## 9.1 Introduction

Waste management and recycling is the essential part of environmental sanitation. Increasing population, economic, and industrial growth have been moved out together with global environmental pollution crisis through the production of vast quantities of solid, liquid, and gaseous waste materials. According to the environment protection rule, aiming at limiting the use of the dump, the development of new recycling techniques capable to utilize the wastes into new marketable products gains an ever-increasing importance. As a consequence of the complex chemical composition and the presence of several useful elements (other than hazardous elements) in the solid wastes, the production of waste-based glasses and glass-ceramics stands for a very good approach for the utilization of the solid waste materials. This is due to the fact that the vitrification and devitrification processes are capable to increase the homogeneity and keep constancy of the chemical composition of the product

Functional Glasses and Glass-Ceramics. http://dx.doi.org/10.1016/B978-0-12-805056-9.00009-X

and to adjust the properties in order to address the reutilization [1–3]. The major advantages of these vitrification and devitrification processes are flexibility of the process, which allows treatment of various types of waste; excellent stabilization of hazardous inorganic substances; and substantial reduction in volume of the treated waste (from 20% to 97%, depending on the category of waste). In addition, the products of glasses and glass-ceramics have several beneficial properties suitable for practical utilization such as capability of the glass network to embed the heavy metals, high chemical durability, high surface hardness, and good mechanical and thermal properties [4]. These glasses and glass-ceramics, obtained from solid waste materials, are used as construction and architectural components; thermal, sound and electric insulators; wall tiles; tile glazes; filtering and wear-resistant parts; etc. [5–8]. Consequently, these processes are now widely accepted for the safe processing of a number of solid waste materials. In this chapter, different features of glasses and glass-ceramics prepared by using different kinds of solid waste materials are described.

An attempt was made using the Web of Science (version 5.17) as described in Chapter 1 to provide a global publication trend and quantitative growth in the area of waste materials. Here, year-wise data were extracted for the last 25 years (1990–2014) by searching, using the topic keyword *glass* followed by refining the document type as *article* published in the journals and meetings. It was further refined using the term *waste materials*. This plot is shown in Fig. 9.1. It gives us a clear publication trend and quantitative growth of published papers on waste materials during

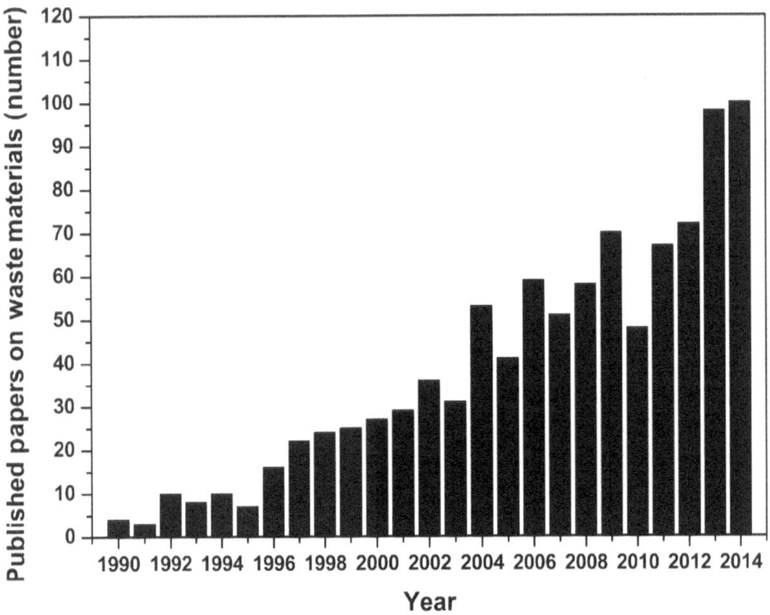


**Fig. 9.1** Year-wise published papers on glasses and glass-ceramics obtained from waste materials during the last 25 years (1990–2014).

the last 25 years (1990–2014). It is seen that there is a rapid growth of the published papers during the period 1990–2014. The rapid growth of the published papers during the last 15 years clearly signifies the importance of this subject matter.

## 9.2 Different types of solid waste materials

There is no universally recognized classification of solid waste materials. Hence, the most important types of solid waste materials are shown in Fig. 9.2 and considered for discussion in the present chapter. These are (i) industrial waste, (ii) coal fly ash, (iii) e-waste, (iv) incinerator fly ash, and (v) rice husk ash. To foresee their suitability for the preparation of glasses and glass-ceramics, it is necessary to know their chemical compositions. For this reason, chemical composition (wt.%) ranges of different solid waste materials are provided in Table 9.1. It is seen that there is wide differences (intra and inter) in the chemical compositions. The chemical composition of the final glasses and glass-ceramics, therefore, must be worked out judiciously to achieve the desired properties. Sometimes, it may require addition of other types of waste materials or chemicals.

To find out the total number of published papers on glasses and glass-ceramics derived from individual waste material during the last 25 years (1990–2014), the total publication of 969 articles on waste materials (see Fig. 9.1) was further refined by term of *individual waste material* (e.g., industrial waste, coal fly ash, and e-waste). This plot is depicted in Fig. 9.3. It gives us a clear quantitative idea of the papers published on the glasses and glass-ceramics derived from the respective waste material and a comparison among them. The decreasing order of publication trend of the individual waste material derived glasses and glass-ceramics is as follows:

Industrial waste > Coal fly ash > e-Waste > Incinerator fly ash ≫ Rice husk ash

It is seen that during the last 25 years (1990–2014) among the 969 published articles on glasses and glass-ceramics derived from waste materials, the highest number (223) of papers was published using industrial waste, the second highest (190) is with coal fly ash, the third position (70) is with e-waste (cathode ray tube (CRT), thin film transistor liquid crystal display (TFT-LCD), and mobile waste glasses), and the lowest position (7 only) is with rice husk ash. This trend of publications also gives us the idea about the extent of attention paid to the individual solid waste materials

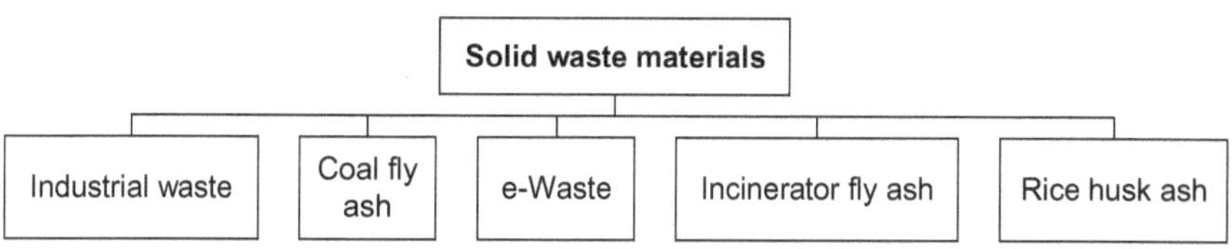


**Fig. 9.2** Important types of solid waste materials.

Table 9.1 **Chemical composition (wt.%) of different solid waste materials**

| | Industrial waste | | | | e-Waste | | | Municipal solid waste incinerator | | |
|---|---|---|---|---|---|---|---|---|---|---|
| **Oxide** | **Low-iron blast furnace slag [9,10]** | **High-iron blast furnace slag [4]** | **Bottle and window glass cullet [4,11]** | **Coal fly ash [5]** | **CRT panel glass [4]** | **CRT funnel glass [4]** | **TFT-LCD glass [4,12]** | **Fly ash [4]** | **Bottom fly ash [4]** | **Rice husk ash [4,8]** |
| $SiO_2$ | 33.41–37.63 | 10.3–13.7 | 70.8–72.60 | 15.17–66.15 | 57.87–60.7 | 51.5–54.1 | 61.20–72.84 | 7.3–27.5 | 30.3–47.4 | 89.4–94.5 |
| CaO | 36.36–38.09 | 38.7–40.4 | 9.4–10.33 | 0.36–23.71 | 0.10 | 3.5–3.77 | 1.5–20.06 | 16.6–19.5 | 18.8–23.1 | 0.17–2.55 |
| $Al_2O_3$ | 9.54–14.45 | 1.1–3.9 | 0.97–2.4 | 7.14–29.60 | 1.7–3.76 | 1.80–3.21 | 0–16.3 | 3.2–11.0 | 9.9–13.0 | 0.06–3.81 |
| $Na_2O$ | | | 13.0–14.0 | 0.15–5.06 | 7.5–9.89 | 6.20–10.21 | 0–0.3 | 13.1 | 1.9–4.5 | 0.50–0.78 |
| $K_2O$ | | | 0.25–1.1 | 0.49–2.72 | 6.9–7.29 | 8.2–9.47 | | 11.2 | 0.9–1.0 | 1.1–1.6 |
| MgO | 9.64–10.25 | 7.4–8.2 | 1.84–2.1 | 0.60–8.98 | | 1.43 | 0–1.16 | 2.6–3.1 | 4.3–10.2 | 0.23–1.30 |
| BaO | | | | | 7.95–9.90 | 0.8–1.28 | | | | |
| SrO | | | | | 8.06 | 0.7–0.89 | | | | |
| $B_2O_3$ | | | | | | | 0–10.72 | | | |
| $TiO_2$ | 0.79–3.09 | | | 0.13–1.66 | | | | | | |
| FeO/$Fe_2O_3$ | 0.38–0.45 | 9.9–11.2 | 0.28–0.3 | 3.61–15.80 | 0.22 | 0.13 | | 1.4–5.0 | 4.3–10.2 | 0.54–1.95 |
| MnO | 0.26–0.56 | | | 0.05–0.27 | | | | | | 0.03–1.09 |
| $P_2O_5$ | | 2.0 | | 0.12–0.56 | | | | 1.7 | 1.2–1.9 | 0.53–3.6 |
| ZnO | | | 0–0.12 | 0.06–1.4 | 0.63 | 0.41 | | | | |
| PbO | | | | 0.05 | 0.01–0.02 | 18.40–22.00 | | | | |
| $Cl^-$ | | | | | | | | 10.3–22.0 | | |

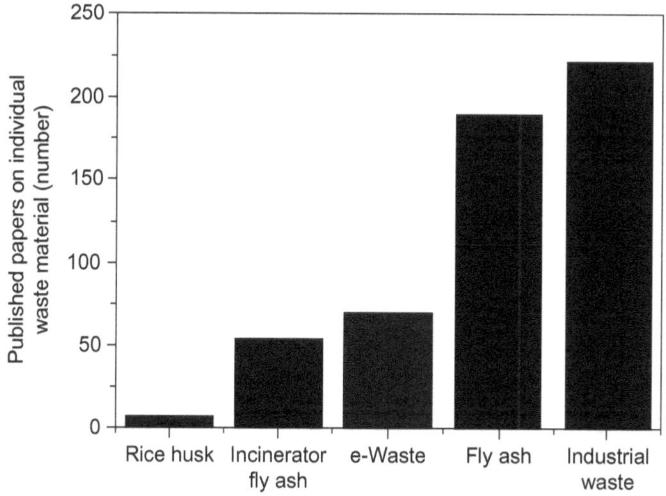


**Fig. 9.3** Paper published on glasses and glass-ceramics obtained from different types of solid waste materials during the last 25 years (1990–2014).

toward their processing and utilization in the form of glasses and glass-ceramics. Therefore, the industrial waste and rice husk ash are the highest and least attended areas, respectively.

## 9.3 Processing techniques of solid waste materials

There are various processing techniques of solid waste materials; the widely used techniques are (i) vitrification and crystallization, (ii) foaming, and (iii) sintering. These are illustrated in Fig. 9.4. These are briefly described below.

### *9.3.1 Vitrification and crystallization technique*

The vitrification and crystallization techniques yield dense glasses and glass-ceramics, respectively. These are the well-established techniques for converting various kinds of solid wastes into several reusable materials with excellent chemical stability [1,3,5]. This method consists of vitrification of solid waste materials (along with glass forming additives if necessary) by melting in the range of 1300–1500°C and followed by casting to obtain glass. The glass is subjected to controlled nucleation (sometimes nucleating agents are added with the starting solid waste materials if necessary) and crystallization processes, induced through a heat-treatment (devitrification) protocol, to transform glass into glass-ceramic (a polycrystalline material along with residual glass). The major disadvantage of this method is an energy-intensive process involving relatively high costs. However, its use can be fully acceptable if high quality products with necessary properties are produced, which can compete with existing materials for practical applications.

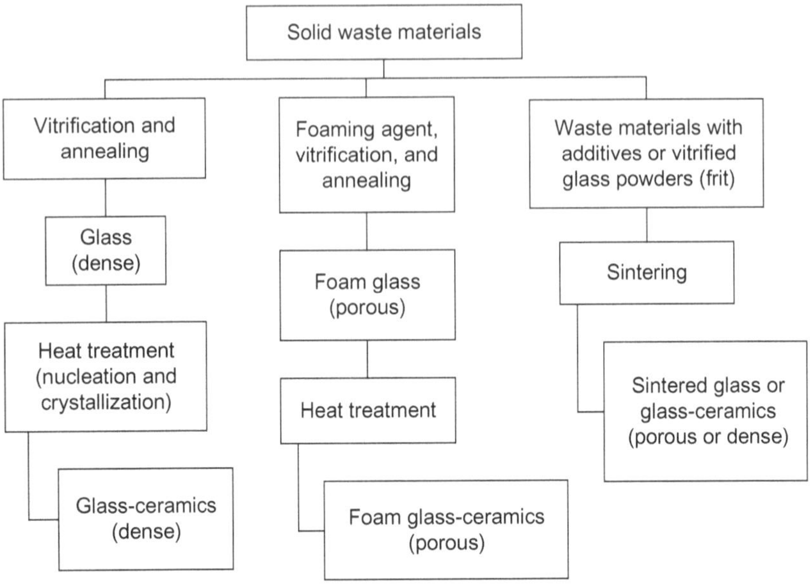


**Fig. 9.4** Different techniques of processing of solid waste materials.

## *9.3.2 Foaming technique*

Foam glasses and glass-ceramics are prepared by sintering a mixture of powders of solid waste materials and foaming agents upon heating [7,13,14]. These foaming agents are pore-forming or gas-forming agents such as dolomite ($CaMg(CO_3)_2$), calcite ($CaCO_3$), soda ash ($Na_2CO_3$), silicon carbide (SiC), titanium nitride (TiN), $SiC+MnO_2$, and coke. These additives are introduced into the glass foam batch in small quantity (up to around 20 wt.%). Quite a lot of processes take place under thermal treatment of the glass foam batch, which resulting in the foam formation. When the temperature of the batch exceeds its softening temperature, the glass batch starts sintering and forms a continuous sintered body. Particles of the foaming agent become insulated by softening glass. After a certain temperature is reached, they start emitting gases, due to combustion or decomposition, frothing the batch melts. Some gas-forming reactions are shown below:

$$CaCO_3 \rightarrow CaO + CO_2 \uparrow \tag{9.1}$$

$$SiC + 2O_2 \rightarrow SiO_2 + CO_2 \uparrow \tag{9.2}$$

$$2MnO_2 \rightarrow 2MnO + O_2 \uparrow \tag{9.3}$$

$$C + 2O_2 \rightarrow CO_2 \uparrow \tag{9.4}$$

As a result of gas release, the pores emerge in the entire sintered body where the particles of the foaming agent are entrapped. The shape of pores and the properties of foam glass achieved largely depend on the concentration and type of the foaming agent used. Foam glass-ceramics are obtained through the nucleation and crystallization processes induced by a suitable heat-treatment protocol. Controlling the processing parameters, it is possible to produce foam glass with porosity, apparent density, and compressive strength values of about 85–95 vol.%, 0.1–0.3 g $cm^3$, and 0.4–6 MPa, respectively.

### 9.3.3 Sintering technique

This sintering technique is similar to that of the conventional ceramic technology. The final products of dense or porous sintered glasses and glass-ceramics can be obtained by two ways such as *direct sintering* and *frit sintering* processes [9,15]. In the *direct sintering* technique, one type of waste is mixed up with other types or additives and is subjected to direct sintering at elevated temperature. The viscous flow sintering process leads to sintered glasses and glass-ceramics avoiding the high temperature melting stage (1300–1500°C). The dissolution of inorganic waste in the liquid phase forms the new matrices of glasses and glass-ceramics. The major advantage of this technique is that it is an energy saving process, thus cost effective. However, the homogeneity of the product is inadequate. In the *frit sintering* technique, initially, frit is prepared by melting the waste mixture and followed by pouring in either air or water. This frit is ground to the desired particle sizes. Final products are obtained by pressing the glass powders followed by sintering at the predetermined (DAT or DSC method) temperatures. As the final shape and size of the products are controlled by pressing in mold of desired shape and size, it avoids the cutting and grinding operations. Homogeneity is also relatively good. This technique is largely suitable for the deposition of a glaze (see Chapter 4, Section 4.3.4).

## 9.4 Processing and properties of solid waste materials derived glasses and glass-ceramics

The end use of the solid waste material derived glasses and glass-ceramics depends on their properties, which are largely dependent on the composition of the solid waste materials and their processing technique. Consequently, the processing and properties of glasses and glass-ceramics obtained from different solid waste materials are briefly described individually as follows:

### 9.4.1 Industrial waste

There are various kinds of industrial wastes, for example, slag of ferrous and nonferrous metallurgical industries, by-products, and ashes and wastes from mining and chemical industries The most important industrial wastes are low- and high-iron blast furnace slag, bottle, and window glass cullet. Their chemical compositions are provided in Table 9.1.

Francis [16] described the conversion of low-iron blast furnace slag into glass-ceramics by melting at 1350°C and subsequent heat treatment at 900°C. These glass-ceramics contain three different crystalline phases of gehlenite ($Ca_2Al_2SiO_7$), barium aluminum silicate ($BaAl_2Si_2O_8$), and diopside pyroxene ($Ca(Mg,Al)(Si,Al)_2O_6$). Fig. 9.5 shows the SEM photomicrographs of these glass-ceramics heat-treated at (A) 900°C (exhibits needle-like crystals) and (B) 1050°C (exhibits rod-like-shaped crystals). The white phases of needle- and rod-like crystals consist of both gehlenite and barium aluminum silicate, whereas the light gray dendritic phase is of pyroxene.

Fernandes et al. [13] described the preparation of foam glass from waste sheet glass and fly ash using carbonates as foaming agents using a low sintering process at 850°C. Fig. 9.6 depicts the photograph of foam glasses obtained from mixture of glass cullet-fly ash calcite ($GAC_1$) and glass cullet-fly ash dolomite ($GAD_1$). The foam glasses have the porosity, apparent density, and compressive strength value ranges

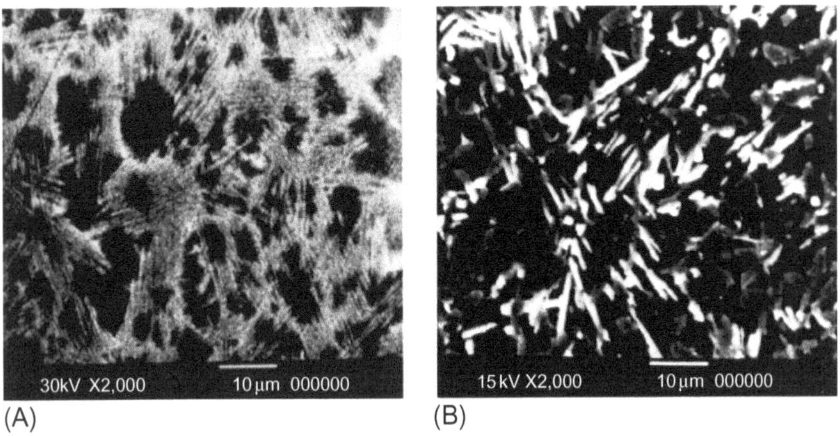


**Fig. 9.5** SEM photomicrographs of blast furnace slag derived glass-ceramics heat-treated at (A) 900°C (exhibits needle-like crystals) and (B) 1050°C (exhibits rod-like-shaped crystals). Reproduced with permission from A.A. Francis, Conversion of blast furnace slag into new glass-ceramic material, J. Euro. Ceram. Soc. 24 (2004) 2819–2824, Copyright © Elsevier.

**Fig. 9.6** Photograph of foam glasses obtained from mixture of glass cullet-fly ash calcite ($GAC_1$) and glass cullet-fly ash dolomite ($GAD_1$).
Reproduced with permission from H.R. Fernandes, D.U. Tulyaganov, J.M.F. Ferreira, Preparation and characterization of foams from sheet glass and fly ash using carbonates as foaming agents, Ceram. Int. 35 (2009) 229–235, Copyright © Elsevier.

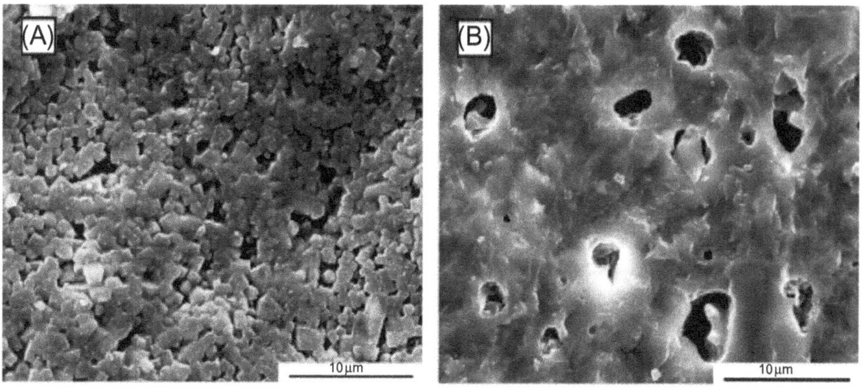


**Fig. 9.7** SEM photomicrographs of glass-ceramics obtained by a conventional ceramic-sintering route from the mixture of ground blast furnace slag and (A) 5 wt.% potash feldspar powder and (B) 8 wt.% potash feldspar powder.
Reproduced with permission from H. Liu, H. Lu, D. Chen, H. Wang, H. Xu, R. Zhang, Preparation and properties of glass–ceramics derived from blast-furnace slag by a ceramic-sintering process, Ceram. Int. 35 (2009) 3181–3184, Copyright © Elsevier.

of 75%–84%, 0.36–0.41 g $cm^{-3}$, and 2.40–2.80 MPa, respectively. The foams also bear good correlations between the compressive strength, apparent density, and microstructure (pore size, struts' thickness, and internal porosity).

Liu et al. [9] reported the preparation of blast furnace slag derived glass-ceramics by a ceramic-sintering process. The green body was prepared from the pulverized low-iron blast furnace slag (10–20 μm) blending with 5–10 wt.% potash feldspar powder. The sintering was carried out in air at the nucleation temperature of 720–760°C and crystallization temperature of 800–900°C for different durations (20–60 min), with the heating rates of 2–5°C $min^{-1}$, followed by a high temperature treatment at 1200°C. Fig. 9.7 shows the SEM photomicrographs of the glass-ceramics from the mixture of ground blast furnace slag and (A) 5 wt.% potash feldspar powder and (B) 8 wt.% potash feldspar powder. It is interesting to note that the glass-ceramic sample obtained with 5 wt.% potash feldspar contains both the gehlenite ($Ca_2Al_2SiO_7$, in large amount) and akermanite ($Ca_2MgSi_2O_7$, in small amount) crystal phases, whereas the glass-ceramic sample obtained with 8 wt.% potash feldspar contains only the gehlenite ($Ca_2Al_2SiO_7$) crystal phase in very small amount. The glass-ceramic sample obtained with 5 wt.% potash feldspar exhibits the highest microhardness (5.2 GPa), bending strength (>85 MPa), and chemical resistance (nearly 100%). Therefore, 5 wt.% potash feldspar is the optimum additive in the case of low-iron blast furnace slag to obtain good glass-ceramics.

### 9.4.2 Coal fly ash

Coal fly ash is a waste material produced in coal-burning thermal power plants. The increasing demand of energy production creates about 600 million tons of coal fly ashes in the thermal power plants, and this value tends to increase in the near future.

Conversion of this fly ash into functional glasses and glass-ceramics is one of the best methods for its reutilization and keeping the environment clean. Its chemical composition ranges are given in Table 9.1.

Kim and Kim [17] reported the processing and properties of a glass-ceramic obtained from coal fly ash through an economic process. They prepared the glass from the mixture of fly ash and additives (CaO and $TiO_2$) in Pt-Rh crucibles at 1400°C followed by casting and annealing near $T_g$. The glass was transformed into anorthite ($CaAl_2Si_2O_8$) crystal phase containing glass-ceramics by heat treatment in the range of 950–1050°C. Fig. 9.8 represents the SEM photomicrographs showing the alteration of crystal size of glass-ceramics, obtained from coal fly ash derived glasses, by a single-step heat treatment (simultaneous nucleation and crystal growth) protocol at (A) 950, (B) 980, (C) 1000, and (D) 1020°C for 60 min. The mechanical properties (hardness, strength, fracture toughness, elastic constant, and wear rate) of the glass-ceramics indicated good possibilities for their use as structural materials.

Peng et al. [18] reported the preparation of nanocrystalline glass-ceramics by crystallization of vitrified coal fly ash. In this work, the authors added two kinds of mixed fluxing additives ($Na_2O+CaO$ and $BaO+CaO$) with the fly ash. The glasses were

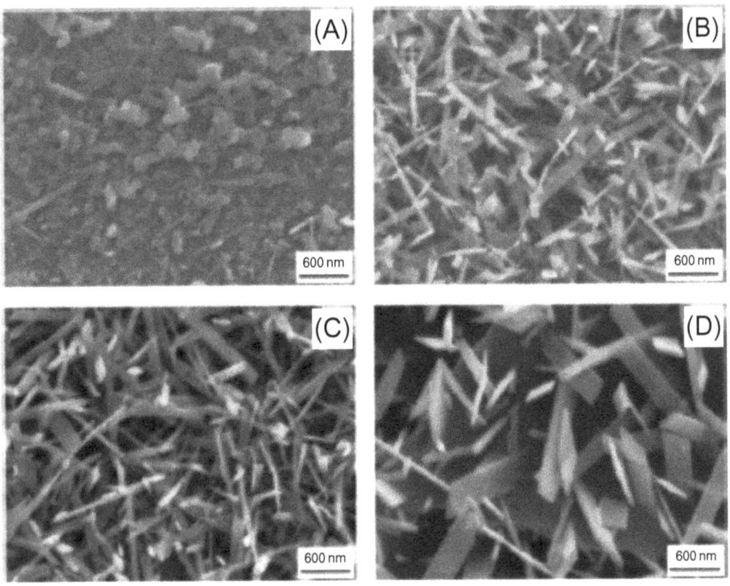


**Fig. 9.8** SEM photomicrographs showing the alteration of crystal size of glass-ceramics, obtained from coal fly ash derived glasses, by a single step heat treatment (simultaneous nucleation and crystal growth) protocol at (A) 950°C, (B) 980°C, (C) 1000°C, and (D) 1020°C for 60 min.
Reproduced with permission from J.M. Kim, H.S. Kim, Processing and properties of a glass-ceramic from coal fly ash from a thermal power plant through an economic process, J. Euro. Ceram. Soc. 24 (2004) 2825–2833, Copyright © Elsevier.

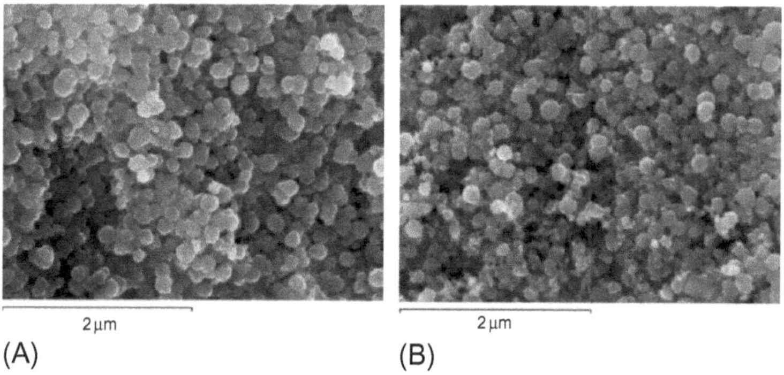


**Fig. 9.9** SEM photomicrographs of nanocrystalline glass-ceramics obtained from glasses of (A) coal fly ash+8.0 wt.% $Na_2O$+9.2 wt.% CaO by heat-treating at 825°C/60 min and (B) coal fly ash+9.0 wt.% BaO+9.1 wt.% CaO by heat treating at 840°C/90 min.
Reproduced with permission from F. Peng, K. Liang, A. Hu, H. Shao, Nano-crystal glass-ceramics obtained by crystallization of vitrified coal fly ash, Fuel 83 (2004) 1973–1977, Copyright © Elsevier.

prepared by melting in platinum crucible at 1550°C, and the glass-ceramics were obtained by heat treating the glass at different temperatures. Fig. 9.9 shows the SEM photomicrograph of nanocrystalline glass-ceramics obtained from glasses of (A) coal fly ash+8.0 wt.% $Na_2O$+9.2 wt.% CaO by heat-treating at 825°C/60 min and (B) coal fly ash+9.0 wt.% BaO+9.1 wt.% CaO by heat treating at 840°C/90 min. In both cases, the wollastonite ($CaSiO_3$) is the only crystal phase.

### 9.4.3 e-Waste

Disposal electronic-waste (e-waste) is a serious global environmental problem. These e-wastes contain many materials as discarded along with several glasses, which are originated mainly from CRT, TFT-LCD, and mobile phone. Among these, the CRT glasses comprise the largest fragment of the recent e-waste in many countries. For example, in the European Union (in 2001) 80% [19] and in the United States (in 2013), 43% [20] of the total e-waste discarded was CRT glasses. Therefore, processing of e-waste glasses into reusable glasses and glass-ceramics would be a very successful waste management technique. In view of this, the chemical composition ranges of different e-waste glasses are provided in Table 9.1.

Bernardo et al. [21] reported the fabrication of sanidine glass-ceramics using CRT glass following the sintering route. They prepared the precursor glass by melting the mixture of CRT glass, lime, and mining residue of feldspar at 1300°C. The glass was dry ball milled to the powder of sizes <37 μm(400 mesh), and the pressed powder compacts were sintered in air at 880°C to obtain glass-ceramics. The major crystal phase of these glass-ceramics is sanidine, $(K,Na)(AlSi_3)O_8$. Fig. 9.10A depicts the photograph of the wollastonite glass-ceramics like sintered sanidine glass-ceramics obtained from a mixed composition of 28 wt.% CRT panel glass, 25 wt.% lime,

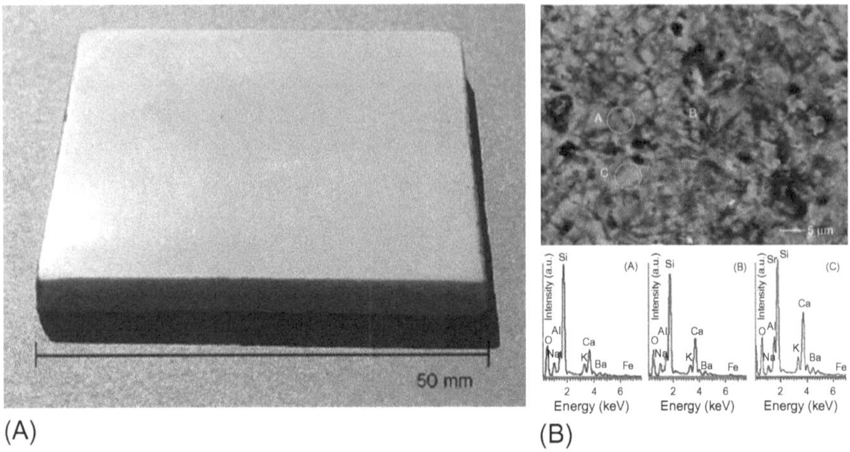


**Fig. 9.10** (A) Photograph of wollastonite glass-ceramics like sintered sanidine glass-ceramics obtained from a mixed composition of 28 wt.% CRT panel glass, 25 wt.% lime, and 47 wt.% mining residues (aluminosilicate). (B) EDS spectra of selected areas (marked as A, B, and C) of the sintered sanidine glass-ceramics.
Reproduced with permission from E. Bernardo, R. Castellan, S. Hreglich, I. Lancellotti, Sintered sanidine glass-ceramics from industrial wastes, J. Euro. Ceram. Soc. 26 (2006) 3335–3341, Copyright © Elsevier.

and 47 wt.% mining residues (aluminosilicate). Fig. 9.10B shows the EDS spectra of selected areas (marked as A, B, and C) of the sintered sanidine glass-ceramics. The zone B of fibrous crystals is attributed to sanidine. These glass-ceramics are useable as construction materials.

Glass foams are other highly valuable products for use as thermal and acoustic insulator, which can also be produced from waste CRT glasses. Méar et al. [14] described fabrication of macroporous foam glasses from waste funnel glasses. They prepared the glass foam by sintering the powder compacts of the powder mixtures of CRT funnel glass and foaming agents, SiC or TiN, in the temperature range of 750–950°C. Fig. 9.11 shows SEM photomicrographs of the foam glasses obtained by mixing CRT funnel glass powder with 5 wt.% SiC heat-treated at (A) 750°C/90 min and (B) 950°C/90 min and with 4 wt.% TiN heat-treated at (A) 750°C/90 min and (B) 950°C/90 min. The pore diameters of the foam glasses vary in the range of 3.6–360.0 μm.

## 9.4.4 Incinerator fly ash

Incineration of the municipal waste gives rise to considerable amounts of solid residues such as the bottom, boiler, and filter ashes, which are commonly known as municipal solid waste incinerator (MSWI) fly ash or shortly as incinerator fly ash.

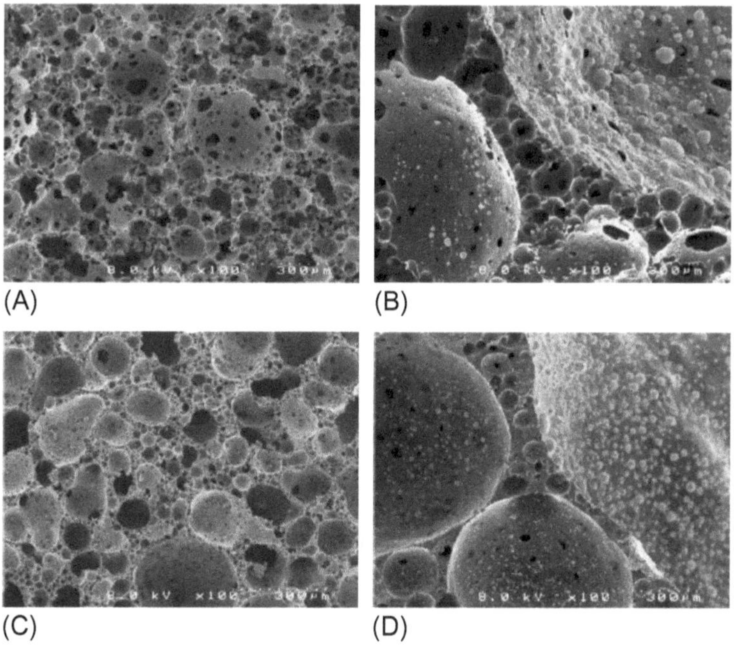


**Fig. 9.11** SEM photomicrographs of foam glass obtained by mixing CRT funnel glass powder with 5 wt.% SiC heat-treated at (A) 750°C/90 min and (B) 950°C/90 min and with 4 wt.% TiN heat-treated at (C) 750°C/90 min and (D) 950°C/90 min.
Reproduced with permission from F. Méar, P. Yot, M. Ribes, Effects of temperature, reaction time and reducing agent content on the synthesis of macroporous foam glasses from waste funnel glasses, Mater. Lett. 60 (2006) 929–934, Copyright © Elsevier.

The volume of this fly ash is huge and creates many environmental problems. For example, burning 1000 kg of municipal waste, the obtained products are around 300 kg of bottom ash and 30 kg of fly ash. However, it may vary from place to place and country to country. A chemical composition range of incinerator fly ash is provided in Table 9.1. Conversion of incinerator fly ash into reusable glasses and glass-ceramics is one of the most promising solutions, of incinerator fly ash created environmental problems, among the various available technologies.

Romero et al. [22] reported the production of sintered glass-ceramics from vitrified urban incinerator waste. They prepared the precursor glass by melting fly ash in a platinum crucible at 1500°C followed by casting onto a cold metal plate. The resulting glass was crushed and sieved to a particle size of 75 μm. Cylindrical compacts were fabricated by uniaxial cold pressing of the powder at 250 MPa without using binder. Glass-ceramics were prepared by sintering in the temperature range of 800–925°C. Fig. 9.12 shows the SEM photomicrographs and XRD patterns of MSWI fly ash derived sintered glass-ceramics heat-treated at (A) 750, (B) 800, and (C) 850°C for 30 min. The XRD patterns indicate the formation of both monoclinic and triclinic

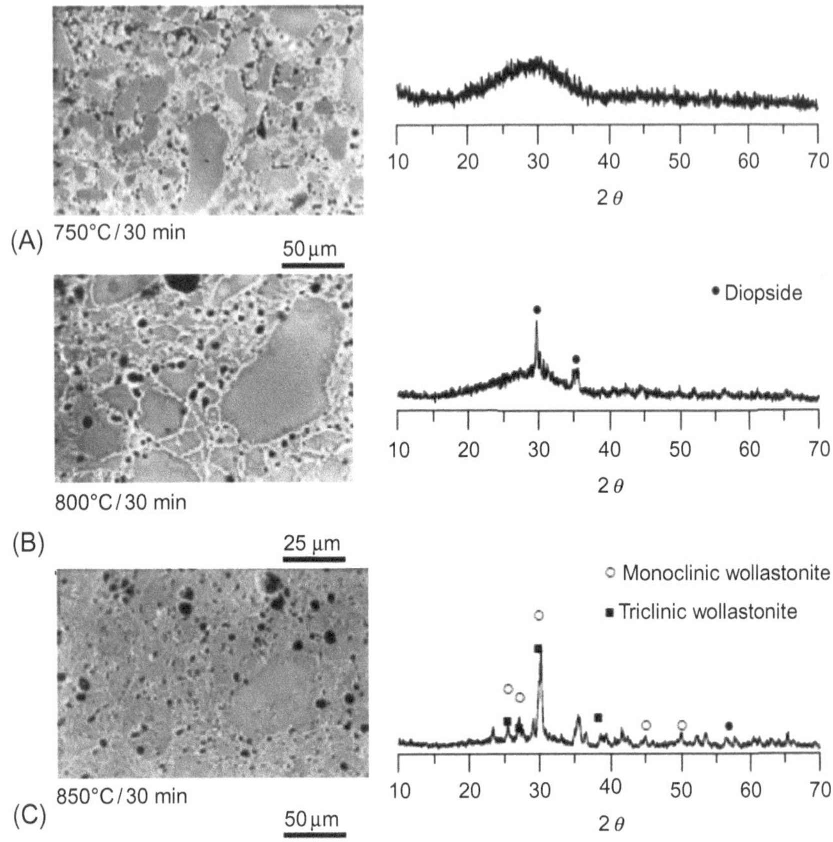


**Fig. 9.12** SEM photomicrographs and XRD patterns of municipal solid waste incinerator (MSWI) fly ash derived sintered glass-ceramics heat-treated at (A) 750°C, (B) 800°C, and (C) 850°C for 30 min.
Reproduced with permission from M. Romero, J.M. Rincón, R.D. Rawlings, A.R. Boccaccini, Use of vitrified urban incinerator waste as raw material for production of sintered glass-ceramics, Mater. Res. Bull. 36 (2001) 383–395, Copyright © Elsevier.

wollastonite ($CaSiO_3$) crystals via the intermediate formation of diopside ($CaMg(SiO_3)_2$) crystal phase in the glass-ceramics.

Ferraris et al. [23] reported the fabrication of glass matrix composites by sintering the mixture of vitrified MSWI bottom ash and solid wastes of aluminum alloy industry. The composites were prepared in air by a low temperature (740–830°C) pressureless viscous-phase sintering procedure. Fig. 9.13A shows the photograph of (a) a MSWI bottom ash derived glass, (b) glass matrix composite obtained from mixture of MSWI bottom ash glass and aluminum foundry waste, and (c) glass matrix composite obtained from mixture of a different MSWI bottom ash glass and aluminum foundry waste. Fig. 9.13B displays the SEM photomicrograph of the glass matrix

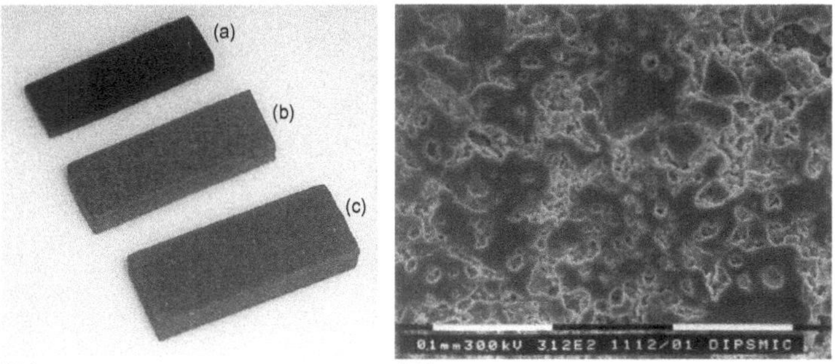


(A) (B)

**Fig. 9.13** (A) Photograph of (a) a municipal solid waste incinerator (MSWI) bottom ash derived glass, (b) glass matrix composite obtained from mixture of MSWI bottom ash glass and aluminum foundry waste, and (c) glass matrix composite obtained from mixture of a different MSWI bottom ash glass and aluminum foundry waste. (B) SEM photomicrograph of the glass matrix composite obtained from mixture of a different MSWI bottom ash glass and aluminum foundry waste (as shown in (c)) sintered at 810°C/1 h.
Reproduced with permission from M. Ferraris, M. Salvo, F. Smeacetto, L. Augier, L. Barbieri, A. Corradi, I. Lancellotti, Glass matrix composites from solid waste materials, J. Euro. Ceram. Soc. 21 (2001) 453–460, Copyright © Elsevier.

composite obtained from mixture of a different MSWI bottom ash glass and aluminum foundry waste (as shown in (c)) sintered at 810°C/1 h. Gehlenite ($Ca_2Al(AlSiO_7)$) and pyroxene ($CaMgSi_2O_6$) are the major crystal phases of the composites. The room temperature bending strength of the composite was found to be around 50 MPa; hence, its most promising functional applications are in the area of tiles.

### 9.4.5 Rice husk ash

Rice husk is the outer covering of the paddy grain, which produces 25% ash on burning. Annually, 33 million tons of rice husk ash (RHA) are being produced worldwide [24]. This ash contains 89–95% of amorphous silica ($SiO_2$) along with other oxides. Its chemical compositional range is given in Table 9.1. Therefore, the RHA is a huge source of silica, which can be utilized for the production of various functional glasses and glass-ceramics.

Andreola et al. [24] described the fabrication of glass-ceramic tiles by using RHA as silica precursor. The RHA glass frit was prepared from a composition (wt.%) of 46.52, RHA; 13.84, $Al_2O_3$; 13.16, MgO; 22.17, $Na_2CO_3$; and 4.33, $B_2O_3$ by melting at 1450°C followed by quenching in water. Glass powder of size <63 μm was obtained by crushing and grinding. The glass-ceramic tile samples were fabricated by uniaxial pressing of the powder at 40 MPa in a steel die followed by the sintering-crystallization process in the temperature range of 650–1000°C. Fig. 9.14 depicts the SEM photomicrographs of glass-ceramic tiles obtained using rice husk ash as silica precursor heat-treated at (A) 700 and (B) 800°C for 40 min. These glass-ceramics were found to produce nepheline ($Na_2O{\cdot}Al_2O_3{\cdot}2SiO_2$) and forsterite

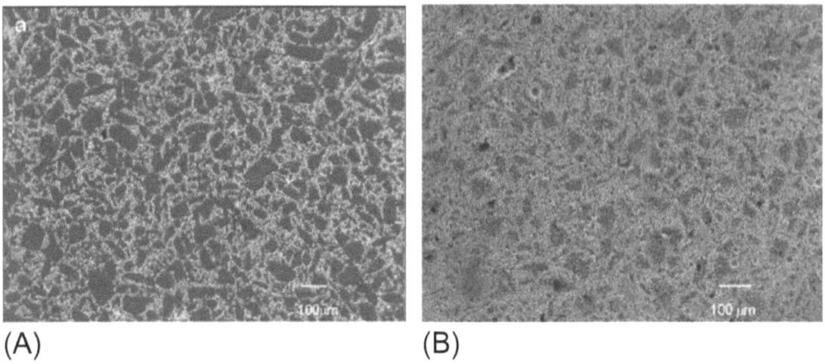


**Fig. 9.14** SEM photomicrographs of glass-ceramic tiles obtained using rice husk ash as silica precursor heat-treated at (A) 700°C and (B) 800°C for 40 min.
Reproduced with permission from F. Andreola, M.I. Martín, A.M. Ferrari, I. Lancellotti, F. Bondioli, J.M. Rincón, M. Romero, L. Barbieri, Technological properties of glass-ceramic tiles obtained using rice husk ash as silica precursor, Ceram. Int. 39 (2013) 5427–5435, Copyright © Elsevier.

($2MgO{\cdot}SiO_2$) crystal phases. Their bending strength, Young's modulus, shear modulus, Poisson's ratio, and Mohs hardness were found to vary in the ranges of 24–39 MPa, 43–61 GPa, 17–23 GPa, 0.14–0.30, and 6–9, respectively.

Wu et al. [25] reported the preparation of porous 45S5 Bioglass derived glass-ceramic scaffolds by using rice husk as a porogen additive. The scaffolds were prepared by mixing 45S5 Bioglass (45.0, $SiO_2$; 24.5, CaO; 24.5, $Na_2O$; and 6.0, $P_2O_5$ wt.%) powders with different percentages of RHA. The scaffold green body was fabricated by pressing the mixture along with polyvinyl alcohol (PVA) binder at 100 kg $cm^{-2}$. The pressed compact was heat-treated initially at 450°C for 1 h to burn out the PVA and finally sintered at 1050°C for 1 h. Fig. 9.15 illustrates the SEM

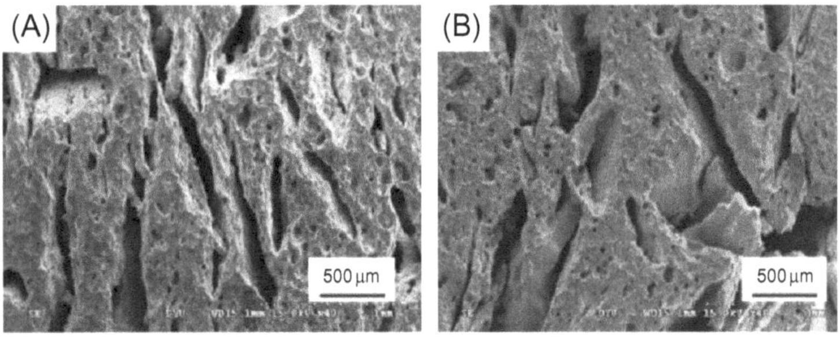


**Fig. 9.15** SEM photomicrographs of sintered (at 1050°C/1 h) porous glass-ceramic scaffolds of composition (A) 20 wt.% 45S5 Bioglass and 80 wt.% rice husk ash and (B) 30 wt.% 45S5 Bioglass and 70 wt.% rice husk ash.
Reproduced with permission from S.-C. Wu, H.-C. Hsu, S.-H. Hsiao, W.-F. Ho, Preparation of porous 45S5 Bioglass-derived glass–ceramic scaffolds by using rice husk as a porogen additive, J. Mater. Sci. Mater. Med. 20 (2009) 1229–1236, Copyright © Springer.

photomicrographs of sintered (at 1050°C/1 h) porous glass-ceramic scaffolds of composition (A) 20 wt.% 45S5 Bioglass and 80 wt.% rice husk ash and (B) 30 wt.% 45S5 Bioglass and 70 wt.% rice husk ash. Both the samples exhibit elongated macro (>420 μm in length and >100 μm in breadth) and meso (25–80 μm in spherical diameter) porous structures. Major crystalline phases of these glass-ceramic scaffolds are $Na_2Ca_2Si_3O_9$ and apatite. These glass-ceramic scaffolds were found to be exhibited in the increasing trend of formation of apatite layer with time on immersion in simulated body fluid (SBF) solution.

## 9.5 Functional applications

From the preceding discussion, it is clear that various dense and porous glasses and glass-ceramics with exploitable properties can be prepared from different solid waste materials. These glasses and glass-ceramics can be used for different functional applications. A summarized list of the functional applications of different solid waste material derived glasses and glass-ceramics is provided in Table 9.2 [5–8]. Their major uses are as construction and architectural components; building blocks; wall tiles; thermal, electric, and sound insulators; grinding and abrasive medium; anticorrosive container lining; wear-resistant parts; etc.

## 9.6 Summary

Here, different dense and porous glasses and glass-ceramics fabricated from various solid waste materials were described. Solid waste materials such as industrial waste, coal fly ash, e-waste, MSWI fly ash, and rice husk ash with their chemical composition were considered for discussion. Processing of these wastes through vitrification, nucleation and crystallization, and foaming and sintering procedures was described. Processing and properties of the glasses and glass-ceramics obtained from individual solid waste materials were deliberated with reference to industrial waste, coal fly ash, e-waste, incinerator fly ash, and rice husk ash. Various functional applications of various solid waste derived glasses and glass-ceramics were offered in details. Finally, future outlook and exercises on glasses and glass-ceramics obtained from various solid waste materials were provided.

## 9.7 Future outlook

On the basis of the chemical composition of different types of solid waste materials and tuning possibilities by addition of external chemicals or other category of waste, it is possible to tailor the batches most suitable for vitrification and subsequent crystallization to monitor the properties of the waste-based glasses and glass-ceramics for reutilization. Therefore, the challenges are to be addressed to produce components possessing the physicochemical, mechanical, and esthetic characteristics required by the specific applications. The solid wastes contain some toxic elements; therefore, wide application and commercial exploitation of the products made from wastes should be appropriately addressed and clarified to ensure their acceptance by the

**Table 9.2 Functional applications of different solid waste material derived glasses and glass-ceramics [5–8]**

| Solid waste materials | Solid waste material derived glasses and glass-ceramics | Functional applications |
|---|---|---|
| Industrial waste | Glass (dense) | Building blocks, wall tiles |
| | Foam glass (porous) | Thermal and sound insulator |
| | Glass-ceramics (dense) | Construction and architectural components, wall tiles, high-resistant wall cladding, grinding and abrasive medium, anticorrosive container lining, wear-resistant parts |
| | Glass-ceramics (porous) | Thermal and sound insulator |
| Coal fly ash | Glass (dense) | Radiation shielding materials, construction materials, building blocks, wall tiles |
| | Glass-ceramics (dense) | Construction and architectural components, electric insulators, grinding and abrasive medium, anticorrosive container lining, wear-resistant parts |
| | Glass-ceramics (porous) | Thermal and sound insulator |
| e-Waste | Glass (dense) | Tile glaze |
| | Foam glass (porous) | Thermal and sound insulator |
| | Glass-ceramics (dense) | Construction and architectural components, tile glaze |
| | Glass-ceramics (porous) | Thermal and sound insulator |
| | Micro- and macro-size open cell structured glass-ceramics | Filtering materials |
| Municipal solid waste incinerator fly ash | Glass (dense) | Building blocks and wall tiles |
| | Glass (porous) | Thermal and sound insulator |
| | Glass-ceramics (dense) | Construction and architectural components, building blocks, wall tiles, grinding and abrasive medium, anticorrosive container lining and wear-resistant parts |
| | Glass-ceramics (porous) | Thermal and sound insulator |
| Rice husk ash | Glass (porous) | Catalyst support, 45S5 Bioglass scaffolds |
| | Glass-ceramics (dense) | Cordierite glass-ceramics, lithium disilicate bioglass-ceramics |
| | Glass-ceramics (porous) | Bioactive glass-ceramic scaffolds |

public. The preparation procedure should be established in a cost effective manner to reduce the production cost in industries [3,5,26]. The major weakness of glass and glass-ceramic industries is the high "embodied energy" (refers to the energy used in the manufacturing). This is the burning challenge to the researchers and industries to act for significant decreasing of the embodied energy. One of the best approaches is to develop glasses with lower melting and glass transition temperature ($T_g$) [27]. Another important technique is the direct sintering of the solid wastes or mixture of solid wastes [1]. Attention should be paid to develop waste-based glasses and glass-ceramics with new or complex properties and functionalities, which cannot be achieved by using virgin raw materials [4].

## Exercises

**9.1** What is the significance of the development of functional glasses and glass-ceramics from solid waste materials?

**9.2** What are the important types of solid waste materials?

**9.3** Describe the chemical compositions of the important types of solid waste materials. What conclusion can you draw from their chemical compositions?

**9.4** Describe the different important processing techniques of fabrication of glasses and glass-ceramics from solid waste materials.

**9.5** Describe the processing technique and important properties of the glasses and glass-ceramics obtained from industrial wastes.

**9.6** Describe the processing technique and important properties of the glasses and glass-ceramics obtained from coal fly ashes.

**9.7** Describe the processing technique and important properties of the glasses and glass-ceramics obtained from e-wastes.

**9.8** Describe the processing technique and important properties of the glasses and glass-ceramics obtained from MSWI fly ashes.

**9.9** Describe the processing technique and important properties of the glasses and glass-ceramics obtained from rice husk ash.

**9.10** Describe the different functional applications of the glasses and glass-ceramics obtained from various solid waste materials.

## References

[1] W.E. Lee, A.R. Boccaccini, J.A. Labrincha, C. Leonelli, C.H. Drummond III, C.R. Cheeseman, Green engineering—ceramic technology and sustainable development, Am. Ceram. Soc. Bull. 86 (2007) 18–25.

[2] E. Gomez, D.A. Rani, C.R. Cheeseman, D. Deegan, M. Wise, A.R. Boccaccini, Thermal plasma technology for the treatment of wastes: a critical review, J. Hazard. Mater. 161 (2009) 614–626.

[3] L. Barbieri, A. Corradi, I. Lancellotti, Wastes based glasses and glass-ceramics, Mater. Constr. 51 (2001) 197–208.

[4] A. Rincón, M. Marangoni, S. Cetin, E. Bernardo, Recycling of inorganic waste in monolithic and cellular glass-based materials for structural and functional applications, J. Chem. Technol. Biotechnol. 91 (2016) 1946–1961.

[5] R.D. Rawlings, J.P. Wu, A.R. Boccaccini, Glass-ceramics: their production from wastes—a review, J. Mater. Sci. 41 (2006) 733–761.
[6] S. Singh, A. Kumar, D. Singh, K.S. Thind, G.S. Mudahar, Barium–borate–fly ash glasses: as radiation shielding materials, Nucl. Instr. Meth. Phys. Res. B 266 (2008) 140–146.
[7] J. König, R.R. Petersen, Y. Yue, Fabrication of highly insulating foam glass made from CRT panel glass, Ceram. Int. 41 (2015) 9793–9800.
[8] S. Chandrasekhar, K.G. Satyanarayana, P.N. Pramada, P. Raghavan, T.N. Gupta, Review. Processing, properties and applications of reactive silica from rice husk—an overview, J. Mater. Sci. 38 (2003) 3159–3168.
[9] H. Liu, H. Lu, D. Chen, H. Wang, H. Xu, R. Zhang, Preparation and properties of glass–ceramics derived from blast-furnace slag by a ceramic-sintering process, Ceram. Int. 35 (2009) 3181–3184.
[10] Y. Zhao, D. Chen, Y. Bi, M. Long, Preparation of low cost glass–ceramics from molten blast furnace slag, Ceram. Int. 38 (2012) 2495–2500.
[11] Y. Pontikes, L. Esposito, A. Tucci, G.N. Angelopoulos, Thermal behaviour of clays for traditional ceramics with soda–lime–silica waste glass admixture, J. Eur. Ceram. Soc. 27 (2007) 1657–1663.
[12] K.-L. Lin, W.-K. Chang, T.-C. Chang, C.-H. Lee, C.-H. Lin, Recycling thin film transistor liquid crystal display (TFT-LCD) waste glass produced as glass–ceramics, J. Clean. Prod. 17 (2009) 1499–1503.
[13] H.R. Fernandes, D.U. Tulyaganov, J.M.F. Ferreira, Preparation and characterization of foams from sheet glass and fly ash using carbonates as foaming agents, Ceram. Int. 35 (2009) 229–235.
[14] F. Méar, P. Yot, M. Ribes, Effects of temperature, reaction time and reducing agent content on the synthesis of macroporous foam glasses from waste funnel glasses, Mater. Lett. 60 (2006) 929–934.
[15] I. Ponsot, R. Detsch, A.R. Boccaccini, E. Bernardo, Waste derived glass ceramic composites prepared by low temperature sintering/sinter-crystallisation, Adv. Appl. Ceram. 114 (2015) S17–S25.
[16] A.A. Francis, Conversion of blast furnace slag into new glass-ceramic material, J. Eur. Ceram. Soc. 24 (2004) 2819–2824.
[17] J.M. Kim, H.S. Kim, Processing and properties of a glass-ceramic from coal fly ash from a thermal power plant through an economic process, J. Eur. Ceram. Soc. 24 (2004) 2825–2833.
[18] F. Peng, K. Liang, A. Hu, H. Shao, Nano-crystal glass-ceramics obtained by crystallization of vitrified coal fly ash, Fuel 83 (2004) 1973–1977.
[19] C. Gable, B. Shireman, Computer and electronic product stewardship: are we ready for the challenge? Environ. Qual. Manag. 11 (2001) 35–45.
[20] An Analysis of the Demand for CRT Glass Processing in the U.S., https://www.recyclingtoday.com/FileUploads/file/ An Analysis of the Demand for CRT Glass Processing in the US.pdf, 2013, 1–47 (accessed 01.07.16).
[21] E. Bernardo, R. Castellan, S. Hreglich, I. Lancellotti, Sintered sanidine glass-ceramics from industrial wastes, J. Eur. Ceram. Soc. 26 (2006) 3335–3341.
[22] M. Romero, J.M. Rincón, R.D. Rawlings, A.R. Boccaccini, Use of vitrified urban incinerator waste as raw material for production of sintered glass-ceramics, Mater. Res. Bull. 36 (2001) 383–395.
[23] M. Ferraris, M. Salvo, F. Smeacetto, L. Augier, L. Barbieri, A. Corradi, I. Lancellotti, Glass matrix composites from solid waste materials, J. Eur. Ceram. Soc. 21 (2001) 453–460.

[24] F. Andreola, M.I. Martín, A.M. Ferrari, I. Lancellotti, F. Bondioli, J.M. Rincón, M. Romero, L. Barbieri, Technological properties of glass-ceramic tiles obtained using rice husk ash as silica precursor, Ceram. Int. 39 (2013) 5427–5435.
[25] S.-C. Wu, H.-C. Hsu, S.-H. Hsiao, W.-F. Ho, Preparation of porous 45S5 Bioglass®-derived glass–ceramic scaffolds by using rice husk as a porogen additive, J. Mater. Sci. Mater. Med. 20 (2009) 1229–1236.
[26] H. Isa, A review of glass-ceramics production from silicate wastes, Int. J. Phys. Sci. 6 (2011) 6781–6790.
[27] E. Axinte, Glasses as engineering materials: a review, Mater. Des. 32 (2011) 1717–1732.

# Part Three

# Functional glasses and glass-ceramics of nonoxide systems

# Functional chalcogenide glasses and glass-ceramics

# 10

## Chapter Outline



## 10.1 Introduction

Oxide glasses have the inadequacy of transmission up to 4 μm in the electromagnetic spectrum. As a consequence, chalcogenide glasses were discovered in quest of glasses, which can transmit in the long infrared region up to 25 μm. Chalcogenide glasses are made up from mixtures of elements containing at least one or more chalcogen elements (S, Se, and Te) of the group 16 or VIA (except O) of the periodic table of elements as shown in Fig. 10.1A. In addition, the mixture may contain single or multiple electropositive elements such as As, Ge, Ga, Sb, Sn, In, P, Si, Al, Pb, Ag, La, Na, K, Rb, and Cs. Chalcogenide glasses in conjunction with halogen elements (Cl, Br, and I) create the chalcohalide glasses. According to the band theory, they possess the band gap of semiconductors; as a result, chalcogenide glasses are considered as "*semiconducting or semiconductor glasses*" [1]. Since these glasses transmit up to the long infrared region, chalcogenide glasses are also regarded as "*infrared glasses*" [2]. The nature of bonding in chalcogenide glasses is predominantly covalent while that in oxide glasses is ionic. Chalcogenide glasses possess several unique properties such as capability to transmit up to the long infrared region, low softening point and glass transition temperature, less mechanical stability, and inability to react with water. These glasses are used in several passive and active functional applications such as thermal imaging, night vision, infrared laser power delivery, nonlinear optics,


Functional Glasses and Glass-Ceramics. http://dx.doi.org/10.1016/B978-0-12-805056-9.00010-6

| Partial periodic table of elements | | | | |
|---|---|---|---|---|
| Group | | | | |
| 13 | 14 | 15 | 16 | 17 |
| III-A | IV-A | V-A | VI-A | VII-A |
| B<br>Boron | C<br>Carbon | N<br>Nitrogen | O<br>Oxygen | F<br>Fluorine |
| Al<br>Aluminium | Si<br>Silicon | P<br>Phosphorous | **S**<br>**Sulphur** | Cl<br>Chlorine |
| Ga<br>Gallium | Ge<br>Germanium | As<br>Arsenic | **Se**<br>**Selenium** | Br<br>Bromine |
| In<br>Indium | Sn<br>Tin | Sb<br>Antimony | **Te**<br>**Tellurium** | I<br>Iodine |

(A)

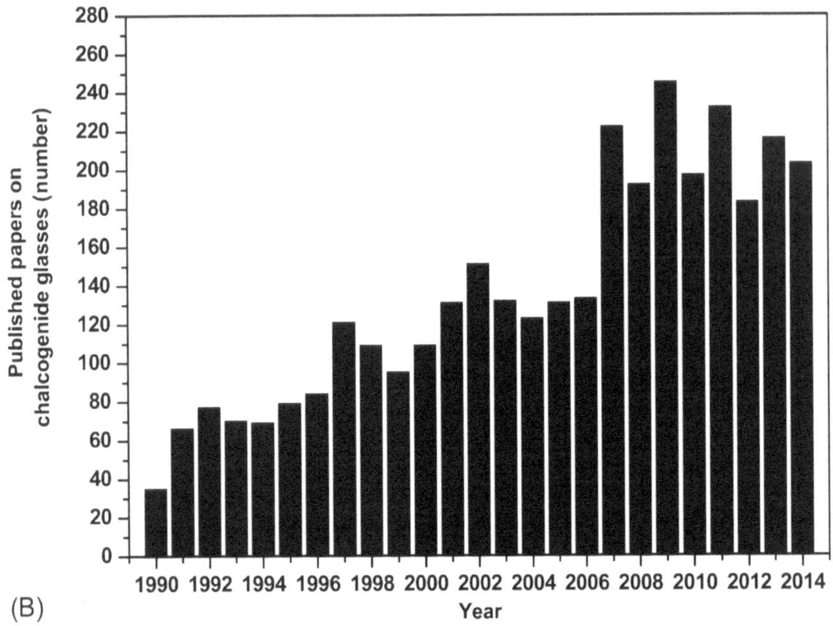


(B)

**Fig. 10.1** (A) Partial periodic table of elements showing the position of chalcogen elements S, Se, and Te *(shaded)*. (B) Year-wise published papers on chalcogenide glasses during the last 25 years (1990–2014).

chemical sensing, and memory switching. In this chapter, various features of chalcogenide glasses and glass-ceramics will be described.

An attempt was made using the Web of Science (version 5.17) as described in Chapter 1 to provide a global publication trend and quantitative growth in the area of chalcogenide glasses and glass-ceramics. Here, year-wise data were extracted for the last 25 years (1990–2014) by searching, using the topic keyword "*glass*" followed by refining the document type as "*article*" published in the journals and meetings. It was further refined using the term "*chalcogenide glass*." This plot is shown in Fig. 10.1B. It gives us a clear publication trend and quantitative growth of published papers on chalcogenide glasses and glass-ceramics during the last 25 years (1990–2014).

## 10.2 History of development of chalcogenide glasses

Some chronological events of development of chalcogenide glasses are summarized in Table 10.1. Perhaps, Frerichs [3] was the first in 1953 to report arsenic trisulfide ($As_2S_3$) as the new optical glass in useful thickness with good transparency in the

**Table 10.1 Some chronological events of development of chalcogenide glasses**

| Year | Event | Author(s) | Ref. |
|---|---|---|---|
| 1953 | Perhaps, arsenic trisulfide ($As_2S_3$) was reported as the new optical glass in useful thickness with good transparency in the infrared region up to 12–13 μm | Frerichs | [3] |
| 1963 | Developed glasses of the Si–As–Te system with infrared transmission from 1 to 20 μm and high softening points (around 500°C) | Hilton and Brau | [4] |
| 1964 | Reported glasses of the As–Ge–Se system | Nemilov; Savage and Nielsen | [5–7] |
| 1965 | Reported glasses of the S–Se–Te system | Pearson | [8] |
| 1966 | Developed glasses of the IVA–VA–VIA ternary systems, for example, Ge–As–Te and Ge–P–Te | Hilton et al. | [9] |
| 1970 | Reported chalcohalide glasses of the Ge–Se–I and Si–Se–I systems | Dembovsky and Popova | [10] |
| 1972 | Reported chalcohalide glasses of the As–Te–I system | Quin-Rod and Johnson | [11] |
| 1975 | First report on purification of glasses by distillation applied to the $Ge_{28}Sb_{12}Se_{60}$ and $Ge_{33}As_{12}Se_{55}$ glasses | Hilton, Hayes, and Rechtin | [12] |
| 1975 | First report on purification of glasses by distillation applied to the $As_2Se_3$ glass | Moynihan et al. | [13] |
| 1992 | Preparation of high purity chalcogenide glasses by distillation using Mg strip as an oxygen getter | Reitter et al. | [14] |

infrared region up to 12–13 μm. He also reported the extension of the infrared absorption edge up to 20 μm using selenium. In 1963, Hilton and Brau [4] reported the development of glasses of the Si–As–Te system with infrared transmission from 1 to 20 μm and high softening points (around 500°C). In 1964, Nemilov [5,6] and Savage and Nielsen [7] simultaneously reported the development of glasses of the As–Ge–Se system along with studies on viscosity, structure, and transmission properties. In 1965, Pearson [8] described the glasses of the S–Se–Te system. In 1966, Hilton et al. [9] developed glasses of the IVA–VA–VIA ternary systems, for example, Ge–As–Te and Ge–P–Te. Dembovsky and Popova [10] reported the chalcohalide glasses of the Ge–Se–I and Si–Se–I systems in 1970, whereas Quin-Rod and Johnson [11] reported the development of chalcohalide glasses of the As–Te–I system in 1972. Hilton, et al. [12] in 1975 were the first to report the purification of chalcogenide glasses by distillation. They demonstrated this with the examples of $Ge_{28}Sb_{12}Se_{60}$ and $Ge_{33}As_{12}Se_{55}$ glasses. While Moynihan et al. [13] simultaneously reported the purification of glasses by distillation with the example of the $As_2Se_3$ glass in 1975. In 1992, Reitter et al. [14] reported the preparation of high purity chalcogenide glasses by distillation using Mg strip as an oxygen getter.

## 10.3 Structure of chalcogenide glasses

The structures of chalcogenide glasses are described by inorganic networks with one-, two-, and three-dimensional (1D, 2D, and 3D) space connectivity or in some cases in between of them [15]. These are shown in Fig. 10.2. In these structures, the chemical bonds have high degree of freedom permitting bending and rotations without affecting the total bond energy. Fig. 10.2A shows the 1D structure of Se or S glass that is made up of infinite chains. Fig. 10.2B shows the 2D distorted planar structure of $As_2S_3$ or $As_2Se_3$ glass that is made up of 2-coordinated S (or Se) and 3-coordinated As.

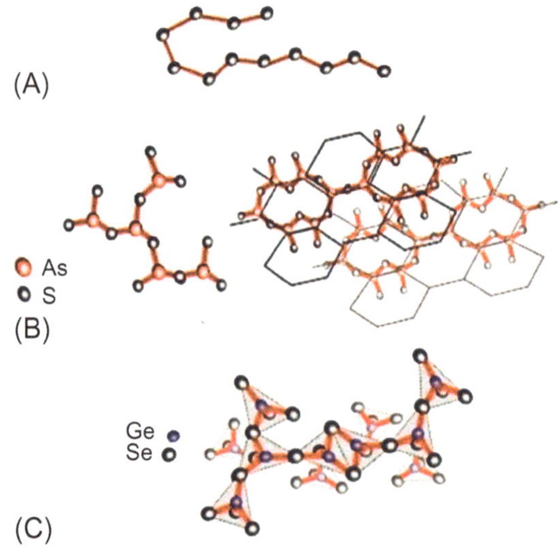


**Fig. 10.2** Structure of different chalcogenide glasses. (A) 1D Se glass made of infinite chains, (B) 2D distorted planar $As_2S_3$ glass made of 2-coordinated S and 3-coordinated As, and (C) 3D $GeSe_2$ glass made of $GeSe_4$ tetrahedral connections. Courtesy: X.H. Zhang, B. Bureau, P. Lucas, C. Boussard-Pledel, J. Lucas, Glasses for seeing beyond visible, Chem. Eur. J. 14 (2008) 432–442.

Fig. 10.2C shows the 3D structure of $GeSe_2$ or $GeS_2$ glass that is made up of $GeSe_4$ tetrahedral connections, that is, the Ge atom is 4-coordinated. In the structures as shown in Figs. 10.2B and C, the glass network is linked by the bridges of the S or Se atom.

Fig. 10.3 shows the room temperature polarized (HH) and depolarized (HV) Raman spectra of (A) $Ge_{30}As_{10}Se_{25}Te_{35}$ and (B) $Ge_{30}As_{10}Se_{30}Te_{10}I_{20}$ glasses [16]. The broad Raman bands of the $Ge_{30}As_{10}Se_{25}Te_{35}$ glass extended in the ranges of 140–190, 190–230, and 230–280 $cm^{-1}$ are assigned to the $[GeSe_{4-x}Te_x]$ mixed tetrahedral units, $[AsSe_{3-x}Te_x]$ mixed pyramids, and combination of $[GeSe_{4-x}Te_x]$ mixed tetrahedral units and Ge–Te–Ge chains, respectively, whereas, in the $Ge_{30}As_{10}Se_{30}Te_{10}I_{20}$ glass, the broad Raman bands widened over the ranges of 150–220, 160–190, 190–230, and 230–280 $cm^{-1}$ are attributed to the combination of $[AsSe_{3-y}I_y]$ mixed pyramids and $[GeSe_{4-y}I_y]$ mixed tetrahedral units, $[GeSe_{4-y}I_y]$ mixed tetrahedral units, $[AsSe_{3-y}Te_y]$ mixed pyramids, and $[GeSe_{3/2}I]$ mixed tetrahedral units, respectively. Schematic two-dimensional representations of the structures of the $Ge_{30}As_{10}Se_{25}Te_{35}$ and $Ge_{30}As_{10}Se_{30}Te_{10}I_{20}$ glasses are illustrated in Fig. 10.4A and B, respectively [16]. In these structures, the Ge atom is 4-coordinated, the As atom is 3-coordinated, while both the Se and Te atoms are 2-coordinated. Hence, both the Se and Te atoms act as the link propagating groups, whereas the I atom acts as the link terminating species.

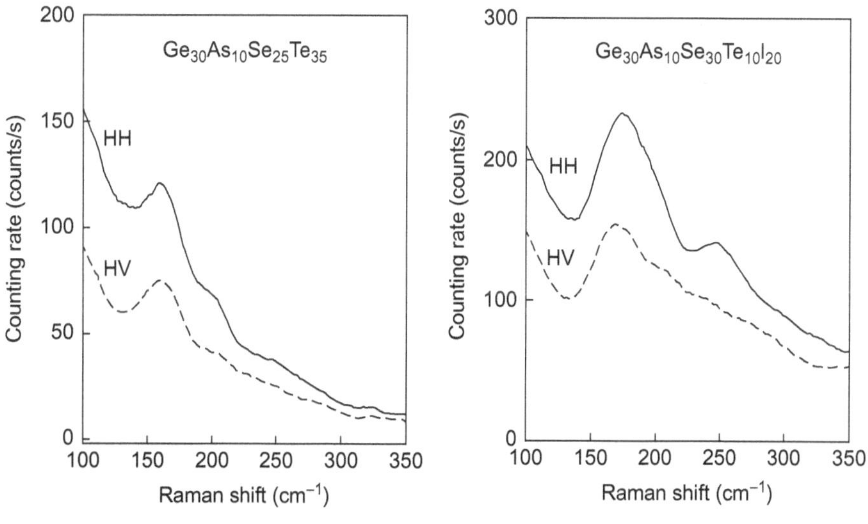


**Fig. 10.3** Room temperature polarized (HH) and depolarized (HV) Raman spectra of (A) $Ge_{30}As_{10}Se_{25}Te_{35}$ and (B) $Ge_{30}As_{10}Se_{30}Te_{10}I_{20}$ glasses.
Reproduced with permission from V.Q. Nguyen, J.S. Sanghera, J.A. Freitasc, I.D. Aggarwal, I.K. Lloyd, Structural investigation of chalcogenide and chalcohalide glasses using Raman spectroscopy, J. Non-Cryst. Solids 248 (1999) 103–114, Copyright © Elsevier.

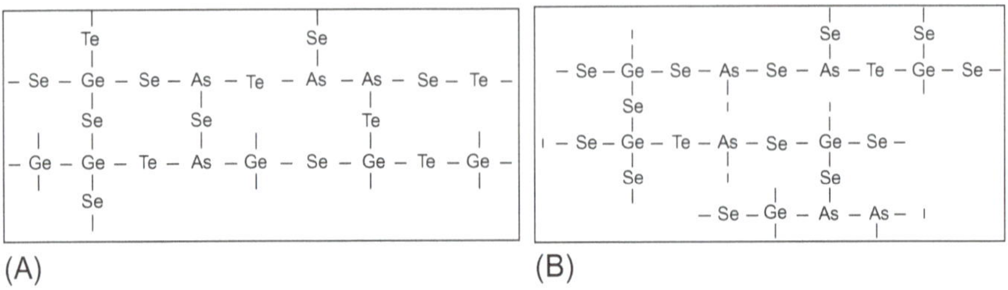


**Fig. 10.4** Schematic two-dimensional representation of structure of glasses in the systems (A) Ge–As–Se–Te and (B) Ge–As–Se–Te–I.
Reproduced with permission from V.Q. Nguyen, J.S. Sanghera, J.A. Freitasc, I.D. Aggarwal, I.K. Lloyd, Structural investigation of chalcogenide and chalcohalide glasses using Raman spectroscopy, J. Non-Cryst. Solids 248 (1999) 103–114, Copyright © Elsevier.

## 10.4 Preparation of chalcogenide glasses

Bulk chalcogenide glasses are prepared by the conventional melt-quenching technique using high purity (99.999%–99.9999%) granular (2–8 mm) either elemental (e.g., As, Ge, S, Se, and Te) or compound (e.g., $As_2S_3$ and $As_2Se_3$) or both as raw materials. All the raw materials are weighed and loaded in precleaned silica glass tube in a glove box, as shown in Fig. 10.5A, maintaining oxygen content around 1 ppm. The silica glass tube is cleaned by etching with dilute aqueous HF solution followed by washing with distilled water and then with ethanol or isopropanol. The silica tube is dried at 120–150°C for several hours. The silica tube with the raw material mixture is then connected to an evacuation system and evacuated down to $10^{-4}$ to $10^{-5}$ mbar. The mixture is baked around 100°C for 3–5 h under vacuum to remove surface-adsorbed oxygen and moisture before the tube is sealed. At the end of this degassing procedure, the tube is sealed by an oxy-hydrogen torch flame. Subsequently, the sealed tube is firmly fixed into a rocking furnace, as shown in Fig. 10.5B. It is then heated in the temperature range of 800–1150°C (depends on glass components) raising the temperature at a slow rate of 1°C $min^{-1}$ to avoid possible

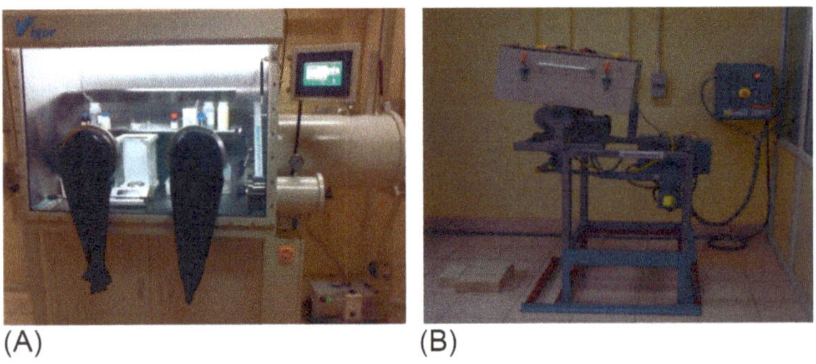


**Fig. 10.5** Photographs of (A) glove box and (B) rocking furnace used in chalcogenide glass preparation.

explosion and held at this temperature for 5–20 h (depends on glass composition). Finally, the glass is obtained by quenching the melt in either air or water or cold water (depends on glass composition) followed by annealing at 20°C below the glass transition temperature ($T_g$) for 4–10 h to remove residual thermal stress. Finally, the annealed glass is sized by cutting and polished to mirror smoothness.

## 10.5 Preparation of high purity chalcogenide glasses

The transparency region (the section in between the short and long wavelength cut-offs) of chalcogenide glasses is limited by the absorption bands of impurities. These impurities built up in the glass network and their bands are called extrinsic absorption. These impurities are of two classes—oxide and hydride origins. A summary of the absorption bands of oxide and hydride impurities is provided in Table 10.2 [17,18]. This gives us a clear picture of how the transmissions of different chalcogenide glasses are impaired by the presence of these bands. With the given best effort as described in Section 10.4, it is very difficult to remove these impurities from the glasses and obtain impurity-induced absorption band free good transmission of the chalcogenide glasses. Let us consider Fig. 10.6A that displays the transmission spectral profiles of the $Ge_{30}As_{10}Se_{30}Te_{30}$ glasses containing (1) 103.4, (2) 66.7, (3) 17.4, (4) 5.6, (5) 0.8, and (6) 0.6 ppmw oxide [19]. The curves (5) and (6) point out that the impurity content should be less than 1 ppm to obtain for good transmission of the chalcogenide glasses.

Accordingly, various approaches have been reported to prepare high purity chalcogenide glasses. One good technique is distillation of chalcogenide glass components

**Table 10.2 Absorption bands of oxide and hydride impurities [17,18]**

| Bond | Wavelength (μm) |
|---|---|
| ***Oxide impurities*** | |
| As–O | 7.90s, 9.50w, 12.34vs |
| Sb–O | 10.55m, 14.53s, 17.60s |
| Si–O | 9.27vs, 12.50m |
| Ge–O | 10.40m, 10.77s, 11.40vs, 12.40vs, 18.18m |
| Se–O | 10.67w, 11.06vs, 10.96s |
| Te–O | 12.95s, 15.50vs |
| ***Hydride impurities*** | |
| O–H | 1.45w, 2.24m, 2.78s, 2.93 vs, 6.30s |
| S–H | 2.05w, 2.54w, 3.11m, 3.70m, 4.01s |
| Se–H | 2.32w, 3.53w, 4.12m, 4.57s |
| Te–C | 5.07m |
| C–H | 3.41m, 3.50m |

Intensity: *vs*, very strong; *s*, strong; *m*, medium; *w*, weak; *sh*, shoulder.

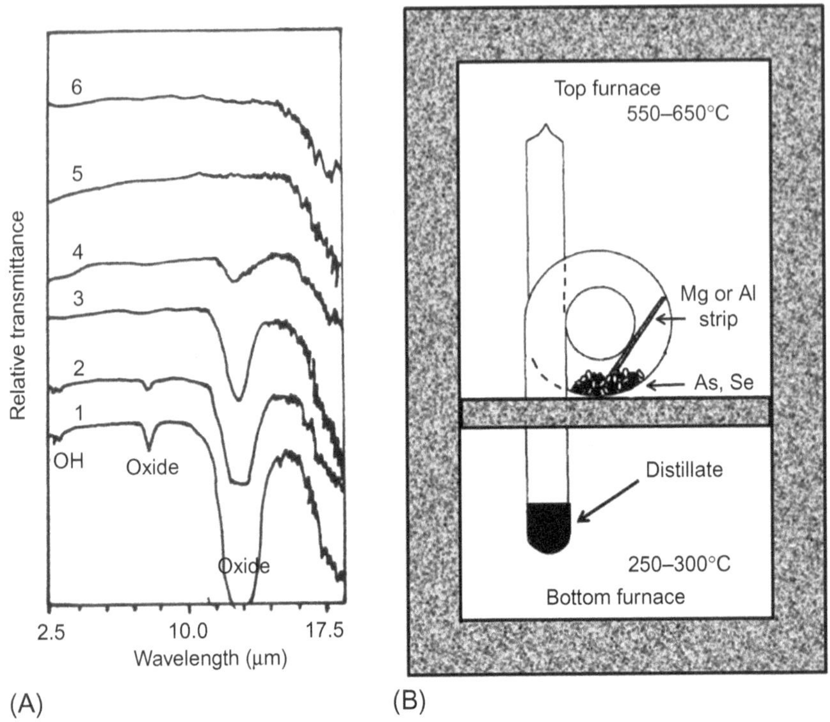


**Fig. 10.6** (A) Transmission spectral profiles of $Ge_{30}As_{10}Se_{30}Te_{30}$ glasses containing (1) 103.4, (2) 66.7, (3) 17.4, (4) 5.6, (5) 0.8, and (6) 0.6 ppmw oxide. (B) Purification is carried out in a sealed looped silica tube by distillation of components of chalcogenide glasses in the presence of Mg or Al strip as oxygen getter in a two-level (top, higher-temperature, and bottom, lower-temperature) furnace.
(A) Reproduced with permission from J.S. Sanghera, V.Q. Nguyen, R. Miklos, I.D. Aggralwal, Measurement of bulk absorption coefficients of chalcogenide and chalcohalide glasses at 10.6 μm using $CO_2$ laser calorimetry. J. Non-Cryst. Solids, 161 (1993) 320–322, Copyright © Elsevier.

along with oxygen getter strips such as Ti, Ce, La, Mg, Ba, and Al. In this technique, oxygen is removed from the glass batch according to the following simple reaction:

$$Mg + O_2 \rightarrow 2MgO \qquad (10.1)$$

Fig. 10.6B shows how the purification is carried out in a sealed looped silica tube by distillation of components of chalcogenide glasses in the presence of Mg or Al strip as oxygen getter in a two-level (top, higher temperature, and bottom, lower temperature) furnace [20]. The glass batch is placed into the looped part of the tube along with 100–300 mg oxygen getter of Mg or Al strip and sealed under vacuum. Distillation is done by heating the top part of the furnace in the range of 550–650°C (depends on the melting or boiling point of the components) and the lower part in the range of 250–300°C. After completion of distillation that takes nearly 24 h, the lower part of the tube is sealed, and the glass preparation is finished in a similar way as described in Section 10.4.

## 10.6 Properties of chalcogenide glasses and glass-ceramics

Photographs of different chalcogenide glasses of various dimensions developed in the author's laboratory are shown in Fig. 10.7. It is very exciting to exhibit the photographs of Ga–Sb–S chalcogenide glasses as depicted in Fig. 10.8 [21]. It is seen that the photograph taken by a camera working in the visible light (Fig. 10.8A) appears as black, that is, opaque. While the photograph taken by a microcamera operating at 940 nm (Fig. 10.8B) of the longer wavelength infrared light appears as highly transparent and the glasses show signs of excellent visible homogeneity. A comparison of spectral transmission windows of silica, fluoride, and different chalcogenide glasses is illustrated in Fig. 10.9 [22]. It is clear from this figure that the infrared absorption edges (cutoffs) of silica and fluoride glasses are around 4.5 and 6.5 μm, respectively, whereas it is extended toward longer infrared region from 10 to 21 μm for chalcogenide glasses, which depends on glass composition.

A comparison of compositions and different properties of some commercial chalcogenide glasses are represented in Table 10.3 [23–25]. It is observed that these commercial chalcogenide glasses consist of different chalcogens (e.g., S, Se, and Te) in conjunction with other elements (e.g., As, Ge, and Sb). Their properties such as density, glass transition temperature ($T_g$), refractive index ($n$), and transmission window vary in the ranges of 3.20–4.84 g $cm^{-3}$, 185–368°C, 2.3807–2.7869, and 0.6–8.0 to 0.8–15.0 μm, respectively. These properties are exclusively dependent on glass composition.

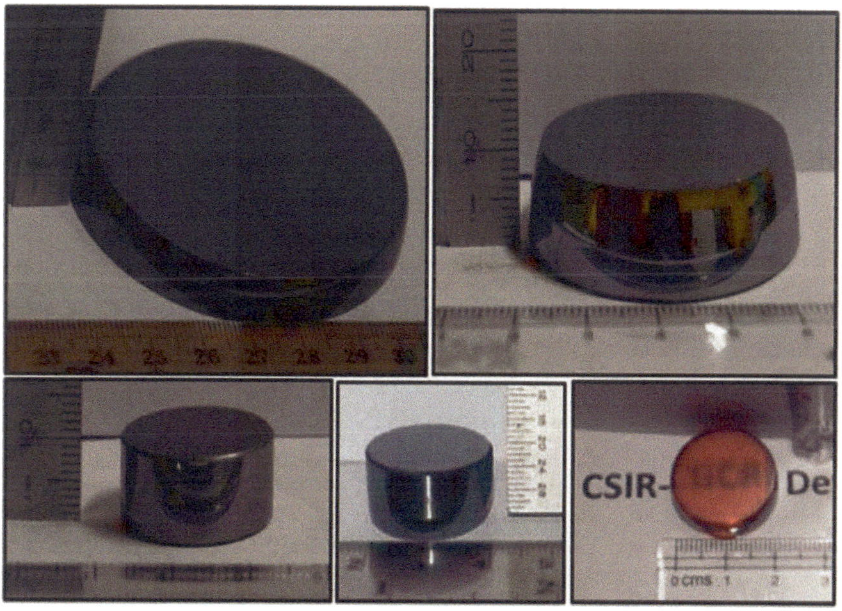


**Fig. 10.7** Photographs of different chalcogenide glasses of various dimensions developed in the author's laboratory.

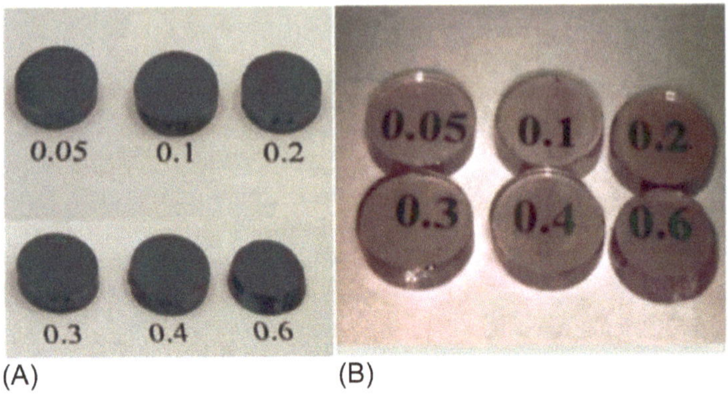


**Fig. 10.8** Photographs of Ga–Sb–S chalcogenide glasses containing different concentrations of $Dy^{3+}$ ions taken by a (A) camera working in the visible light and (B) microcamera operating at 940 nm.

Reproduced with permission from M. Zhang, A. Yang, Y. Peng, B. Zhang, H. Ren, W. Guo, Y. Yang, C. Zhai, Y. Wang, Z. Yang, D. Tang, $Dy^{3+}$-doped Ga–Sb–S chalcogenide glasses for mid-infrared lasers, Mater. Res. Bull. 70 (2015) 55–59, Copyright © Elsevier.

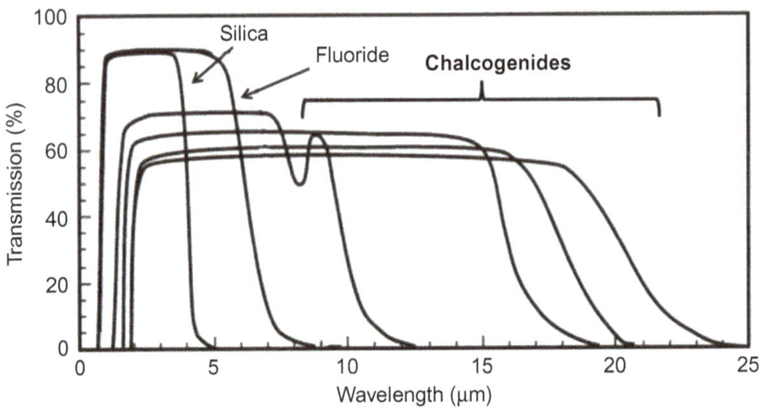


**Fig. 10.9** Comparison of spectral transmission windows of silica, fluoride, and different chalcogenide glasses.

Reproduced with permission from J. Lucas, Infrared glasses, Curr. Opin. Solid State Mater. Sci. 4 (1999) 181–187, Copyright © Elsevier.

The most striking property of the chalcogenide glasses is their high resistance against attack of chemical reagents [2]. They are not soluble in water or in aqueous solution of acids or salts. As a result, they are not attacked by atmospheric moisture and sea water. Moreover, they cannot be destructed by hydrofluoric acid. All these chemical durability properties are quite different from that of oxide glasses. In addition, they also do not react with organic solvents like ethanol, iso-propanol, and toluene.

**Table 10.3 Comparison of compositions and properties of some commercial chalcogenide glasses**

| Property | AMTIR-6[a] [23] | IG6[b] [24] | AMTIR-1[a] [23] | GASIR1[c] [25] | GASIR2[c] [25] | IG3[b] [24] |
|---|---|---|---|---|---|---|
| Composition (at.%) | $As_{20}S_{60}$ | $As_{20}Se_{60}$ | $Ge_{33}As_{12}Se_{55}$ | $Ge_{22}As_{20}Se_{58}$ | $Ge_{20}Sb_{15}Se_{65}$ | $Ge_{30}As_{13}Se_{32}Te_{25}$ |
| Density (g $cm^{-3}$) | 3.20 | 4.63 | 4.40 | 4.40 | 4.70 | 4.84 |
| Glass transition temperature, $T_g$ (°C) | 187 | 185 | 368 | 292 | 264 | 275 |
| Softening temperature, $T_s$ (°C) | 210 | – | 405 | – | – | – |
| Upper-use temperature (°C) | 150 | – | 300 | 250 | 200 | – |
| Coefficient of thermal expansion, $\alpha$ ($\times 10^{-6}$°$C^{-1}$) | 21.6 | 20.8 | 12.0 | 17.0 | 16.0 | 10.4 |
| Knoop Hardness | 109 | 1.04 GPa | 170 | 170 (Vickers) | 160 (Vickers) | 1.36 GPa |
| Young's modulus ($\times 10^6$ psi) | 2.3 | 18.3 GPa | 3.2 | 17.89 GPa | 19.11 GPa | 22.0 GPa |
| Poisson's ratio | 0.24 | – | 0.27 | 0.28 | 0.27 | – |
| Transmission range (μm) | 0.6–8.0 | 0.85–12.0 | 0.7–12.0 | 0.8–15.0 | 1.0–14.0 | 1.2–12.0 |
| Refractive index, $n$ (at 10 μm) | 2.3807 | 2.7781 | 2.4981 | 2.4944 | 2.5842 | 2.7869 |
| $dn/dT$ ($\times 10^{-6}$°$C^{-1}$ at 10 μm) | <1 (at 5 μm) | 32.2 | 72 | 55 | 58 | 102.8 |
| Dispersion | | | | | | |
| 3–5 μm | 155 | – | 202 | – | – | – |
| 8–12 μm | – | – | 109 | – | – | – |

[a]AMTIR-1 and AMTIR-6. Branded products of Amorphous Materials, Inc., TX, the United States.
[b]IG3 and IG6. Branded products of Schott AG, Mainz, Germany.
[c]GASIR1 and GASIR2. Branded products of Umicore Electro-Optic Materials, Olen, Belgium.

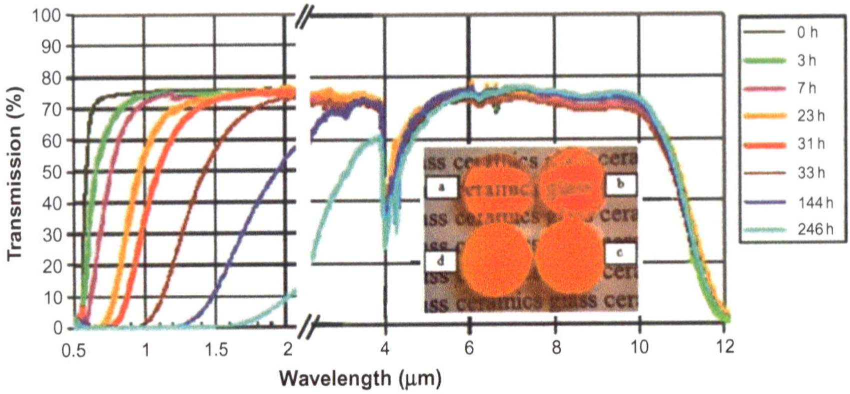


**Fig. 10.10** Transmission spectra of $5GeS_2$–$1Sb_2S_3$–2CsCl glass and glass-ceramics obtained by heat-treating at 290°C for different durations. Inset shows the photographs of glass-ceramic samples obtained by heat-treating at 290°C for (a) 0 h, (b) 7 h, (c) 73 h, and (d) 144 h. Reproduced with permission from X. Zhang, M.A. Hongli, J. Lucas, A new class of infrared transmitting glass-ceramics based on controlled nucleation and growth of alkali halide in a sulphide based glass matrix, J. Non-Cryst. Solids 337 (2004) 130–135, Copyright © Elsevier.

Fig. 10.10 shows the transmission spectra of $5GeS_2$–$1Sb_2S_3$–2CsCl glass and glass-ceramics obtained by heat-treating at 290°C for different durations [26]. Inset shows the photographs of glass-ceramic samples obtained by heat-treating at 290°C for (a) 0 h, (b) 7 h, (c) 73 h, and (d) 144 h. The visual transparency of the glass-ceramic samples gradually becomes opaque with progress of the heat treatment time. As a result of this growth, the short wavelength visible absorption edge of the resultant glass-ceramics gradually shifted toward longer wavelength near infrared (NIR) region without affecting the transmission in the mid-infrared (MIR) range and the infrared absorption edges (cutoffs). This has happened due to Mie scattering caused by the size (submicron) limit of the crystallite particles with respect to the wavelength of visible to NIR light. The presence of submicron-size crystallite particles of the glass-ceramics does not have considerable effect on the transmission beyond the wavelength of 5 μm and the infrared absorption edges (cutoffs). For that reason, most of these glass-ceramics can be used for transmission beyond 5 μm.

## 10.7 Properties of transition metal and rare-earth doped chalcogenide glasses and glass-ceramics

To find out the total number of published papers on individual transition metal and rare-earth metal ion containing chalcogenide glasses and glass-ceramics during the last 25 years (1990–2014), the total publication of 3405 articles on chalcogenide glasses and glass-ceramics was further refined by term of "*individual transition metal or rare-earth metal*." This evaluation was carried out according to the technique described in Section 10.1 earlier. These plots are illustrated in Fig. 10.11A and B. It gives us a clear quantitative idea of paper published on individual transition metal and rare-earth metal ion containing chalcogenide glasses and glass-ceramics as well as a comparison among them. The decreasing order of publication trends of the individual transition metal and rare-earth metal ion doped chalcogenide glasses and glass-ceramics are as follows:

$$Cu \gg Fe > Cr > Mn > Ti \approx Co > V$$

$$E \gg Dy > Pr > Nd > Yb > Sm > Tb \approx Ho > Tm > Gd$$

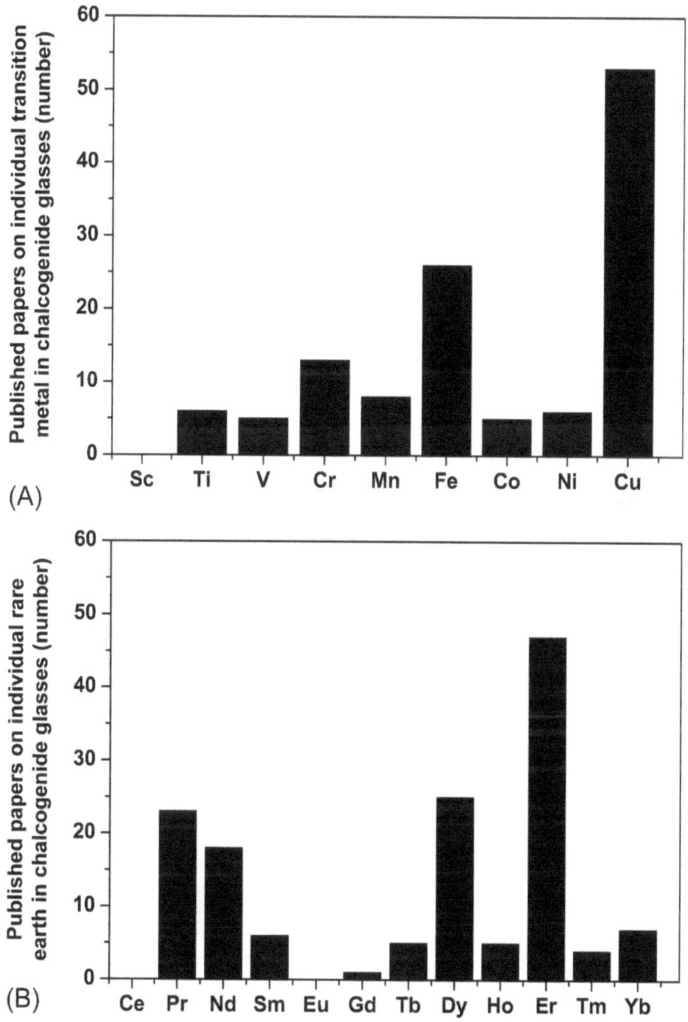


**Fig. 10.11** Total numbers of published papers during the last 25 years (1990–2014) on individual (A) transition and (B) rare-earth metal ion in chalcogenide glasses.

It is seen that during the last 25 years (1990–2014), among the 3405 published articles, the highest number (53) of papers was published using copper (Cu) and the second highest (26) is with iron (Fe) in the transition metal family and the highest number (47) of papers was published using erbium (Er) and the second highest (25) is with dysprosium (Dy) in the rare-earth metal family. These trends of publications give us the idea about the importance of doping the chalcogenide glasses with transition metal and rare-earth metal ions.

Fig. 10.12A illustrates the absorption spectra of undoped and 0.1 mol.% vanadium (III) doped Ga–La–S glasses [27]. Inset shows the enlarged region of the band peaking at 1100 nm due to vanadium. Fig. 10.12B shows the excitation and photoluminescence (PL) spectra of 0.01 mol.% vanadium (III) doped Ga–La–S glasses [27]. The Pl spectra taken by exciting at 514, 808, and 1064 nm exhibit similar characteristic PL band peaking around 1500 nm with a full width at half maximum of about 500 nm. This indicates that all the three excitation bands belong to the same oxidation state of $V^{3+}$ likewise the PL band. The broadness of the PL spectra designates that the $V^{3+}$ ion is in a low crystal field site.

Fig. 10.13 displays the (A) Vis-NIR transmission and (B) mid-IR emission (excitation at 1.32 μm) spectra of Ga–Sb–S glasses doped with different concentrations of $Dy^{3+}$ ions [21]. There are six absorption bands of the $Dy^{3+}$ ion, which are credited to the electron transitions from the ground state $^6H_{15/2}$ to the excited states as shown in Fig. 10.13A. The emission spectra exhibit four emission bands peaking at 2.95, 3.59, 4.17, and 4.40 μm. The bands at 2.95 and 4.40 μm are assigned to the emission transitions $^6H_{13/2} \rightarrow {}^6H_{15/2}$ and $^6H_{11/2} \rightarrow {}^6H_{13/2}$, respectively. The $Dy^{3+}$ ion-doped Ga–Sb–S glasses exhibit high quantum efficiency (75.9–88.1%) and large emission cross section ($0.38\text{-}1.11 \times 10^{-20}$ cm$^2$), which advocate their promising use as MIR lasers.

Fig. 10.14A shows the infrared luminescence spectra of $NdCl_3$-doped Ga–Ge–As–S glass excited at 0.8 μm [28]. It displays three PL emission bands centered at 890, 1078, and 1360 nm, which assigned to the electronic emission transitions of $^4F_{3/2} \rightarrow {}^4I_{9/2}$, $^4F_{3/2} \rightarrow {}^4I_{11/2}$, and $^4F_{3/2} \rightarrow {}^4I_{13/2}$, respectively. The excited-state lifetime is $110 \pm 5$ μs. As a result, within the accuracy of the Judd's equations approximation, the quantum yield of luminescence is found to be nearly 100%. These results advocate that the Nd-doped chalcogenide glasses are very promising for the development of active optical materials for use in the infrared spectral region.

Fig. 10.14B illustrates the mid-IR emission (excited at 0.8 μm) spectra of $80GeS_2$–$20Ga_2S_3 + 0.5Tm_2S_3$ glass-ceramics heat-treated at 458°C for 0, 10, 17, and 25 h [29]. Inset shows the fluorescence lifetime at 3.8 μm. As a result of crystallization, a nearly fivefold increase of intensity of $Tm^{3+}$ emission transition $^3H_5 \rightarrow {}^3F_4$ at 3.8 μm and about 76 μs increase in lifetime have been achieved. These increments are attributed to the formation of β-$GeS_2$ nanocrystals in glass-ceramics as a result of heat treatment of precursor glasses. Thus, rare-earth doped chalcogenide glass-ceramics are found to be very promising for use as MIR lasers.

Fig. 10.15A shows the upconversion emission spectra of $Er_2S_3$-doped $GeS_2$–$Ga_2S_3$ glasses excited at (A) 1550 and (B) 813 nm [30]. Owing to relatively low phonon energy of the range of 300–400 cm$^{-1}$, these glasses exhibit efficient upconversion

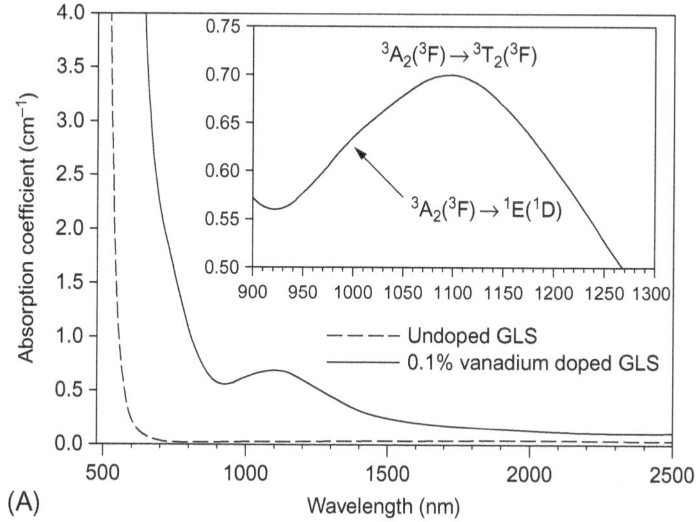


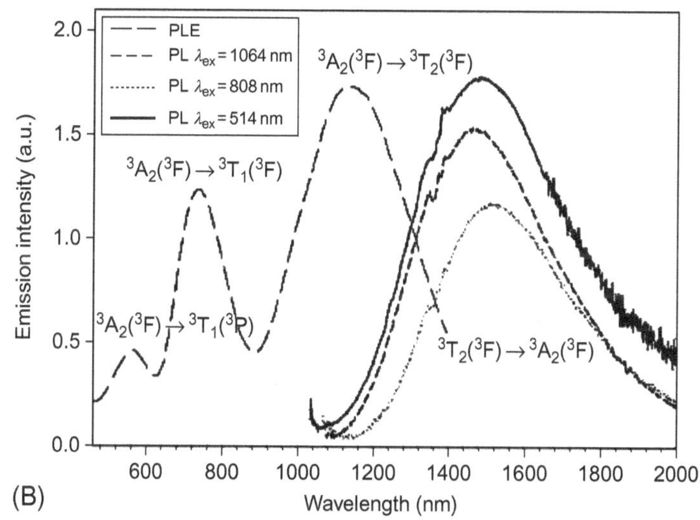


**Fig. 10.12** (A) Absorption spectra of undoped and 0.1 mol.% vanadium (III) doped Ga–La–S glasses. Inset shows the enlarged region of the band peaking at 1100 nm due to vanadium. (B) Excitation and photoluminescence spectra of 0.01 mol.% vanadium (III) doped Ga–La–S glasses.

Reproduced with permission from M. Hughes, H. Rutt, D. Hewak, R.J. Curry, Spectroscopy of vanadium (III) doped gallium lanthanum sulphide chalcogenide glass, Appl. Phys. Lett. 90 (2007) 031108 (3 pages), Copyright © AIP.

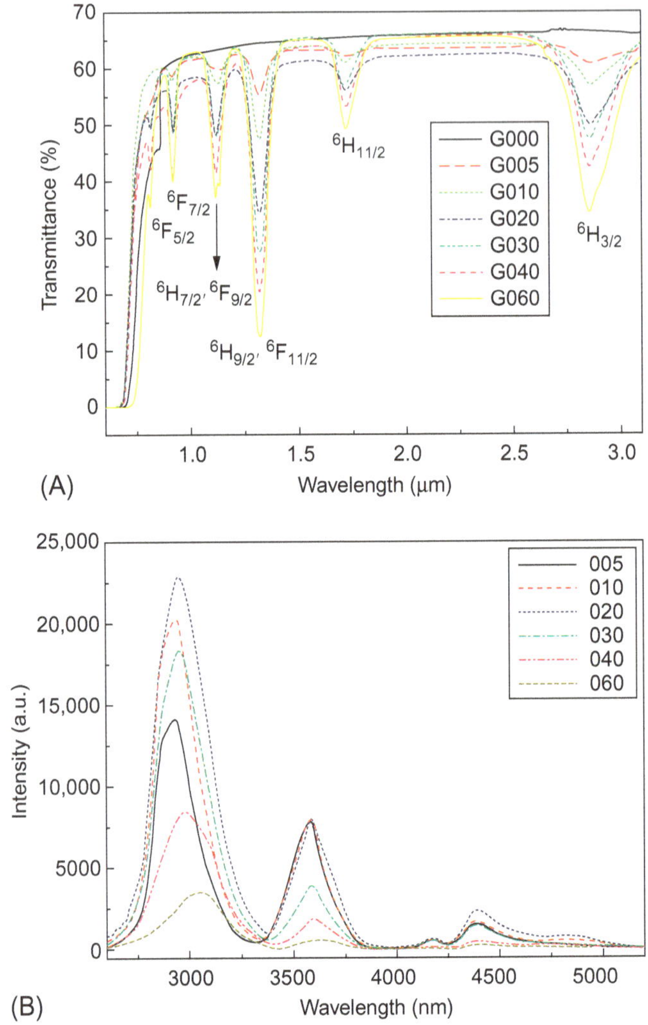


**Fig. 10.13** (A) Vis-NIR transmission and (B) mid-IR emission (excitation at 1.32 μm) spectra of Ga–Sb–S glasses doped with different concentrations of $Dy^{3+}$ ions.
Reproduced with permission from M. Zhang, A. Yang, Y. Peng, B. Zhang, H. Ren, W. Guo, Y. Yang, C. Zhai, Y. Wang, Z. Yang, D. Tang, $Dy^{3+}$-doped Ga–Sb–S chalcogenide glasses for mid-infrared lasers, Mater. Res. Bull. 70 (2015) 55–59, Copyright © Elsevier.

emission in the visible and NIR region (around 550, 670, and 1000 nm). The low phonon energy of the glass matrix suppresses the nonradiative losses as a result of multiphonon relaxation. Fig. 10.15B displays the upconversion emission spectra (excited at 1550 nm) of $25GeS_2$–$35Ga_2S_3$–40CsCl + 0.3Er base glass and glass-ceramics heat-treated at 350°C for 20, 23, and 30 h [31]. The base glass luminescence is

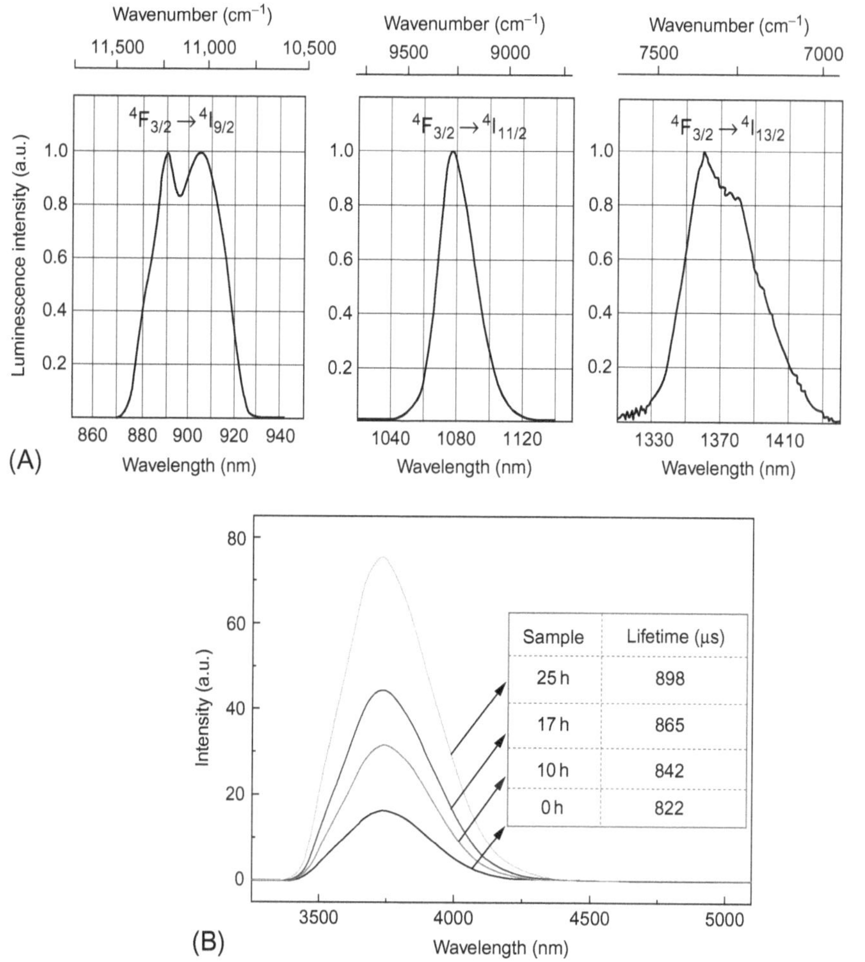


**Fig. 10.14** (A) Infrared luminescence spectra of $NdCl_3$-doped Ga–Ge–As–S glass excited at 0.8 μm. (B) Mid-IR emission (excited at 0.8 μm) spectra of $80GeS_2$–$20Ga_2S_3$+$0.5Tm_2S_3$ glass-ceramics heat-treated at 458°C for 0, 10, 17, and 25 h. Inset shows the fluorescence lifetime at 3.8 μm.

(A) Reproduced with permission from A. Belykh, L. Glebov, C. Lerminiaux, S. Lunter, M. Mikhailov, A. Plyukhin, M. Prassas, A. Przhevuskii, Spectral and luminescence properties of neodymium in chalcogenide glasses, J. Non-Cryst. Solids 213 and 214 (1997) 238–244, Copyright © Elsevier. (B) Reproduced with permission from C. Lin, S. Dai, C. Liu, B. Song, Y. Xu, F. Chen, J. Heo, Mechanism of the enhancement of mid-infrared emission from $GeS_2$–$Ga_2S_3$ chalcogenide glass-ceramics doped with $Tm^{3+}$, Appl. Phys. Lett. 100 (2012) 231910 (4 pages), Copyright © AIP.

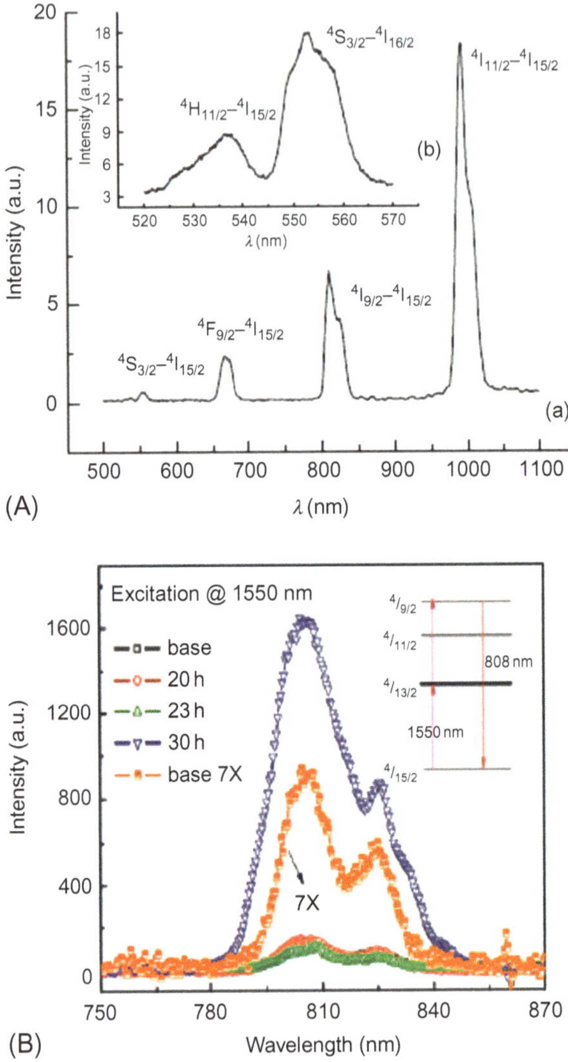


**Fig. 10.15** (A) Upconversion emission spectra of $Er_2S_3$-doped $GeS_2$–$Ga_2S_3$ glasses excited at (a) 1550 and (b) 813 nm. (B) Upconversion emission spectra (excited at 1550 nm) of $25GeS_2$–$35Ga_2S_3$–$40CsCl+0.3Er$ base glass and glass-ceramics heat-treated at 350°C for 20, 23, and 30 h. The base glass luminescence is multiplied by seven times. Inset shows the partial energy level diagram illustrating the upconversion emission transition.
(A) Reproduced with permission from A. Tverjanovich, Ya.G. Grigoriev, S.V. Degtyarev, A.V. Kurochkin, A.A. Man'shina, Yu.S. Tver'yanovich, Up-conversion fluorescence in Er-doped chalcogenide glasses based on $GeS_2$–$Ga_2S_3$ system, J. Non-Cryst.Solids 286 (2001) 89–91, Copyright © Elsevier. (B) Reproduced with permission from C. Lin, L. Calvez, Z. Li, S. Dai, H. Tao, H. Ma, X. Zhang, B. Moine, X. Zhao, Enhanced up-conversion luminescence in $Er^{3+}$-doped 25GeS2–35Ga2S3–40CsCl chalcogenide glass–ceramics, J. Am. Ceram. Soc. 96 (2013) 816–819, Copyright © The American Ceramic Society.

multiplied by seven times. Inset shows the partial energy level diagram illustrating the upconversion emission transition. It is seen that on heat treatment for 30 h, the intensity of upconversion emission is enhanced by 12-fold.

This has happened due to the change of medium-range order of the base glass as a result of the formation of low phonon energy (269 $cm^{-1}$) $Ga_2S_xCl_{6-x}$ nanocrystals on heat treatment for 30 h.

## 10.8 Functional applications of chalcogenide glasses and glass-ceramics

Various functional applications of chalcogenide glasses and glass-ceramics are depicted in Fig. 10.16 [17,18]. These applications are broadly divided into two categories—passive and active applications. The passive applications include thermal imaging, laser power delivery system, near field microscopy, and chemical sensing, whereas the active applications consist of rare-earth doped fiber, nonlinear optics, and chemical sensing as well. Their details are provided in Fig. 10.16. An example of thermal image obtained with a set of molded chalcogenide glass lenses in combination of an uncooled detector is shown in Fig. 10.17A [32], while a night vision scene obtained from an IR camera (inset shows its fittings) is made up of chalcogenide glass lenses that are used in the car driving assistance system of BMW cars is illustrated in Fig. 10.17B [15]. The schematic representation of the molding process to fabricate chalcogenide glass lenses are represented in Fig. 10.18A [33]. A photograph of the molded aspheric and diffractive lenses of chalcogenide glasses is also shown in Fig. 10.18B [32].

A photograph of a CD-Rewritable memory device made of chalcogenide glasses is shown in Fig. 10.19A. To demonstrate its mechanism, a schematic representation of temperature-time relationship during programming in the phase-change rewritable memory device is illustrated in Fig. 10.19A [34], where $T_a$ and $T_x$ are the amorphization and crystallization temperatures, respectively. The SET and RESET states of the memory correspond to a stored binary 1 or binary 0.

The refractive index of the low $T_g$ (about 50°C) chalcogenide glass of the As-S-I system has been found to possess the refractive index in the range of 2.3–2.4, which is very close to that of diamond (about 2.4). Consequently, this glass is being used for raw diamond processing. In brief, the process is as follows. When the defective raw diamond piece is embedded in these glasses, the defects within diamond become visible, which enable to optimize the cutting area of the raw diamonds to obtain maximum product. Fig. 10.20A illustrates a photograph of chalcogenide glass encapsulated raw diamond showing the defects observed in it [35].

$Li_2S$–$P_2S_5$ glasses and glass-ceramics have high lithium-ion conductivity (about $10^{-4}$ S $cm^{-1}$) and a wide electrochemical window. Consequently, they are used as electrolyte in all solid-state lithium-ion batteries, and the batteries exhibited excellent cycle performance [36–38]. Fig. 10.20B shows the charge-discharge cycle performance of the all solid-state Li-In/$Li_2S$–$P_2S_5$ electrolyte/$Li_4Ti_5O_{12}$ cell [36]. Open and solid circles indicate charge and discharge capacities, respectively, and they

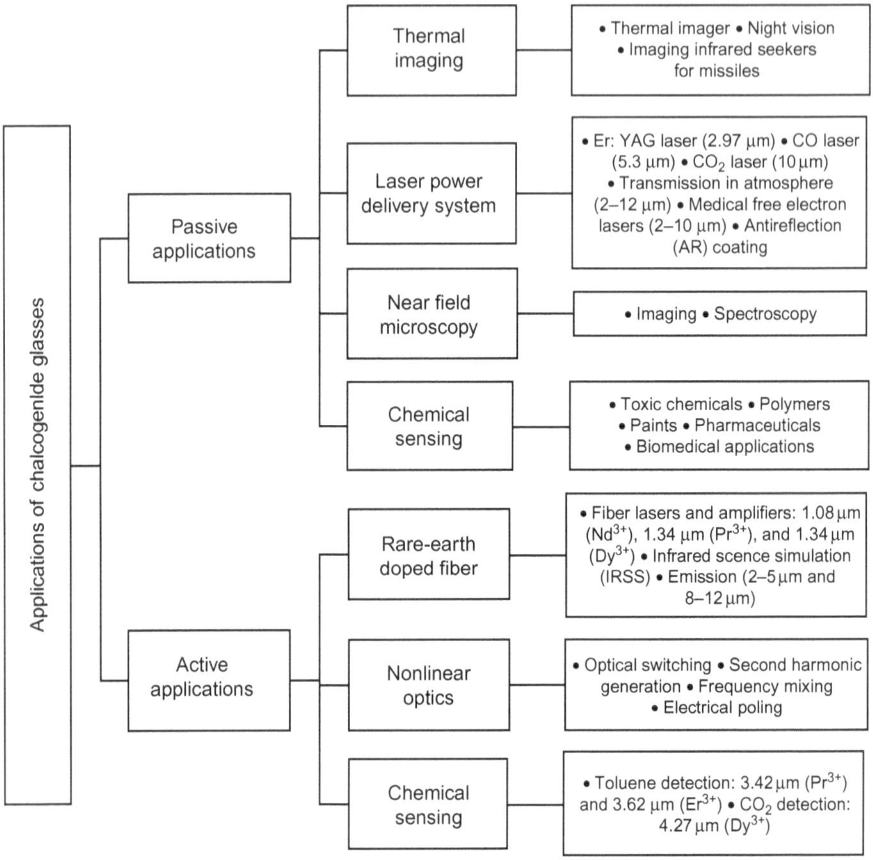


**Fig. 10.16** Various functional applications of chalcogenide glasses and glass-ceramics.

overlap each other. Solid diamonds represent charge-discharge efficiency. Recently, Hayashi et al. [39] have reported the superionic glass-ceramic containing $Na_3PS_4$ crystal phase as the electrolyte for room-temperature rechargeable sodium battery with the cell configuration of (−ve Na-Sn/$Na_3PS_4$/$TiS_2$ +ve). In this electrolyte, cubic $Na_3PS_4$ phase is crystallized in the $75Na_2S$–$25P_2S_5$ (mol.%) glass on heat treatment at 270°C for 1 h. The $Na^+$ ion conductivity of the resultant glass-ceramics is over $10^{-4}$ S cm$^{-1}$.

Recently, Gonçalves et al. [40] have reported chalcogenide glasses as a new class of thermoelectric materials suitable for use as nonconvention power generation. Tellurium-based semiconducting glasses, $Cu_xA_yTe_z$ (A = Ge, As, Ga, and Si), containing high concentrations of copper have shown the dimensionless figure of merit (ZT) values as high as 0.2 at 300 K with high Seebeck coefficients, very low thermal conductivities and tunable electric conductivities. In situ nanocrystallization by appropriate thermal treatment may lead to further improvement of their thermoelectric properties.

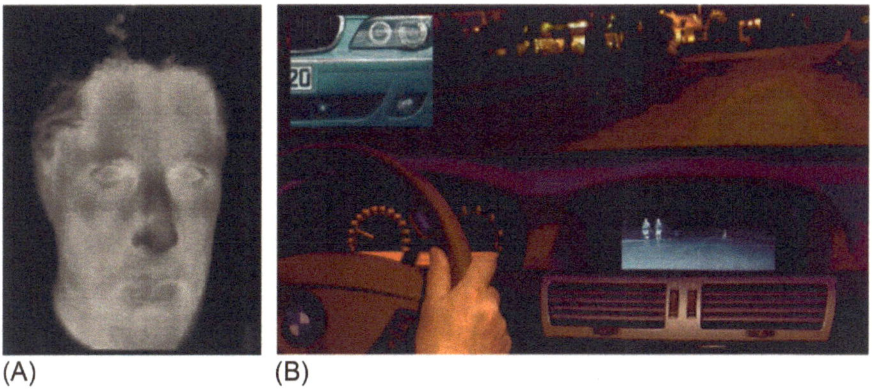


**Fig. 10.17** (A) Thermal image obtained with a set of molded chalcogenide glass lenses in combination of an uncooled detector. (B) Night vision scene obtained from an IR camera (inset shows its fittings) made of chalcogenide glass lenses that are used in the car driving assistance system of BMW cars.
(A) Reproduced with permission from X.H. Zhang, Y. Guimond, Y. Bellec, Production of complex chalcogenide glass optics by molding for thermal imaging, J. Non-Cryst. Solids 326 and 327 (2003) 519–523, Copyright © Elsevier. (B) Reproduced with permission from X.H. Zhang, B. Bureau, P. Lucas, C. Boussard-Pledel, J. Lucas, Glasses for seeing beyond visible, Chem. Eur. J. 14 (2008) 432–442, Copyright © Wiley.

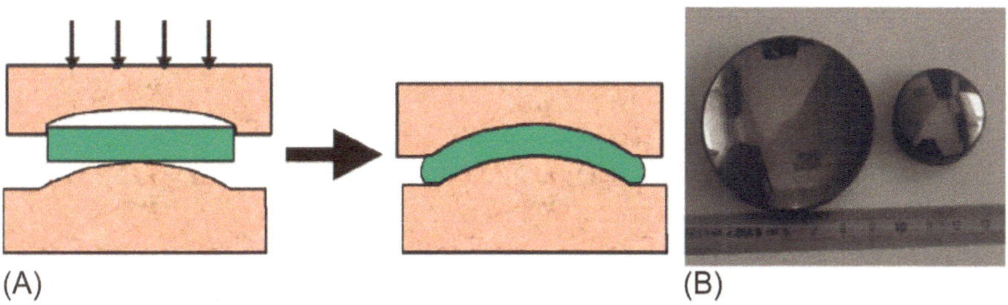


**Fig. 10.18** (A) Schematic representation of molding process to fabricate chalcogenide glass lenses. (B) Molded aspheric and diffractive lenses of chalcogenide glasses.
(A) Reproduced with permission from B. Bureau, X.H. Zhang, F. Smektala, J.-L. Adam, J. Troles, H. Ma, C. Boussard-Plèdel, J. Lucas, P. Lucas, D. L. Coq, M. R. Riley, J. H. Simmons, Recent advances in chalcogenide glasses, J. Non-Cryst. Solids 345 and 346 (2004) 276-283, copyright © Elsevier. (B) Reproduced with permission from X.H. Zhang, Y. Guimond, Y. Bellec, Production of complex chalcogenide glass optics by molding for thermal imaging, J. Non-Cryst. Solids 326 and 327 (2003) 519–523, Copyright © Elsevier.

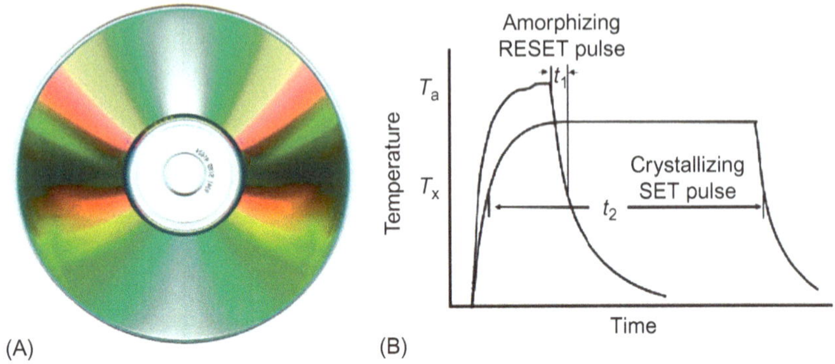


**Fig. 10.19** (A) Photograph of a CD-Rewritable. (B) Schematic representation of temperature-time relationship during programming in a phase-change rewritable memory device. $T_a$ and $T_x$ are the amorphization and crystallization temperatures, respectively. The SET and RESET states of the memory correspond to a stored binary 1 or binary 0.
Courtesy: S. Hudgens, B. Johnson, Overview of phase-change chalcogenide nonvolatile memory technology, MRS Bull. November (2004) 829–832.

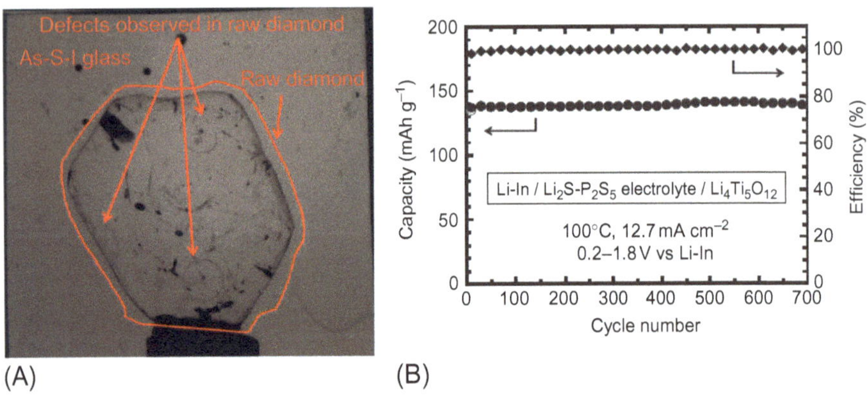


**Fig. 10.20** (A) Photograph of chalcogenide glass encapsulated raw diamond showing the defects observed in it. (B) Charge-discharge cycle performance of the all solid-state Li–In/ Li2S–P2S5 electrolyte/$Li_4Ti_5O_{12}$ cell. Open and solid circles indicate charge and discharge capacities, respectively, and they overlap each other. Solid diamonds represent charge-discharge efficiency.
(A) Reproduced with permission from A.R. Molla, A. Tarafder, N. Sasmal, B. Karmakar, J. Mistry, Synthesis and characterization of low Tg As–S–I chalcohalide glass for processing of raw diamonds, Int. J. Appl. Glass Sci. (2016) 1–5, Copyright © The American Ceramic Society.
(B) Reproduced with permission from M. Tatsumisago, M. Nagao, A. Hayashi, Recent development of sulfide solid electrolytes and interfacial modification for all-solid-state rechargeable lithium batteries, J. Asian Ceram. Soc. 1 (2013) 17–25, Copyright © Elsevier.

## 10.9 Summary and conclusions

Here, various aspects of chalcogenide glasses and glass-ceramics were described. These include history of development and structure of chalcogenide glasses. This is followed by the description of the preparation of normal and high purity chalcogenide glasses. Then, properties of undoped, transition metal, and rare-earth metal doped chalcogenide glasses and glass-ceramics were described with various specific examples including some commercial examples. Finally, functional applications of chalcogenide glasses and glass-ceramics were delineated with reference to several motivating cases. Considering various characteristics of chalcogenide glasses and glass-ceramics, it can be concluded that chalcogenide glasses and glass-ceramics are unique materials.

## 10.10 Future outlook

From foregoing discussion, it is quite clear that the future of chalcogenide glasses is very bright. There is no alternative of chalcogenide glasses for use as infrared material in comparison to their properties, ease of processing, and cost. The development of practical devices requires better understanding the glass formation conditions and structure. Many stimulating developments can be expected through applications of structural understanding, which can create new opportunity in the infrared science. Progress can be made for planar waveguide design that can open the way to more compact and sensitive sensors for use in medical, biological, and industrial appliances. However, the search for perfect infrared glass still remains the subject of much investigation. Future trends would be manufacturing chalcogenide glasses with best possible properties required for production of efficient passive and active devices.

## Exercises

**10.1** Define chalcogenide glasses with examples.

**10.2** Understanding the history of development of chalcogenide glasses is necessary to create new opportunities in the infrared science. Establish this statement with proper justification.

**10.3** Describe the structure with coordination number of the glasses in the systems: As–S, As–Se, Ge–Se, Ge–As–Se, Ge–As–Se–Te, and Ge–As–Se–Te–I.

**10.4** Describe briefly the preparation of high purity chalcogenide glasses.

**10.5** Differentiate silica, fluoride, and chalcogenide glasses with respect to their transmission spectra.

**10.6** What are the effects of heat treatment on the transmission spectra of chalcogenide glass-ceramics? Explain with proper explanation.

**10.7** What are the effects of heat treatment on the mid-IR emission (excitation at 0.8 μm) spectra at 3.8 μm and their lifetime of the $80GeS_2$–$20Ga_2S_3+0.5Tm_2S_3$ glass-ceramics?

**10.8** Describe various functional applications of chalcogenide glasses and glass-ceramics.

**10.9** Describe the functional mechanism of chalcogenide glass-based phase-change rewritable memory devices.

**10.10** Describe the uses of chalcogenide glasses in all solid-state lithium-ion and sodium-ion rechargeable batteries.

## References


[1] R. Fairman, B. Ushkov (Eds.), Semiconducting Chalcogenide Glass III, Elsevier, Philadelphia, PA, 2005.

[2] V.F. Kokorina, Glasses for Infrared Optics, CRC Press, Inc., Boca Raton, FL, 1996.

[3] R. Frerichs, New optical glasses with good transparency in the infrared, J. Opt. Soc. Am. 43 (1953) 1153–1157.

[4] A.R. Hilton, M. Brau, New high temperature infrared transmitting glasses, Infrared Phys. 3 (1963) 69–72.

[5] S.V. Nemilov, Viscosity and structure of glasses in the system AsGeSe in the region with high selenium content, Zh. Prikl. Khim. 37 (1964) 1452–1456.

[6] S.V. Nemilov, Viscosity and structure of glasses in the system As-Ge-Se in the region with low selenium content, Zh. Prikl. Khim. 37 (1964) 1699–1701.

[7] J.A. Savage, S. Nielsen, Preparation of glasses transmitting in the infrared between 8 and 15 microns, Phys. Chem. Glasses 5 (1964) 82–87.

[8] A.B. Pearson, Preparation and properties of sulphide, selenide, and telluride glasses, Glass Ind. 46 (1965) 18–20.

[9] A.R. Hilton, C.E. Jones, M. Brau, New high temperature infrared transmitting glasses-III, Infrared Phys. 6 (1966) 183–194.

[10] S.A. Dembovsky, N.P. Popova, Glass-formation in the systems Ge-Se-I and Si-Se-I, Neorg. Mater. 6 (1970) 138–142.

[11] K. Quin-Rod, R.F. Johnson Jr., Compositional dependence of the thermal, structural, and electrical properties of As-Te-I glasses, J. Non-Cryst. Solids 7 (1972) 53–76.

[12] A.R. Hilton, D.J. Hayes, M.D. Rechtin, Infrared absorption of some high-purity chalcogenide glasses, J. Non-Cryst. Solids 17 (1975) 319–338.

[13] C.T. Moynihan, P.B. Macedo, M.S. Maklad, R.K. Mohr, R.E. Howard, Intrinsic and impurity infrared absorption in $As_2Se_3$ glass, J. Non-Cryst. Solids 17 (1975) 369–385.

[14] A.M. Reitter, A.N. Sreeram, A.K. Varshneya, D.R. Swiler, Modified preparation procedure for laboratory melting of multicomponent chalcogenide glasses, J. Non-Cryst. Solids 139 (1992) 121–128.

[15] X.H. Zhang, B. Bureau, P. Lucas, C. Boussard-Pledel, J. Lucas, Glasses for seeing beyond visible, Chem. Eur. J. 14 (2008) 432–442.

[16] V.Q. Nguyen, J.S. Sanghera, J.A. Freitasc, I.D. Aggarwal, I.K. Lloyd, Structural investigation of chalcogenide and chalcohalide glasses using Raman spectroscopy, J. Non-Cryst. Solids 248 (1999) 103–114.

[17] D. Lezal, Chalcogenide glasses—survey and progress, J. Optoelectron. Adv. Mater. 5 (2003) 23–34.

[18] D. Lezal, J. Pedlikova, J. Zavadila, Chalcogenide glasses for optical and photonics applications, J. Optoelectron. Adv. Mater. 6 (2004) 133–137.

[19] J.S. Sanghera, V.Q. Nguyen, R. Miklos, I.D. Aggralwal, Measurement of bulk absorption coefficients of chalcogenide and chalcohalide glasses at 10.6 μm using $CO_2$ laser calorimetry, J. Non-Cryst. Solids 161 (1993) 320–322.

[20] W.A. King, A.G. Clare, W.C. LaCourse, Laboratory preparation of highly pure $As_2Se_3$ glass, J. Non-Cryst. Solids 181 (1995) 231–237.

[21] M. Zhang, A. Yang, Y. Peng, B. Zhang, H. Ren, W. Guo, Y. Yang, C. Zhai, Y. Wang, Z. Yang, D. Tang, $Dy^{3+}$-doped Ga–Sb–S chalcogenide glasses for mid-infrared lasers, Mater. Res. Bull. 70 (2015) 55–59.

[22] J. Lucas, Infrared glasses, Curr. Opin. Solid State Mater. Sci. 4 (1999) 181–187.

[23] A.R. Hilton Sr., J. McCord, R. Timm, R.A. LeBlanc, Amorphous materials molded IR lens progress report, Proc. SPIE 6940 (2008) 69400Q (6 pages).

[24] A.G. Schott, Mainz, Germany. *www.schott.com/advanced_optics.*

[25] Umicore Electro-Optic Materials, Olen, Belgium.*www.optics.umicore.com.*

[26] X. Zhang, M.A. Hongli, J. Lucas, A new class of infrared transmitting glass-ceramics based on controlled nucleation and growth of alkali halide in a sulphide based glass matrix, J. Non-Cryst. Solids 337 (2004) 130–135.

[27] M. Hughes, H. Rutt, D. Hewak, R.J. Curry, Spectroscopy of vanadium (III) doped gallium lanthanum sulphide chalcogenide glass, Appl. Phys. Lett. 90 (2007) 031108 (3 pages).

[28] A. Belykh, L. Glebov, C. Lerminiaux, S. Lunter, M. Mikhailov, A. Plyukhin, M. Prassas, A. Przhevuskii, Spectral and luminescence properties of neodymium in chalcogenide glasses, J. Non-Cryst. Solids 213 and 214 (1997) 238–244.

[29] C. Lin, S. Dai, C. Liu, B. Song, Y. Xu, F. Chen, J. Heo, Mechanism of the enhancement of mid-infrared emission from $GeS_2$-$Ga_2S_3$ chalcogenide glass-ceramics doped with $Tm^{3+}$, Appl. Phys. Lett. 100 (2012) 231910 (4 pages).

[30] A. Tverjanovich, Ya.G. Grigoriev, S.V. Degtyarev, A.V. Kurochkin, A.A. Man'shina, Yu. S. Tver'yanovich, Up-conversion fluorescence in Er-doped chalcogenide glasses based on $GeS_2$-$Ga_2S_3$ system, J. Non-Cryst. Solids 286 (2001) 89–91.

[31] C. Lin, L. Calvez, Z. Li, S. Dai, H. Tao, H. Ma, X. Zhang, B. Moine, X. Zhao, Enhanced up-conversion luminescence in $Er^{3+}$-Doped $25GeS_2$-$35Ga_2S_3$-$40CsCl$ chalcogenide glass–ceramics, J. Am. Ceram. Soc. 96 (2013) 816–819.

[32] X.H. Zhang, Y. Guimond, Y. Bellec, Production of complex chalcogenide glass optics by molding for thermal imaging, J. Non-Cryst. Solids 326 and 327 (2003) 519–523.

[33] B. Bureau, X.H. Zhang, F. Smektala, J.-L. Adam, J. Troles, H. Ma, C. Boussard-Plèdel, J. Lucas, P. Lucas, D.L. Coq, M.R. Riley, J.H. Simmons, Recent advances in chalcogenide glasses, J. Non-Cryst. Solids 345 and 346 (2004) 276–283.

[34] S. Hudgens, B. Johnson, Overview of phase-change chalcogenide nonvolatile memory technology, MRS Bull. 11 (2004) 829–832.

[35] A.R. Molla, A. Tarafder, N. Sasmal, B. Karmakar, J. Mistry, Synthesis and characterization of low Tg As–S–I chalcohalide glass for processing of raw diamonds, Int. J. Appl. Glas. Sci. 7 (2016) 1–5.

[36] M. Tatsumisago, M. Nagao, A. Hayashi, Recent development of sulfide solid electrolytes and interfacial modification for all-solid-state rechargeable lithium batteries, J. Asian Ceram. Soc. 1 (2013) 17–25.

[37] S.S. Berbano, M. Mirsaneh, M.T. Lanagan, C.A. Randall, Lithium thiophosphate glasses and glass–ceramics as solid electrolytes: processing, microstructure, and properties, Int. J. Appl. Glas. Sci. 4 (2013) 414–425.

[38] A. Hayashi, M. Tatsumisago, Recent development of bulk-type solid-state rechargeable lithium batteries with sulfide glass-ceramic electrolytes, Electron. Mater. Lett. 8 (2012) 199–207.

[39] A. Hayashi, K. Noi, A. Sakuda, M. Tatsumisago, Superionic glass-ceramic electrolytes for room-temperature rechargeable sodium batteries, Nat. Commun. 3 (2012) 856 (5 pages).

[40] A.P. Gonçalves, E.B. Lopes, G. Delaizir, J.B. Vaney, B. Lenoir, A. Piarristeguy, A. Pradel, J. Monnier, P. Ochin, C. Godart, Semiconducting glasses: a new class of thermoelectric materials? J. Solid State Chem. 193 (2012) 26–30.

# Functional halide glasses and glass-ceramics

11

## Chapter Outline



## 11.1 Introduction

The term halide glasses designate those inorganic glasses of which the anionic part is composed of at least one or more element(s) of the halogen family, that is, F, Cl, Br, and I of the group VIIA or 17 of the periodic table, and the cationic part is from the pretransition metal (e.g., Na and Al), transition metal (e.g., Sc, Fe and Mn) or post-transition metal (e.g., Zn, Ga, Zr, Cd, In, and Hf ) series. The halide glasses may be of either single metal halide (e.g., $BeF_2$ or $ZnCl_2$) or multimetal halides (e.g., $ZrF_4$–$BaF_2$–$LaF_3$–$AlF_3$–NaF). Since the majority of the halide glass family is occupied by the fluoride glasses, consequently, the halide glasses frequently mean the fluoride glasses. These fluoride glasses, when they contain $ZrF_4$ as the major component, are known as fluorozirconate glasses. Again, these glasses are composed of fluorides of heavy metals (atomic weight > 90); thus, they are also called heavy metal fluoride glasses (HMFGs). These glasses were developed to circumvent the limitation of the optical properties of the oxide glasses. For example, their infrared (IR) absorption edge can be shifted from around 7 to 30 μm when moving from fluoride to chloride to bromide to iodide glasses. Nevertheless, the halide glasses are noticeably inferior compared with the oxide glasses as a result of their poor mechanical strength, resistance to moisture attack, and devitrification tendency. However, the halide glasses are

Functional Glasses and Glass-Ceramics. http://dx.doi.org/10.1016/B978-0-12-805056-9.00011-8

very important materials and used in various functional applications such as in optics, sensors, IR instrument, medicine, telecommunication, and scintillating glasses for X-ray camera in medical and security devices. In this chapter, various features of halide glasses and their glass-ceramics will be described with special reference to the fluoride family.

An attempt was made using the Web of Science (version 5.17) as described in Chapter 1 to provide a global publication trend and quantitative growth in the area of fluoride glasses and glass-ceramics. Here, year-wise data were extracted for the last 25 years (1990–2014) by searching, using the topic keyword "glass" followed by refining the document type as "article" published in the journals and meetings. It was further refined using the term "fluoride glass." This plot is shown in Fig. 11.1. It gives us a clear publication trend and quantitative growth of published papers on fluoride glasses and glass-ceramics during the last 25 years (1990–2014).

## 11.2 History of development of halide glasses

Some chronological events of the development of halide glasses are provided in Table 11.1. The discovery of fluoride glass dates back to 1898, and the credit goes to Lebeau [1,2]. He first observed the glass formation in beryllium fluoride ($BeF_2$). In 1926, Goldschmidt [3] found out isostructural similarity between $BeF_2$ and $SiO_2$ as the reason of glass formation by $BeF_2$. Vitreous $ZnCl_2$ is the best known chloride glass for a long time as it is isostructural to $SiO_2$ like $BeF_2$, but the exact time of discovery is unknown [4]. In 1933, Heine [5] found out satisfactory glass formation by ternary and quaternary systems of $BeF_2$ with polyvalent metal fluorides. During

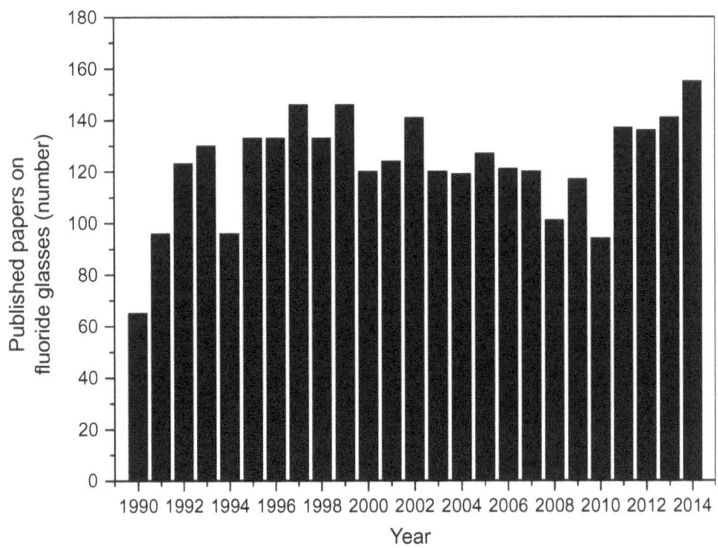


**Fig. 11.1** Year-wise published papers on fluoride glasses during the last 25 years (1990–2014).

**Table 11.1 Some chronological events of development of halide glasses**

| Event | Year | Author | Ref. |
|---|---|---|---|
| First observation of glass formation of beryllium fluoride ($BeF_2$) | 1898 | Lebeau | [1,2] |
| Found out isostructural similarity between $BeF_2$ and $SiO_2$ as the reason of glass formation by $BeF_2$ | 1926 | Goldschmidt | [3] |
| Vitreous $ZnCl_2$ is the best known chloride glass for a long time as it is isostructural to $SiO_2$ like $BeF_2$ | – | – | [4] |
| Found out satisfactory glass formation by ternary and quaternary systems of $BeF_2$ with polyvalent metal fluorides | 1933 | Heine | [5] |
| Developed high ultraviolet transparency glasses with low refractive index and low optical dispersion from combination of beryllium, aluminum, and quadrivalent metal fluorides | 1949, 1950 | Sun et al. | [6,7] |
| First glass formation was noticed in the $BiCl_3$–KCl system | 1960 | Topol et al. | [8] |
| Michel Poulain first discovered heavy metal fluoride glasses (HMFGs) of $ZrF_4$–$BaF_2$–NaF–$NdF_3$ based on $ZrF_4$in 1974 | 1975 | Poulain et al. | [9] |
| Group led by Jacoboni discovered transition metal (M) fluorides ($MF_3$)-based glasses such as $FeF_3$–$MnF_2$–$PbF_2$ | 1979 | Miranday et al. | [10] |
| Group led by Lucas developed new classes of heavy metal fluoride glasses (HMFGs) such as $15BaF_2$–$28ZnF_2$–$28YbF_3$–$28ThF_4$ (BZnYbT) without using $ZrF_4$ | 1980 | Fonteneau et al. | [11] |
| New chloride glass formation was observed in the $ThCl_4$–NaCl–KCl ternary system | 1982 | Hu and Mackenzie | [12] |
| Discovered $CdCl_2$-based chloride glasses in two ternary systems of $CdCl_2$–$BaCl_2$–KCl and $CdCl_2$–$BaCl_2$–NaCl | 1983 | Poulain et al. | [13] |
| Discovered new halide glasses in the AgBr–$PbBr_2$–CuCl–CsBr system | 1985 | Fuding | [14] |
| Group led by Angell developed all-halide superionic glasses in the CsI–AgI and RbI–CuI systems | 1986 | Liu et al. | [15] |
| Discovered tellurium halide (chalcohalide) glasses in binary (e.g., Te–Cl and Te–Br) and ternary (e.g., Te-Cl-S, Te–Cl–Se, Te–Br–S, Te–Br–Se, and Te–I–Se) systems | 1987 | Lucas et al. | [16] |
| Developed far infrared chalcohalide glasses in the Ge–Ga–Te–$ZnI_2$ system possessing infrared cutoff wavelengths beyond 25 μm | 2014 | Xu et al. | [17] |

1949–50, Sun and his coworker [6,7] developed high ultraviolet transparency glasses with low refractive index and low optical dispersion from the combination of beryllium, aluminum, and quadrivalent metal fluorides. These are commercial glasses (Eastman Kodak) and mainly used in manufacturing of compound lenses. Topol et al. [8] first noticed the chloride glass formation in the $BiCl_3$–KCl system.

In actual fact, Michel Poulain first discovered the HMFGs of $ZrF_4$–$BaF_2$–NaF–$NdF_3$ based on $ZrF_4$ in 1974 and published in 1975 [9]. In contrast, the group led by Jacoboni discovered transition metal (M) fluorides ($MF_3$)-based glasses such as $FeF_3$–$MnF_2$–$PbF_2$ in 1979 [10]. Likewise, a group led by Lucas developed new classes of HMFGs such as 15$BaF_2$–28$ZnF_2$–28$YbF_3$–28$ThF_4$ (BZnYbT) without using $ZrF_4$ in 1980 [11]. Hu and Mackenzie [12] observed new chloride glass formation in the $ThCl_4$–NaCl–KCl ternary system. In 1983, Poulain et al. [13] discovered $CdCl_2$-based chloride glasses in two ternary systems of $CdCl_2$–$BaCl_2$–KCl and $CdCl_2$–$BaCl_2$–NaCl. In 1985, Fuding [14] discovered new halide glasses in the AgBr–$PbBr_2$–CuCl–CsBr system. A group led by Angell developed all-halide superionic glasses in the CsI–AgI and RbI–CuI systems in 1986 [15]. In 1987, Lucas et al. [16] discovered tellurium halide (chalcohalide) glasses in binary (e.g., Te–Cl and Te–Br) and ternary (e.g., Te–Cl–S, Te–Cl–Se, Te–Br–S, Te–Br–Se, and Te–I–Se) systems, whereas Xu et al. [17] developed far infrared chalcohalide glasses in the Ge–Ga–Te–$ZnI_2$ system possessing infrared cutoff wavelengths beyond 25 μm in 2011. Recent trend is the exploration of various properties and applications of halide glasses.

## 11.3 Structure of halide glasses

In the family of halide glasses, the mainstreams of the glasses are composed of zirconium fluoride ($ZrF_4$). Thus, the structure of halide glasses is described here with reference to a most simple and stable fluorozirconate $BaF_2$–$ZrF_4$ glass. Kawamoto [18] very resourcefully reviewed the progress of structural studies of $BaF_2$–$ZrF_4$

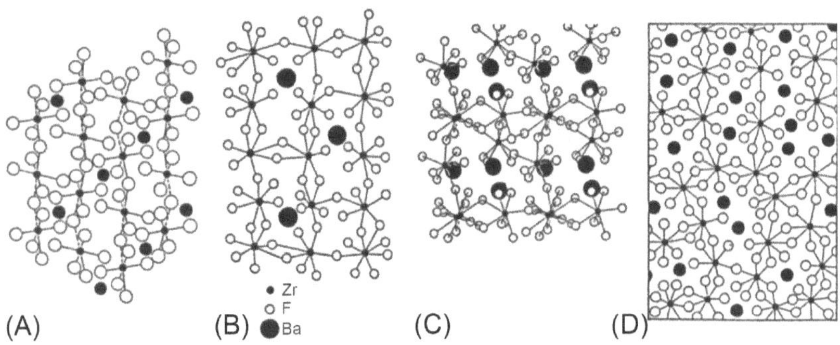


**Fig. 11.2** Various models (A), Almeida et al. [19,20]; (B), Lucas et al. [21,22]; (C), Yasui et al. [23]; and (D), Kawamoto et al. [24–26] for the structure of $BaF_2$–$ZrF_4$ glass. Reproduced with permission from Y. Kawamoto, Progress in structural study of ZrF4-based glasses, Mater. Sci. Forum 6 (1985) 417–426, Copyright © Trans Tech. Publications.

glasses carried out by various investigators. Accordingly, various models proposed for the structure of $BaF_2$–$ZrF_4$ glasses are depicted in Fig. 11.2 [18]. The coordination numbers (CNs) of Zr in $ZrF_n$ polyhedral fluorine environment are found to be 6–8. The first contribution in the study of the structure of $BaF_2$–$ZrF_4$ glasses was made by Almeida and Mackenzie [19] using vibrational (Raman and IR) spectroscopy. This is followed by their X-ray photoemission study [20]. In view of these, their proposed model is shown in Fig. 11.2A. According to this model, the Zr ion makes up six coordinated $[ZrF_6]^{2-}$ polyhedron. These $[ZrF_6]^{2-}$ polyhedrons remain attached with other $[ZrF_6]^{2-}$ polyhedrons through one or more Zr–F–Zr bridge(s), and the $Ba^{2+}$ ion occupies the interstitial (vacant) positions of the structural cases balancing the charge of the $[ZrF_6]^{2-}$ polyhedrons. The structural model proposed by Lucas et al. [21,22] by X-ray scattering and ion dynamic computer simulation studies is illustrated in Fig. 11.2B. Both the experimental radial distribution function and computer simulation indicated the variation of CN of zirconium ion in the range 7.5–7.7. Consequently, in this structural model, they accepted both the $[ZrF_7]^{3-}$ and $[ZrF_8]^{4-}$ polyhedrons. Likewise, the CNs of zirconium ion are 7 and 8 in the model (Fig. 11.2C) proposed by Yasui and Inoue [23] based on molecular dynamic simulations. Similarly, the CNs of zirconium ion are 7 and 8 in the model (Fig. 11.2D) proposed by Kawamoto et al. [24–26] based on Raman spectral analysis.

## 11.4 Synthesis of halide glasses

The halide glasses are prepared from high purity-anhydrous halides as the raw materials. The entire operation is carried out in controlled inert gas atmosphere (argon or nitrogen) dry glove box to exclude moisture ($H_2O$) and oxygen. Sometimes halogenating agents (e.g., ammonium bifluoride for fluoride glasses) are added to reduce oxide or OH impurities. Melting is done in platinum, platinum-gold (5%), gold, or vitreous carbon crucible in the temperature range 800–1000°C (depending on the composition of the glasses) for 1–2 h. Sometimes reactive atmosphere processing is ensured by introducing $NF_3$ or $CCl_4$ in gaseous form into the melting atmosphere to obtain oxide and OH impurity-free glasses. The glass melt is very fluid and can easily be refined and homogenized. After refining, the glass melt is cooled down to such a temperature where the nucleation rate is very low and kept there for a long time (1–2 h). It is cast into a preheated (temperature near glass transition temperature, $T_g$) brass, gold-coated brass, or graphite mold and then annealed below $T_g$ and slowly cooled down to the room temperature.

## 11.5 Properties of halide glasses and glass-ceramics

Compositions and various properties of some typical fluoride glasses are shown in Table 11.2 [27–29]. Fluoride glasses are designated using the first letter of the atomic symbol of the cations of fluorides. For example, the glass of the $ZrF_4$–$BaF_2$–$LaF_3$–$AlF_3$–NaF system is designated as ZBLAN. Thermal properties of fluoride glasses are very important to understand their stability and synthesis. Both the supercooled

**Table 11.2** **Chemical composition and properties of some typical fluoride glasses [27–29]**

| Glass | Composition (mol.%) | $T_g$ (°C) | $T_x$ (°C) | $T_m$ (°C) | $\Delta T_x$ (°C) | $T_{rg}$ | $d$ (g cm$^{-3}$) | $n_D$ | $R_c$ (K min$^{-1}$) |
|---|---|---|---|---|---|---|---|---|---|
| ZBLAN | $53ZrF_4$–$20BaF_2$–$4LaF_3$–$3AlF_3$–20NaF | 262 | 352 | 455 | 90 | 0.5758 | 4.319 | 1.4985 | 1.7 |
| ZBLA | $57ZrF_4$–$34BaF_2$–$5LaF_3$–$4AlF_3$ | 320 | 392 | 513 | 72 | 0.6238 | 4.540 | 1.5190 | 55 |
| HfBL | $62HfF_4$–$33BaF_2$–$5LaF_3$ | 312 | 395 | 530 | 83 | 0.5887 | 5.780 | 1.5150 | |
| YABC | $20YF_3$–$40AlF_3$–$20BaF_2$–$20CaF_2$ | 430 | 535 | 710 | 105 | 0.6056 | 4.000 | 1.4400 | 60 |
| CLAP | $26.1CdF_2$–10LiF–$30.6AlF_3$–$33.3PbF_2$ | 285 | 394 | | 109 | | 5.986 | 1.6060 | 300 |
| BIG | $30BaF_2$–$18InF_3$–$12GaF_3$–$20ZnF_2$–$10YF_3$–$6ThF_4$–$4ZrF_4$ | 332 | 460 | 576 | 128 | 0.5764 | | 1.5050 | |
| PZG | $35.3PbF_2$–$23.5ZnF_2$–$34.3GaF_3$–$4.9YF_3$–$2AlF_3$ | 270 | 325 | 547 | 55 | 0.4936 | | 1.5900 | |
| ZnSB | $47ZnF_2$–$24SrF_2$–$10BaF_2$–$5CdF_2$–$6InF_3$–$4GaF_3$–$2LaF_3$–2NaF | 303 | 377 | 615 | 74 | 0.4927 | | 1.4980 | |
| TLB | $30ThF_4$–60LiF–$10BaF_2$ | 256 | 310 | 560 | 54 | 0.4571 | 5.290 | 1.4950 | |
| SBY | $40ScF_3$–$40BaF_2$–$20YF_3$ | 360 | 410 | 720 | 50 | 0.5000 | 4.380 | 1.4990 | |

$T_g$, glass transition temperature; $T_x$, onset of crystallization temperature; $T_m$, melting temperature; $\Delta T_x$, $(T_x - T_g)$, supercooled liquid region and glass stability; $T_{rg} = T_g/T_m$, reduced glass transition temperature; $d$, density; $n_D$, refractive index at sodium D-line; $R_c$, critical cooling rate.

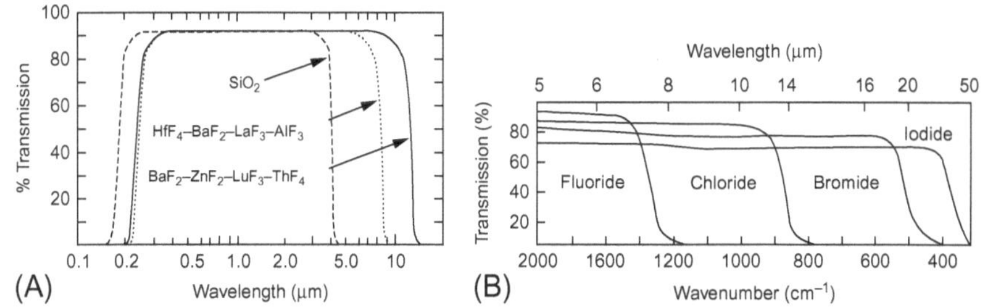


**Fig. 11.3** (A) Comparison of transmission windows of two heavy metal fluoride glasses (HMFGs) with that of silica ($SiO_2$) glass. (B) Infrared cutoff of different halide glasses based on fluoride, chloride, bromide, and iodide.
(A) Courtesy: J.M. Parker, P.W. France, Optical properties of halide glasses, in: M.H. Lewis (Ed.), Glasses and Glass-Ceramics, Chapman and Hall, New York, NY, 1989. (B) Reproduced with permission from J. Lucas, X.H. Zhang, The tellurium halide glasses, J. Non-Cryst. Solid 125 (1990) 1–16, Copyright © Elsevier.

liquid region ($\Delta T_x$) and reduced glass transition temperature ($T_{rg}$) represent the glass forming ability. In general, larger $\Delta T_x$ and $T_{rg}$ values usually exhibit higher stability of the supercooled liquid against crystallization and enable the formation of fluoride glasses with larger dimension. The *critical cooling rate* ($R_c$) is another important parameter (details are provided in Chapter 12), which is characterized as the slowest cooling rate at which a glass is obtained without crystallization from a melt of a given composition. The ZBLAN glass has very low $R_c$ (1.7 K $min^{-1}$) value, whereas the $R_c$ value of the CLAP glass is very high (300 K $min^{-1}$). The significance is that the ZBLAN glass can be prepared by employing moderate cooling, whereas the CLAP glass requires very high cooling rate particularly during casting.

Fig. 11.3A shows a comparison of transmission windows of two HMFGs with that of silica ($SiO_2$) glass [30]. It is seen that the infrared absorption edge of silica glass is extended from 4–8 to 11 μm in fluoride glasses based on their composition. These infrared absorption edges of halide glasses can further be extended up to 30 μm in the glasses of heavier halides of chloride, bromide, and iodide. Fig. 11.3B shows a comparison of the infrared cutoff of different halide glasses based on fluoride, chloride, bromide, and iodide [31].

## 11.6 Properties of transition metal and rare-earth ion doped halide glasses and glass-ceramics

To find out the total number of published papers on individual transition metal and rare-earth metal ion containing fluoride glasses and glass-ceramics during the last 25 years (1990–2014), the total publication of 3074 articles on fluoride glasses and glass-ceramics was further refined by term of *individual transition metal or rare-earth metal*. This evaluation was carried out according to the technique described in Section 11.1 earlier. These plots are illustrated in Fig. 11.4A and B. It gives us a clear

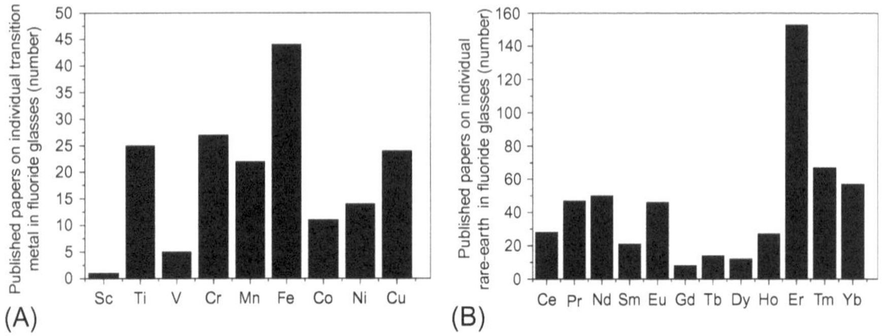


**Fig. 11.4** Total numbers of published papers during the last 25 years (1990–2014) on individual (A) transition and (B) rare-earth metal ion in fluoride glasses.

quantitative idea of paper published on individual transition metal and rare-earth metal ion containing fluoride glasses and glass-ceramics and a comparison among them. The decreasing order of publication trends of the individual transition metal and rare-earth metal ion doped fluoride glasses and glass-ceramics are as follows:

$$Fe > Cr > Ti > Cu > Mn > Ni > Co > V > Sc$$

$$Er \gg Tm > Yb > Nd > Pr > Eu > Ce > Ho > Sm > Tb > Dy > Gd$$

It is seen that during the last 25 years (1990–2014), among the 3074 published articles, the highest number (44) of papers was published using iron (Fe), and the second highest (27) is with chromium (Cr) in the transition metal family, and the highest number (153) of papers was published using erbium (Er), and the second highest (67) is with thulium (Tm) in the rare-earth metal family. These trends of publications give us the idea about the importance of doping the fluoride glasses with transition metal and rare-earth metal ions.

Fig. 11.5 illustrates the room temperature (A) photoluminescence (PL) and (B) excitation spectra of a $Fe^{3+}$-doped fluorozirconate glass of composition (mol.%) $51.8ZrF_4$–$20.0BaF_2$–$19.6LiF$–$3.3AlF_3$–$5.3LaF_3$+$0.4FeF_3$ [32]. When excited with a 476.2 nm laser light at 300 K, the broad PL band was found to peak around 780 nm. The band was assigned to the Stokes-shifted radiative transitions between the lowest crystal field split levels ($^4T_1 \rightarrow {}^6A_1$) of the $Fe^{3+}$ ion as a result of $d^5$ electronic configuration. The PL excitation spectrum is a measure of the magnitude of the integrated PL intensity as a function of excitation wavelength. The excitation spectrum has well defined peaks at 650 and 450 nm with a shoulder around 500 nm.

Fig. 11.6 depicts the mid-infrared emission spectrum of $Ho^{3+}$-ion doped $BaF_2$–$InF_3$–$ZnF_2$–$YF_3$–$ThF_4$ (BIZnYT) glass at 290 K (excitation by xenon lamp) [33]. Three emission bands are found to peak around 1.2, 2.0, and 3.0 μm. They are assigned to the $^5I_6 \rightarrow {}^5I_8$, $^5I_7 \rightarrow {}^5I_8$, and $^5I_6 \rightarrow {}^5I_7$ transitions, respectively.

Fig. 11.7A demonstrates the IR and UV excited RGB emission (white light) spectrum of $Tm^{3+}$, $Yb^{3+}$, and $Er^{3+}$ triply doped fluoroaluminate glass, and Fig. 11.7B illustrates the schematic energy-level diagram of $Tm^{3+}$, $Yb^{3+}$, and $Er^{3+}$ ions showing the

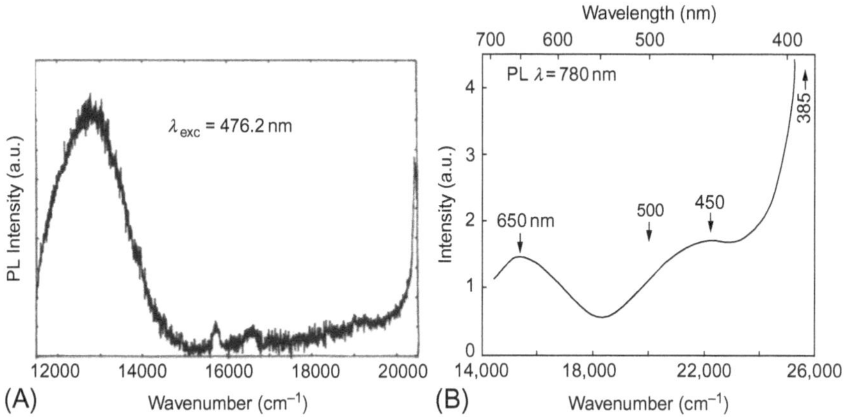


**Fig. 11.5** Room temperature (A) photoluminescence and (B) excitation spectra of a $Fe^{3+}$-doped fluorozirconate glass of composition (mol.%) 51.8$ZrF_4$–20.0$BaF_2$–19.6LiF–3.3$AlF_3$–5.3$LaF_3$ + 0.4$FeF_3$.
Reproduced with permission from J.A. Freitas Jr., U. Strom, D.C. Tran, Photoluminescence studies of fluorozirconate glasses, Mater. Sci. Forum 6 (1985) 627–634, Copyright © Trans Tech. Publications.

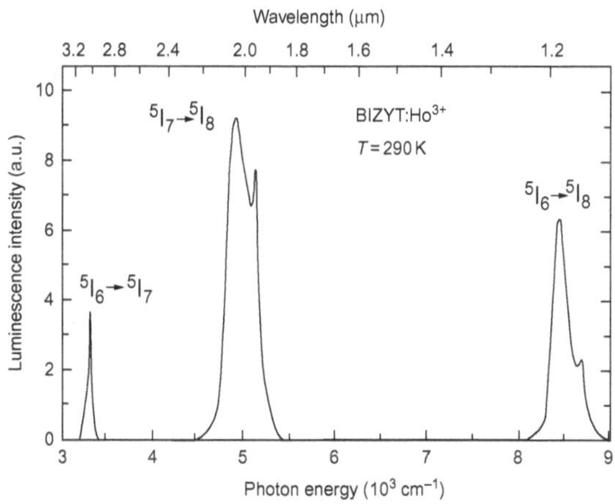


**Fig. 11.6** Mid-infrared emission spectrum of $Ho^{3+}$-ion doped $BaF_2$–$InF_3$–$ZnF_2$–$YF_3$–$ThF_4$ (BIZnYT) glass at 290 K (excitation by xenon lamp).
Reproduced with permission from J.-C. Adam, C. Guéry, J. Lucas, J. Rubin, B. Moine, G. Boulon, Infrared optical transitions of $Ho^{3+}$ ions in fluoride glasses, Mater. Sci. Forum, 19–20 (1987) 573–580, Copyright © Trans Tech. Publications.

mechanism for RGB (white light) generation [34]. Blue emission (around 450 nm) is obtained as a result of UV (360 nm) excitation, whereas the green (around 550 nm) and red (650 nm) emission bands are obtained as a result of upconversion (UC) emission from IR (1000 nm) to visible. The UC was facilitated owing to low phonon energy (450–500 cm$^{-1}$) of the fluoride glass compared with that of the oxide glasses.

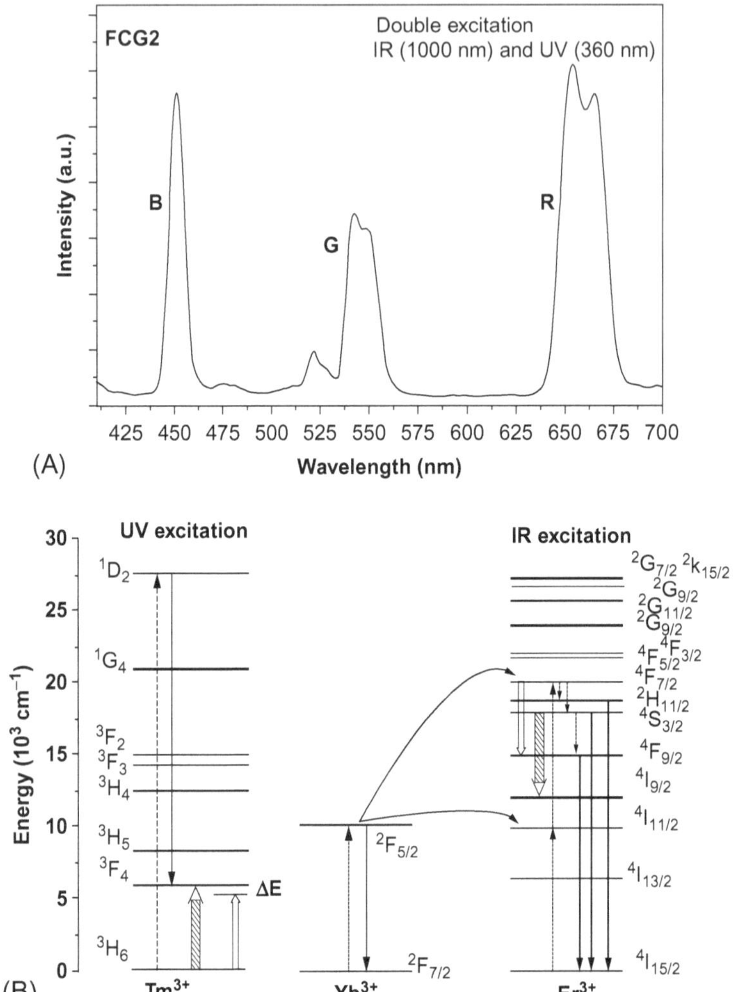


**Fig. 11.7** (A) IR and UV excited RGB emission spectrum of $Tm^{3+}$, $Yb^{3+}$, and $Er^{3+}$ triply doped fluoroaluminate glass. (B) Schematic energy-level diagram of $Tm^{3+}$, $Yb^{3+}$, and $Er^{3+}$ ions showing the mechanism for RGB generation.
Reproduced with permission from J.E.C. Silva, G.F. de Sa', P.A. Santa-Cruz, Red, green and blue light generation in fluoride glasses controlled by double excitation, J. Alloys Compd. 323–324 (2001) 336–339, Copyright © Elsevier.

A comparison of phonon energy of different glasses is provided in Table 11.3. Low phonon energy suppresses the nonradiative emission as a result of reduced multiphonon emission rate. A comparison of multiphonon emission rate of fluoride glasses ($BaF_2$-$ThF_2$ based and ZBLA) with that of different oxide glasses is shown in Fig. 11.8 [35]. It is seen that the fluoride glasses have the lowest multiphonon emission rate compared with that of different oxide glasses.

**Table 11.3 Phonon energy of some rare-earth host glasses and halide crystals**

| Host | Phonon energy, $\hbar_{ph}$ ($cm^{-1}$) |
|---|---|
| Borate glass | 1400 |
| Phosphate glass | 1260 |
| Silica/silicate glass | 1100 |
| Germanate glass | 900 |
| Tellurite glass | 700 |
| Antimony oxide glass | 602 |
| Fluoride glass | 450–500 |
| $SrF_2$ crystal | 360 |
| Chalcogenide glass | 350 |
| $LaF_3$ crystal | 350 |
| $LaCl_3$ crystal | 260 |
| $LaBr_3$ crystal | 175 |

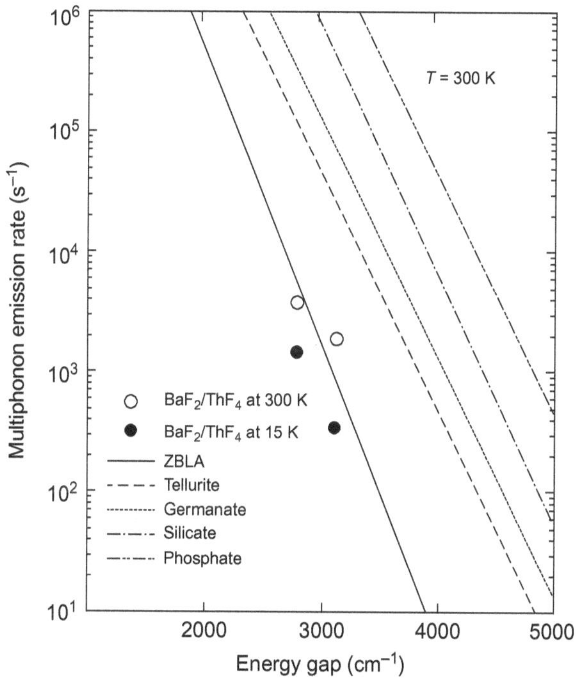


**Fig. 11.8** Comparison of multiphonon emission rate of fluoride glasses ($BaF_2$-$ThF_2$ based and ZBLA) with that of different oxide glasses.
Reproduced with permission from W.A. Sibley, D.C. Yeh, M. Suscavage, M.G. Drexhage, Multiphonon relaxation and infrared to visible conversion of $Er^{3+}$ and $Yb^{3+}$ ions in barium-thorium fluoride glass, Mater. Sci. Forum 19–20 (1987) 567–572, Copyright © Trans Tech. Publications.

The above facts can be recognized if we keep in mind the following theoretical considerations. The fluorescence lifetime ($\tau_f$) can be expressed by the following relation:

$$\frac{1}{\tau_f} = \frac{1}{\tau_R} + W_{nr} \tag{11.1}$$

where $\tau_R$ and $W_{nr}$ are the radiative lifetime and nonradiative relaxation rate, respectively. Hence, the quantum efficiency ($\eta$) is expressed by

$$\eta = \tau_f / \tau_f \tag{11.2}$$

and nonradiative relaxation rate ($W_{nr}$) is given by

$$W_{nr} = W_{mp} + W_{et} \tag{11.3}$$

where $W_{mp}$ and $W_{et}$ are the multiphonon relaxation rate and energy transfer rate, respectively. The multiphonon relaxation process arises from the interaction of the rare-earth ion with the fluctuating electric field due to vibration of the matrix lattice. The number ($p$) of phonon of energy ($\hbar_{ph}$) required to bridge the energy gap ($E$) between an excited level, and the next lower energy level can be evaluated using the following relation:

$$p = \Delta E / \hbar\omega_{ph} \tag{11.4}$$

Therefore, the lower the phonon energy ($\hbar_{ph}$), the lesser the rate of multiphonon relaxation process and the loss of excitation energy as a result of thermalization. This in turn increases population in the excited state and radiation transition probability.

Fig. 11.9 shows the (A) extinction coefficient of 0.5% Eu-doped fluorochlorozirconate glasses heat-treated at different temperatures for 70 min and (B) plot of photostimulated luminescence (PSL) efficiencies and transmittance (1.45 mm thick sample) at 633 nm of 1000 ppm Eu-doped fluorobromozirconate (FBZ) glasses as function of heat treatment temperature [36]. The extinction coefficient of the FCZ glass-ceramics gradually decreases with rise in heat-treatment temperature as a result of scattering of the generated $BaCl_2$ crystallites in the resultant glass-ceramics. Similarly, the loss of transmittance at 633 nm, as shown in Fig. 11.9A, can be interpreted. There is a trade-off between optical transparency at 633 nm and relative PSL over a narrow heat-treatment temperature range (270–280°C). This coincidence clearly signifies that as soon as the $BaCl_2$ crystallites are formed, the europium (Eu) ion enters into the low phonon energy ($BaCl_2 \approx 350\ cm^{-1}$ and fluoride glass $\approx 500\ cm^{-1}$) environment of $BaCl_2$ crystals, enhancing the PSL efficiency. As an evidence, the TEM images of Eu-doped FCZ glass-ceramic heat-treated at 290°C for 5 min (left) and 10 min (right) are shown in Fig. 11.10 [37].

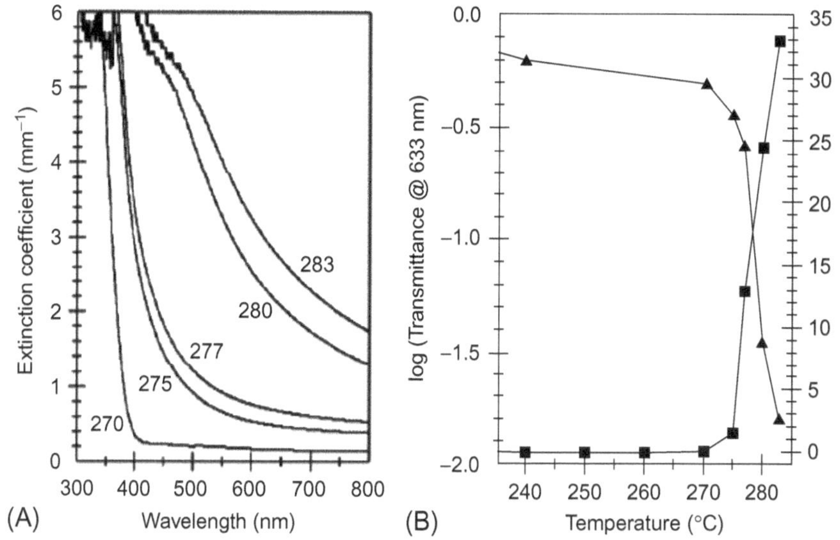


**Fig. 11.9** (A) Extinction coefficient of 0.5% Eu-doped fluorochlorozirconate glasses heat-treated at different temperatures for 70 min. (B) Plot of photostimulated luminescence efficiencies and transmittance (1.45 mm thick sample) at 633 nm of 1000 ppm Eu-doped fluorobromozirconate glasses as a function of heat-treatment temperature.
Reproduced with permission from S. Schweizer, J.-M. Spaeth, M. Secu, A. Edgar, G.V.M. Williams, Photostimulable defects in nano-crystallites in fluorozirconate glasses, Phys. Stat. Sol. 202 (2005) 243–249, Copyright © Wiley.

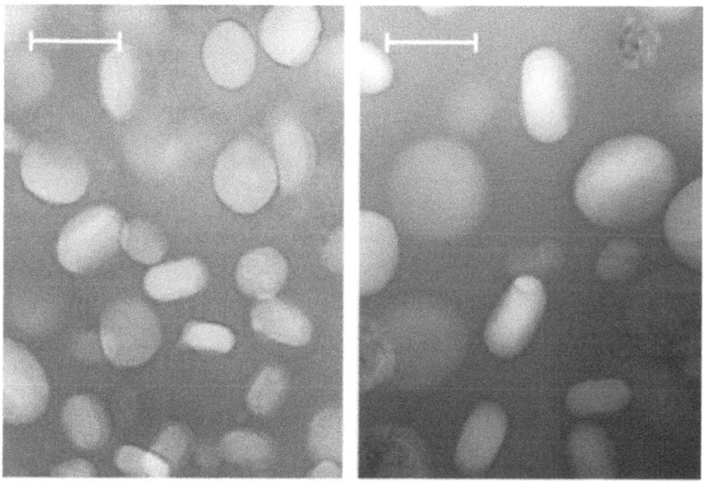

**Fig. 11.10** TEM images of Eu-doped FCZ glass-ceramic heat-treated at 290°C for 5 min (left) and 10 min (right). Scale = 100 nm.
Reproduced with permission from S. Schweizer, L W. Hobbs, M. Secu, J.-M. Spaeth, A. Edgar, G.V.M. Williams, Photostimulated luminescence in Eu-doped fluorochlorozirconate glass-ceramics, Appl. Phys. Lett. 83 (2003) 449–451, Copyright © AIP.

## 11.7 Functional applications of halide glasses and glass-ceramics

The major advantages of halide glasses over chalcogenide glasses are that they can be used in the UV and visible regions as well, and they are useable in the infrared regions in comparison with the oxide glasses. Various functional applications of halide (fluoride) glasses are illustrated in Fig. 11.11. They are used in passive and active applications. Some of their notable applications are infrared laser beam delivery systems, sensing, infrared lasers, scintillation, and laser cooling. Some of these applications are briefly described as follows.

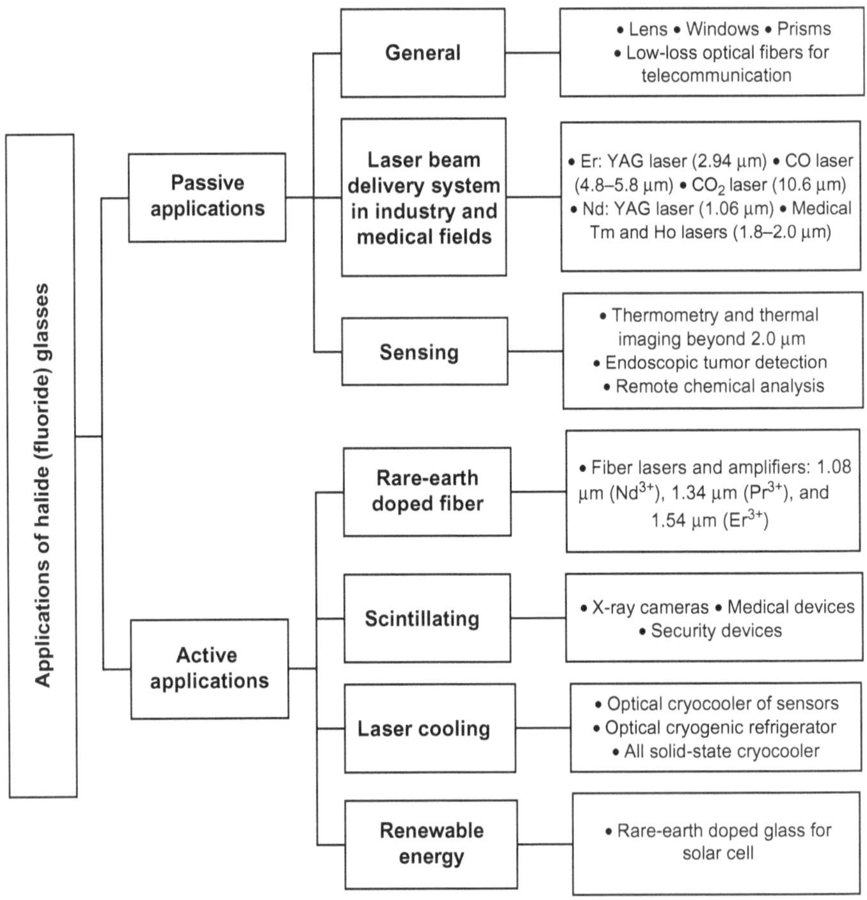


**Fig. 11.11** Functional applications of halide (fluoride) glasses.

### 11.7.1 Scintillation

Fig. 11.12A shows the scintillation (X-ray excited luminescence) spectra of $EuF_2$-doped FCZ glass-ceramics containing hexagonal $BaCl_2$ nanoparticles [38]. It is seen that there are two X-ray luminescence bands peaking at 407 and 485 nm. The band at 407 nm is assigned to the 5d-4f transition of $Eu^{2+}$ ions, whereas the origin of the 485 nm band is unknown; however, its intensity increases linearly with increasing the concentration of $EuF_2$ doping. The scintillation efficiency of the glass-ceramic plate and image quality are of paramount importance for medical and security applications. Fig. 11.12B shows the scintillation image of a mouse joint, recorded with a 2.0% $EuF_2$-doped FCZ glass-ceramic plate [38]. The plate shown has very high resolution of 110 lp $mm^{-1}$ at a MTF of 0.2, which is close to the quality obtained using single crystal of cadmium tungstate. Therefore, the FCZ glass-ceramic scintillator could be used as an alternative material for image formation resulting from scintillation. The advantages of these glass-ceramics are that they can be scaled to any shape and size with excellent resolution.

### 11.7.2 Laser cooling

In 1995, Epstein and coworkers [39] first convincingly demonstrated laser cooling in an $Yb^{3+}$-doped HMFG, ZBLANP ($ZrF_4$–$BaF_2$–$LaF_3$–$AlF_3$–NaF–$PbF_2$). Laser cooling of rare-earth doped solids is based on the anti-Stokes fluorescence, where the emitted photons have a mean energy higher than that of the absorbed photons. However, until

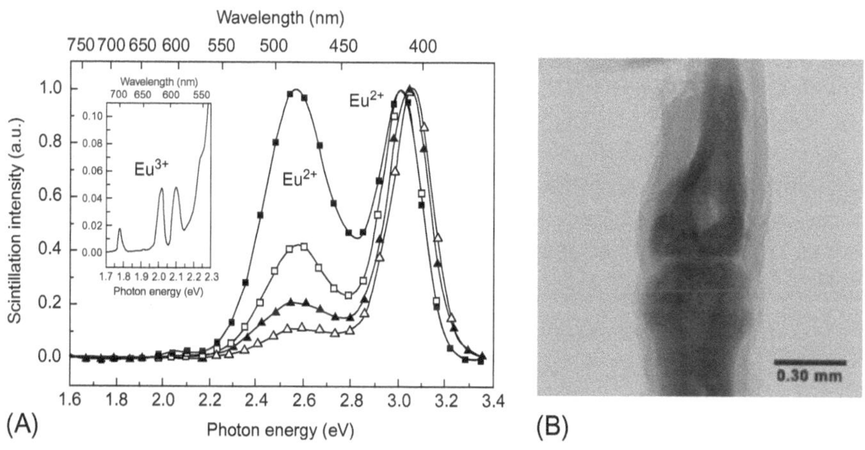


**Fig. 11.12** (A) Scintillation (X-ray excited luminescence) spectra of $EuF_2$-doped FCZ glass-ceramics (△, 0.2% $EuF_2$; ▲, 0.5% $EuF_2$; □, 1.0% $EuF_2$; and ■, 2.0% $EuF_2$). The inset shows the absorption bands of $Eu^{3+}$ ions in this glass. (B) Scintillation image of a mouse joint, recorded with a 2.0% $EuF_2$-doped FCZ glass-ceramic plate.
Reproduced with permission from J.A. Johnson, S. Schweizer, B. Henke, G. Chen, J. Woodford, P.J. Newman, D.R. MacFarlane, Eu-activated fluorochlorozirconate glass-ceramic scintillators, J. Appl. Phys. 100 (2006) 034701 (5 pages), Copyright © AIP.

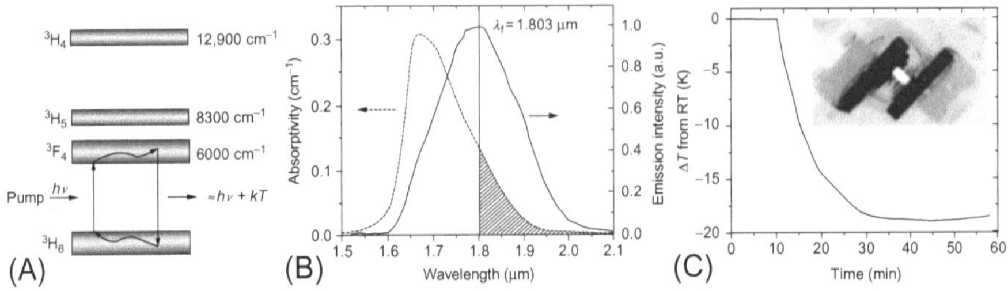


**Fig. 11.13** (A) Energy-level diagram of $Tm^{3+}$:ZBLANP glass showing the mechanism of laser cooling. (B) Absorption *(dotted curve)* and fluorescence *(solid curve)* spectra of $Tm^{3+}$:ZBLANP glass showing the shaded area where cooling is expected. (C) Temperature as a function of time showing the extent of cooling for multiple pump passes through $Tm^{3+}$:ZBLANP glass sample. Inset shows photograph of both the cold sample *(white)* and the hot mirror *(black)*. Reproduced with permission from C.W. Hoyt, M.P. Hasselbeck, M. Sheik-Bahae, R.I. Epstein, S. Greenfield, J. Thiede, J. Distel, J. Valencia, Advances in laser cooling of thulium-doped glass, J. Opt. Soc. Am. B20 (2003) 1066–1074, Copyright © OSA.

now, the achieved efficiency of temperature difference has increased from 0.3 to 92 K, which is much lower than the theoretical limit. Here, we describe the laser cooling of $Tm^{3+}$-doped fluorozirconate (ZBLANP) glass down to 19 K below the ambient temperature with a multiple-pass pump scheme [40]. Fig. 11.13A illustrates the energy-level diagram of $Tm^{3+}$:ZBLANP glass showing the mechanism of laser cooling. The dopant $Tm^{3+}$ ion is excited by the pump from the top of the ground-state manifold ($^3H_6$) to the bottom of the excited-state manifold ($^3F_4$). The atoms thermalize in both manifolds by absorbing vibrational energy from the host glass, and the subsequent fluorescence removes an energy $h\nu_f - h\nu$ for each absorbed photon. Fig. 11.13B displays the absorption (dotted curve) and fluorescence (solid curve) spectra of $Tm^{3+}$:ZBLANP glass showing the shaded area where cooling is expected. The vertical solid line marks the mean fluorescent wavelength at 1.803 µm. Fig. 11.13C shows the reduction of temperature as a function of time showing the extent of cooling for multiple pump passes through the $Tm^{3+}$:ZBLANP glass sample. Inset shows photograph of both the cold sample (white) and the hot mirror (black).

### 11.7.3 Solar cell

Recently, Martín and coworkers [41] have reported the experimental enhancement of the photocurrent in a solar cell using UC process in $Yb^{3+}$ and $Er^{3+}$ ions codoped fluoroindate glasses exciting at 1480 nm. Fig. 11.14A displays the photograph showing a fluoroindate glass codoped with $Er^{3+}$ and $Yb^{3+}$ placed on the top of a c-Si solar cell. The excitation is carried out from the top with a 1480 nm laser diode (not visible in the picture). Fig. 11.14B solar cell photocurrent and external quantum efficiency (EQE) as a function of laser power at 1480 nm as measured with the setup shown in (A) and glasses containing different concentrations of $Er^{3+}$ ion at a fixed

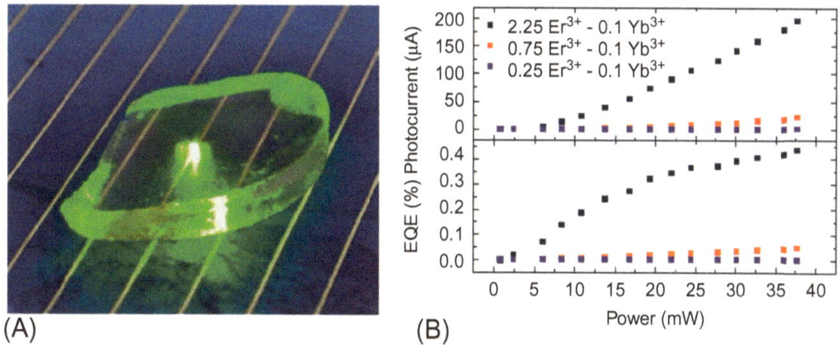


**Fig. 11.14** (A) Photograph showing a fluoroindate glass codoped with $Er^{3+}$ and $Yb^{3+}$ placed on the top of a c-Si solar cell. The excitation is carried out from the top with a 1480 nm laser diode. (B) Solar-cell photocurrent and external quantum efficiency (EQE) as a function of laser power at 1480 nm as measured with the setup shown in (A) and glasses containing different concentrations of $Er^{3+}$ ion at a fixed concentration of $Yb^{3+}$ ion.

Reproduced with permission from M.A. Hernández-Rodríguez, M.H. Imanieh, L.L. Martín, I.R. Martín, Experimental enhancement of the photocurrent in a solar cell using upconversion process in fluoroindate glasses exciting at 1480 nm, Sol. Energy Mater. Sol. Cells 116 (2013) 171–175, Copyright © Elsevier.

concentration of $Yb^{3+}$ ion. It is seen that a photocurrent of nearly 200 μA can be obtained at 37 mW of pump power for the sample codoped with 2.25 mol.% of $Er^{3+}$ and 0.1 mol.% of $Yb^{3+}$. This experiment, therefore, clearly evidences the enhanced photocurrent generation by the UC processes of $Yb^{3+}$ and $Er^{3+}$ ions codoped fluoroindate glasses. Other investigators also obtained improved efficiency of silicon solar cells using rare-earth doped upconverting fluoride glasses [42,43].

## 11.8 Summary

Various features of halide glasses and glass-ceramics particularly of the fluoride family are briefly described in this chapter. Initially, the publication trend and quantitative growth of published papers on fluoride glasses and glass-ceramics during the last 25 years (1990–2014) were presented. This was followed by the depiction of history of development, structure, synthesis, and properties of the halide glasses. Then, the properties of transition metal and rare-earth ion-doped halide glasses and glass-ceramics were described with specific instances. Finally, the functional applications of halide glasses and glass-ceramics were elaborated with definite examples.

## Exercises

**11.1** Compare halide, chalcogenide, and oxide glasses.
**11.2** Name the first fluoride glass. Compare its structure with that of silica glass.
**11.3** Who and when HMFG was discovered? What was the glass system?

**11.4** Describe the different models of structure of $BaF_2$–$ZrF_4$ glass. What are the basic differences of these models?
**11.5** Describe briefly the procedure of synthesis of fluoride glass.
**11.6** Describe the importance of critical cooling rate ($R_c$) in manufacturing of fluoride glasses?
**11.7** Describe the UV and infrared cutoffs of fluoride, chloride, bromide, and iodide glasses. Compare these with that of silica glass.
**11.8** Describe the PL and excitation spectra (with schematic profiles) of a $Fe^{3+}$-doped fluorozirconate glass of the $ZrF_4$–$BaF_2$–LiF–$AlF_3$–$LaF_3$–$FeF_3$ system.
**11.9** Describe the UC phenomenon with schematic spectral profile and partial energy diagram of $Yb^{3+}$ and $Er^{3+}$ doubly doped fluoroaluminate glass. Also describe their associated theories.
**11.10** Why is there a trade-off between the optical transparency and PSL efficiencies in Eu-doped FBZ glasses as function of heat-treatment temperature? Explain the reason.
**11.11** Describe various functional applications of halide (fluoride) glasses.
**11.12** Describe short notes on (i) scintillation, (ii) laser cooling, and (iii) efficiency enhancement of silicon solar cells by using fluoride glasses or glass-ceramics.

# References

[1] P. Lebeau, Anhydrous beryllium fluoride and oxyfluoride of beryllium, Chem. News 77 (1898) 288–291.
[2] P. Lebeau, Beryllium fluoride glass by heating ammonium fluoroberyllate, C. R. Geosci. 126 (1898) 1418–1421.
[3] V.M. Goldschmidt, Geochemische verteilungsgesetzeder elemente. VII. Die gesetzeder kristallchemie, Skr. Nor. Vidensk.-Akad. (Oslo) 8 (1926) 127–137.
[4] H. Rawson, Inorganic Glass Forming Systems, Academic Press, New York, NY, 1976.
[5] G. Heine, Über die darstellung und eigenschaften einiger beryllium fluoridgläser, Angew. Chem. 46 (1933) 473–477.
[6] K.-H. Sun, T.E. Callear, US patent 2,466,506, 1949.
[7] K.-H. Sun, M.L. Huggins, US patent 2,511,224, 1950.
[8] L.A. Topol, S.W. Mayer, L.D. Ransom, The heat of fusion of bismuth trichloride—a comparison of calorimetric and cryoscopic determinations, J. Phys. Chem. 64 (1960) 862–865.
[9] M. Poulain, M. Poulain, J. Lucas, Verresfluores au tetrafluorure de zirconium proprietes optiques d'un verre dope au $Nd^{3+}$, Mater. Res. Bull. 10 (1975) 243–246.
[10] J. Miranday, C. Jacoboni, R. de Pape, Rev. Chim. Min. 16 (1979) 277–282 (J. Miranday, C. Jacoboni, R. de Pape, New transition metal fluoride glasses isolated in the systems, J. Non-Cryst. Solids 43 (1981) 393–401.).
[11] G. Fonteneau, H. Slim, F. Lahaie, J. Lucas, Nouveaux verres fluorestrans metteursdansl' infrarougedans les systemes $LnF_3$-$BaF_2$-$ZnF_2$, Mater. Res. Bull. 15 (1980) 1425–1432.
[12] H. Hu, J.D. Mackenzie, New chloride glasses in the NaCl-KCl-$ThCl_4$ system, J. Non-Cryst. Solids 51 (1982) 269–272.
[13] M. Poulain, M. Matecki, J.-L. Mouric, M. Poulain, Cadmium halide glasses. II.—chloride glasses, Mater. Res. Bull. 18 (1983) 631–636.
[14] M. Fuding, New halide glasses in the AgBr-$PbBr_2$-CuCl-CsBr system, Mater. Sci. Forum 5 (1985) 127–134.

[15] C. Liu, H.G.K. Sundar, C.A. Angell, All-halide superionic glasses, Solid State Ionics 18–19 (1986) 442–448.
[16] J. Lucas, X.H. Zhang, G. Fonteneau, Tellurium halide glasses-new materials for IR fibers. in: Proc. SPIE 0843, 1987. http://dx.doi.org/10.1117/12.968206.
[17] H. Xu, X. Wang, Q. Nie, Y. He, P. Zhang, T. Xu, S. Dai, X. Zhang, Glass formation and properties of Ge–Ga–Te–$ZnI_2$ far infrared chalcohalide glasses, J. Non-Cryst. Solids 383 (2014) 212–215.
[18] Y. Kawamoto, Progress in structural study of $ZrF_4$-based glasses, Mater. Sci. Forum 6 (1985) 417–426.
[19] R.M. Almeida, J.D. Mackenzie, Vibrational spectra and structure of fluorozirconate glasses, J. Chem. Phys. 74 (1981) 5954–5961.
[20] R.M. Almeida, J. Lau, J.D. Mackenzie, X-ray photoemission study of fluorozirconate glass and related crystals, J. Non-Cryst. Solids 69 (1984) 161–165.
[21] J. Lucas, C.A. Angel, S. Tamaddon, Fluoride bridging modes in fluorozirconate glasses by X-ray and computer simulation studies, Mater. Res. Bull. 19 (1984) 945–951.
[22] J. Lucas, D. Louer, C.A. Angel, A structural model for fluorozirconate glass, Mater. Sci. Forum 6 (1985) 449–456.
[23] I. Yasui, H. Inoue, Molecular dynamic simulations of changes in structure in $ZrF_4$ based glasses, J. Non-Cryst. Solids 71 (1985) 39–47.
[24] Y. Kawamoto, F. Sakaguchi, Thermal properties and Raman spectra of crystalline and vitreous $BaZrF_6$, $PbZrF_6$, and $SrZrF_6$, Bull. Chem. Soc. Jpn. 56 (1983) 2138–2141.
[25] Y. Kawamoto, T. Horisaka, Short-range structures of barium, lead, and meta-fluorozirconate glasses, J. Non-Cryst. Solids 56 (1983) 39–44.
[26] Y. Kawamoto, The Raman spectra of barium fluorozirconate glasses and their interpretation, Phys. Chem. Glasses 25 (1984) 88–91.
[27] M. Poulain, A. Soufiane, Y. Messaddeq, M.A. Aegerter, Fluoride glasses: synthesis and properties, Braz. J. Phys. 22 (1992) 205–217.
[28] J.L. Adam, Fuoride glass research in France: fundamentals and applications, J. Fluor. Chem. 107 (2001) 265–270.
[29] G. Rault, J.L. Adam, F. Smektala, J. Lucas, Fuoride glass compositions for waveguide applications, J. Fluor. Chem. 110 (2001) 165–173.
[30] J.M. Parker, P.W. France, Optical properties of halide glasses, in: M.H. Lewis (Ed.), Glasses and Glass-Ceramics, Chapman and Hall, New York, NY, 1989.
[31] J. Lucas, X.H. Zhang, The tellurium halide glasses, J. Non-Cryst. Solids 125 (1990) 1–16.
[32] J.A. Freitas Jr., U. Strom, D.C. Tran, Photoluminescence studies of fluorozirconate glasses, Mater. Sci. Forum 6 (1985) 627–634.
[33] J.-C. Adam, C. Guéry, J. Lucas, J. Rubin, B. Moine, G. Boulon, Infrared optical transitions of $Ho^{3+}$ ions in fluoride glasses, Mater. Sci. Forum 19–20 (1987) 573–580.
[34] J.E.C. Silva, G.F. de Sa', P.A. Santa-Cruz, Red, green and blue light generation in fluoride glasses controlled by double excitation, J. Alloys Compd. 323–324 (2001) 336–339.
[35] W.A. Sibley, D.C. Yeh, M. Suscavage, M.G. Drexhage, Multiphonon relaxation and infrared to visible conversion of $Er^{3+}$ and $Yb^{3+}$ ions in barium-thorium fluoride glass, Mater. Sci. Forum 19–20 (1987) 567–572.
[36] S. Schweizer, J.-M. Spaeth, M. Secu, A. Edgar, G.V.M. Williams, Photostimulable defects in nano-crystallites in fluorozirconate glasses, Phys. Status Solidi 202 (2005) 243–249.
[37] S. Schweizer, L.W. Hobbs, M. Secu, J.-M. Spaeth, A. Edgar, G.V.M. Williams, Photostimulated luminescence in Eu-doped fluorochlorozirconate glass-ceramics, Appl. Phys. Lett. 83 (2003) 449–451.

[38] J.A. Johnson, S. Schweizer, B. Henke, G. Chen, J. Woodford, P.J. Newman, D.R. MacFarlane, Eu-activated fluorochlorozirconate glass-ceramic scintillators, J. Appl. Phys. 100 (2006). 034701 (5 pages).

[39] R.I. Epstein, M.I. Buchwald, B.C. Edwards, T.R. Gosnell, C.E. Mungan, Observation of laser-induced fluorescent cooling of a solid, Nature 377 (1995) 500–503.

[40] C.W. Hoyt, M.P. Hasselbeck, M. Sheik-Bahae, R.I. Epstein, S. Greenfield, J. Thiede, J. Distel, J. Valencia, Advances in laser cooling of thulium-doped glass, J. Opt. Soc. Am. B20 (2003) 1066–1074.

[41] M.A. Hernández-Rodríguez, M.H. Imanieh, L.L. Martín, I.R. Martín, Experimental enhancement of the photocurrent in a solar cell using upconversion process in fluoroindate glasses exciting at 1480 nm, Sol. Energy Mater. Sol. Cells 116 (2013) 171–175.

[42] F. Pellé, S. Ivanova, J.-F. Guillemoles, Upconversion of 1.54 μm radiation in $Er^{3+}$ doped fluoride-based materials for c-Si solar cell with improved efficiency, EPJ Photovolt. 2 (2011). 20601 (10 pages).

[43] B. Henke, B. Ahrens, J.A. Johnson, P.-T. Miclea, S. Schweizer, Upconverting glasses for high-efficiency solar cells. SPIE Newsroom (2009). http://dx.doi.org/10.1117/2.1200910.1811 3 pages.

# Functional bulk metallic glasses

**12**

## Chapter Outline



## 12.1 Introduction

Bulk metallic glasses (BMGs) are metallic alloys with a disordered atomic-scale structure. The term "*bulk*" defines at least of millimeter scale. They are different from the conventional alloys that are crystalline and therefore have highly ordered atomic arrangements. Glass is defined as "*a noncrystalline solid, which exhibits the glass transformation (transition) range in the volume or enthalpy-temperature relationship*" [1]. Therefore, the BMG can be defined as "*a noncrystalline metallic alloy, which exhibits the glass transformation (transition) range in the volume or enthalpy-temperature relationship and can be prepared at least in millimeter scale.*" Hence, the BMGs have the glass transition temperature ($T_g$) that is not obtainable in the case of conventional alloys. The BMGs can be cast in molds to near-net shapes. The atoms of BMGs are "frozen" in random disordered structures with short-range order and without any long-range order like crystalline counterparts. This noncrystalline structure lacking in grain boundaries provides BMGs high strength, toughness, hardness, elasticity, corrosion, and wear resistance. As a result of these beneficial outstanding properties of BMGs, they are used in a wide range of functional applications such as striking face plate in golf clubs, casing in cellular phones, watch parts, optical mirrors, connecting parts for optical fibers, microelectromechanical systems (MEMS), high sensitivity small pressure sensors, high density information-storage material, slat truck cover for aircrafts, and jewelry. As a consequence of this fact, there are a huge number of publications in this domain.

An attempt was made using the Web of Science (version 5.17) as described in Chapter 1 to provide a global publication trend and quantitative growth in the area of BMGs. Here, year-wise data were extracted for the last 25 years (1990–2014)

**Functional Glasses and Glass-Ceramics. http://dx.doi.org/10.1016/B978-0-12-805056-9.00012-X**

by searching, using the topic keyword "*glass*" followed by refining the document type as "*article*" published in the journals and meetings. It was further refined using the term "*bulk metallic glass.*" This plot is shown in Fig. 12.1. It gives us a clear publication trend and quantitative growth of published papers on BMGs during the last 25 years (1990–2014). It is seen that there is an exponential increase (from 3 to 583) in paper publication during the period of 1990–2007 and then after there is a decreasing trend during the period of 2007–14. Nevertheless, the publication is still enormous in the tune of 380 in the year 2014.

Various aspects of BMGs will be described briefly in the following sections.

## 12.2 History of development of BMGs

Some chronological events of the development of BMGs are summarized in Table 12.1 [2–19]. Duwez and his coworkers [2] discovered the metallic glass by cooling the metallic liquid of $Au_{80}Si_{20}$ at a very high rate of $1 \times 10^6$ °C $s^{-1}$ at Caltech, the United States, in 1960. As a result of very high cooling rate, the glass dimension was a few microns. In 1969, Chen and Turnbull [3] successfully prepared amorphous microspheres of the ternary Pd–Si–N (where N = Ag, Cu, or Au) system. These glasses cannot be regarded as BMGs because of their micron-size dimension.

In fact, Chen [4] gave birth of BMGs in 1974. He prepared the first bulk metallic-glass wire of 0.5–1.0 mm diameter of $Pd_{77.5}Cu_6Si_{16.5}$ at lower cooling rate of $10^3$K $s^{-1}$. In 1982, Turnbull and his coworkers [6] were successful to fabricate

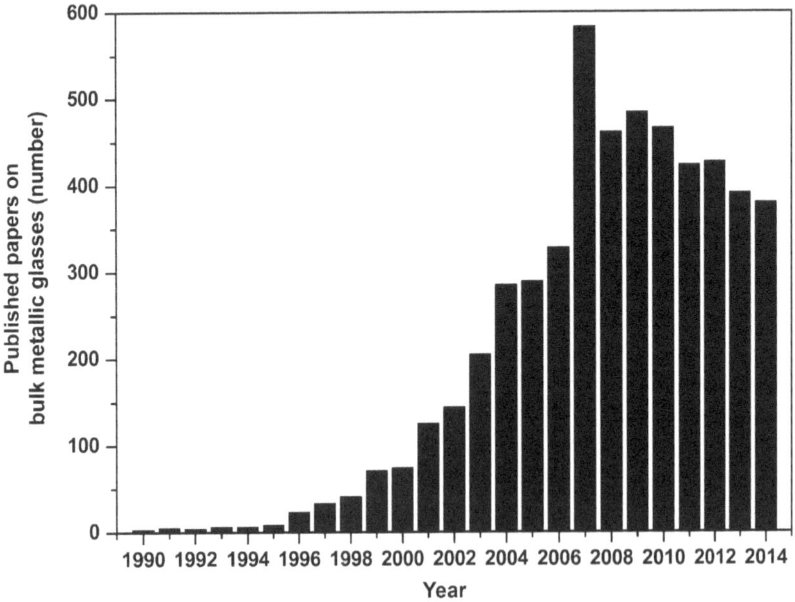


**Fig. 12.1** Year-wise published papers on bulk metallic glasses (BMGs) during the last 25 years (1990–2014).

**Table 12.1 Some chronological events of development of bulk metallic glasses**

| Year | Event | Author(s) | Ref. |
|---|---|---|---|
| 1960 | The first metallic glass was discovered by rapid quenching (splat cooling) ($1 \times 10^6$ °C $s^{-1}$) of $Au_{80}Si_{20}$ liquid | Klement, Willens, and Duwez | [2] |
| 1969 | Successfully prepared amorphous spheres of the ternary Pd–Si–N (where N = Ag, Cu, or Au) system | Chen and Turnbull | [3] |
| 1974 | Birth of bulk metallic glass (BMG). The first bulk metallic glass of 0.5–1.0 mm diameter of $Pd_{77.5}Cu_6Si_{16.5}$ at lower cooling rate of $10^3$K $s^{-1}$ | Chen | [4] |
| 1982 | Little higher diameter spheres of the $Au_{55}Pb_{22.5}Sb_{22.5}$ metallic glass were prepared | Lee, Kendall, and Johnson | [5] |
| 1982 | Glassy spheroids of 0.53 cm minor diameter of $Pd_{40}Ni_{40}$ $P_{20}$ were slowly cooled (1.4 K $s^{-1}$) under $10^{-6}$ Torr vacuum | Drehman, Greer, and Turnbull | [6] |
| 1984 | Obtained $Pd_{40}Ni_{40}P_{20}$ BMG ingot of cm size at the cooling rate of 10 K $s^{-1}$ using boron oxide ($B_2O_3$) to purify the melt and eliminate heterogeneous nucleation | Kui, Greer, and Turnbull | [7] |
| 1989 | Obtained exceptionally wide glass formation range (3–83 at.% and 0–60 at.% Ni) in the Al–La–Ni system | Inoue, Zhang, and Masumoto | [8] |
| 1991 | Cylinders of diameter up to 4.0 mm of $Mg_{65}Cu_{25}Y_{10}$ BMG were obtained with critical cooling rate of 93–391 K $s^{-1}$ | Inoue, Kato, Zhang, Kim, and Masumoto | [9] |
| 1991 | Wide glass formation ranges were obtained in $Zr_{65}Al_{7.5}Cu_{2.5}(Co_{(1-x-y)}Ni_xCu_y)$ ($x$ = 0–15; $y$ = 10–23) system by melt spinning | Zhang, Inoue, and Masumoto | [10] |
| 1993 | Rods of diameter up to 14 mm of $Zr_{41.2}Ti_{13.8}Cu_{12.5}Ni_{10.0}Be_{22.5}$ BMG were obtained at the critical cooling rate of 10 K $s^{-1}$ or less. It is the first commercial BMG known as Vitreloy (Vit1) and also known as *Liquidmetal* | Peker and Johnson | [11] |

*Continued*

Table 12.1 **Continued**

| Year | Event | Author(s) | Ref. |
|---|---|---|---|
| 1997 | Using $B_2O_3$ flux treatment cylinders of diameter up to 75 mm of $Pd_{40}Cu_{30}Ni_{10}P_{20}$ BMG were obtained at the critical cooling rate of 0.167 K $s^{-1}$ | Inoue, Nishiyama, and Kimura | [12] |
| 1998 | Established minor alloying additives dramatically enhance BMG formation as per 0.5–1 at.% silicon addition increased Cu–Ti–Zr–Ni ingot thickness from 4 to 7 mm | Choi-Yim, Busch, and Johnson | [13] |
| 2001 | Vit105 ($Zr_{52.5}Cu_{17.9}Ni_{14.6}Al_{10}Ti_5$) was established as one of the best BMG-forming alloys | Kündig, Löffler, Johnson, Uggowitzer, and Thiyagarajan | [14] |
| 2007 | BMG of high fracture strength of around 2000 MPa was obtained in $Cu_{45}Zr_{45}Al_5Ag_5$ | Zhang, Jia, Zhang, and Inoue | [15] |
| 2003–09 | Rare-earth-based bulk metallic glasses (REBMGs) were established as multifunctional materials. | Luo and Wang | [16] |
| 2009 | $Mg_{60+x}Zn_{35-x}Ca_5$ $(0 \leq x \geq 7)$ BMGs were established as a new generation biodegradable implants | Zberg, Uggowitzer, and Löffler | [17] |
| 2011 | Developed $Pd_{79}Ag_{3.5}P_6Si_{9.5}Ge_2$ BMG as the highest damage tolerant (fracture toughness and strength) material | Demetriou, Launey, Garrett, Schramm, Hofmann, Johnson, and Ritchie | [18] |
| 2016 | Developed impurity-insensitive $Zr_{50}Ti_4Y_1Al_{10}Cu_{25}Ni_7Co_2Fe_1$ BMG along with rod fabrication of 16 mm diameter suitable for low-cost industrial production and commercialization | Chen, Cheng, and Zhang | [19] |

glassy spheroids of 0.53 cm (minor axis diameter) of $Pd_{40}Ni_{40}P_{20}$ by a process of slow cooling (1.4 K $s^{-1}$) under $10^{-6}$ Torr vacuum. In 1984, this group obtained $Pd_{40}Ni_{40}P_{20}$ BMG ingot of cm size at the cooling rate of 10 K $s^{-1}$ using boron oxide ($B_2O_3$) to purify the melt and eliminate heterogeneous nucleation [7]. Unfortunately, these revolutionary contributions [3,4,6,7] on the development of BMGs did not make a huge impact in the materials science community because of the use of very expensive palladium, which resulted in unsuitability for use in a wide range of applications.

In the late 1980s, the breakthrough came in the development of BMGs with the discovery of mixing of many glass-forming metals of different sizes and expanding the glass formation range by Inoue and coworkers at Tohoku University, Japan

[8–10]. Schematic illustrations of the characteristics of different components of various BMGs are shown in Fig. 12.2A [20]. These glasses contained a minimum of three metals and occasionally reached up to five. It was proved from these studies that noble metals were no longer the indispensable components in BMGs [8–10]. Consequently, it opened up the suitability for their use in a wide range of applications because of cost effectiveness. In 1993, Peker and Johnson [11] obtained rods of diameter up to 14 mm of $Zr_{41.2}Ti_{13.8}Cu_{12.5}Ni_{10.0}Be_{22.5}$ BMG at the critical cooling rate of $10\ K\ s^{-1}$ or less. It was the first commercial BMG known as Vitreloy (Vit1) and also known as *Liquidmetal*. It is noteworthy that the glass contains 22.5 at.% beryllium (Be). It is the smallest metal atom and found to be the most befitting to fill the empty spaces of the glass matrix. As a result, it very efficiently stabilizes the glassy phase. Up until now, this glass is one of the best BMGs and represents the densest packing in zirconium-based compositions [21]. In 1997, Inoue and coworkers [12] obtained cylinders of diameter up to 75 mm of $Pd_{40}Cu_{30}Ni_{10}P_{20}$ BMG at the critical cooling rate of $0.167\ K\ s^{-1}$ using $B_2O_3$ flux treatment. It is observed that during the period of 1960–97, the major attention was given to find out new BMG system and increase casting thickness of the glass. A plot of critical casting thickness of glass formation as a function of glass discovery year and glass system is shown in Fig. 12.2B [22]. Thus, the casting thickness raised by more than three orders of magnitude in the last four decades in comparison with the discovery of the first metallic lass.

In 1998, Choi-Yim et al. [13] have established that minor alloying additives dramatically enhance BMG formation as per 0.5–1 at.% silicon addition increased Cu–Ti–Zr–Ni ingot thickness from 4 to 7 mm. Evidence in support of this fact is

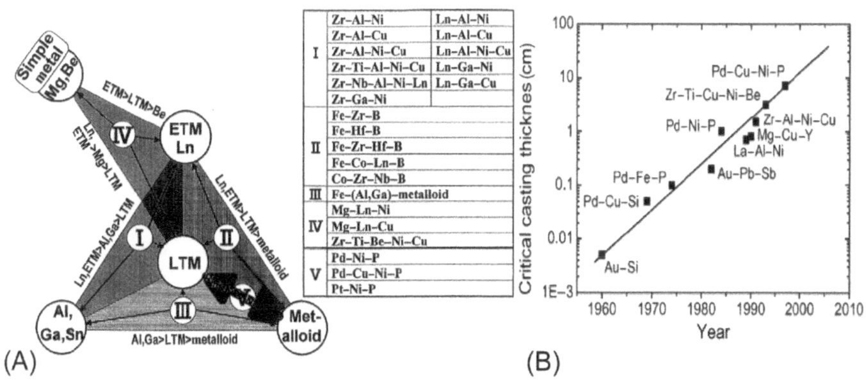


**Fig. 12.2** (A) Schematic illustrations of the characteristics of different components of various BMGs. ETM and LTM stand for the transition metals of IV–VI and VII–VIII groups in the periodic table, respectively. (B) Critical casting thickness of glass formation as a function of glass discovery year and glass system.

(A) Reproduced with permission from A. Inoue, Stabilization of metallic supercooled liquid and bulk amorphous alloys, Acta Mater. 48 (2000) 279–306, Copyright © Elsevier. (B) Reproduced with permission from J. F. Löffler, Bulk metallic glasses, Intermetallics, 11 (2003) 529–540, Copyright © Elsevier.

illustrated with the photomicrographs as shown in Fig. 12.3A and B [23]. Fig. 12.3C depicts the plot of relaxation time as a function of $T_g/T$ indicating different fragility of $Cu_{46}Zr_{54}$ and $Cu_{46}Zr_{47}Al_7$ alloys [24]. It is seen that the relaxation dynamics is considerably sluggished down (by one order of magnitude around 800 K) because of aluminum alloying. Thus, aluminum addition alters the fragility by making the $Cu_{46}Zr_{47}Al_7$ ternary glass stronger than that of binary $Cu_{46}Zr_{54}$ glass. In 2001, Kündig et al. [14] established $Zr_{52.5}Cu_{17.9}Ni_{14.6}Al_{10}Ti_5$ (Vit105) glass as one of the best BMG-forming alloys and commercialized. In 2007, Zhang et al. [15] obtained high fracture strength of around 2000 MPa in $Cu_{45}Zr_{45}Al_5Ag_5$ BMG. During the period of 2003–09, Wang and coworkers [16] developed various rare-earth-based bulk metallic glasses (RE BMGs) and established as multifunctional materials.

In 2009, Löffler and coworkers [17] established $Mg_{60+x}Zn_{35-x}Ca_5$ $(0 \leq x \geq 7)$ BMGs as a new generation biodegradable implants. In 2011, Demetriou et al. [18]

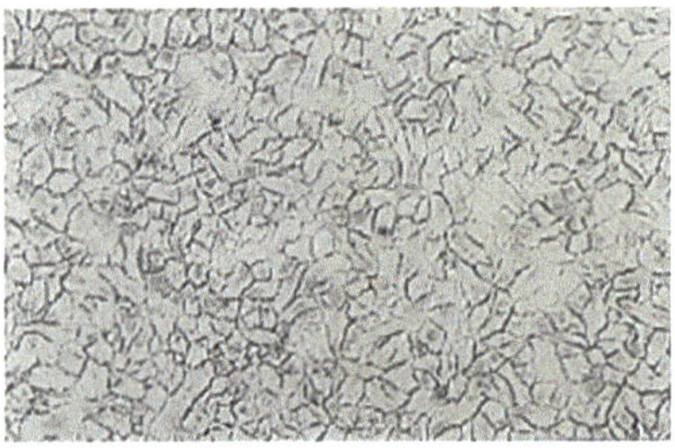
(A)

(B)

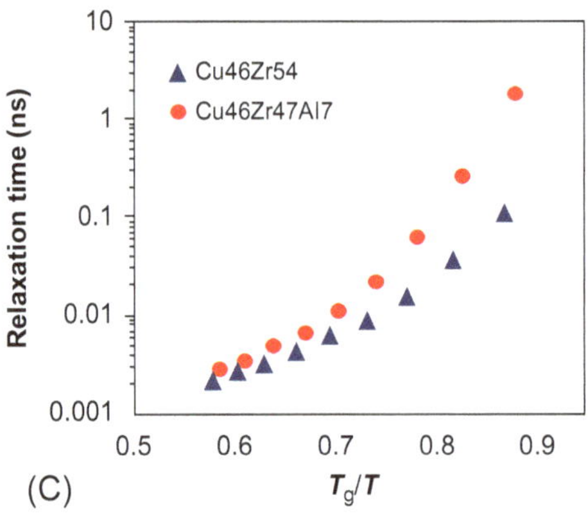

(C)

**Fig. 12.3** Photomicrographs of as-cast rods (6.4 mm diameter) $Zr_{52.5}Al_{10}Ti_5Cu_{17.5}Ni_{14.6}$ alloy (A) without and (B) with dopant of 0.1% B + 0.25% Si + 0.1% Pb. (C) Plot of relaxation time as a function of $T_g/T$ indicating different fragility of $Cu_{46}Zr_{54}$ and $Cu_{46}Zr_{47}Al_7$ alloys.
(A) and (B) Reproduced with permission from Z.P. Lu, C.T. Liu, Role of minor alloying additions in formation of bulk metallic glasses: a review, J. Mater. Sci. 39 (2004) 3965–3974, Copyright © Springer. (C) Reproduced with permission from Y.Q. Cheng, E. Ma, Atomic-level structure and structure-property relationship in metallic glasses, Prog. Mater. Sci. 56 (2011) 379–473, Copyright © Elsevier.

developed $Pd_{79}Ag_{3.5}P_6Si_{9.5}Ge_2$ BMG as the highest damage tolerant (fracture toughness and strength) material. The map of damage tolerance (fracture toughness vs yield strength) of different material is shown in Fig. 12.4 [18]. The encircled zone indicated by arrow suggests that the damage tolerance of metallic glasses extends beyond traditional benchmarks toward levels previously inaccessible to any materials. Earlier, we have witnessed that impurity removal from the metallic alloys enhances the BMG formation and increases the casting thickness [7,12]. In 2016, Chen et al. [19] have developed the impurity-insensitive $Zr_{50}Ti_4Y_1Al_{10}Cu_{25}Ni_7Co_2Fe_1$ BMG composition along with rod fabrication of 16 mm diameter, which is found to be suitable for low-cost industrial production and commercialization. It is suggested that the glass composition with a large number (eight) of components obstructs the heterogeneous crystallization induced by nanoscale nuclei resulted from impurities, which influences positively on the maintenance of glass-forming ability (GFA), stability and mechanical properties.

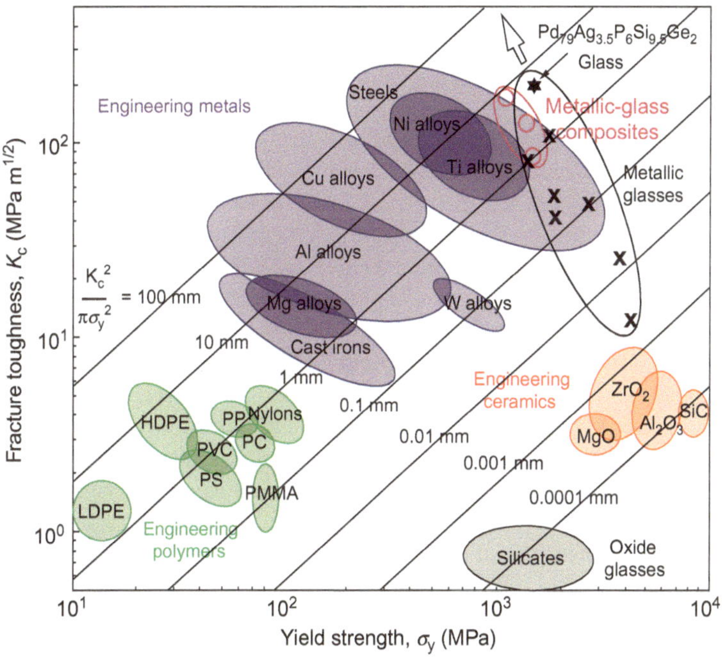


**Fig. 12.4** Map of damage tolerance (fracture toughness vs yield strength) of different materials. The encircled zone indicated by *arrow* suggests that the damage tolerance of metallic glasses extends beyond traditional benchmarks toward levels previously inaccessible to any materials. Reproduced with permission from M.D. Demetriou, M.E. Launey, G. Garrett, J.P. Schramm, D.C. Hofmann, W.L. Johnson, R.O. Ritchie, A damage-tolerant glass, Nat. Mater. 10 (2011) 123–128, Copyright © Nature Group, McMillan.

## 12.3 Kinetic behavior of BMG formation

In favor of glass formation, the cooling process of the melt (liquid) is to be controlled in such a manner to prevent crystal formation. Both the nucleation and crystal occur simultaneously during cooling of a melt. Their rates change continuously as the temperature decreases during cooling. Therefore, the glass formation must deal with the interaction between these two processes. The rates of nucleation ($I$) and crystal growth ($U$) are given by the following two expressions:

$$I = \left(n_v kT/3\,\pi\alpha_0^3\eta\right)\exp\left(-W^*/kT\right) \tag{12.1}$$

$$U = \left(kT/3\pi\alpha_0^2\eta\right)\left[1 - \exp\left(\Delta G/kT\right)\right] \tag{12.2}$$

where $n_v$ is the number of formula unit of the crystal per unit volume of the melt, $k$ is the Boltzmann constant, $T$ is the absolute temperature in Kelvin, $\alpha_0$ is the average distance of atomic jump (approximately equal to the interatomic separation distance), $\eta$ is the melt viscosity, $W^*$ is the thermodynamic barrier of nucleation, and $\Delta G$ is the thermodynamic barrier of crystal growth. The rate at which a melt can be transformed into a crystal is expressed by

$$X_t = 1 - \exp\left(-\pi U^3 I t^4/3\right) \tag{12.3}$$

where $X_t$ is the volume fraction crystallized after time $t$. Glass formation, therefore, requires $X_t$ to reside below the detectable limit. To maintain this condition, both $I$ and $U$ are to be very low.

To assess the possibility of glass formation for a given substance, it is necessary to understand the influence of kinetic factors in glass formation by using the time-temperature-transformation (*T-T-T*) curves [25,26]. In a *T-T-T* curve, the time taken to crystallize to a certain volume fraction ($X_t$), as determined by using Eq. (12.3), is plotted against the under cooling temperature, that is, from the melting or liquidus temperature ($T_m$ or $T_l$) down to the glass transition temperature ($T_g$). The *T-T-T* curve shows a typical "C" (the tip is called the "nose") shape due to the fact that the thermodynamic driving force for crystallization increases with increasing undercooling, while the atomic mobility in the liquid reduces. At the nose temperature ($T_N$), the time ($\tau_N$) requires to crystallize a particular volume fraction is minimum. This temperature represents a critical value of crystallization, which must be avoided to achieve glass formation. It is, therefore, necessary to evaluate the critical cooling rate ($R_c$) that is expressed by the following relation:

$$R_c \approx (T_m - T_N)/\tau_N = \Delta T_N/\tau_N \tag{12.4}$$

The maximum glass thickness ($t_{max}$) that could be obtained without crystallization is given by

$$t_{max} \approx \left(D_T \cdot \tau_N\right)^{0.5} \tag{12.5}$$

where $D_T$ is the thermal diffusivity.

Fig. 12.5A shows the schematic illustration of comparison of time-temperature-transformation (*T-T-T*) diagrams of glassy, amorphous, and crystalline alloys along with their critical cooling rates ($R_c$) [27]. The critical cooling rate 0.067 K $s^{-1}$ is achieved for glassy alloy formation, which is $\sim 10^8$ times smaller than that for ordinary amorphous alloy. The supercooled liquid state of glassy alloy, therefore, requires long periods of several thousand seconds to achieve high stability. Such a spectacular increase in the thermal stability of supercooled liquid has enabled the production of large-scale BMG with various outer shapes and massy ingot of 80 mm in diameter and 85 mm high.

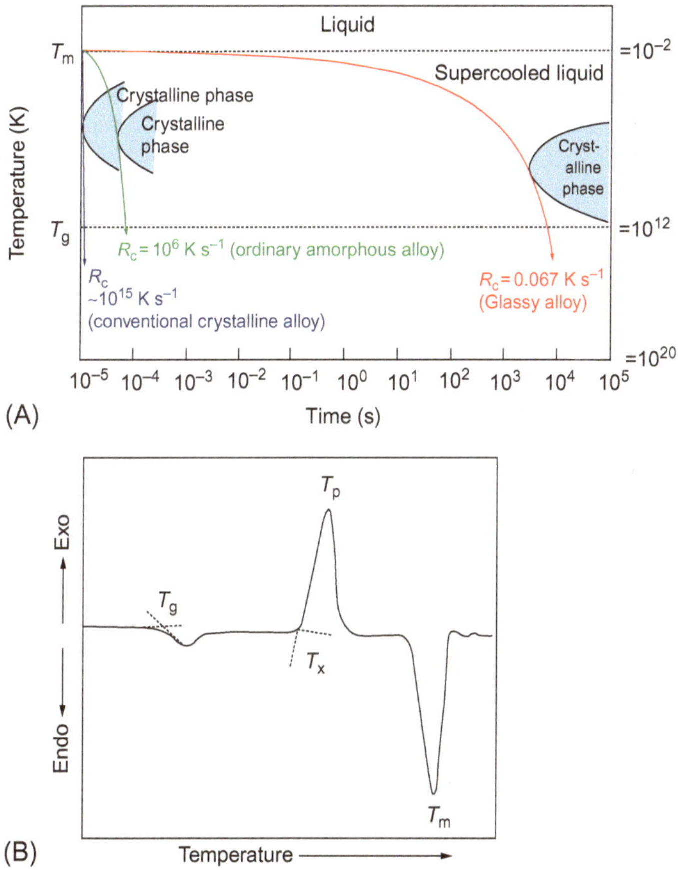


**Fig. 12.5** (A) Schematic illustration of comparison of time-temperature-transformation (*T-T-T*) diagrams of glassy, amorphous, and crystalline alloys along with their critical cooling rates ($R_c$). The supercooled liquid state of glassy alloy requires long periods of several thousand seconds to achieve high stability. (B) Schematic illustration of a typical DTA or DSC curve of a crystallizable glass ($T_g$, glass transition temperature; $T_x$, onset of crystallization exotherm; $T_p$, exotherm peak of crystallization; and $T_m$, endothermic peak of first melting). Reproduced with permission from A. Inoue, A. Takeuchi, Recent development and application products of bulk glassy alloys, Acta Mater. 59 (2011) 2243–2267, Copyright © Elsevier.

Turnbull [28] had forecasted that the reduced glass transition temperature ($T_{rg}$) can be used as a decisive factor for determining the GFA of any material. The reduced glass transition temperature ($T_{rg}$) is defined as the ratio of the glass transition temperature ($T_g$) to the liquidus temperature ($T_l$), that is, $T_{rg} = T_g/T_l$. In accordance with this criterion, a liquid with $T_{rg} \geq 2/3$ becomes very sluggish to crystallization and can readily be undercooled into the glassy state with a low cooling rate. Hitherto, this criterion is regarded as one of the best "thumb rule" for deciding the GFA of any melt (liquid), and it played a solution role for the development of many glasses together with BMGs.

A schematic illustration of a typical DTA or DSC curve of a crystallizable glass is depicted in Fig. 12.5B for better understanding of the glass formation of BMGs, where $T_g$, $T_x$, $T_p$, and $T_m$ are the temperatures of the glass transition, onset of crystallization exotherm, exothermic peak of crystallization, and endothermic peak of first melting, respectively. The $T_m$ may also be regarded as liquidus temperature ($T_l$). The supercooled liquid region ($\Delta T_x$) is defined as the difference between $T_x$ and $T_g$, that is, $\Delta T_x = (T_x - T_g)$. A comparison of critical cooling rates ($R_c$) versus reduced glass transition temperature ($T_{rg}$) of BMGs with that of silicate and conventional metallic glasses is shown in Fig. 12.6A [29]. It is seen that some BMGs has the GFA very close to that of silicate glasses. The change of critical cooling rate ($R_c$) and maximum glass thickness ($t_{max}$) of BMGs as a function of temperature of supercooled liquid region ($\Delta T_x$) is shown in Fig. 12.6B [20]. It is seen that the $R_c$ decreases and $t_{max}$ increases with increase in $\Delta T_x$, that is, the GFA increases with expansion of $\Delta T_x$.

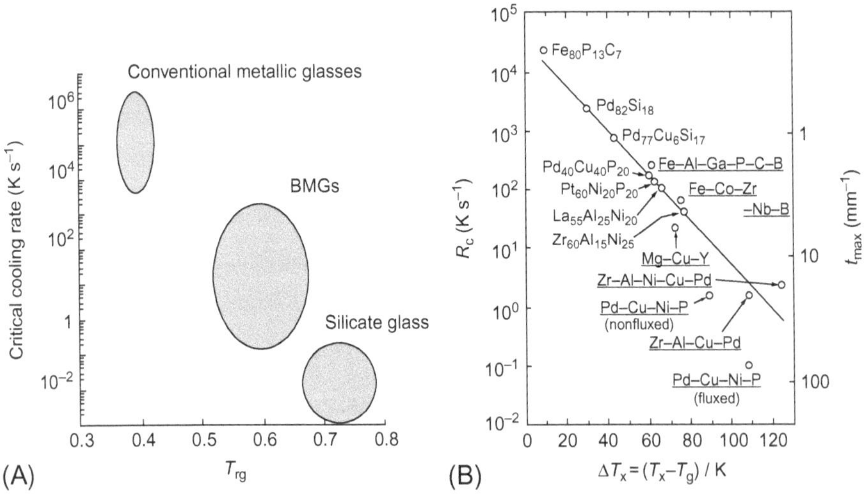


**Fig. 12.6** (A) A comparison of $R_c$ versus reduced glass transition temperature ($T_{rg}$) of BMGs with that of silicate and conventional metallic glasses. (B) Change of $R_c$ and $t_{max}$ of BMGs as a function of temperature of supercooled liquid region ($\Delta T_x$).

(A) Reproduced with permission from W.H. Wang, C. Dong, C.H. Shek, Bulk metallic glasses, Mater. Sci. Eng. R Rep. 44 (2004) 45–89, Copyright © Elsevier. (B) Reproduced with permission from A. Inoue, Stabilization of metallic supercooled liquid and bulk amorphous alloys, Acta Mater. 48 (2000) 279–306, Copyright © Elsevier.

## 12.4 Preparation of BMGs

The preparation of BMGs is generally carried out in two steps. In the first step, the homogeneous alloy ingot is prepared in a refractory crucible by induction or arc melting the mixture of high purity (99.9%–99.999%) elements of metals and/or metalloids in appropriate proportion under a high purity argon atmosphere. This ingot is remelted four to five times to ensure the compositional homogeneity. The temperature selection depends on the melting point of the elements used. Schematic illustration of a typical setup for initial alloy ingot preparation by induction heating is shown in Fig. 12.7A [30]. The second step consists of pressure injection process of molten alloy ingot into water-cooled copper mold in argon atmosphere. Schematic illustration of bulk metallic glass preparation by induction heating and pressure-die casting arrangement is shown in Fig. 12.7B [30]. BMGs of different shapes and sizes can also be prepared by using copper-mold injection-casting method or copper-mold tilt-pour-casting method. Photographs of BMGs of various shapes and sizes are shown in Fig. 12.8A and B [31,32].

## 12.5 Properties of BMGs

Fig. 12.9A shows the XRD patterns of as-cast rods and melt-spun ribbon of $Ti_{47}Cu_{38}Zr_{7.5}Fe_{2.5}Sn_2Si_1Ag_2$ BMG [33]. The glasses of both the 7 mm diameter rod and ribbon are X-ray amorphous as a consequence of their XRD patterns display only a main halo without any other Bragg's reflection peaks. Whereas the 8 mm diameter glass rod contains the crystalline intermetallic compounds of $Ti_2Cu$, $Ti_3Cu_4$, and TiCu as revealed in their sharp peaks in addition to the main halo of amorphous character in its XRD patterns. The high resolution TEM (HRTEM) image with selected area electron diffraction (SAED) pattern (the inset) of 7 mm diameter rod of the above glass is shown

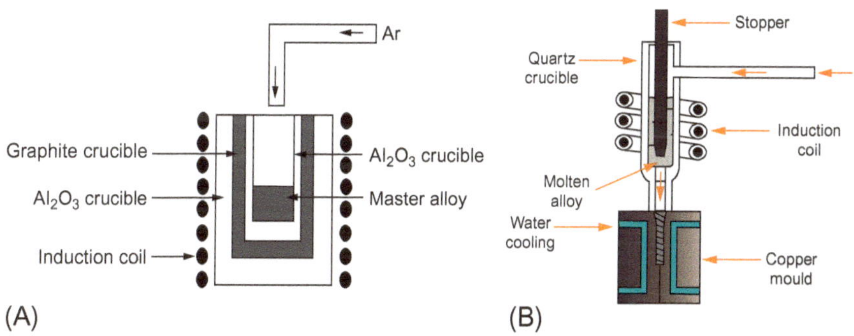


**Fig. 12.7** Schematic illustration of (A) setup for initial alloy ingot preparation by induction heating and (B) bulk metallic glass preparation by induction heating and pressure-die casting arrangement.
Reproduced with permission from W. Pilarczyk, Preparation and characterization of Zr-based bulk metallic glasses in form of plate, J. Alloys Compd. 615 (2014) S132–S135, Copyright © Elsevier.

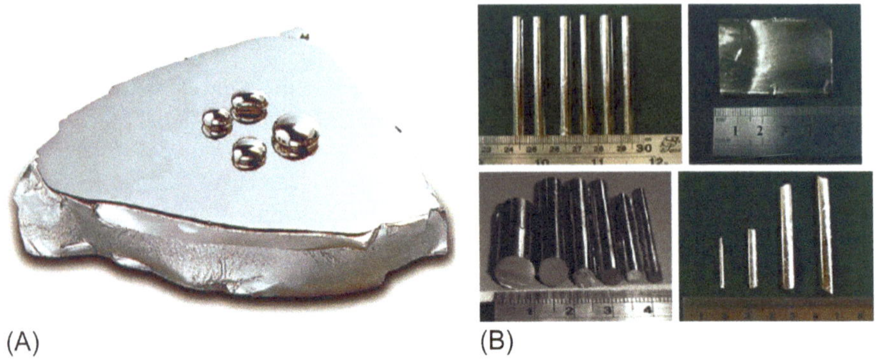


**Fig. 12.8** (A) Photograph of Zirconium (Zr)-based metallic glass of "Liquidmetal Technologies." (B) Photographs of rods and plate of Zr and Mg-based BMGs.
(A) Reproduced with permission from M. Telford, The case for bulk metallic glass, Mater. Today 7 (2004), 36–43, Copyright © Elsevier. (B) Reproduced with permission from J.C. Huang, J.P. Chu, J.S.C. Jang, Recent progress in metallic glasses in Taiwan, Intermetallics, 17 (2009) 973–987, Copyright © Elsevier.

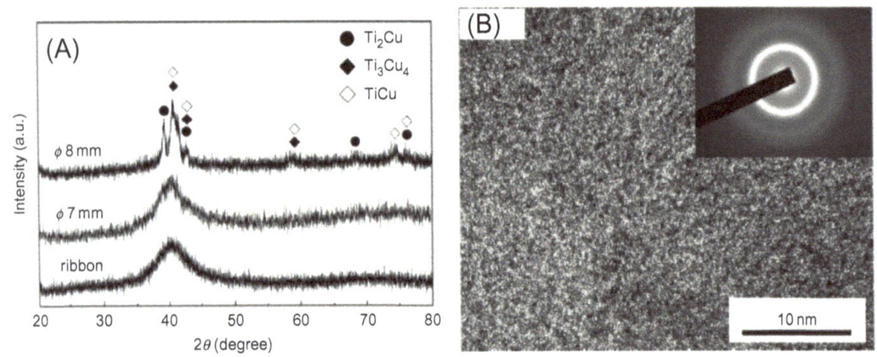


**Fig. 12.9** (A) XRD patterns of as-cast rods and melt-spun ribbon and (B) The high resolution TEM (HRTEM) image with selected area electron diffraction (SAED) pattern (inset) of rod (7 mm diameter) of $Ti_{47}Cu_{38}Zr_{7.5}Fe_{2.5}Sn_2Si_1Ag_2$ BMG.
Reproduced with permission from S. Pang, Y. Liu, H. Li, L. Sun, Y. Li, T. Zhang, New Ti-based Ti–Cu–Zr–Fe–Sn–Si–Ag bulk metallic glass for biomedical applications, J. Alloys Compd. 625 (2015) 323–327, Copyright © Elsevier.

in Fig. 12.9B. The HRTEM image exhibits a homogeneous maze-like pattern without any crystalline fringe contrast, and the corresponding SAED pattern shows diffuse halo rings, which are distinguishing of amorphous structure.

Compositions and thermal properties of some BMGs are shown in Table 12.2. Both the supercooled liquid region ($\Delta T_x$) and reduced glass transition temperature ($T_{rg}$) represent the GFA. In general, larger $\Delta T_x$ and $T_{rg}$ values usually exhibit higher stability of the supercooled liquid against crystallization and enable the formation of BMGs with larger critical diameter ($D_c$).

**Table 12.2** **Compositions and thermal properties of some bulk metallic glasses**

| Composition (at.%) | $D_c$ (mm) | $T_g$ (K) | $T_x$ (K) | $\Delta T_x$ (K) | $T_m$ (K) | $T_l$ (K) | $T_{rg}$ | Ref. |
|---|---|---|---|---|---|---|---|---|
| ***Iron (Fe) based*** | | | | | | | | |
| $Fe_{78}B_{22}$ | 2.0 | 675 | 690 | 15 | 1406 | 1468 | 0.460 | [34] |
| $Fe_{72}Y_6B_{22}$ | 2.0 | 897 | 940 | 43 | 1385 | 1434 | 0.630 | [34] |
| $Fe_{68}Nb_4Y_6B_{22}$ | 2.0 | 920 | 983 | 63 | 1374 | 1427 | 0.640 | [34] |
| $Fe_{55}Cr_{20}Mo_5P_{13}C_7$ | 6.0 | 782 | 810 | 28 | 1215 | 1289 | 0.607 | [35] |
| $Fe_{67.5}Mo_{7.5}P_{10}C_{10}B_5$ | 3.0 | 749 | 845 | 96 | – | 1337 | 0.560 | [36] |
| $Fe_{62.5}Co_5Mo_{7.5}P_{10}C_{10}B_5$ | 4.0 | 738 | 837 | 99 | – | 1332 | 0.554 | [36] |
| $Fe_{60}Ni_{7.5}Mo_{7.5}P_{10}C_{10}B_5$ | 4.0 | 727 | 824 | 97 | – | 1326 | 0.548 | [36] |
| $Fe_{60}Co_5Ni_{2.5}Mo_{7.5}P_{10}C_{10}B_5$ | 3.0 | 733 | 835 | 102 | – | 1375 | 0.533 | [36] |
| $Fe_{48}Cr_{15}Mo_{14}C_{15}B_6Gd_2$ | 2.0 | 856 | 887 | 31 | 1393 | 1519 | 0.564 | [37] |
| $Fe_{48}Cr_{15}Mo_{14}C_{15}B_6Y_2$ | 2.0 | 841 | 885 | 44 | 1389 | 1460 | 0.576 | [37] |
| ***Zirconium (Zr) based*** | | | | | | | | |
| $Zr_{55}Cu_{30}Ni_5Al_{10}$ | 1.0 (plate) | 703 | 790 | 87 | – | 1210 | 0.581 | [30] |
| $Zr_{41.2}Ti_{13.8}Cu_{12.5}Ni_{10.0}Be_{22.5}$(Vit1) | 14.0 | 623 | 672 | 49 | 932 | 996 | 0.670 | [11] |
| $Zr_{52.5}Cu_{17.9}Ni_{14.6}Al_{10}Ti_5$ (LM105) | – | 672 | 741 | 69 | – | – | – | [38] |
| ***Palladium (Pd) based*** | | | | | | | | |
| $Pd_{40}Cu_{30}Ni_{10}P_{20}$ | 72.0 | 575 | 670 | 95 | 804 | 840 | 0.720 | [12] |
| $Pd_{79}Ag_{3.5}P_6Si_{9.5}Ge_2$ | 1.5 | 613 | 644 | 31 | 967 | 1065 | 0.580 | [18] |
| ***Titanium (Ti) based*** | | | | | | | | |
| $Ti_{47}Cu_{38}Zr_{7.5}Fe_{2.5}Sn_2Si_1Ag_2$ | 7.0 | 641 | 693 | 52 | – | – | – | [33] |
| $Ti_{46}Cu_{27.5}Zr_{11.5}Co_7Sn_3Si_1Ag_4$ | 5.0 | 640 | 691 | 51 | 1068 | 1207 | 0.530 | [39] |

*Continued*

Table 12.2 **Continued**

| Composition (at.%) | $D_c$ (mm) | $T_g$ (K) | $T_x$ (K) | $\Delta T_x$ (K) | $T_m$ (K) | $T_l$ (K) | $T_{rg}$ | Ref. |
|---|---|---|---|---|---|---|---|---|
| ***Magnesium (Mg) based*** | | | | | | | | |
| $Mg_{65}Ni_{21}Pr_{14}$(in argon) | 3.0 | 452 | 494 | 42 | 748 | 794 | 0.570 | [40] |
| $Mg_{65}Ni_{21}Pr_{14}$ (in air) | 3.0 | 454 | 495 | 41 | 753 | 790 | 0.570 | [40] |
| $Mg_{66}Zn_{30}Ca_3Sr_1$ | 6.0 | 369 | 383 | 14 | 613 | 654 | 0.564 | [41] |
| ***Rare-earth (RE) based*** | | | | | | | | |
| $Gd_{55}Al_{25}Co_{20}$ | 2.0 | 585 | 657 | 72 | 949 | 971 | 0.602 | [42] |
| $Er_{55}Al_{25}Co_{20}$ | 5.0 | 663 | 722 | 59 | 1056 | 1079 | 0.628 | [42] |
| $La_{68}Al_{10}Cu_{20}Co_2$ | 3.0 | 374 | 439 | 65 | – | – | – | [43] |
| $Ce_{68}Al_{10}Cu_{20}Co_2$ | 3.0 | 351 | 427 | 76 | – | – | – | [43] |
| $(Ce_{50}La_{50})_{68}Al_{10}Cu_{20}Co_2$ | 3.0 | 362 | 419 | 57 | – | – | – | [43] |
| $Dy_{46}Y_{10}Al_{24}Co_{18}Fe_2$ | – | 627 | 677 | 50 | 991 | 1023 | 0.633 | [44] |
| ***Cobalt (Co) based*** | | | | | | | | |
| $Co_{71}Mo_9P_{14}B_6$ | 4.5 | 742 | 770 | 28 | 1290 | 1374 | 0.540 | [45] |
| ***Yttrium (Y) based*** | | | | | | | | |
| $Y_{55}Al_{25}Co_{20}$ | 2.0 | 633 | 694 | 61 | 1035 | 1060 | 0.597 | [42] |

$D_c$, critical diameter (in mm) for complete glass formation; $K$, unit of temperature in Kelvin; $T_g$, glass transition temperature; $T_x$, onset crystallization temperature; $\Delta T_x = (T_x - T_g)$, supercooled liquid region; $T_m$, melting temperature; $T_l$, liquidus temperature; and $T_{rg} = T_g/T_l$, reduced glass transition temperature.

BMGs are the answers for many technological solutions because of their several outstanding properties such as high strength, high fracture toughness, high hardness, and high fatigue strength. A galaxy of these properties of BMGs is provided in Table 12.3 [20,27,29]. A comparison of mechanical properties between some classical BMGs and oxide glasses is also provided in Table 12.4 [29,38]. Here, we have considered Young's modulus ($E$), Poisson's ratio ($\nu$), bulk modulus ($B$), shear modulus ($G$), and the ratio between $B$ and $G$ ($B/G$). The fragility of a glass forming liquid can be correlated with the bulk ($B$) and shear moduli ($G$), that is, Poisson's ratio ($\nu$) or the ratio between $B$ and $G$ ($B/G$) of the corresponding glass. A small value of $\nu$ means that atoms or molecules can scarcely rearrange themselves to shear strains without a severe disturbance in bonding configurations, and the large $\nu$ indicates the easiness of atomic rearrangement. This designates that the GFA has a close relation with the microstructure of a glass forming system. A decreasing value of $\nu$ indicates high GFA of the glass forming system [46].

The values of $\nu$ for a variety of BMGs range from 0.35 to 0.40, which are close to that of crystalline metal, for instance, 0.37 for copper, whereas the $\nu$ values of typical oxide glasses vary in the range of 0.17–0.24. The oxide glasses are brittle, since the atoms or molecules can scarcely rearrange themselves to shear strains without a drastic disturbance in bonding configurations. On the contrary, the metallic glasses with higher value of $\nu$ have larger plastic deformation demonstrating the easiness of atomic rearrangement in the materials. The comparison designates that $\nu$ bears a certain relationship with the atomic configuration in these noncrystalline materials. The GFA also has relation with the value of $\nu$ of a glass forming system. The usual metallic glasses have poor GFA, and their critical cooling rate ($R_c$) for the glass formation varies in the range of 104–107 K $s^{-1}$. Whereas the $R_c$ of the BMGs ranges from 1 to 100 K $s^{-1}$, for example, the $R_c$ of Vit1 is lower than 1 K $s^{-1}$. It is, therefore, clear that the $R_c$ of the BMGs approaches to that of the oxide glasses ($R_c$ is less than 1 K $s^{-1}$) [29]. A similar correlation can also be drawn with the ratio between $B$ and $G$ ($B/G$), which is in the ranges of 3.0–4.5 and 1.2–1.6 for classical BMGs and typical oxide glasses, respectively.

Fig. 12.10 represents a comparison of strength versus elastic limit of glassy metal alloys with that of crystalline metal alloys, silica, wood, and polymers [31]. It is observed that the glassy metal alloys possess higher strength and elastic limit than that of crystalline metal alloys (e.g., steels and titanium alloys), silica, and wood. This fact undoubtedly demonstrates the uniqueness of the glassy metal alloys. Fig. 12.11 illustrates the comparison of tensile strength and Vickers hardness as a function of

**Table 12.3 Properties of bulk metallic glasses [20,27,29]**

| *Properties* |
|---|
| • High strength • High fracture toughness • High hardness • High fatigue strength • High impact fracture energy • High wear resistance • High elastic energy • High corrosion resistance • High acoustic attenuation • High reflection ratio • High frequency permeability • High hydrogen storage capability • High viscous flow ability • High manufacturability • Good soft magnetism • High magnetostriction • Efficient electrode (chlorine gas) • Self-sharping property |

**Table 12.4** **Comparison of mechanical properties between some classical bulk metallic glasses and oxide glasses [29,38]**

| Glass composition (at.%)/system | Density, $\rho$ (g $cm^{-3}$) | Young's modulus, $E$ (GPa) | Poisson's ratio, $\nu$ | Bulk modulus, $B$ (GPa) | Shear modulus, $G$ (GPa) | $B/G$ |
|---|---|---|---|---|---|---|
| ***Bulk metallic glasses*** | | | | | | |
| $Zr_{52.5}Cu_{17.9}Ni_{14.6}Al_{10}Ti_5$ (LM105) | 6.570 | 92.7 | 0.380 | – | – | – |
| $Zr_{41}Ti_{14}Cu_{12.5}Ni_{10}Be_{22.5}$ (Vit1) | 6.125 | 101.0 | 0.350 | 114.10 | 37.4 | 3.05 |
| $Zr_{48}Nb_8Cu_{12}Fe_8Be_{24}$ | 6.436 | 95.7 | 0.359 | 113.60 | 35.2 | 3.23 |
| $Zr_{50.56}Ti_{5.14}Cu_{18.85}Ni_{11.15}Al_{14.3}$ | 6.608 | 92.7 | 0.363 | 112.60 | 34.0 | 3.31 |
| $Cu_{60}Zr_{20}Hf_{10}Ti_{10}$ | 8.315 | 101.1 | 0.368 | 128.20 | 36.9 | 3.47 |
| $Pr_{60}Cu_{20}Ni_{10}Al_{10}$ | 6.900 | 37.2 | 0.363 | 45.20 | 13.6 | 3.32 |
| $Pd_{39}Ni_{10}Cu_{30}P_{21}$ | 9.152 | 98.2 | 0.400 | 159.10 | 35.1 | 4.53 |
| ***Oxide glasses*** | | | | | | |
| Soda lime silicate glass (window), $Na_2O–CaO–SiO_2$ | 2.421 | 67.2 | 0.211 | 38.70 | 27.7 | 1.40 |
| Silica glass (fused quartz), $SiO_2$ | 2.201 | 72.7 | 0.170 | 36.90 | 31.0 | 1.19 |
| Borosilicate glass, $Na_2O–B_2O_3–SiO_2$ | 2.320 | 61.9 | 0.240 | 40.52 | 24.9 | 1.63 |

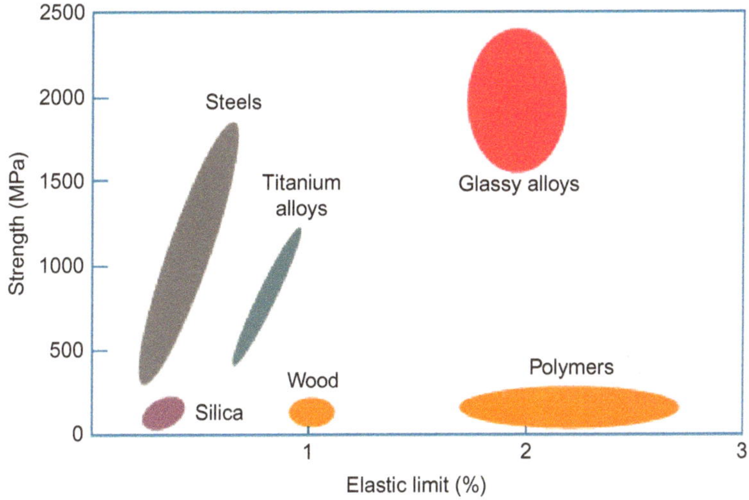


**Fig. 12.10** A comparison of strength versus elastic limit of glassy metal alloys with that of crystalline metal alloys, silica, wood, and polymers. Glassy metal alloys possess higher strength and elastic limit than that of crystalline metal alloys.
Reproduced with permission from M. Telford, The case for bulk metallic glass, Mater. Today 7 (2004), 36–43, Copyright © Elsevier.

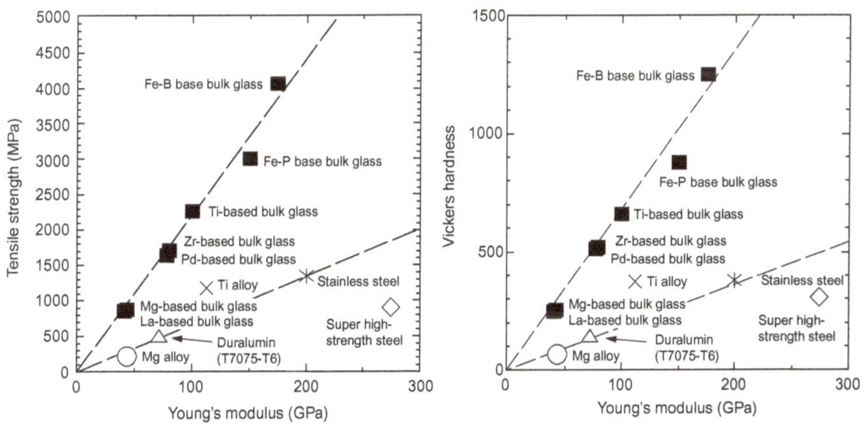


**Fig. 12.11** Comparison of tensile strength and Vickers hardness as a function of Young's modulus of various BMGs and crystalline alloys.
Reproduced with permission from A. Inoue, Stabilization of metallic supercooled liquid and bulk amorphous alloys, Acta Mater. 48 (2000) 279–306, copyright © Elsevier.

Young's modulus of various BMGs and crystalline alloys [20]. The BMGs have high tensile strength of 840–4000 MPa and high Vickers hardness of 250–1250 combined with Young's modulus of 47–170 GPa depending on the compositions. Similarly, a comparison of fracture strength as a function of Young's modulus of various BMGs and crystalline metallic alloys is presented in Fig. 12.12A [27]. It is seen that the

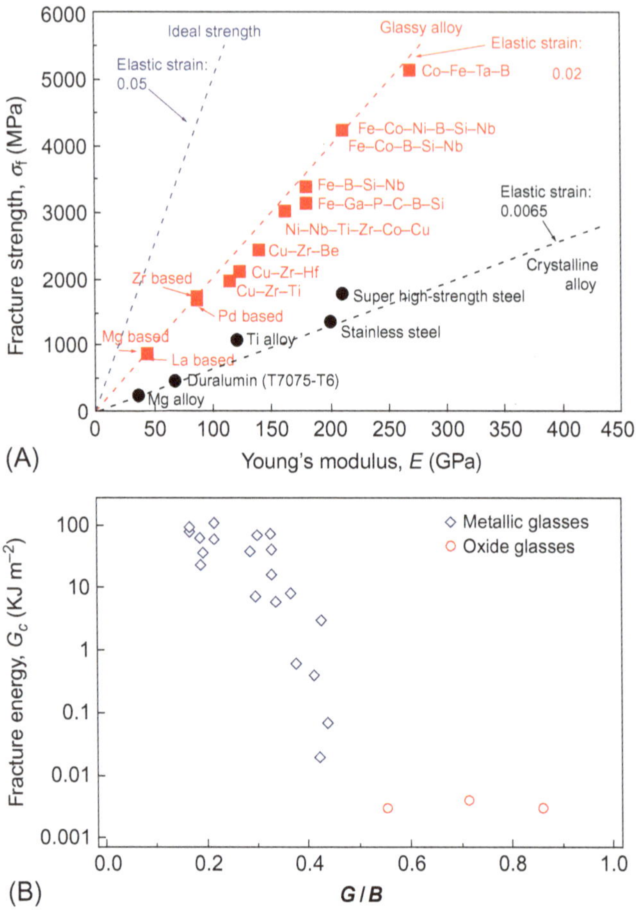


**Fig. 12.12** (A) Comparison of fracture strength as a function of Young's modulus of various BMGs and crystalline metallic alloys. (B) Comparison of fracture energy as a function of ratio of shear and bulk moduli (*G*/*B*) of metallic and oxide glasses.
(A) Reproduced with permission from A. Inoue, A. Takeuchi, Recent development and application products of bulk glassy alloys, Acta Mater. 59 (2011) 2243–2267, Copyright © Elsevier. (B) Reproduced with permission from C.A. Schuh, T.C. Hufnagel, U. Ramamurty, Mechanical behavior of amorphous alloys, Acta Mater. 55 (2007) 4067–4109, Copyright © Elsevier.

fracture strength of the BMGs vary in the range of 850–5200 MPa when Young's modulus fluctuate in the range of 47–270 GPa. Whereas in the case of crystalline metallic alloys, the fracture strength varies in the range of 200–1800 MPa when Young's modulus fluctuates in the range of 40–215 GPa. Thus, the BMGs display higher tensile strength, Vickers hardness, and fracture strength than those of crystalline alloys.

Likewise, the fracture energy of the BMGs can be correlated with their elastic properties. A comparison of fracture energy ($G_c$) as a function of ratio of shear and bulk moduli ($G/B$) of metallic and oxide glasses is illustrated in Fig. 12.12B [47]. It is observed that the fracture energy ($G_c$) of the metallic glasses vary in the range of 0.01–100 kJ $m^{-2}$ when their ratio of shear and bulk moduli ($G/B$) fluctuates in the range of 0.18–0.42. Whereas for oxide glasses, their fracture energy ($G_c$) varies around 0.005 kJ $m^{-2}$ when their ratio of shear and bulk moduli ($G/B$) fluctuate in the range of 0.56–0.86. Thus, the metallic glasses with ratio of shear and bulk moduli less than 0.41 ($G/B < 0.42$) are tough, while the oxide glasses with $G/B > 0.55$ are brittle. Similar correlation can also be expressed in terms of Poison's ratio ($\nu$). Therefore, all these mechanical properties make BMGs a different class of materials, which are different from the crystalline alloys and oxide glasses.

## 12.6 Functional applications of BMGs

We have witnessed that the BMGs possess unique, unconventional, and functional properties, which have resulted from their long-range disordered atomic configurations and multicomponent characteristics. With respect to preparation and fabrication of functional articles, the unique advantages of BMGs are near-net shape casting, simple production techniques, viscous flow formability and nanoscale imprint ability. One of the major advantages of BMGs for use as functional materials is that there are many degrees of freedom to tailor their various properties such as electric, mechanical, and magnetic properties as a result of their flexibility in composition, microstructural manipulation, shape, and dimensions. All these properties and advantages have given BMGs a high level of technological, environmental, and economical importance. They are used as functional materials in many fields such as structural, sensor, sporting goods, nanotechnology, information data storage, and biomedical. The present and future application potentials of BMGs are summarized in Table 12.5 [20,27,29,48]. Fig. 12.13 illustrates the functional uses of BMGs with group of application area and BMG articles and devices [31,49]. Examples of various types of hardware of BMGs are presented in Fig. 12.14 [32,50]. Fig. 12.15 displays the schematic illustration of construction of the world smallest size (1.5 mm diameter) microgeared motor and Ni-based bulk metallic glass fabricated gear parts [48]. Photographs of (A) pressure sensor devices and (B) slat truck cover of aircrafts made of BMGs are displayed in Fig. 12.16 [48].

Wang [51] has described various functional properties of BMGs, which are elastic moduli, hard magnetic properties, magnetocaloric effect, spin dynamics, heavy-fermion behavior, polyamorphism, polymer-like thermoplastic behavior, and spontaneous formation of nanostructures by fracture. He has focused on a variety of BMGs that have potential applications as functional glassy materials. Nishiyama et al. [52] have reported several high performance industrial products of BMGs, for example, magnetic yoke for linear actuator. Hasan et al. [53] have reported the functionalization of metallic glasses through hierarchical patterning, which is critical for use in electronics and photonics.

Table 12.5 **Applications of bulk metallic glasses [20,27,29,48]**

| |
|---|
| ***Application area [20,27]*** |
| • Structural • Sensor • Precision machinery • Optical • Ornamental • Spring • Sporting goods • Wear-resistant coating • Precision nozzle • Corrosion-resistant • Magnetic • Microtechnology • Nanotechnology • Information data storage • Biomedical implant • Medical instrument • Fuel-cell separator • Microelectromechanical systems (MEMS) • Actuator • Magnetic refrigerant • Biosensors |
| ***Application as material [20,29]*** |
| • Machinery structural materials • Cutting materials • Die materials • Tool materials • Bonding materials • Sporting goods materials • Corrosion resistance materials • Writing appliance materials • Optical precision materials • Hydrogen storage materials • Soft magnetic materials • High magnetostrictive materials • Electrode materials • Composite materials • Acoustic absorption materials • Penetrator • Medical devices materials • Spin-glass (SG) materials • Nanoimprinting for optoelectronic devices • Hologram patterns |
| ***Articles of applications [48]*** |
| • Striking face plate in golf clubs • Frame in tennis rackets • Optical mirrors • Casing in cellular phones • Casing in electromagnetic instruments • Connecting parts for optical fibers • Shot penning balls • Electromagnetic shielding plates • Soft magnetic choke coils • Soft magnetic high-frequency power coils • High torque-geared motor parts • High corrosion-resistant coating plates • Vessels for lead-free soldering • High-sensitivity small pressure sensors for vehicles • Colliori-type liquid flowmeter • Spring • Slat truck cover for airplane • In-printing plate • High density information-storage material • Yoke material for linear actuator • High-frequency type antenna material • Magnetic iron core for high speed motor • Biomedical endoscope parts • Tool • Hard disk drives • Compact headphones • Hard magnets • Watch parts |

Pang et al. [33] have observed that the $Ti_{47}Cu_{38}Zr_{7.5}Fe_{2.5}Sn_2Si_1Ag_2$ BMG possesses a good combination of high GFA, good mechanical properties, corrosion resistance and biocompatibility, which are found to be suitable for biomedical implant applications. Wang et al. [54] have reported biodegradable CaMgZn BMG for potential skeletal application based on the evaluation of both in vitro tests on ion release and cytotoxicity and in vivo implantation. Cao et al. [55] have reported a series of bioresorbable BMGs consisting of biocompatible elements Ca, Mg, and Zn for the use in orthopedic applications. The interesting property of these BMGs is that their biocorrosion can be tailored through alloy composition. Metallic glasses possess a unique combination of magnetostriction and extremely soft magnetic properties, which make them ideal candidates for developing high performance biosensors. Li et al. [56] have demonstrated the detection of pathogen (e.g., *Salmonella*) in liquid using the metallic glass-based biosensors and found a detection limit better than 50 CFU $mL^{-1}$. The ability of BMGs to fabricate into net shape complex geometries in a single processing step with patterning the surface enables to program desirable and predictable cellular response into a 3D biomaterial. Generally, considering the

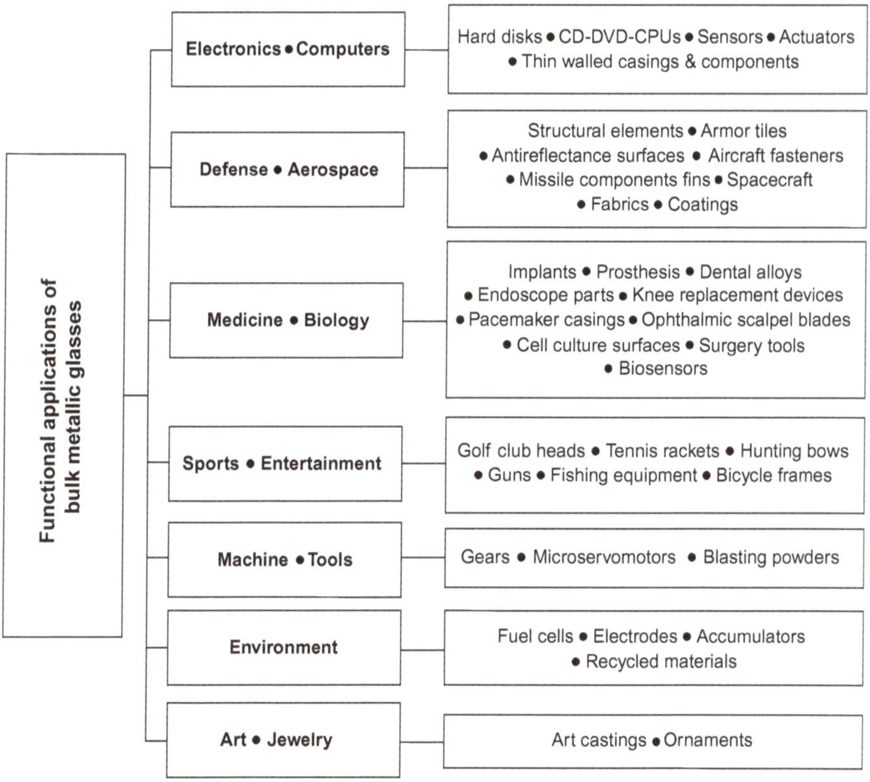


**Fig. 12.13** Functional uses of BMGs with group of application area and BMG articles and devices.

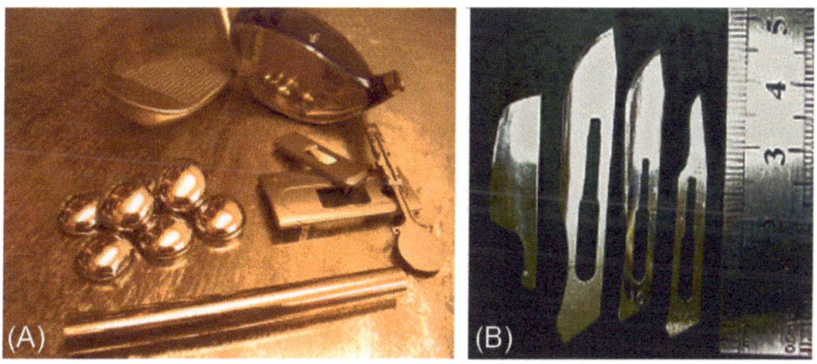


**Fig. 12.14** (A) Examples of bulk metallic glass hardware of Liquidmetal Technologies: golf clubs, ingots, electronic casings, rod (12 mm diameter), and large plate (under the sample). (B) Bulk metallic glass surgical razor.
(A) Reproduced with permission from D.C. Hofmann, Bulk metallic glasses and their composites: a brief history of diverging fields, J. Mater (2013) Article ID 517904 (8 pages), Copyright © Hindwai Publishing Corporation. (B) Reproduced with permission from J.C. Huang, J.P. Chu, J.S.C. Jang, Recent progress in metallic glasses in Taiwan, Intermetallics, 17 (2009) 973–987, Copyright © Elsevier.

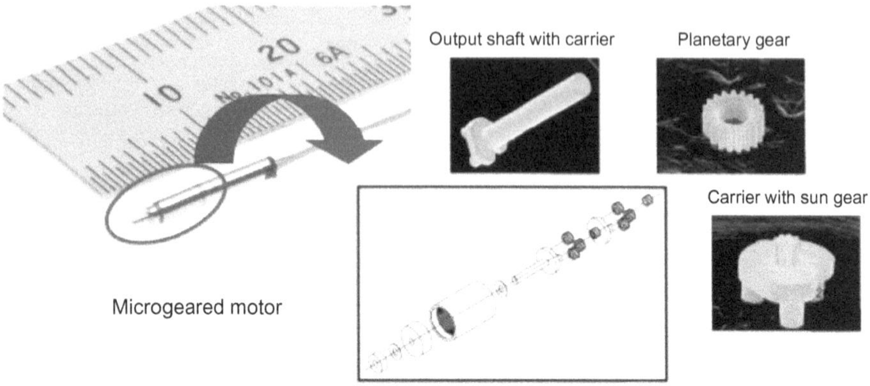


**Fig. 12.15** Schematic illustration of construction of the world smallest size (1.5 mm diameter) microgeared motor and Ni-based bulk metallic glass fabricated gear parts.
Courtesy: A. Inoue, X.M. Wang, W. Zhang, Developments and applications of bulk metallic glasses, Rev. Adv. Mater. Sci. 18 (2008) 1–9.

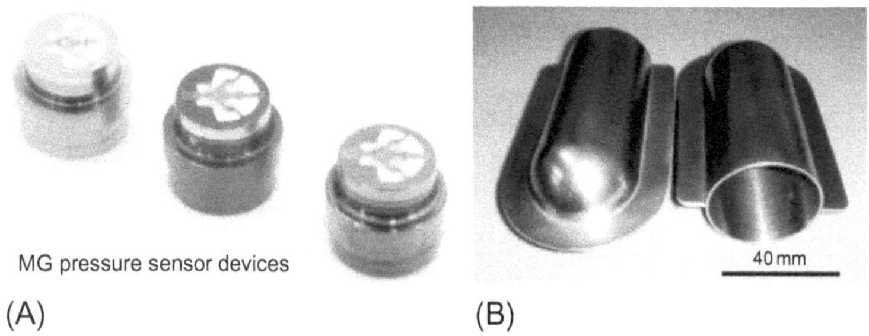


**Fig. 12.16** Photographs of (A) pressure sensor devices and (B) slat truck cover of aircrafts made of BMGs.
Courtesy: A. Inoue, X. M. Wang, W. Zhang, Developments and applications of bulk metallic glasses, Rev. Adv. Mater. Sci. 18 (2008) 1–9.

processing capabilities and properties, it is concluded that BMGs are useful for many medical devices and implants [57].

## 12.7 Summary

Here, with a brief introduction of BMGs and publication trends for the last 25 years, their various aspects were described. Chronological development of BMGs was described in the history section with specific examples and illustrations.

Kinetic behavior of bulk metallic glass formation was described with reference to the GFA, nucleation rate ($I$), crystallization rate ($U$), critical cooling rate ($R_c$), reduced glass transition temperature ($T_{rg}$), and maximum glass thickness ($t_{max}$). Preparation of BMGs was described with schematic illustrations of setup necessary for melting and casting. In the property section, various thermal and mechanical properties were described with specific glass composition and illustrations. Various functional applications of BMGs were described with current information and future outlook.

## 12.8 Future outlook

BMGs possess unique and unconventional properties. For this reason, they are the choice of materials for many functional applications. At present, there are some difficulties in their preparation such as melting in air, obtaining in larger size, and requiring high purity raw materials. Therefore, it is necessary to develop BMG compositions that could be prepared in air in larger sizes using relatively less pure raw materials. Other future directions would be on the road of increasing GFA, avoiding the use of toxic or costly elements, and functionalizing by patterning through surface engineering. For biomedical applications, it is necessary to develop BMG compositions with good combination of superior mechanical properties and excellent biocompatibility and biofunctionality along with satisfactory clinical performance. Besides their use in high technology applications, for their use in everyday appliances, it is necessary to develop BMG compositions, which can yield larger sizes using cheaper raw materials.

## Exercises

**12.1** What are the differences between the amorphous metallic glasses and BMGs?

**12.2** How do you define BMGs?

**12.3** What are the effects of $B_2O_3$ as fluxing agent in manufacturing of BMGs?

**12.4** What the effects of critical cooling rate ($R_c$) and reduced glass transition temperature ($T_{rg}$) on GFA of BMGs? How do they control the manufacturing of BMGs?

**12.5** Multicomponent BMGs or glasses containing elements of different sizes are more stable than that of less component or glasses containing similar size elements. Justify the statement with proper reasons.

**12.6** Describe the preparation of BMGs in brief with schematic illustration of experimental setup.

**12.7** BMGs possess unique and unconventional properties. Establish this statement with some experimental evidences.

**12.8** BMGs are tough, whereas oxide glasses are brittle. Explain this fact with experimental data.

**12.9** Describe several functional applications of BMGs.

**12.10** Name several articles that are used in medicine and biology.

# References


[1] B. Karmakar, Fundamentals of glass and glass nanocomposites, in: B. Karmakar, K. Rademann, A.L. Stepanov (Eds.), Glass Nanocomposites: Synthesis, Properties and Applications, Elsevier, Amsterdam, 2016, pp. 3–53.

[2] W. Klement, R.H. Willens, P. Duwez, Non-crystalline structure in solidified gold-silicon alloys, Nature 187 (1960) 869–870.

[3] H.S. Chen, D. Turnbull, Formation, stability and structure of palladium-silicon based alloy glasses, Acta Metall. 17 (1969) 1021–1031.

[4] H.S. Chen, Thermodynamic considerations on the formation and stability of metallic glasses, Acta Metall. 22 (1974) 1505–1511.

[5] M.C. Lee, J.M. Kendall, W.L. Johnson, Spheres of the metallic glass $Au_{55}Pb_{22.5}Sb_{22.5}$ and their surface characteristics, Appl. Phys. Lett. 40 (1982) 382–384.

[6] A.L. Drehman, A.L. Greer, D. Turnbull, Bulk formation of a metallic glass: $Pd_{40}Ni_{40}P_{20}$, Appl. Phys. Lett. 41 (1982) 716.

[7] H.W. Kui, A.L. Greer, D. Turnbull, Formation of bulk metallic glass by fluxing, Appl. Phys. Lett. 45 (1984) 615–616.

[8] A. Inoue, T. Zhang, T. Masumoto, Al–La–Ni amorphous alloys with a wide supercooled liquid region, Mater. Trans. JIM 30 (1989) 965–972.

[9] A. Inoue, A. Kato, T. Zhang, S.G. Kim, T. Masumoto, Mg–Cu–Y amorphous alloys with high mechanical strengths produced by a metallic mold casting method, Mater. Trans. JIM 32 (1991) 609–916.

[10] T. Zhang, A. Inoue, T. Masumoto, Amorphous Zr–Al–TM (TM = Co, Ni, Cu) alloys with significant supercooled liquid region of over 100 K, Mater. Trans. JIM 32 (1991) 1005–1010.

[11] A. Peker, W.L. Johnson, A highly processable metallic glass–$Zr_{41.2}Ti_{13.8}Cu_{12.5}Ni_{10.0}Be_{22.5}$, Appl. Phys. Lett. 63 (1993) 2342–2344.

[12] A. Inoue, N. Nishiyama, H. Kimura, Preparation and thermal stability of bulk amorphous $Pd_{40}Cu_{30}Ni_{10}P_{20}$ alloy cylinder of 72 mm in diameter, Mater. Trans. JIM 38 (1997) 179–183.

[13] H. Choi-Yim, R. Busch, W.L. Johnson, The effect of silicon on the glass forming ability of the bulk metallic glass forming alloy during processing of composites, J. Appl. Phys. 83 (1998) 7993–7998.

[14] A.A. Kündig, J.F. Löffler, W.L. Johnson, P.J. Uggowitzer, P. Thiyagarajan, Influence of decomposition on the thermal stability of undercooled Zr-Ti-Cu-Ni-Al alloys, Scr. Mater. 44 (2001) 1269–1273.

[15] W. Zhang, F. Jia, Q. Zhang, A. Inoue, Effects of additional Ag on the thermal stability and glass-forming ability of Cu–Zr binary glassy alloys, Mater. Sci. Eng. A 459 (2007) 330–336.

[16] Q. Luo, W.H. Wang, Rare earth based bulk metallic glasses, J. Non-Cryst. Solids 355 (2009) 759–775.

[17] B. Zberg, P.J. Uggowitzer, J.F. Löffler, MgZnCa glasses without clinically observable hydrogen evolution for biodegradable implants, Nat. Mater. 8 (2009) 887–891.

[18] M.D. Demetriou, M.E. Launey, G. Garrett, J.P. Schramm, D.C. Hofmann, W.L. Johnson, R.O. Ritchie, A damage-tolerant glass, Nat. Mater. 10 (2011) 123–128.

[19] C. Chen, Y. Cheng, T. Zhang, Synthesis of impurity-insensitive Zr-based bulk metallic glass, J. Non-Cryst. Solids 439 (2016) 1–5.

[20] A. Inoue, Stabilization of metallic supercooled liquid and bulk amorphous alloys, Acta Mater. 48 (2000) 279–306.

[21] M. Chen, A brief overview of bulk metallic glasses, NPG Asia Mater. 3 (2011) 82–90.
[22] J.F. Löffler, Bulk metallic glasses, Intermetallics 11 (2003) 529–540.
[23] Z.P. Lu, C.T. Liu, Role of minor alloying additions in formation of bulk metallic glasses: a review, J. Mater. Sci. 39 (2004) 3965–3974.
[24] Y.Q. Cheng, E. Ma, Atomic-level structure and structure-property relationship in metallic glasses, Prog. Mater. Sci. 56 (2011) 379–473.
[25] P.W. McMillan, Glass-Ceramics, second ed., Academic Press, London, 1979.
[26] J.E. Shelby, Introduction to Glass Science and Technology, The Royal Society of Chemistry, Cambridge, 1997.
[27] A. Inoue, A. Takeuchi, Recent development and application products of bulk glassy alloys, Acta Mater. 59 (2011) 2243–2267.
[28] D. Turnbull, Under what conditions can a glass be formed? Contemp. Phys. 10 (1969) 473–488.
[29] W.H. Wang, C. Dong, C.H. Shek, Bulk metallic glasses, Mater. Sci. Eng. R. Rep. 44 (2004) 45–89.
[30] W. Pilarczyk, Preparation and characterization of Zr-based bulk metallic glasses in form of plate, J. Alloys Compd. 615 (2014) S132–S135.
[31] M. Telford, The case for bulk metallic glass, Mater. Today 7 (2004) 36–43.
[32] J.C. Huang, J.P. Chu, J.S.C. Jang, Recent progress in metallic glasses in Taiwan, Intermetallics 17 (2009) 973–987.
[33] S. Pang, Y. Liu, H. Li, L. Sun, Y. Li, T. Zhang, New Ti-based Ti–Cu–Zr–Fe–Sn–Si–Ag bulk metallic glass for biomedical applications, J. Alloys Compd. 625 (2015) 323–327.
[34] Q. Yu, X.D. Wang, H.B. Lou, Q.P. Cao, J.Z. Jiang, Atomic packing in Fe-based metallic glasses, Acta Mater. 102 (2016) 116–124.
[35] S. Li, Q. Wei, Q. Li, B. Jiang, Y. Chen, Y. Sun, Development of Fe-based bulk metallic glasses as potential biomaterials, Mater. Sci. Eng. C 52 (2015) 235–241.
[36] X. Jia, Y. Li, H. Wang, G. Xie, S. Yamaura, W. Zhang, Synthesis and properties of ferromagnetic Fe-based (Fe, Ni, Co)–Mo–P–C–B bulk metallic glasses with large supercooled liquid region, Physica B 476 (2015) 141–146.
[37] G.C. Lavorato, G. Fiore, A. Castellero, M. Baricco, J.A. Moya, Preparation and characterization of Fe-based bulk metallic glasses in plate form, Physica B 407 (2012) 3192–3195.
[38] Liquidmetal® Technologies—Corporate, Rancho Santa Margarita, CA, USA, *http://www.liquidmetal.com*.
[39] T. Wang, Y.D. Wu, J.J. Si, Y.H. Cai, X.H. Chen, X.D. Hui, Novel Ti-based bulk metallic glasses with superior plastic yielding strength and corrosion resistance, Mater. Sci. Eng. A 642 (2015) 297–303.
[40] Y.X. Wei, X.K. Xi, D.Q. Zhao, M.X. Pan, W.H. Wang, Formation of MgNiPr bulk metallic glasses in air, Mater. Lett. 59 (2005) 945–947.
[41] H. Li, S. Pang, Y. Liu, L. Sun, P.K. Liaw, T. Zhang, Biodegradable Mg–Zn–Ca–Sr bulk metallic glasses with enhanced corrosion performance for biomedical applications, Mater. Des. 67 (2015) 9–19.
[42] S. Li, R.J. Wang, M.X. Pan, D.Q. Zhao, W.H. Wang, Formation and properties of $RE_{55}Al_{25}Co_{20}$ (RE=Y, Ce, La, Pr, Nd, Gd, Tb, Dy, Ho and Er) bulk metallic glasses, J. Non-Cryst. Solids 354 (2008) 1080–1088.
[43] X.F. Liu, R.J. Wang, D.Q. Zhao, M.X. Pan, W.H. Wang, Bulk metallic glasses based on binary cerium and lanthanum elements, Appl. Phys. Lett. 91 (2007). 041901 (3 pages).
[44] S. Li, R.J. Wang, M.X. Pan, D.Q. Zhao, W.H. Wang, Bulk metallic glasses based on heavy rare earth dysprosium, Scr. Mater. 53 (2005) 1489–1492.

[45] L. Bie, Q. Li, D. Cao, H. Li, J. Zhang, C. Chang, Y. Sun, Preparation and properties of quaternary CoMoPB bulk metallic glasses, Intermetallics 71 (2016) 7–11.
[46] W.H. Wang, Correlations between elastic moduli and properties in bulk metallic glasses, J. Appl. Phys. 99 (2006). 093506 (10 pages).
[47] C.A. Schuh, T.C. Hufnagel, U. Ramamurty, Mechanical behavior of amorphous alloys, Acta Mater. 55 (2007) 4067–4109.
[48] A. Inoue, X.M. Wang, W. Zhang, Developments and applications of bulk metallic glasses, Rev. Adv. Mater. Sci. 18 (2008) 1–9.
[49] E. Axinte, Metallic glasses from "alchemy" to pure science: present and future of design, processing and applications of glassy metals, Mater. Des. 35 (2012) 518–556.
[50] D.C. Hofmann, Bulk metallic glasses and their composites: a brief history of diverging fields, J. Mater. (2013). Article ID 517904 (8 pages).
[51] W.H. Wang, Bulk metallic glasses with functional physical properties, Adv. Mater. 21 (2009) 4524–4544.
[52] N. Nishiyama, K. Amiya, A. Inoue, Bulk metallic glasses for industrial products, Mater. Trans. 45 (2004) 1245–1250.
[53] M. Hasan, J. Schroers, G. Kumar, Functionalization of metallic glasses through hierarchical patterning, Nano Lett. 15 (2015) 963–968.
[54] Y.B. Wang, X.H. Xie, H.F. Li, X.L. Wang, M.Z. Zhao, E.W. Zhang, Y.J. Bai, Y.F. Zheng, L. Qin, Biodegradable CaMgZn bulk metallic glass for potential skeletal application, Acta Biomater. 7 (2011) 3196–3208.
[55] J.D. Cao, N.T. Kirkland, K.J. Laws, N. Birbilis, M. Ferry, Ca–Mg–Zn bulk metallic glasses as bioresorbable metals, Acta Biomater. 8 (2012) 2375–2383.
[56] S. Li, S. Horikawa, M. Park, Y. Chai, V.J. Vodyanoy, B.A. Chin, Amorphous metallic glass biosensors, Intermetallics 30 (2012) 80–85.
[57] J. Schroers, G. Kumar, T.M. Hodges, S. Chan, T.R. Kyriakides, Bulk metallic glasses for biomedical applications, JOM 61 (2009) 21–29.

# Index

Note: Page numbers followed by *f* indicate figures, and *t* indicate tables.

Kolibri 360 GmbH
Ferdinand-Jühlke-Straße 7
99095 Erfurt, Deutschland
produktsicherheit@kolibri360.de